Progress in Brain Research
Volume 201

Functional Neural Transplantation III

Primary and Stem Cell Therapies for Brain Repair, Part II

Advisory Editors

Progress in Brain Research
Volume 201

Functional Neural Transplantation III

Primary and Stem Cell Therapies for Brain Repair, Part II

Edited by

Stephen B. Dunnett
Brain Repair Group
School of Biosciences
Cardiff University
Cardiff, United Kingdom

Anders Björklund
Wallenberg Neuroscience Center
Division of Neurobiology
Lund University
Lund, Sweden

ELSEVIER

AMSTERDAM – BOSTON – HEIDELBERG – LONDON – NEW YORK – OXFORD
PARIS – SAN DIEGO – SAN FRANCISCO – SYDNEY – TOKYO

Elsevier
Radarweg 29, PO Box 211, 1000 AE Amsterdam, The Netherlands
The Boulevard, Langford Lane, Kidlington, Oxford, OX5 1GB, UK

First edition 2012

Library of Congress Cataloging-in-Publication Data
A catalog record for this book is available from the Library of Congress

British Library Cataloguing in Publication Data
A catalogue record for this book is available from the British Library

ISBN: 978-0-444-59544-7
ISSN: 0079-6123

Printed and bound by CPI Group (UK) Ltd, Croydon, CR0 4YY

Transferred to digital print 2013

Contributors

Rojin Abraham
Stem Cell Institute, KU Leuven, Onderwijs & Navorsing V, Leuven, Belgium

Joanna Adamczak
Max-Planck-Institute for Neurological Research, In vivo NMR, Cologne, Germany

Melissa R. Andrews[1]
Centre for Brain Repair, University of Cambridge, Forvie Site, Robinson Way, Cambridge, UK

Anders Björklund
Wallenberg Neuroscience Center, Lund University, Lund, Sweden

Marina Boido
Neuroscience Institute of the Cavalieri Ottolenghi Foundation, University of Torino, Torino, Italy

Cesario V. Borlongan
Center of Excellence for Aging & Brain Repair, Department Neurosurgery and Brain Repair, University of South Florida Morsani College of Medicine, Tampa, FL, USA

Mary Bartlett Bunge
The Miami Project to Cure Paralysis, University of Miami Miller School of Medicine, The Lois Pope LIFE Center, Miami, FL, USA; Department of Cell Biology, University of Miami Miller School of Medicine, Rosenthiel Medical Sciences Building, Miami, FL, USA; Department of Neurological Surgery, University of Miami Miller School of Medicine, The Lois Pope LIFE Center, Miami, FL, USA

Roberto Cantello
ALS Centre, Department of Neurology, Eastern Piedmont University, "Maggiore della Carità" Hospital, Novara, Italy

Daniel J. Chew
Centre for Brain Repair, University of Cambridge, Forvie Site, Robinson Way, Cambridge, UK

Peter H. Chipman
Department of Anatomy and Neurobiology, Dalhousie University, Halifax, NS, Canada

Stephen B. Dunnett
Brain Repair Group, School of Biosciences, Cardiff University, Cardiff, Wales, UK

[1]Current address: School of Medicine University of St Andrews, Medical and Biological Sciences Building, North Haugh, St Andrews, UK.

David J. Eve
Center of Excellence for Aging & Brain Repair, Department Neurosurgery and Brain Repair, University of South Florida Morsani College of Medicine, Tampa, FL, USA

Franca Fagioli
Stem Cell Transplantation and Cellular Therapy Unit, Pediatric Onco-Hematology Division, "Regina Margherita" Children's Hospital, University of Torino, Torino, Italy

James W. Fawcett
Centre for Brain Repair, University of Cambridge, Forvie Site, Robinson Way, Cambridge, UK

Ivana Ferrero
Stem Cell Transplantation and Cellular Therapy Unit, Pediatric Onco-Hematology Division, "Regina Margherita" Children's Hospital, University of Torino, Torino, Italy

Robin J.M. Franklin
Wellcome Trust and MRC, Cambridge Stem Cell Institute, University of Cambridge, Cambridge, UK; Department of Veterinary Medicine, University of Cambridge, Cambridge, UK; Department of Biology, Georgetown University, Washington, DC, USA

Fred H. Gage
Laboratory of Genetics, Salk Institute for Biological Studies, La Jolla, CA, USA

Mathias Hoehn
Max-Planck-Institute for Neurological Research, In vivo NMR, Cologne, Germany

Osamu Honmou
Department of Neurology, Yale University, School of Medicine, New Haven, CT, USA; Department of Neural Regenerative Medicine, Sapporo Medical University, Sapporo, Japan

Jeffrey K. Huang
Wellcome Trust and MRC, Cambridge Stem Cell Institute, University of Cambridge, Cambridge, UK; Department of Veterinary Medicine, University of Cambridge, Cambridge, UK; Department of Biology, Georgetown University, Washington, DC, USA

Ahmed Ibrahim
Institute of Neurology, University College London, London, UK

Zi-Bing Jin
Laboratory for Stem Cell & Retinal Regeneration, Center for Vision Research, The Eye Hospital, School of Ophthalmology & Optometry, Wenzhou Medical College, Wenzhou, China

Hans S. Keirstead
Reeve-Irvine Research Center, University of California at Irvine, Irvine, CA, USA; Sue and Bill Gross Stem Cell Research Center, University of California at Irvine,

Irvine, CA, USA; Department of Anatomy & Neurobiology, School of Medicine, University of California at Irvine, Irvine, CA, USA; Department of Neurological Surgery, School of Medicine, University of California at Irvine, Irvine, CA, USA

Jeffery D. Kocsis
Department of Neurology, Yale University, School of Medicine, New Haven, CT, USA; Center for Neuroscience and Regeneration Research, VA Connecticut Healthcare System, West Haven, CT, USA

Zaal Kokaia
Laboratory of Stem Cells and Restorative Neurology, Lund Stem Cell Center, University Hospital, Lund, Sweden

Daqing Li
Institute of Neurology, University College London, London, UK

Ying Li
Institute of Neurology, University College London, London, UK

Olle Lindvall
Laboratory of Stem Cells and Restorative Neurology, Lund Stem Cell Center, University Hospital, Lund, Sweden

Letizia Mazzini
ALS Centre, Department of Neurology, Eastern Piedmont University, "Maggiore della Carità" Hospital, Novara, Italy

Christopher Metcalf
Center of Excellence for Aging & Brain Repair, Department Neurosurgery and Brain Repair, University of South Florida Morsani College of Medicine, Tampa, FL, USA

Keith W. Muir
Institute of Neurosciences and Psychology, University of Glasgow, Southern General Hospital, Glasgow, Scotland, UK

Victor F. Rafuse
Department of Anatomy and Neurobiology, Dalhousie University, Halifax, NS, Canada

Geoffrey Raisman
Institute of Neurology, University College London, London, UK

Paul R. Sanberg
Center of Excellence for Aging & Brain Repair, Department Neurosurgery and Brain Repair, University of South Florida Morsani College of Medicine, Tampa, FL, USA

John D. Sinden
ReNeuron Limited, Surrey Research Park, Guildford, Surrey, UK

Masayo Takahashi
Laboratory for Retinal Regeneration, RIKEN Center for Developmental Biology, Kobe, Japan; Center for iPS Research and Application, Kyoto University, Kyoto, Japan

Jeremy S. Toma
Department of Anatomy and Neurobiology, Dalhousie University, Halifax, NS, Canada

Pierre Vanderhaeghen
IRIBHM (Institute for Interdisciplinary Research), Université Libre de Bruxelles (U.L.B.), Brussels, Belgium; Welbio, Université Libre de Bruxelles (U.L.B.), Brussels, Belgium

Alessandro Vercelli
Neuroscience Institute of the Cavalieri Ottolenghi Foundation, University of Torino, Torino, Italy

Catherine M. Verfaillie
Stem Cell Institute, KU Leuven, Onderwijs & Navorsing V, Leuven, Belgium

Indira Vishnubhatla
ReNeuron Limited, Surrey Research Park, Guildford, Surrey, UK

Ryan R. Wiliams
The Miami Project to Cure Paralysis, University of Miami Miller School of Medicine, The Lois Pope LIFE Center, Miami, FL, USA

Lindsey A. Wyatt
Reeve-Irvine Research Center, University of California at Irvine, Irvine, CA, USA; Sue and Bill Gross Stem Cell Research Center, University of California at Irvine, Irvine, CA, USA; Department of Anatomy & Neurobiology, School of Medicine, University of California at Irvine, Irvine, CA, USA; Department of Neurological Surgery, School of Medicine, University of California at Irvine, Irvine, CA, USA

Contents (Part II)

Contents (Part I)

SECTION

Introduction

1

CHAPTER

Introduction (Part II)

1

Stephen B. Dunnett*, Anders Björklund†

*Brain Repair Group, School of Biosciences, Cardiff University, Cardiff, Wales, and MRC Centre for Neuropsychiatric Genetics and Genomics, School of Medicine, Cardiff University, Cardiff, UK

†Wallenberg Neuroscience Center, Lund University, Lund, Sweden

Functional neural transplantation has come a long way in the past four decades. Pioneering studies during the1970s and early1980s established basic methods for achieving effective survival and integration of embryonic neuronal tissue grafts in the mammalian central nervous system so that, in the early 1980s, three major review volumes covering this new field appeared in close succession (Björklund and Stenevi, 1985; Sladek and Gash, 1984; Wallace and Das, 1983). Their focus was largely on the principles of defining the technical aspects of delivery, demonstrating neuronal survival, characterizing anatomical growth of connections between grafted cells and the host brain, and providing the first biochemical, physiological, and simple assessments of function. From the outset, there was an interest in whether functional recovery in simple animal models might presage the way to eventual clinical applications of novel surgical strategies to treat hitherto untreatable neurological diseases.

By the early 1990s, the capacity of neural transplants to alleviate functional deficits in animal models of a range of neurodegenerative conditions was well established, and the feasibility of translation dramatically established by the first demonstration of unequivocal functional benefit in Parkinson's disease, at least for some patients (Lindvall et al., 1990). This led us to coordinate the first major review volume, *Functional Neural Transplantation*, focusing explicitly on the range of animal models that were then being used to evaluate the functional efficacy of neural grafts, exploring the mechanisms and conditions for functional recovery, and summarizing the first clinical applications, most notably in Parkinson's disease, but also looking forward to applications in a range of other conditions, including Huntington's disease, aging, ischemia, spinal cord injury, and metabolic disease (Dunnett and Björklund, 1994).

By the time of the second review volume, *Functional Neural Transplantation II*, 8 years later, alternative sources of cells were now being considered, including immortalized cell lines, cell encapsulation, and xenotransplantation (Dunnett and Björklund, 2000); although the first human embryonic stem cell lines were clearly appearing on the horizon, they had not yet translated to functional transplantation even in animal models. Clinical studies had by now commenced in several other

Progress in Brain Research, Volume 201, ISSN 0079-6123, http://dx.doi.org/10.1016/B978-0-444-59544-7.00017-2

diseases, including Huntington's disease, retinal disease, ischemia and spinal cord injury, albeit with patchy evidence for efficacy, but the focus was still on using primary fetal cells.

The past decade has seen major swings in the fortunes of functional cell transplantation, in both negative and positive directions. On the one hand, the first double-blind placebo-controlled study of cell transplantation not only found only modest evidence for efficacy but more seriously also reported the emergence of significant side effects in the form of graft-induced dyskinesias in several of the grafted patients (Freed et al., 2001). Although this has cast a significant cautionary cloud over the field of using primary fetal neural tissues for cell therapy, in particular for Parkinson's disease, it has not inhibited the emergence of more substantive trials in Huntington's disease, stroke, pain control, spinal cord injury, ALS, and macular disease. On the other hand, dramatic advances have been made in understanding normal neuronal differentiation in the developing embryo, now allowing the controlled direction of stem cell differentiation toward a range of specific neuronal fates, as required to provide alternative neuronal populations for transplantation. Moreover, other sources of pluripotent cells have been identified, including adult somatic stem cells, the generation of induced pluripotent stem cells, and even the direct transdifferentiation of neuronal cells from differentiated cells of other nonneuronal lineages. With these advances, stem cells now offer a realistic prospect as an alternative source of cells for cell replacement therapy that circumvent the difficult ethical, practical, immunological, and quality control issues surrounding the use of primary fetal tissues.

In the light of recent developments and new opportunities, we now consider that a further review of our field is timely, and the present volume, *Functional Neural Transplantation III* (Dunnett and Björklund, 2012), constitutes the realization of this project. First and foremost, we thank our authors for their uniformly excellent contributions to the volume, bringing their considerable insights and expertise to bear on the tasks we set and bearing with fortitude our pressures to maintain stringent deadlines in order to ensure that time-sensitive material can be brought rapidly into the public domain. In spite of the many technical difficulties, we see rapid and significant progress in the development of protocols that hold promise for translation to clinical application, and we find a remarkable degree of optimism that the challenges are soluble and the prospects for functional repair in as yet incurable neurodegenerative and traumatic conditions may for the first time be achievable.

As the previous volume, *Functional Neural Transplantation III* is again published in Elsevier's *Progress in Brain Research* series, and we wish also to thank and acknowledge the support of the Elsevier production team in enabling us to bring the project to realization. In the course of production, we realized that the breadth, depth, and quality of the individual submissions exceed the overall length that can be accommodated in a single volume. Therefore, the final project is now published over two volumes of the new format *Progress in Brain Research* series; although the division is somewhat arbitrary, dictated by natural groupings of chapters that go together, it is our wish that the two parts should be considered conceptually as a unitary intellectual whole, rather than as separate volumes. With this comprehensive focus in mind, both parts include the contents listing and indexing of the integrated project.

After a similar generic introduction, this second part picks up the themes of the first part, now addressing the power of neural transplantation and related cell therapies in a range of CNS systems outside of the basal ganglia (which was the focus of the first part). Many of the same challenges arise in the need to find preclinical solutions to cell replacement and repair for a range of devastating neurological conditions, and to identify suitable cells that can deliver effective functional recovery to patients. However, this is not just "more of the same." Both stroke and spinal cord research, in particular, have adopted a broader focus on alternative mechanisms for functional recovery than just neuronal replacement and circuit repair, with neuroprotective and trophic functions making an important—and in some cases, primary—contribution. We also see a much greater emphasis on the role of the glial environment within which neurons survive, connect, and function. And, finally, there has been a far broader vision of the possibilities for developing alternative and novel cell sources of cells for transplantation, including the challenges of differentiating functional neural cells from autologous, somatic, and adult-derived sources, in particular, with neuroprotection in stroke as a target.

We believe that the resulting volumes now provide the most up-to-date survey of the current status of this fast-developing field. We hope that it provides a valuable resource and receives widespread interest within our cell therapy community.

References

Björklund, A., Stenevi, U., 1985. Neural Grafting in the Mammalian CNS. Elsevier, Amsterdam.

Dunnett, S.B., Björklund, A., 1994. Functional Neural Transplantation. Raven Press, New York.

Dunnett, S.B., Björklund, A., 2000. Functional Neural Transplantation II. Novel Cell Therapies for CNS Disorders. Elsevier Science, Amsterdam.

Dunnett, S.B., Björklund, A., 2012. Functional Neural Transplantation III. Primary and Stem Cell Transplantation for Brain Repair. Elsevier Science, Amsterdam.

Freed, C.R., Greene, P.E., Breeze, R.E., Tsai, W.Y., DuMouchel, W., Kao, R., Dillon, R.N., Winfield, H., Culver, S., Trojanowski, J.Q., Eidelberg, D., Fahn, S., 2001. Transplantation of embryonic dopamine neurons for severe Parkinson's disease. N. Engl. J. Med. 344, 710–719.

Lindvall, O., Brundin, P., Widner, H., Rehncrona, S., Gustavii, B., Frackowiak, R., Leenders, K.L., Sawle, G., Rothwell, J.C., Marsden, C.D., Björklund, A., 1990. Grafts of fetal dopamine neurons survive and improve motor function in Parkinson's disease. Science 247, 574–577.

Sladek, J.R., Gash, D.M., 1984. Neural Transplants, Development and Function. Plenum Press, New York.

Wallace, R.B., Das, G.D., 1983. Neural Tissue Transplantation Research. Springer-Verlag, New York.

CHAPTER

Transplantation in the future

2

Fred H. Gage[1]

Laboratory of Genetics, Salk Institute for Biological Studies, La Jolla, CA, USA

[1]Corresponding author. Tel.: +1-858-453-4100 x1013, e-mail address: gage@salk.edu

Abstract

Cellular transplantation to the brain and spinal cord remains a promising approach both for probing basic biological questions and as a potential therapy for neurological diseases. This chapter summarizes some of the main constraints that continue to limit general biological applications and, specifically, clinical applications. These constraints include the critical features of the successful donor cell as well as those of the receptive host tissue and organism. In addition, we explore future directions, with special emphasis on genetic engineering, combinations of novel cell types combined with trophic factors, and training of the host organism to improve the accurate integration of grafted cells. Some speculations are made regarding universal donor cells, but these advances will depend on additional basic work to bring this approach to the clinic. The convergence of advanced molecular and cellular methods together with improved methods of *in vivo* imaging adds to our optimism for significant advances in cellular transplantation in the near future.

Keywords

donor cell, host, future directions, safety, transplantation, universal donor

Cellular transplantation to the nervous system has developed into one of the most useful tools for developmental neurobiology and has served to demonstrate the remarkable structural plasticity and reorganization in the adult brain. However, the routine or successful transition to clinical applications has been rocky. The editors of this volume, Stephen Dunnett and Anders Björklund, asked me to write a perspective for this comprehensive update on neural transplantation and they gave me latitude with regard to what I would cover. I will deconstruct some of the constraints and assumptions that those of us working in this field share and will project into the future to see where we might be heading. The other outstanding chapters in this volume will certainly provide very useful, up-to-date, and cutting-edge summaries of

Progress in Brain Research, Volume 201, ISSN 0079-6123, http://dx.doi.org/10.1016/B978-0-444-59544-7.00001-9

results from the most current areas of activity in the field, so I will not review the literature exhaustively but rather provide some opinions and speculations.

1 CONSTRAINTS

1.1 Role of the cellular graft

One of the constraints when designing a transplantation study is to determine what the goal of the experiment is, and for CNS transplantation studies, the options are diverse. For many of the basic biological questions, where transplants are tools for discovery, the constraints are related to the age of the donor and the host as well as the purity of the donor cells. In most cases, these studies rely on accuracy of integration in the transplanted site, for which one can use the existing host circuitry as the blueprint for what accurate integration should look like. For preclinical or translational studies, we generally need to decide whether we are seeking to replace dead cells or protect dying cells. In other studies, we may be using the graft to augment a regenerative response by providing a cellular bridge or scaffold, as in spinal cord injury. Alternatively, the graft may be a source of secreted molecules, such as growth factors or transmitters, or even enzymes to metabolize toxins as in the case of lysosomal storage diseases; in all these latter cases, no or very little integration of the graft into circuitries is required.

As with many areas of research, our goals sometimes change as the results change. If we set out to induce recovery of function in a Huntington's disease model through a cell replacement strategy, for example, then finding that behavioral recovery occurs without long-term graft survival or integration, our interest may change from cell replacement to induced host plasticity.

1.2 Temporal features of the disease model

Another constraint on the use of cellular transplants is the temporal nature of the disease that is being examined. For acute injury models like spinal cord injury or traumatic brain injury, the timing of the transplantation is often critical in that grafting too early or grafting too late can miss a therapeutic window of effectiveness. Chronic late-stage diseases are also complicated because, if one waits too long in the progress of the disease, additional complications may arise, unrelated to the initial symptoms and pathology that were the original targets of the graft. Even more complicated is when cellular grafts are used for protective or prophylactic purposes. One of the many issues one has to be concerned about is whether the cellular graft is changing the course of the disease or disrupting the underlying normal function.

1.3 Safety

As the source of cells becomes more primitive, such as embryonic stem (ES) cells, the chance of continued proliferation and tumorigenicity increases. These potential setbacks continue to be a problem and will likely lead to the development of genetically engineered safety valves, like genetically inducible toxins built into every cell of the transplant but only activated upon cell mitosis. This theme of molecular engineering of cells will be a common thread in the solutions I propose and the directions I think we are likely to be heading in.

The other major safety issue is immunocompatibility of the graft and the host. While the brain is "immunoprivileged," it is clear now that some level of rejection can occur when histocompatibility is breached—yet no clear solution exists. Perhaps transient immune suppression, giving time for the blood–brain barrier to heal, is adequate in some cases, but we do not really know whether it encourages an ongoing, low-grade inflammation that persists and inhibits the optimal graft–host interaction. Again I would argue that developing some reliable way—using zinc fingers or TALENS or other Zinc finger technology—to molecularly engineer the donor cells to homologously insert the host MHC into the immune locus of the donor cells would go a long way toward addressing this issue. Currently, these methods are expensive and complicated, but both of these issues of cost and complexity are changing at a remarkable pace. In general, these issues of safety suggest the value of a universal donor (UD) cell for each application that has an easy cloning site inserted in its deleted MHC locus so the host MHC can be inserted for each unique patient. An alternative already being considered is banking enough immunologically diverse donor cells to cover the expected diversity of the disease. This approach may be more effective in more genetically homogeneous populations with more homogeneous genomes.

Another thought or concern about safety is the possible malfunction of a well-integrated graft. Even if a cellular graft does not form a tumor, is not immunologically rejected, and in fact makes synaptic connections, how can we be sure the correct connections will be made? But, maybe it does not matter; we just do not know enough about required connectivity at this point. Some investigators (such as Stephen Dunnett) have speculated and provided data that neural graft integration is activity dependent, so training the graft as it matures in the host to perform behaviors or tasks in which the grafted cells will be expected to participate may be required or may at least enhance successful and accurate graft integration.

1.4 Donor cell type: Current

Currently, the cell types being used most effectively in preclinical studies are still fresh fetal tissue (though still poorly characterized) as well as isolated, cloned, and propagated fetal cells that are differentiated to expandable neural progenitor cells. Encapsulation is still being explored but less than in years past for CNS applications, and oncogene immortalized fetal cells, while still useful experimentally, generate concern when envisioned clinically. More recently, induced pluripotent (iPS) cells and induced neurons or neural progenitors (iN) have been extensively researched. In all cases, what is of critical importance is finding the age of the donor cell that optimizes its chances of differentiating into the specific cell phenotype that represents the aim of the experiment (dopamine neurons, astroglia, oligodendroglia or GABAergic interneurons, or even some combination of cell phenotypes). When fetal tissue has been the primary tissue source, there has been a consensus strategy of dissecting fetal tissue at specific developmental stages from specific brain areas, when the cells are committed to a lineage but not mature yet. The hope has been that this approach would optimize accurate integration, and in many, if not most, cases there was supporting evidence for this (see Section 1.5).

Due in large part to extraordinary advances in the understanding of the basic cellular and molecular mechanisms of neurodevelopment, progress is being made to direct primitive cells into very specific subclasses of cells. More application of this developmental knowledge to ES cell and iPS cell studies is needed to generate the appropriate environment in a dish that assures that the induced derived progenitor cells have their cell fate patterned before transplantation, to optimize the chances of generating the appropriate and integrated cell types *in vivo*. We still do not have excellent control over these pretransplantation conditions, but even with current technology and knowledge, this is an area where real progress can be made, especially with very primitive cells like ES and iPS cells whose fates are more malleable *in vitro*.

1.5 State of the host upon implantation

While the choice of the cell type is critical, the state of the host upon grafting is in many cases at least as, if not more, important. We already know that the age of the host is important, with the older host being more refractory to successful integration. However, we really do not understand why this is so, and there may be ways to prepare the aged host to be more youthful prior to grafting. Diet, exercise, decreased stress, and rest may affect cell survival and integration, but as far as I know, these factors have not been explored in detail.

An obvious variable in graft survival in clinical and preclinical cases is the fact that most cellular grafts are implanted into a diseased brain, whereas most of the parameters that are explored for any particular donor cell are conducted in a healthy brain. Cell loss, inflammation, cavitation, vascular obstruction, and even gliosis all challenge even the best donor cell. Modeling the disease in the animal model clearly helps, and the closer the animal model recapitulates the pathology, the better chance that cell has in a clinical setting. Once again, one could consider some pretransplantation strategies to prepare even the diseased host brain for the implant. Along these lines, it makes sense to pay more attention to the activity of the host after transplantation to optimize the survival and integration of the cells. As mentioned before, this strategy may require different sets of activities for the host that would optimize the development and integration of the implant. Even with our current technology and knowledge, the importance of the activity of the host could be explored and could provide quite doable and inexpensive ways to enhance graft survival and function, not only for the current batch of donor cell candidates but also for any of the cell types that may be generated in the future. Specifically, I am referring again to training the subject (transplant) on tasks for which the transplant is expected to have a function. This approach is especially salient when considering the required youthfulness of the implanted cells and the reasonable assumption that activity-dependent training aids effective development and ultimately function.

2 FUTURE DIRECTIONS

The general neural transplantation strategy makes the assumption that there is no or very little plasticity in the diseased brain, but we know that adult and even aged brains retain remarkable plasticity, both at the synaptic and the cellular level. Thus, it makes sense to use the plasticity of the host brain to improve graft survival and function and to use the grafted cells to increase host plasticity to restore function. An obvious combination is delivering specific neurotrophic factors in a gradient to mimic developmental patterning to both support graft survival and induce directed axonal outgrowth. In addition, behavioral training and/or physical exercise can be used to selectively enhance survival and/or induce, increase, and maintain connections between the graft and the host. In a more engineered approach, the grafted cells could be genetically engineered to express a receptor that could be activated by a compound that is otherwise inactive in the body in order to regulate or enhance some process specific to the graft rather than the host. An analogous approach would be to apply optogenetics to activate or inhibit grafted cell activity selectively with light.

2.1 Cellular transplants to study human disease

It is clear that in the near future much effort will be put into optimizing the use of iPS cells and iN cells as donor cells because they afford the opportunity to study diseased human neural cells and tissues_in an *in vivo* environment. This latter application represents not so much using cellular grafting to repair the damaged brain but rather using cellular grafting of iPS cells to animal models of human disease in search of the underlying mechanisms that cause the disease or to establish more homologous models that can be used to test for new therapies, whether they are cellular, genetic (vector-based), or traditional drug-based therapies. The strategy as it is currently playing out in many labs is to program somatic cells from patients with specific diseases, either sporadic or monogenetic, and then develop methods to specifically differentiate them into the cell types that are presumed to be affected by the disease. By comparing the rate of differentiation and final differentiated fate with the appropriate age-matched, phenotypically and genetically normal cells/individuals, the expectation/hope is that a phenotype specific to the disease can be detected. To this end, there are several levels of outcome that have different values or uses. The optimal outcome is that a phenotype is discovered that makes sense in terms of the disease and includes a phenocopy of some *in vivo* state of the diseased patient. This type of model can be used to search for mechanisms that cause the cellular pathology of a disease. Alternatively, a phenotype may be discovered that is similar to but distinct from the disease; this phenotype could still be used to search for therapies even though this model would be less useful in searching for the mechanism of the disease process. Finally, a phenotype may be discovered that has no relationship with the disease, but there is a robust difference between the diseased cells and the phenotypically normal cells. This model could still be used diagnostically to determine whether a person has the disease, which is not a trivial advance in some cases.

All this *in vivo* modeling becomes relevant to neural transplantation when iPS cells are transplanted to animal models of the disease. Several obvious advantages stand out for using transplantation. First, grafting the iPS cell-derived neural progenitors to a normal immunocompromised mouse brain, in the presumed appropriate brain region, permits monitoring of differentiation in a more realistic setting than in a culture dish. If the same or similar aberrant phenotype can be detected, then it is very likely that the mechanism or cause of the disease is intrinsic to or cell autonomous to the cells that were grafted, given that the host is normal and healthy (albeit immunocompromised). If, however, phenotypically normal, iPS-derived neural progenitor cells are transplanted to the brains of transgenic mice that are carrying a human mutation or mutations that are thought to be causative in the human disease and the grafted cells adopt a disease phenotype that is different from that seen following grafting to a wild-type mouse, then one can infer that there are cell nonautonomous factors critical for the expression of the cellular disease phenotype.

This approach will require considerable development with regard to all the constraints listed above, including age of cell, host, cell type, cell number, immunocompatibility, etc., but its value as an additional modeling tool seems clear. The use of such models for secondary drug screening for effectiveness and for testing toxicity seems obvious, at least to me.

2.2 Synthetic, bioengineered cells

Genetically engineered cells are here to stay and will eventually become the staple for clinical transplantation in order to have more homogeneous, reproducible, and predictable cell lines. This reproducibility will be important to foster the involvement of pharmaceutical companies and even biotech companies, which I believe will be important to make major clinical advances in the field. To generate their attention and commitment of resources and effort, we have to convince industrial partners to believe that neural transplantation therapy works and is commercially viable. Tools are being generated, and enough is known about genetic engineering and homologous recombination to at least envision cells with many of the properties of a UD cell.

Some of the properties that need to be built in are obvious: (1) the capacity to divide *in vitro* on demand and to completely stop dividing without dying *in vivo*; (2) immunocompatibility, or at least a molecular cassette built into the cell to easily insert genetic immunocompatibility. Some more elaborate, but still doable, engineering would include the following. (3) The cell would need to be primitive enough to be able to be differentiated to a state optimal for successful grafting, thus having fluorescent molecule markers that are selectively activated when the cell reaches that state so that the cells could be FACS-sorted for purity and homogeneity. (4) Along these lines of homogeneity and purity, we often think we need a single identified cell type and we attempt to generate purity and homogeneity; however, it may be that, in an adult human diseased brain, additional specific support cells, such as astroglia or oligodendroglia, may be needed to be cotransplanted with the target donor cells at a specific cotransplant density to achieve optimal integration. This approach will be aided by additionally engineering the cell to express multiple, colored molecular indicators of fate that can be used for FACS-sorting of the UD cells. Thus, by differentiating the UD cell to neural precursors of some lineage and other cells to glia precursors and mixing together the correct ratio before grafting, some increased level of optimization can be reached. (5) Another molecular marker that is activated when the cell differentiates and is functional would also be useful, especially if this marker could be detected with PET or some other noninvasive means of monitoring the implanted UD cells. While these UD engineered cells are not available now and some additional impressive work is needed to develop them, they are not unachievable goals and may be crucial to move neural transplantation from an academic exercise to a viable therapy for any of the major neurological or neurodegenerative diseases that are being discussed or targeted for neural transplantation.

One of the reality checks on this UD cell approach is the choice of the starting cell to be engineered. This choice is not clear currently, as many, if not all, cells have their own limitations. An ES cell would be one of the cells to consider. However, something that has not been covered in this brief chapter is the vulnerability or variability of the cellular gene to cell division. It has long been stated that the most dangerous thing that a cell can do is divide. It is becoming clear now that passaging cells results in genomic changes/mutations, due to incomplete or overwhelmed genome repair systems. These genetic changes, mutations, or modifications quickly lead to changes that can lead to aberrant behavior of some cells with passage. It can be argued that most of the time these changes have no impact on the cell's behavior, but when growing millions of cells for several passages—which would likely be needed for some transplants even for a single individual—these genetic changes need to be considered. Enhanced gene repair machinery could be engineered to provide an enhanced surveillance of the genome, but this possibility is moving into incompletely explored areas. A more speculative approach would be to engineer, from elemental components, a synthetic cell or at least one that has a gene designed from the beginning to be a UD cell. This idea is not completely fanciful. Past year, the successful construction of the first self-replicating, synthetic bacterial cell was reported. The 1.08 million base pair chromosome of a modified *Mycoplasma mycoides* genome was synthesized. The synthetic cell was named *Mycoplasma mycoides* JCVI-syn1.0 and is the proof of principle that genomes can be designed in the computer, chemically made in the laboratory, and transplanted into a recipient cell to produce a new, self-replicating cell controlled only by the synthetic genome. It will be a while before a eukaryotic cell is synthesized, but once a proof of principle is demonstrated and a need exists, humans have a remarkable capacity to achieve previously unattainable goals. I think neural transplantation may be at a similar stage. Proof of principle has been demonstrated in several instances for successful neural transplantation, and the clinical need is clear. We are now on the cusp of achieving the previously unattainable.

SECTION

2

Stroke

CHAPTER

Neural differentiation and support of neuroregeneration of non-neural adult stem cells

3

Rojin Abraham, Catherine M. Verfaillie*

Stem Cell Institute, KU Leuven, Onderwijs & Navorsing V, Leuven, Belgium

**Corresponding author. Tel.: +32-16-330295, Fax: +32-16-330293, e-mail address: catherine.verfaillie@med.kuleuven.be*

Abstract

Although it is well established that neural stem cells (NSCs) or neural stem/progenitor cells differentiated from pluripotent stem cells can generate neurons, astrocytes, and oligodendrocytes, a number of other cell populations are also being considered for therapy of central nervous system disorders. Here, we describe the potential of (stem) cells from other postnatal tissues, including bone marrow, (umbilical cord) blood, fat tissue, or dental pulp, which themselves do not (robustly) generate neural progeny. However, these non-neuroectoderm derived cell populations appear to capable of inducing endogenous neurogenesis and angiogenesis. As these "trophic" effects are also, at least partly, responsible for some of the beneficial effects seen when NSC are grafted in the brain, these non-neuroectodermal cells may exert beneficial effects when used to treat neurodegenerative disorders.

Keywords

mesenchymal stem/stromal cells, endothelial progenitor cells, neural stem cells, trophic factors

Abbreviations

AFS	amniotic fluid stem
AMI	acute myocardial infarction
BMC	bone marrow cell
CNS	central nervous system
DAMP	damage-associated molecular patterns

Progress in Brain Research, Volume 201, ISSN 0079-6123, http://dx.doi.org/10.1016/B978-0-444-59544-7.00002-0

DC	dendritic cell
EC	endothelial cells
EO	early outgrowth
EPC	endothelial progenitor cell
FCS	fetal calf serum
HMGB1	high-mobility group box 1
HSC	hematopoietic stem cell
IL-10	interleukin-10
IL-1β	interleukin-1β
IL-6	interleukin-6
LO	late outgrowth
MAPC	multipotent adult progenitor cell
MASC	multiptotent adult stem cell
MIAMI	marrow isolated adult multilineage inducible
MSC	mesenchymal stem cell
NK	natural killer
NPC	neuroprogenitor cell
NSC	neural stem cell
OB	olfactory bulb
PB	peripheral blood
SIDS	stroke-induced immunodepression
SVZ	subventricular zone
UCB	umbilical cord blood
USSC	unrestricted somatic stem cell
VSEL	very small embryonic stem cell like cell

1 INTRODUCTION

Adult stem cells are currently being evaluated and used in clinical trials for a number of central nervous system (CNS) disorders, among which are multiple sclerosis, Parkinson disease, and stroke (Koch et al., 2009). As ischemic stroke incorporates the most varied number of target cells to be "treated"/"regenerated" in the setting of CNS disorders, we will, in this chapter, use stroke as the model disease to discuss the possible role adult stem cell grafts can play in the setting of CNS diseases.

Stroke is the third leading cause of death behind heart disease and cancer (American Heart Association, 2009). In ischemic stroke, which comprises 87% of all cases, decreased cellular respiration secondary to oxygen depletion causes neuronal death in central areas of the infarct (Bramlett and Dietrich, 2004). Decreased intracellular ATP supply leads to loss of membrane potential, influx of calcium, and loss of membrane integrity resulting in rapid necrosis. The area surrounding the primary infarct is known as the penumbra and receives some perfusion from adjacent areas. This area survives the initial insult but ultimately dies if blood flow is not restored (Dirnagl et al., 1999). Further neural cell death occurs via a number of mechanisms, including widespread depolarization in response to loss of ATP causing release of glutamate, leading to excitotoxic injury. In addition, inflammatory and immune responses further worsen the lesion.

Damage-associated molecular patterns (DAMPs) released from dying cells, for example, heat-shock proteins, high-mobility group box 1 (HMGB1), and hyaluronan, induce inflammatory mediators as DAMPs are sensed by Toll-like receptors and expressed on cells of the innate immune system (Arslan et al., 2011; Kong and Le, 2011). Of note, although HMGB1 increases in an initial phase the proinflammatory effect of ischemia, it appears also to have a regenerative effect, by improving neurogenesis and angiogenesis (Hayakawa et al., 2011). As a result of DAMP signalling, inflammatory mediators such as cytokines (IL-1β, TNF, IL-6), chemokines, and nitric oxide are produced, worsening cell death and causing leakiness of the blood brain barrier (Maroso et al., 2011; Sozen et al., 2009). Cytokines and chemokines also induce upregulation of adhesion molecules on endothelial cells (ECs), which cause further recruitment of circulating leucocytes into the cerebral parenchyma (Wang et al., 2007).

Neutrophils first migrate into the brain parenchyma, followed by macrophages and lymphocytes a few days after injury (Wang et al., 2007). These play a crucial role in the establishment of the inflammatory process which has a deleterious effect in stroke. Indeed, the extent of stroke is severely decreased when mice deficient in T and B cells are used (Hurn et al., 2007). However, a subset of T-lymphocytes, namely, CD4/CD25/FOXP3-positive regulatory T cells, appear to suppress the production of proinflammatory cytokines such as TNFα, interleukin-1β, and interferon, exerting neuroprotective effects (Planas and Chamorro, 2009).

There is bidirectional communication between the CNS and the immune system to maintain homoeostasis and CNS lesions disturb this homeostasis. Cytokines and neurotransmitters produced as a result of the lesion influence chemosensitive brain areas involved in immune control, such as the hypothalamus (Meisel and Meisel, 2012; Trakhtenberg and Goldberg, 2012). Furthermore, inflammatory mediators, released into the systemic circulation, influence peripheral immune organ function. Therefore, quickly following brain injury, stroke-induced immunodepression (SIDS) is seen, which can last for several weeks. SIDS is associated with infectious complications, which, due to the release of additional inflammatory signals, may worsen the brain lesion. However, there may also be a positive effect associated with this generalized immunodeficiency, as it may decrease autoimmune-like responses against for instance neoantigens released following brain injury (Gee et al., 2007; Madinier et al., 2009).

In adult brain, neurogenesis occurs mainly in the subventricular zone (SVZ) of the lateral ventricles and in the hippocampal dentate gyrus. Neural stem cells (NSCs) in the SVZ have been identified as a subset of subependymal astrocyte-like cells (type B cells), that generate rapidly proliferating neuroprogenitor cells (NPCs; type C cells), which then give rise to migrating progenitor cells (type A cells; Gage, 2000). NPCs originating from the SVZ migrate tangentially toward the olfactory bulb (OB) through the rostral migratory stream. Upon reaching the subependymal zone in the OB, NPCs migrate radially to the overlying granule and periglomerular cell layers where they differentiate mainly into interneurons. Enhanced neurogenesis occurs in response to brain injury (Arvidsson et al., 2002). However, in most cases, this response is insufficient in magnitude or duration to induce significant functional recovery. As NSC can differentiate into neurons, astrocytes, and oligodendrocytes, their ability to replace lost neural cells in CNS disorders has been tested. A fraction of NSCs survive after transplantation, migrate toward areas of infarct (Kelly et al., 2004), and generate neurons (Englund et al., 2002) that may form connections with host cells (Park et al., 2002). However, NSCS are clearly not able to regenerate all lost neural cells. Nevertheless, functional improvement is observed, which has been ascribed to neuroprotective (Lee et al., 2007; Ourednik et al., 2002) and immunomodulatory (Fujiwara et al., 2004; Pluchino et al., 2005) effects of NSCs. A number of other non-neural cell types have similar functions and are therefore being evaluated for their ability to improve functionality following an ischemic event or in other neurodegenerative disorders (Ramos-Cabrer et al., 2010).

2 ADULT NON-NEURAL STEM CELLS

In this review, we first describe the phenotype and properties of several non-neural adult (stem) cell populations that have been/are being tested in preclinical models of stroke, and some in clinical studies, including bone marrow cells (BMCs), umbilical cord blood (UCB) cells, endothelial progenitor cells (EPCs), and different types of mesenchymal stem cells (MSCs). Subsequently, a review of the results from preclinical and clinical studies will be given.

3 BONE MARROW CELLS/UMBILICAL CORD BLOOD CELLS

This population of cells is highly heterogeneous, including myeloid and lymphoid cells, a small proportion of hematopoietic stem cells (HSCs) as well as a small population of EPCs and MSCs. Likewise, unfractionated UCB cells contain a mixture of HSC, more mature hematopoietic cells, EPCs and MSCs. The possible beneficial effects of BMCs/UCB cells in ischemic diseases come from the production of trophic factors that decrease neural cell death, improve endogenous neurogenesis and neural plasticity, and perhaps direct generation of blood vessels, but not from direct regeneration of neural cells. An advantage is that therapies can be commenced at any time following the ischemic event.

BMCs have been most extensively used in the setting of acute myocardial infarction (AMI). These studies have shown that moderate beneficial effects on heart function ensue from the therapy (measured as left ventricular ejection fraction; Passier et al., 2008). Administration of granulocyte colony stimulating factor results in the increased frequency of progenitors/stem cells in the peripheral blood (PB) and has been used either alone as therapy for AMI or combined with BMCs, infused in the coronary arteries, again with mixed results in the clinical setting (Minatoguchi, 2011). In addition, studies in peripheral limb ischemia models have demonstrated that injection of BMCs may worsen the inflammation also seen in this ischemia model (Aranguren et al., 2008). Indeed, compared with saline controls, excess donor as well as host-derived T-lymphocytes and dendritic cells (DCs) were found infiltrating the ischemic muscle, leading to a worsening of the ischemia and subsequent fibrosis in the treated group compared with the saline control group. This observation should serve as a note of caution, in that grafting of such cell populations in the ischemic, necrotic, and inflammatory disease may contribute to and exacerbate the inflammatory response with potentially detrimental consequences later on.

4 ENDOTHELIAL CELLS/ENDOTHELIAL PROGENITOR CELLS

It was long believed that regeneration of endothelial beds postnatally occurred nearly exclusively by endoangiogenic mechanisms, that is, the sprouting of mature ECs from preexisting vessels, likely in response to angiogenic growth factors. However, there is now very solid evidence that EPCs derived from the BM that reach organs via the circulation (Asahara et al., 1999; Lin et al., 2000) or from within the organ itself (Balasubramaniam et al., 2010) partake in revascularization in a process termed postnatal vasculogenesis. However, the field of EPC has been confusing chiefly because two kinds of cell populations have been named EPCs. EPCs are cultured from PB. During these cultures, colonies appear early, containing cells that express not only CD31 but also CD45 (Yoder et al., 2007). These cells cannot make vascular tubes, but instead form macrophage colonies in methylcellulose assays. These so-termed early outgrowth (EO)-EPCs are hematopoietic in origin and may contribute to revascularization by the production of trophic factors that affect both ECs and late outgrowth (LO)-EPCs (Rehman et al., 2003; Zentilin et al., 2006). A second type of colony also grows from cultures of PB. They can be detected after 2–3 weeks, can be multiply replated, have enormous expansion potential, and have been classified as LO-EPC (Asahara et al., 1999; Ingram et al., 2005; Lin et al., 2000; Sieveking et al., 2008; Yoder et al., 2007). LO-EPCs are CD45 negative; express vWF, KDR, Flt-1, Tie-2, VE-cadherin, and CD31; uptake acLDL; and form robust vascular tubes *in vitro* and *in vivo*. Moreover, upon transplantation *in vivo*, they incorporate in existing endothelial linings (Balasubramaniam et al., 2010; Hendrickx et al., 2010; Lin et al., 2000). LO-EPCs can also be isolated from UCB. Compared with postnatal derived LO-EPCs, UCB-derived LO-EPCs have significantly greater proliferation potential (Droetto et al., 2004).

In considering EPCs for vascular repair in ischemic disorders, three limitations need to be kept in mind. (i) EO EPCs in elderly sick patients are defective. Their expansion potential is limited and they support repair of the vascular bed significantly less well (Schmidt-Lucke et al., 2005). (ii) ECs derived from EPCs are very immunogenic, making allogeneic transplantation more difficult. (iii) Finally, when considering autologous therapies, the fact that EPC expansion *in vitro* requires several weeks, signifies that therapy can only be commenced several weeks after the ischemic event.

5 MESENCHYMAL STEM/STROMAL CELLS

5.1 Phenotype and origin of MSCs

Mesenchymal stem (stromal) cells were first described by Friedenstein et al., who isolated from bone marrow non-hematopoietic cells that formed colonies of cells with fibroblast-like morphology. These cells, initially named fibroblast colony forming cells, could be expanded adherent to culture plates and differentiate into osteoblasts, adipocytes, and chondroblasts. They were later renamed MSC by Caplan and colleagues (Caplan, 1991). Since the initial description of MSCs from BM, MSCs have also been isolated from many other tissues, including adipose tissue, UCB, Wharton's jelly, placenta, amniotic fluid, dental pulp, and many "solid" organs.

The fact that isolation of MSCs relies on their adherence properties and depends on high concentrations of fetal calf serum (FCS) with large batch to batch variation, the phenotypic and functional characteristics of MSCs isolated in different laboratories varies. Cultured MSCs are identified by a combination of morphologic, phenotypic, and functional properties. Several cell surface antigens have been used to identify cultured MSCs, including CD10, CD13, CD29, CD44, CD49a–f, CD63, CD90, CD105, CD106, CD140b, STRO-1, SH2, SH3, and SB-10 and the absence of endothelial (progenitor) cells markers, CD31, CD45, CD3, CD14, CD11b, CD19, CD38, and CD66b (Dominici et al., 2006). Bianco and colleagues determined that CD146, which marks the adventitial reticular cells in the subendothelial reticular cells, allows direct isolation of human MSC from BM (Sacchetti et al., 2007). CD146 positive cells differentiate into osteoblasts, chondrocytes, adipocytes, and smooth muscle cells, a property shared with MSC. In addition, when grafted *in vivo*, they form bone structures, wherein host hematopoiesis develops. CD146+ cells can be re-isolated from these ectopic bone structures, thereby demonstrating true stem cell properties of such MSCs.

Prolonged culture of MSCs *in vitro* is associated with changes in their expressed gene profile including among others, increased transcript levels of cytoskeleton proteins and decreased levels of cell cycle inhibitory genes (Boquest et al., 2005), changes that can also be seen at the epigenetic level (Bork et al., 2010). As mentioned above, variability in MSC function is also caused by relatively minor changes to the culture conditions such different lots of FCS versus autologous serum, inclusion of platelet lysate or addition of growth factors (for instance, FGF2). All these changes are also associated with differences in expressed gene profile and epigenomic state of the cells (Bieback et al., 2009; Boquest et al., 2005; Holzwarth et al., 2010; Sanchez et al., 2009).

5.2 Effects of MSCs on host tissues

One characteristic of MSCs is that they have significant immunomodulatory capacities, whilst not eliciting an immune response. First, MSCs inhibit the proliferation and function of T cells. This inhibition is not MHC-restricted as it can be mediated by both syngeneic/autologous and allogeneic MSC (Krampera et al., 2003; Le Blanc et al., 2003). Contact-dependent mechanisms (Augello et al., 2005) as well as the release of soluble factors are thought to operate in MSC-mediated T cell suppression (Nauta and Fibbe, 2007). Examples of immunosuppressive factors involved are (interferon-γ) induced nitric oxide (NO), and constitutively produced factors such as transforming growth factor-β1, hepatocyte growth factor, PGE2, IL-10, and others (Krampera et al., 2006). Due to the absence of co-stimulatory molecules such as CD80 (B7-1) and CD86 (B7-2), MSCs are believed to render T cells anergic in a reversible manner (Di Nicola et al., 2002; Krampera et al., 2003). Finally, MSCs modulate adaptive T cell immunity indirectly through the induction of suppressive regulatory T cells (Aggarwal and Pittenger, 2005; Maccario et al., 2005).

Furthermore, MSCs inhibit the maturation of monocytes and $CD34^+$ cells into mature DCs *in vitro*, as well as the antigen presenting function of DCs (Nauta and Fibbe, 2007; Uccelli et al., 2008). MSCs are also reported to inhibit the cytotoxic function of human NK (natural killer) cells by downregulating the expression of activating NK receptors such as NKG2D (Spaggiari et al., 2008). Inhibitory effects on B cells are controversial (Nauta and Fibbe, 2007; Uccelli et al., 2008), but since B cell responses are mainly T cell dependent, MSCs have an indirect effect on B cells by inhibiting T cell function.

MSCs are known to be immunoprivileged, displaying low expression levels of MHC molecules, and no expression of co-stimulatory molecules. Thus, they can be used over MHC barriers in humans. Hence, in contrast to, for instance, EPCs or NSCs, MSCs can be used as an "off-the-shelf" product, and may therefore be used to treat the acute phase of stroke.

Aside from modulating the immune system, MSCs also produce trophic cytokines and growth factors that can influence host cells. These factors are in part responsible for the "regeneration" seen following MSC transplantation in a number of disorders, including AMI, peripheral ischemia, or stroke. The precise nature of factors produced by MSCs that enhance, for instance, angiogenesis and neurogenesis are unknown. Although proteomics approaches have identified factors produced by MSCs *in vitro*, studies are still needed to address which factors are produced once

MSCs are introduced *in vivo*, as it is now very clear that the environment wherein MSCs reside influences their phenotype and function.

When transplanted intravenously, MSCs are trapped chiefly by the extensive vascular bed of the lung, and only small numbers of MSC migrate to lesioned tissues (Gao et al., 2001). Even when syngeneic MSCs are infused, or introduced in sites of injury, only very small numbers of cells can be detected long term (Wagner et al., 2009). The beneficial effects found in a variety of disorders as a consequence of MSC transplantation must therefore be due to the short-term production of trophic, anti-inflammatory or immunomodulatory factors, or cell–cell interactions. How the effects occur remains still not fully understood. However, as only short-term presence of MSCs is sufficient to exert beneficial effects, the fast disappearance of MSCs could be considered beneficial, as late problems, such as tumor formation or tissue nonspecific differentiation may not be of (or be of lesser) concern.

5.3 Direct contribution of MSCs to tissues

It is well established that MSCs have the ability to differentiate into various tissues of mesodermal origin (osteocytes, chondrocytes, adipocytes; Caplan, 2007). Less clear is whether they themselves also differentiate into ECs, neural cells, and cells of endodermal origin. This will be further addressed in the section below.

6 ADULT STEM CELLS WITH GREATER POTENCY

6.1 Phenotype and origin of cells with greater potency

Since the 2000s, a number of cells have been described with broader differentiation ability than MSCs, and which are isolated by plastic adherence (as is done for MSCs) from human BM or other tissues, such amniotic fluid and UCB. These include pre-MSCs, which are murine in origin, and the human multipotent adult progenitor cells (MAPCs), very small embryonic stem cell like (VSEL) cells, marrow isolated adult multilineage inducible (MIAMI) cells, multiptotent adult stem cells (MASCs), amniotic fluid stem (AFS) cells, and unrestricted somatic stem cells (USSCs), among others (Anjos-Afonso and Bonnet, 2007; Beltrami et al., 2007; D'Ippolito et al., 2004; Kucia et al., 2007, 2008; Roobrouck et al., 2011a). The only two cell populations that are directly isolated from BM or UCB are VSEL cells and pre-MSCs. All other cell types have been isolated by plating single cell suspensions of the different tissues under conditions that differ in a number of features from the culture conditions used for the expansion of MSCs. Whether these differences in culture conditions are responsible for the differences between the different cell populations, and between these cell populations and "classical" MSCs is unknown. The culture conditions and characteristics of these cell populations have been recently summarized by Roobrouck et al. (2011b). Roobrouck et al. also recently reported that the phenotype and function of human MSCs and human MAPCs is at least in part due to differences in the culture conditions used (Roobrouck et al., 2011a).

A number of these human cell populations are reported to express the ESC-specific transcription factors OCT4 and NANOG. This notion is, however, controversial. The facts that no Oct4-positive cells can be detected in different tissues of Oct4-GFP mice and that conditional tissue specific knockout of Oct4 does not affect tissue regeneration following damage to multiple organs (Lengner et al., 2007), strongly suggest that Oct4 is not required for the adult organism and is silenced during development (Boiani et al., 2004; Kehler et al., 2004). It is, however, possible that *in vitro* culture of adult stem cells obtained from somatic tissue may lead to the reactivation of OCT4. Further complicating this issue in human cells is the fact that a number of OCT4-pseudogenes exist, which can be detected at the transcript or protein level if care is not taken that primers and antibodies specific for the "stemness" isoform, OCT4a, are being used (Atlasi et al., 2008; Liedtke et al., 2007; Zuk, 2009). Ideally, these results should also be confirmed by evaluation of the methylation state of the OCT4 promoter (Cantz et al., 2008).

As for OCT4, at least 11 human pseudogenes have been identified for NANOG (Booth and Holland, 2004; Fairbanks and Maughan, 2006). Aside from pseudogenes, alternative splice variants have been described. In contrast to OCT4a, NANOG is not expressed in later postnatal germline stem cells. Although NANOG expression has been reported in some adult somatic stem cells, no good tools are available to discriminate between NANOG itself and its pseudogenes, or isoforms. Therefore, it is unclear whether the detection of NANOG in some cell population but not in others represents true differences between the populations or is an artifactual detection of NANOG. Promoter methylation studies could help clarify whether NANOG is expressed in adult somatic cells.

6.2 Therapeutic effects of cells with greater potency

As has been shown for MSCs, many of the cells with greater potency (MAPCs, MIAMI cells, USSCs, AFS cells) produce trophic factors that affect endogenous repair mechanisms in the setting of vascular ischemia and have similar immunomodulatory effects while not inducing an immune response.

6.3 Direct contribution to tissues

To what extent any of these cells can themselves contribute to tissues *in vivo* remains unclear. As is true for MSCs, several of the populations have been grafted *in vivo*. In contrast to MSCs, human MAPCs can generate ECs, creating functional vessels in matrigel plugs, as do pre-MSCs, MASCs, and USSCs. Some of these studies have suggested direct contribution to cardiac muscle and neurons. However, levels of contribution were, in general, low, and the differentiated cell types were only qualified by immunostaining, and not function.

7 ADULT STEM CELL GRAFTS FOR STROKE

Here, we will try to summarize the results obtained in preclinical studies that have tested the ability of some of these non-neural (stem) cell populations described above for their ability to mitigate functional impairment following stroke. The preclinical models most often used young adult rats and fewer mice, and only few studies have used aged animals or animals with underlying pathologies, such as hyperlipidemia or hypertension, as is also found in patients with ischemic brain disorders. Interestingly, a significant number of studies grafted the cells intravenously and not directly into sites near the lesion in the brain. Moreover, a significant number of studies grafted human cells in immunocompetent animal models.

Aside from neural repair, revascularization is also required following an ischemic insult to the brain, which might be achieved by grafting EPCs. Only few studies have tested the effect of enriched EPCs in the setting of ischemia. Following systemic transplantation, some grafted cells can be detected near the ischemic area, even though direct contribution of the grafted cells to EC lining of vessels was rare. Nevertheless, as is true for NSCs and MSCs, EPCs also reduce infarct size and improve functional recovery (Fan et al., 2011; Moubarik et al., 2011; Nakamura et al., 2011).

Aside from EPCs, a large number of studies have tested the effect of crude BMCs, UCB cells, or MSCs derived from multiple tissues on functional recovery after stroke. Most but not all studies demonstrated (some) functional improvement, whether cells were transplanted within hours to days after the stroke, or after several weeks. The reasons for the functional improvement quoted in the different studies differ, however, quite extensively. Although some studies have suggested that BMCs, UCB cells, or MSCs may generate neurons both *in vitro* and *in vivo* (Deng and Daley, 2001; Hess et al., 2002a,b; Ikegame et al., 2011; Kopen et al., 1999; Mezey et al., 2003; Yang et al., 2011; Zhao et al., 2002), it is, in general, accepted that this transdifferentiation is minimal if it occurs at all. Nevertheless, substantial functional benefits have been observed after both intracranial and intravascular delivery (Giraldi-Guimaraes et al., 2009; Honma et al., 2006; Liao et al., 2009; Pavlichenko et al., 2008; Sugiyama et al., 2011). However, grafted cells appear survive poorly after transplantation into the brain (Coyne et al., 2006; Keimpema et al., 2009) and whether or not grafted cells that are infused systemically enter the injured brain is also unclear, with some studies identifying donor cells near the infarct (Nomura et al., 2005; Yang et al., 2012), while others do not (Steiner et al., 2012).

It is believed that the grafted cells exert anti-apoptotic and neuroproliferative signals, which lead to enhanced neo-neurogenesis and improved survival of existing and newborn neurons (Alder et al., 2012; Bao et al., 2011; Chen et al., 2012; Deng et al., 2011; Koh et al., 2008; Leu et al., 2011; Steiner et al., 2012; Voulgari-Kokota et al., 2012; Wakabayashi et al., 2011; Yoo et al., 2008). Similarly, grafted cells enhance angiogenesis (Moisan et al., 2012) and neural plasticity (Andrews et al., 2008). In addition, the ability of MSCs to influence T cell function, DC maturation, and NK cell function may be associated with reduced monocyte/microglia activation and scar formation (Bacigaluppi et al., 2009; Ekdahl et al., 2009).

Notably, a number of studies have demonstrated that even BM- or UCB-derived MSC-xenografts can improve functional outcome. This is possibly due to the fact that MSCs are to some extent immunoprotective and may inhibit the innate immune system, including the function of NK cells (Spaggiari et al., 2008), which is responsible for the hyperacute rejection associated with xenografts (Li et al., 2009). As discussed above, stroke itself is associated with SIDS (Meisel and Meisel, 2012; Trakhtenberg and Goldberg, 2012). Although it has not been extensively studied, some data are available to indicate that treatment with immunosuppressive MSCs may not aggravate the immunosuppressed state (Scheibe et al., 2012).

8 CONCLUSIONS

In this chapter, we tried to discuss a number of adult (stem) cells that are currently being evaluated for their possible neuroregenerative properties. We hope that we convinced the readers that aside from NSCs, no strong evidence exists that (stem) cells from other postnatal tissues, including bone marrow, (umbilical cord) blood, fat tissue, or dental pulp, have the capacity to (robustly) generate neural progeny. Nevertheless, these non-neuroectoderm-derived cell populations appear significantly to activate endogenous neurogenesis and angiogenesis. Such "trophic" effects are also, at least partly, responsible for some of the beneficial effects seen when NSCs are grafted in the brain. Therefore, even these non-neuroectodermal cells may exert beneficial effects when used to treat neurodegenerative disorders. However, they, themselves, will be unable to regenerate the lost neural tissue.

References

Aggarwal, S., Pittenger, M.F., 2005. Human mesenchymal stem cells modulate allogeneic immune cell responses. Blood 105, 1815–1822.

Alder, J., Kramer, B.C., Hoskin, C., Thakker-Varia, S., 2012. Brain-derived neurotrophic factor produced by human umbilical tissue-derived cells is required for its effect on hippocampal dendritic differentiation. Dev. Neurobiol. 72, 755–765.

American Heart Association, 2009. Heart disease and stroke statistics update: a report from the Statistics Committee and American Heart Association Stroke Statistics Subcommittee. Circulation 22, e21–e181.

Andrews, E.M., Tsai, S.Y., Johnson, S.C., Farrer, J.R., Wagner, J.P., Kopen, G.C., Kartje, G.L., 2008. Human adult bone marrow-derived somatic cell therapy results in functional recovery and axonal plasticity following stroke in the rat. Exp. Neurol. 211, 588–592.

Anjos-Afonso, F., Bonnet, D., 2007. Nonhematopoietic/endothelial SSEA-1 + cells define the most primitive progenitors in the adult murine bone marrow mesenchymal compartment. Blood 109, 1298–1306.

Aranguren, X.L., McCue, J.D., Hendrickx, B., Zhu, X.H., Du, F., Chen, E., Pelacho, B., Penuelas, I., Abizanda, G., Uriz, M., et al., 2008. Multipotent adult progenitor cells sustain function of ischemic limbs in mice. J. Clin. Invest. 118, 505–514.

Arslan, F., Keogh, B., McGuirk, P., Parker, A.E., 2011. TLR2 and TLR4 in ischemia reperfusion injury. Mediators Inflamm. 2010, 704202.

Arvidsson, A., Collin, T., Kirik, D., Kokaia, Z., Lindvall, O., 2002. Neuronal replacement from endogenous precursors in the adult brain after stroke. Nat. Med. 8, 963–970.

Asahara, T., Takahashi, T., Masuda, H., Kalka, C., Chen, D., Iwaguro, H., Inai, Y., Silver, M., Isner, J., 1999. VEGF contributes to postnatal neovascularization by mobilizing bone marrow-derived endothelial progenitor cells. EMBO J. 18, 3964–3972.

Atlasi, Y., Mowla, S.J., Ziaee, S.A., Gokhale, P.J., Andrews, P.W., 2008. OCT4 spliced variants are differentially expressed in human pluripotent and nonpluripotent cells. Stem Cells 26, 3068–3074.

Augello, A., Tasso, R., Negrini, S.M., Amateis, A., Indiveri, F., Cancedda, R., Pennesi, G., 2005. Bone marrow mesenchymal progenitor cells inhibit lymphocyte proliferation by activation of the programmed death 1 pathway. Eur. J. Immunol. 35, 1482–1490.

Bacigaluppi, M., Pluchino, S., Peruzzotti-Jametti, L., Kilic, E., Kilic, U., Salani, G., Brambilla, E., West, M.J., Comi, G., Martino, G., et al., 2009. Delayed post-ischaemic neuroprotection following systemic neural stem cell transplantation involves multiple mechanisms. Brain 132, 2239–2251.

Balasubramaniam, V., Ryan, S.L., Seedorf, G.J., Roth, E.V., Heumann, T.R., Yoder, M.C., Ingram, D.A., Hogan, C.J., Markham, N.E., Abman, S.H., 2010. Bone marrow-derived angiogenic cells restore lung alveolar and vascular structure after neonatal hyperoxia in infant mice. Am. J. Physiol. Lung Cell. Mol. Physiol. 298, L315–L323.

Bao, X., Wei, J., Feng, M., Lu, S., Li, G., Dou, W., Ma, W., Ma, S., An, Y., Qin, C., et al., 2011. Transplantation of human bone marrow-derived mesenchymal stem cells promotes behavioral recovery and endogenous neurogenesis after cerebral ischemia in rats. Brain Res. 1367, 103–113.

Beltrami, A.P., Cesselli, D., Bergamin, N., Marcon, P., Rigo, S., Puppato, E., D'Aurizio, F., Verardo, R., Piazza, S., Pignatelli, A., et al., 2007. Multipotent cells can be generated in vitro from several adult human organs (heart, liver and bone marrow). Blood 110, 3438–3446.

Bieback, K., Hecker, A., Kocaomer, A., Lannert, H., Schallmoser, K., Strunk, D., Kluter, H., 2009. Human alternatives to fetal bovine serum for the expansion of mesenchymal stromal cells from bone marrow. Stem Cells 27, 2331–2341.

Boiani, M., Kehler, J., Scholer, H.R., 2004. Activity of the germline-specific Oct4-GFP transgene in normal and clone mouse embryos. Methods Mol. Biol. 254, 1–34.

Booth, H.A., Holland, P.W., 2004. Eleven daughters of NANOG. Genomics 84, 229–238.

Boquest, A.C., Shahdadfar, A., Fronsdal, K., Sigurjonsson, O., Tunheim, S.H., Collas, P., Brinchmann, J.E., 2005. Isolation and transcription profiling of purified uncultured human stromal stem cells: alteration of gene expression after in vitro cell culture. Mol. Biol. Cell 16, 1131–1141.

Bork, S., Pfister, S., Witt, H., Horn, P., Korn, B., Ho, A.D., Wagner, W., 2010. DNA methylation pattern changes upon long-term culture and aging of human mesenchymal stromal cells. Aging Cell 9, 54–63.

Bramlett, H.M., Dietrich, W.D., 2004. Pathophysiology of cerebral ischemia and brain trauma: similarities and differences. J. Cereb. Blood Flow Metab. 24, 133–150.

Cantz, T., Key, G., Bleidissel, M., Gentile, L., Han, D.W., Brenne, A., Scholer, H.R., 2008. Absence of OCT4 expression in somatic tumor cell lines. Stem Cells 26, 692–697.

Caplan, A.I., 1991. Mesenchymal stem cells. J. Orthop. Res. 9, 641–650.

Caplan, A.I., 2007. Adult mesenchymal stem cells for tissue engineering versus regenerative medicine. J. Cell. Physiol. 213, 341–347.

Chen, J., Shehadah, A., Pal, A., Zacharek, A., Cui, X., Cui, Y., Roberts, C., Lu, M., Zeitlin, A., Hariri, R., et al., 2012. Neuroprotective effect of human placenta-derived cell treatment of stroke in rats. Cell Transplant. Mar 28. [Epub ahead of print].

Coyne, T.M., Akiva, J.M., Woodbury, D., Black, I.B., 2006. Marrow stromal cells transplanted to the adult brain are rejected by an inflammatory response and transfer donor labels to host neurons and glia. Stem Cells 24 (11), 2483–2492. Epub 2006 Jul 27.

Deng, M., Daley, G.Q., 2001. Expression of interferon consensus sequence binding protein induces potent immunity against BCR/ABL-induced leukemia. Blood 97, 3491–3497.

Deng, Y.B., Ye, W.B., Hu, Z.Z., Yan, Y., Wang, Y., Takon, B.F., Zhou, G.Q., Zhou, Y.F., 2011. Intravenously administered BMSCs reduce neuronal apoptosis and promote neuronal proliferation through the release of VEGF after stroke in rats. Neurol. Res. 32, 148–156.

Di Nicola, M., Carlo-Stella, C., Magni, M., Milanesi, M., Longoni, P.D., Matteucci, P., Grisanti, S., Gianni, A.M., 2002. Human bone marrow stromal cells suppress T-lymphocyte proliferation induced by cellular or nonspecific mitogenic stimuli. Blood 99, 3838–3843.

D'Ippolito, G., Diabira, S., Howard, G.A., Menei, P., Roos, B.A., Schiller, P.C., 2004. Marrow-isolated adult multilineage inducible (MIAMI) cells, a unique population of postnatal young and old human cells with extensive expansion and differentiation potential. J. Cell Sci. 117, 2971–2981.

Dirnagl, U., Iadecola, C., Moskowitz, M.A., 1999. Pathobiology of ischaemic stroke: an integrated view. Trends Neurosci. 22, 391–397.

Dominici, M., Le Blanc, K., Mueller, I., Slaper-Cortenbach, I., Marini, F., Krause, D., Deans, R., Keating, A., Prockop, D., Horwitz, E., 2006. Minimal criteria for defining multipotent mesenchymal stromal cells. The International Society for Cellular Therapy position statement. Cytotherapy 8, 315–317.

Droetto, S., Viale, A., Primo, L., Jordaney, N., Bruno, S., Pagano, M., Piacibello, W., Bussolino, F., Aglietta, M., 2004. Vasculogenic potential of long term repopulating cord blood progenitors. FASEB J. 18, 1273–1275.

Ekdahl, C.T., Kokaia, Z., Lindvall, O., 2009. Brain inflammation and adult neurogenesis: the dual role of microglia. Neuroscience 158, 1021–1029.

Englund, U., Bjorklund, A., Wictorin, K., Lindvall, O., Kokaia, M., 2002. Grafted neural stem cells develop into functional pyramidal neurons and integrate into host cortical circuitry. Proc. Natl. Acad. Sci. U.S.A. 99, 17089–17094.

Fairbanks, D.J., Maughan, P.J., 2006. Evolution of the NANOG pseudogene family in the human and chimpanzee genomes. BMC Evol. Biol. 6, 12.

Fan, Y., Shen, F., Frenzel, T., Zhu, W., Ye, J., Liu, J., Chen, Y., Su, H., Young, W.L., Yang, G.Y., 2011. Endothelial progenitor cell transplantation improves long-term stroke outcome in mice. Ann. Neurol. 67, 488–497.

Fujiwara, Y., Tanaka, N., Ishida, O., Fujimoto, Y., Murakami, T., Kajihara, H., Yasunaga, Y., Ochi, M., 2004. Intravenously injected neural progenitor cells of transgenic rats can migrate to the injured spinal cord and differentiate into neurons, astrocytes and oligodendrocytes. Neurosci. Lett. 366, 287–291.

Gage, F.H., 2000. Mammalian neural stem cells. Science 287, 1433–1438.

Gao, J., Dennis, J.E., Muzic, R.F., Lundberg, M., Caplan, A.I., 2001. The dynamic in vivo distribution of bone marrow-derived mesenchymal stem cells after infusion. Cells Tissues Organs 169, 12–20.

Gee, J.M., Kalil, A., Shea, C., Becker, K.J., 2007. Lymphocytes: potential mediators of post-ischemic injury and neuroprotection. Stroke 38, 783–788.

Giraldi-Guimaraes, A., Rezende-Lima, M., Bruno, F.P., Mendez-Otero, R., 2009. Treatment with bone marrow mononuclear cells induces functional recovery and decreases neurodegeneration after sensorimotor cortical ischemia in rats. Brain Res. Feb 9. [Epub ahead of print].

Hayakawa, K., Qiu, J., Lo, E.H., 2011. Biphasic actions of HMGB1 signaling in inflammation and recovery after stroke. Ann. N. Y. Acad. Sci. 1207, 50–57.

Hendrickx, B., Verdonck, K., Van den Berge, S., Dickens, S., Eriksson, E., Vranckx, J.J., Luttun, A., 2010. Integration of blood outgrowth endothelial cells in dermal fibroblast sheets promotes full thickness wound healing. Stem Cells 28, 1165–1177.

Hess, D.C., Hill, W.D., Martin-Studdard, A., Carothers, J., Brailer, J., Carroll, J., 2002a. Blood into brain after stroke. Trends Mol. Med. 8, 452–453.

Hess, D.C., Hill, W.D., Martin-Studdard, A., Carroll, J., Brailer, J., Carothers, J., 2002b. Bone marrow as a source of endothelial cells and NeuN-expressing cells after stroke. Stroke 33, 1362–1368.

Holzwarth, C., Vaegler, M., Gieseke, F., Pfister, S.M., Handgretinger, R., Kerst, G., Muller, I., 2010. Low physiologic oxygen tensions reduce proliferation and differentiation of human multipotent mesenchymal stromal cells. BMC Cell Biol. 11, 11.

Honma, T., Honmou, O., Iihoshi, S., Harada, K., Houkin, K., Hamada, H., Kocsis, J.D., 2006. Intravenous infusion of immortalized human mesenchymal stem cells protects against injury in a cerebral ischemia model in adult rat. Exp. Neurol. 199, 56–66.

Hurn, P.D., Subramanian, S., Parker, S.M., Afentoulis, M.E., Kaler, L.J., Vandenbark, A.A., Offner, H., 2007. T- and B-cell-deficient mice with experimental stroke have reduced lesion size and inflammation. J. Cereb. Blood Flow Metab. 27, 1798–1805.

Ikegame, Y., Yamashita, K., Hayashi, S., Mizuno, H., Tawada, M., You, F., Yamada, K., Tanaka, Y., Egashira, Y., Nakashima, S., et al., 2011. Comparison of mesenchymal stem cells from adipose tissue and bone marrow for ischemic stroke therapy. Cytotherapy 13, 675–685.

Ingram, D.A., Mead, L.E., Moore, D.B., Woodard, W., Fenoglio, A., Yoder, M.C., 2005. Vessel wall-derived endothelial cells rapidly proliferate because they contain a complete hierarchy of endothelial progenitor cells. Blood 105, 2783–2786.

Kehler, J., Tolkunova, E., Koschorz, B., Pesce, M., Gentile, L., Boiani, M., Lomeli, H., Nagy, A., McLaughlin, K.J., Scholer, H.R., et al., 2004. Oct4 is required for primordial germ cell survival. EMBO Rep. 5, 1078–1083.

Keimpema, E., Fokkens, M.R., Nagy, Z., Agoston, V., Luiten, P.G., Nyakas, C., Boddeke, H.W., Copray, J.C., 2009. Early transient presence of implanted bone marrow stem cells reduces lesion size after cerebral ischaemia in adult rats. Neuropathol. Appl. Neurobiol. 35, 89–102.

Kelly, S., Bliss, T.M., Shah, A.K., Sun, G.H., Ma, M., Foo, W.C., Masel, J., Yenari, M.A., Weissman, I.L., Uchida, N., et al., 2004. Transplanted human fetal neural stem cells survive, migrate, and differentiate in ischemic rat cerebral cortex. Proc. Natl. Acad. Sci. U.S.A. 101, 11839–11844.

Koch, P., Kokaia, Z., Lindvall, O., Brustle, O., 2009. Emerging concepts in neural stem cell research: autologous repair and cell-based disease modelling. Lancet Neurol. 8, 819–829.

Koh, S.H., Kim, K.S., Choi, M.R., Jung, K.H., Park, K.S., Chai, Y.G., Roh, W., Hwang, S.J., Ko, H.J., Huh, Y.M., et al., 2008. Implantation of human umbilical cord-derived mesenchymal stem cells as a neuroprotective therapy for ischemic stroke in rats. Brain Res. 1229, 233–248.

Kong, Y., Le, Y., 2011. Toll-like receptors in inflammation of the central nervous system. Int. Immunopharmacol. 11, 1407–1414.

Kopen, G., Prockop, D., Phinney, D., 1999. Marrow stromal cells migrate throughout forebrain and cerebellum, and they differentiate into astrocytes after injection into neonatal mouse brains. Proc. Natl. Acad. Sci. U.S.A. 96, 10711–10716.

Krampera, M., Glennie, S., Dyson, J., Scott, D., Laylor, R., Simpson, E., Dazzi, F., 2003. Bone marrow mesenchymal stem cells inhibit the response of naive and memory antigen-specific T cells to their cognate peptide. Blood 101, 3722–3729.

Krampera, M., Cosmi, L., Angeli, R., Pasini, A., Liotta, F., Andreini, A., Santarlasci, V., Mazzinghi, B., Pizzolo, G., Vinante, F., et al., 2006. Role for interferon-gamma in the immunomodulatory activity of human bone marrow mesenchymal stem cells. Stem Cells 24, 386–398.

Kucia, M., Halasa, M., Wysoczynski, M., Baskiewicz-Masiuk, M., Moldenhawer, S., Zuba-Surma, E., Czajka, R., Wojakowski, W., Machalinski, B., Ratajczak, M.Z., 2007. Morphological and molecular characterization of novel population of CXCR4+ SSEA-4+ Oct-4+ very small embryonic-like cells purified from human cord blood: preliminary report. Leukemia 21, 297–303.

Kucia, M., Wysoczynski, M., Ratajczak, J., Ratajczak, M.Z., 2008. Identification of very small embryonic like (VSEL) stem cells in bone marrow. Cell Tissue Res. 331, 125–134.

Le Blanc, K., Tammik, L., Sundberg, B., Haynesworth, S.E., Ringden, O., 2003. Mesenchymal stem cells inhibit and stimulate mixed lymphocyte cultures and mitogenic responses independently of the major histocompatibility complex. Scand. J. Immunol. 57, 11–20.

Lee, J.P., Jeyakumar, M., Gonzalez, R., Takahashi, H., Lee, P.J., Baek, R.C., Clark, D., Rose, H., Fu, G., Clarke, J., et al., 2007. Stem cells act through multiple mechanisms to benefit mice with neurodegenerative metabolic disease. Nat. Med. 13, 439–447.

Lengner, C.J., Camargo, F.D., Hochedlinger, K., Welstead, G.G., Zaidi, S., Gokhale, S., Scholer, H.R., Tomilin, A., Jaenisch, R., 2007. Oct4 expression is not required for mouse somatic stem cell self-renewal. Cell Stem Cell 1, 403–415.

Leu, S., Lin, Y.C., Yuen, C.M., Yen, C.H., Kao, Y.H., Sun, C.K., Yip, H.K., 2011. Adipose-derived mesenchymal stem cells markedly attenuate brain infarct size and improve neurological function in rats. J. Transl. Med. 8, 63.

Li, S., Waer, M., Billiau, A.D., 2009. Xenotransplantation: role of natural immunity. Transpl. Immunol. 21, 70–74.

Liao, W., Xie, J., Zhong, J., Liu, Y., Du, L., Zhou, B., Xu, J., Liu, P., Yang, S., Wang, J., et al., 2009. Therapeutic effect of human umbilical cord multipotent mesenchymal stromal cells in a rat model of stroke. Transplantation 87, 350–359.

Liedtke, S., Enczmann, J., Waclawczyk, S., Wernet, P., Kogler, G., 2007. Oct4 and its pseudogenes confuse stem cell research. Cell Stem Cell 1, 364–366.

Lin, Y., Weisdorf, D.J., Solovey, A., Hebbel, R.P., 2000. Origins of circulating endothelial cells and endothelial outgrowth from blood. J. Clin. Invest. 105, 71–77.

Maccario, R., Podesta, M., Moretta, A., Cometa, A., Comoli, P., Montagna, D., Daudt, L., Ibatici, A., Piaggio, G., Pozzi, S., et al., 2005. Interaction of human mesenchymal stem cells with cells involved in alloantigen-specific immune response favors the differentiation of CD4+ T-cell subsets expressing a regulatory/suppressive phenotype. Haematologica 90, 516–525.

Madinier, A., Bertrand, N., Mossiat, C., Prigent-Tessier, A., Beley, A., Marie, C., Garnier, P., 2009. Microglial involvement in neuroplastic changes following focal brain ischemia in rats. PLoS One 4, e8101.

Maroso, M., Balosso, S., Ravizza, T., Liu, J., Bianchi, M.E., Vezzani, A., 2011. Interleukin-1 type 1 receptor/Toll-like receptor signalling in epilepsy: the importance of IL-1beta and high-mobility group box 1. J. Intern. Med. 270, 319–326.

Meisel, C., Meisel, A., 2012. Suppressing immunosuppression after stroke. N. Engl. J. Med. 365, 2134–2136.

Mezey, E., Key, S., Vogelsang, G., Szalayova, I., Lange, G.D., Crain, B., 2003. Transplanted bone marrow generates new neurons in human brains. Proc. Natl. Acad. Sci. U.S.A. 100, 1364–1369.

Minatoguchi, S., 2011. Granulocyte colony stimulating factor, peripheral blood stem cells and bone marrow stem cells for cardiac repair after myocardial infarction. Circ. J. 75, 789–790.

Moisan, A., Pannetier, N., Grillon, E., Richard, M.J., de Fraipont, F., Remy, C., Barbier, E.L., Detante, O., 2012. Intracerebral injection of human mesenchymal stem cells impacts cerebral microvasculature after experimental stroke: MRI study. NMR Biomed. doi:http://dx.doi.org/10.1002/nbm.2806. Apr 27. [Epub ahead of print].

Moubarik, C., Guillet, B., Youssef, B., Codaccioni, J.L., Piercecchi, M.D., Sabatier, F., Lionel, P., Dou, L., Foucault-Bertaud, A., Velly, L., et al., 2011. Transplanted late outgrowth endothelial progenitor cells as cell therapy product for stroke. Stem Cell Rev. 7, 208–220.

Nakamura, K., Tsurushima, H., Marushima, A., Nagano, M., Yamashita, T., Suzuki, K., Ohneda, O., Matsumura, A., 2011. A subpopulation of endothelial progenitor cells with low aldehyde dehydrogenase activity attenuates acute ischemic brain injury in rats. Biochem. Biophys. Res. Commun. 418, 87–92.

Nauta, A.J., Fibbe, W.E., 2007. Immunomodulatory properties of mesenchymal stromal cells. Blood 110, 3499–3506.

Nomura, T., Honmou, O., Harada, K., Houkin, K., Hamada, H., Kocsis, J.D., 2005. I.V. infusion of brain-derived neurotrophic factor gene-modified human mesenchymal stem cells protects against injury in a cerebral ischemia model in adult rat. Neuroscience 136, 161–169.

Ourednik, J., Ourednik, V., Lynch, W.P., Schachner, M., Snyder, E.Y., 2002. Neural stem cells display an inherent mechanism for rescuing dysfunctional neurons. Nat. Biotechnol. 20, 1103–1110.

Park, K.I., Teng, Y.D., Snyder, E.Y., 2002. The injured brain interacts reciprocally with neural stem cells supported by scaffolds to reconstitute lost tissue. Nat. Biotechnol. 20, 1111–1117.

Passier, R., van Laake, L.W., Mummery, C.L., 2008. Stem-cell-based therapy and lessons from the heart. Nature 453, 322–329.

Pavlichenko, N., Sokolova, I., Vijde, S., Shvedova, E., Alexandrov, G., Krouglyakov, P., Fedotova, O., Gilerovich, E.G., Polyntsev, D.G., Otellin, V.A., 2008. Mesenchymal stem cells transplantation could be beneficial for treatment of experimental ischemic stroke in rats. Brain Res. 1233, 203–213.

Planas, A.M., Chamorro, A., 2009. Regulatory T cells protect the brain after stroke. Nat. Med. 15, 138–139.

Pluchino, S., Zanotti, L., Rossi, B., Brambilla, E., Ottoboni, L., Salani, G., et al., 2005. Neurosphere-derived multipotent precursors promote neuroprotection by an immunomodulatory mechanism. Nature 436, 266–271.

Ramos-Cabrer, P., Justicia, C., Wiedermann, D., Hoehn, M., 2010. Stem cell mediation of functional recovery after stroke in the rat. PLoS One 5, e12779.

Rehman, J., Li, J., Orschell, C.M., March, K.L., 2003. Peripheral blood "endothelial progenitor cells" are derived from monocyte/macrophages and secrete angiogenic growth factors. Circulation 107, 1164–1169.

Roobrouck, V., Clavel Claver, C., Jacobs, S.A., Ulloa-Montoya, F., Crippa, S., Sohni, A., Roberts, S., Luyten, F.P., Van Gool, S., Sampaolesi, M., et al., 2011a. Differentiation potential of human postnatal MSC, MAB and MAPC reflected in their transcriptome and partially influenced by the culture conditions. Stem Cells 29 (5), 871–882. doi:http://dx.doi.org/10.1002/stem.633.

Roobrouck, V.D., Vanuytsel, K., Verfaillie, C.M., 2011b. Concise review: culture mediated changes in fate and/or potency of stem cells. Stem Cells 29, 583–589.

Sacchetti, B., Funari, A., Michienzi, S., Di Cesare, S., Piersanti, S., Saggio, I., Tagliafico, E., Ferrari, S., Robey, P.G., Riminucci, M., et al., 2007. Self-renewing osteoprogenitors in bone marrow sinusoids can organize a hematopoietic microenvironment. Cell 131, 324–336.

Sanchez, M., Secades, L., Bordallo, C., Meana, C., Rubin, J.M., Cantabrana, B., Bordallo, J., 2009. Role of polyamines and cAMP-dependent mechanisms on 5alpha-dihydrotestosterone-elicited functional effects in isolated right atria of rat. J. Cardiovasc. Pharmacol. 54, 310–318.

Scheibe, F., Ladhoff, J., Huck, J., Grohmann, M., Blazej, K., Oersal, A., Baeva, N., Seifert, M., Priller, J., 2012. Immune effects of mesenchymal stromal cells in experimental stroke. J. Cereb. Blood Flow Metab. 32 (8), 1578–1588. doi:http://dx.doi.org/10.1038/jcbfm.2012.55. Epub 2012 May 2.

Schmidt-Lucke, C., Rossig, L., Fichtlscherer, S., Vasa, M., Britten, M., Kamper, U., Dimmeler, S., Zeiher, A.M., 2005. Reduced number of circulating endothelial progenitor cells predicts future cardiovascular events: proof of concept for the clinical importance of endogenous vascular repair. Circulation 111, 2981–2987.

Sieveking, D.P., Buckle, A., Celermajer, D.S., Ng, M.K., 2008. Strikingly different angiogenic properties of endothelial progenitor cell subpopulations: insights from a novel human angiogenesis assay. J. Am. Coll. Cardiol. 51, 660–668.

Sozen, T., Tsuchiyama, R., Hasegawa, Y., Suzuki, H., Jadhav, V., Nishizawa, S., Zhang, J.H., 2009. Role of interleukin-1beta in early brain injury after subarachnoid hemorrhage in mice. Stroke 40, 2519–2525.

Spaggiari, G.M., Capobianco, A., Abdelrazik, H., Becchetti, F., Mingari, M.C., Moretta, L., 2008. Mesenchymal stem cells inhibit natural killer-cell proliferation, cytotoxicity, and cytokine production: role of indoleamine 2,3-dioxygenase and prostaglandin E2. Blood 111, 1327–1333.

Steiner, B., Roch, M., Holtkamp, N., Kurtz, A., 2012. Systemically administered human bone marrow-derived mesenchymal stem home into peripheral organs but do not induce neuroprotective effects in the MCAo-mouse model for cerebral ischemia. Neurosci. Lett. 513, 25–30.

Sugiyama, M., Iohara, K., Wakita, H., Hattori, H., Ueda, M., Matsushita, K., Nakashima, M., 2011. Dental pulp-derived CD31(-)/CD146(-) side population stem/progenitor cells enhance recovery of focal cerebral ischemia in rats. Tissue Eng. Part A 17, 1303–1311.

Trakhtenberg, E.F., Goldberg, J.L., 2012. Immunology. Neuroimmune communication. Science 334, 47–48.

Uccelli, A., Moretta, L., Pistoia, V., 2008. Mesenchymal stem cells in health and disease. Nat. Rev. Immunol. 8, 726–736.

Voulgari-Kokota, A., Fairless, R., Karamita, M., Kyrargyri, V., Tseveleki, V., Evangelidou, M., Delorme, B., Charbord, P., Diem, R., Probert, L., 2012. Mesenchymal stem cells protect CNS neurons against glutamate excitotoxicity by inhibiting glutamate receptor expression and function. Exp. Neurol. 236 (1), 161–170. Epub 2012 Apr 25.

Wagner, J., Kean, T., Young, R., Dennis, J.E., Caplan, A.I., 2009. Optimizing mesenchymal stem cell-based therapeutics. Curr. Opin. Biotechnol. 20, 531–536.

Wakabayashi, K., Nagai, A., Sheikh, A.M., Shiota, Y., Narantuya, D., Watanabe, T., Masuda, J., Kobayashi, S., Kim, S.U., Yamaguchi, S., 2011. Transplantation of human mesenchymal stem cells promotes functional improvement and increased expression of neurotrophic factors in a rat focal cerebral ischemia model. J. Neurosci. Res. 88, 1017–1025.

Wang, Q., Tang, X.N., Yenari, M.A., 2007. The inflammatory response in stroke. J. Neuroimmunol. 184, 53–68.

Yang, Y.C., Liu, B.S., Shen, C.C., Lin, C.H., Chiao, M.T., Cheng, H.C., 2011. Transplantation of adipose tissue-derived stem cells for treatment of focal cerebral ischemia. Curr. Neurovasc. Res. 8, 1–13.

Yang, K.L., Lee, J.T., Pang, C.Y., Lee, T.Y., Chen, S.P., Liew, H.K., Chen, S.Y., Chen, T.Y., Lin, P.Y., 2012. Human adipose-derived stem cells for the treatment bof intracerebral hemorrhage in rats via femoral intravenous injection. Cell. Mol. Biol. Lett. 17 (3), 376–392. Epub 2012 Apr 28.

Yoder, M.C., Mead, L.E., Prater, D., Krier, T.R., Mroueh, K.N., Li, F., Krasich, R., Temm, C.J., Prchal, J.T., Ingram, D.A., 2007. Redefining endothelial progenitor cells via clonal analysis and hematopoietic stem/progenitor cell principals. Blood 109, 1801–1809.

Yoo, S.W., Kim, S.S., Lee, S.Y., Lee, H.S., Kim, H.S., Lee, Y.D., Suh-Kim, H., 2008. Mesenchymal stem cells promote proliferation of endogenous neural stem cells and survival of newborn cells in a rat stroke model. Exp. Mol. Med. 40, 387–397.

Zentilin, L., Tafuro, S., Zacchigna, S., Arsic, N., Pattarini, L., Sinigaglia, M., Giacca, M., 2006. Bone marrow mononuclear cells are recruited to the sites of VEGF-induced neovascularization but are not incorporated into the newly formed vessels. Blood 107, 3546–3554.

Zhao, L.R., Duan, W.M., Reyes, M., Keene, C.D., Verfaillie, C.M., Low, W.C., 2002. Human bone marrow stem cells exhibit neural phenotypes and ameliorate neurological deficits after grafting into the ischemic brain of rats. Exp. Neurol. 174, 11–20.

Zuk, P.A., 2009. The intracellular distribution of the ES cell totipotent markers OCT4 and Sox2 in adult stem cells differs dramatically according to commercial antibody used. J. Cell. Biochem. 106, 867–877.

CHAPTER

Stem cell repair of striatal ischemia

4

Zaal Kokaia[1], Olle Lindvall

Laboratory of Stem Cells and Restorative Neurology, Lund Stem Cell Center, University Hospital, Lund, Sweden

[1]Corresponding author. Tel.: +46-46-2220276, Fax: +46-46-2220560, e-mail address: Zaal.Kokaia@med.lu.se

Abstract

Stem cell-based approaches have promising prospects as new treatments to repair the stroke-damaged brain and promote efficient functional recovery. Studies in animal models of stroke have shown that transplantation of neural stem/progenitor cells (NSPCs) isolated from different sources can improve function not only by replacing dead neurons but also by trophic actions, modulation of inflammation, promotion of angiogenesis, and neuroprotection. Endogenous NSPCs produce new neurons following stroke and are also potential therapeutic targets. Joint efforts of basic researchers and clinicians are needed to develop safe and reproducible NSPC sources, test them in relevant animal models of stroke, and optimize the recovery of impaired functions.

Keywords

cell replacement, neural stem cell, neurogenesis, stroke, subventricular zone, transplantation

1 INTRODUCTION

Ischemic stroke is caused by occlusion of a cerebral artery due to thromboembolism, leading to focal ischemia. Following ischemic stroke, neurons rapidly die, but additional cell loss occurs hours and days thereafter with limited repair. Cell death is causing sensory, motor, and cognitive impairments. Apart from thrombolysis during the first hours after an ischemic stroke, which can be given only to a few percent of patients, no effective treatment to improve functional recovery exists in the postischemic phase. Therefore, this disease is desperate for a cure. According to the World Health Organization, 15 million people suffer stroke worldwide each year. Of these, 5 million die and another 5 million are permanently disabled. Although most stroke

Progress in Brain Research, Volume 201, ISSN 0079-6123, http://dx.doi.org/10.1016/B978-0-444-59544-7.00003-2

survivors improve to some degree, the majority of patients still suffer substantial impairments. The most commonly occluded vessel in humans is the middle cerebral artery (MCA), which supplies blood to the basal ganglia and cerebral cortex, including temporal, anterolateral frontal, and parietal lobes. The MCA occlusion model in animals closely mimics the pattern and dynamics of the cerebral lesion in humans.

Research has previously focused on the acute and subacute phases of stroke, assuming that the therapeutic window for rescuing degenerating neurons is limited to this period. It is now becoming more and more clear that the spontaneous recovery phase, which lasts for several months, represents a suitable target for treatments that promote regenerative/repair processes. Stem cell-based approaches are of particular interest with the potential to provide completely novel therapies to restore function in stroke. These approaches are based on delivering stem cells from different sources through intracerebral transplantation or systemic injection, or to target the adult brain's own neural stem/progenitor cells (NSPCs) and modulate their function. In this chapter, we summarize current knowledge regarding these two main stem cell-based approaches for functional restoration in the stroke-damaged brain. We describe recent findings, indicating that stem cells can improve function not only by replacing dead neurons but also by trophic actions, modulation of inflammation, stimulation of angiogenesis, and neuroprotection.

2 TRANSPLANTATION OF NSPCs IN STROKE

Conceptually, the idea behind stem cell therapy is rather straightforward: stem cells are isolated and expanded in culture, directed toward the desired phenotype, and subsequently transplanted into the diseased human recipient (Fig. 1). The grafted cells may replace damaged or dead endogenous neural cells, but can also facilitate plastic changes and provide cellular support and reorganization in the stroke-affected part of the brain.

NSPCs with self-renewal capacity and multipotentiality within the neural lineage can be derived from the fetal brain, embryonic stem (ES) cells, and induced pluripotent stem (iPS) cells. Proof-of-principle that neuronal replacement and symptomatic improvement of impaired function is possible in the human brain has been obtained in patients with Parkinson's disease (Lindvall and Björklund, 2004). By analogy, the new neurons generated from the transplanted NSPCs in the stroke-damaged brain may replace the ones which have died due to the ischemia, establish synaptic connections with the host brain, and contribute to the restoration of injured circuitry. In the striatum, neurons generated from grafted NSPCs could potentially be used to replace GABAergic striatopallidal medium spiny projection neurons or striatal interneurons. In the cortex, newly formed neurons might substitute dead or damaged corticostriatal, corticothalamic, corticospinal, and transcallosal glutamatergic projection neurons, or different types of GABAergic interneurons.

Following an ischemic insult, the remaining surviving neurons participate in the restoration of the disrupted neuronal circuitries. However, the plasticity of these

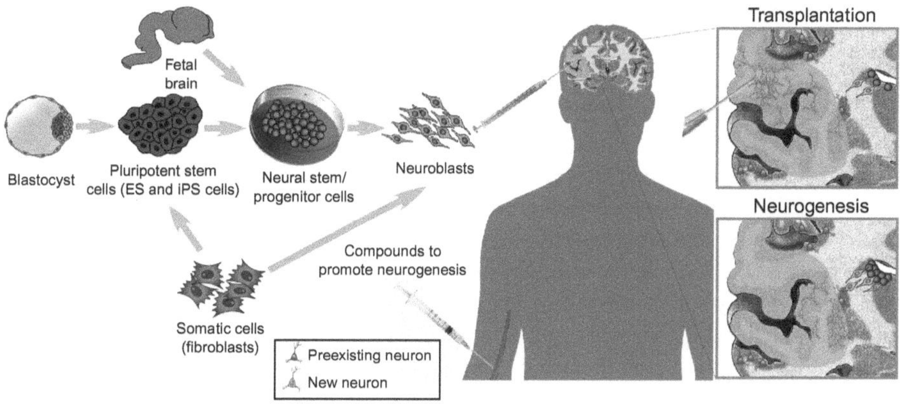

FIGURE 1

Cell replacement therapies for stroke. Human neural stem and progenitor cells (NSPCs) for replacement of dead and damaged neurons through transplantation in stroke-damaged brain can be derived from the fetal brain, pluripotent embryonic stem (ES) cells of blastocyst, and from somatic cells (e.g., skin fibroblasts) through generation of induced pluripotent stem (iPS) cells. Neuroblasts can also be generated through direct reprogramming of somatic cells. Finally, neuronal replacement and enhancement of functional recovery can be achieved by systemic delivery of compounds which promote different steps of stroke-induced neurogenesis from endogenous NSPCs. (For color version of this figure, the reader is referred to the online version of this chapter.)

mature, functionally integrated neurons and their capacity to form new synaptic connections is limited, especially in the aged brain. Therefore, supplying the brain with NSPCs having strong neurogenic potential and capacity to form new synaptic connections and to become incorporated in the host neural circuitries, make these cells very attractive for the regeneration of damaged neuronal networks after stroke.

A large number of studies have used NSPCs isolated from different sources and tested their ability to survive, grow, and reverse functional impairments after transplantation in different models of stroke (Andres et al., 2011; Darsalia et al., 2007; Horie et al., 2011; Kelly et al., 2004; Oki et al., 2012). Studies in which the cells were of human origin are the most interesting ones because it is highly likely that only human-derived cells will be approved for use in the clinical setting. All experimental studies are based on xenotransplantation and require immunosuppression or use of immune-deficient animals to prevent from rejection. Even in humans under allotransplantation conditions, immunosuppression has been used in the majority of clinical trials performed so far. Understanding the properties of human-derived NSPCs when implanted in the stroke-damaged brain of animals and the capacity of the grafts to improve stroke-impaired function will be particularly important for their clinical translation.

Several studies have demonstrated survival, neuronal differentiation, and even functional recovery after transplantation of NSPCs of human origin in animal models of stroke. Human fetal NSPCs migrated toward the ischemic lesion in rodents and differentiated into mature neurons after transplantation in stroke-damaged rodent cortex (Kelly et al., 2004) and striatum (Darsalia et al., 2007). The intracortically transplanted neurospheres survived robustly in naive and ischemic brains. Survival was influenced by the proximity of the graft to the stroke lesion and correlated negatively with the number of inflammatory cells (Kelly et al., 2004). Importantly, the human NSPCs demonstrated their capacity to migrate long distances (about 1.2 mm) predominantly toward the lesion. Migrating cells expressed the neuroblast marker doublecortin (DCX), whereas human cells at the lesion border expressed the immature neuronal marker βIII-tubulin. The follow-up study demonstrated that the positive effect of the cells on behavioral recovery was mediated through human vascular endothelial growth factor (hVEGF) secreted from the grafted cells (Horie et al., 2011). The hVEGF-induced effects included early improvement in blood–brain barrier integrity, suppression of inflammation, and increased neovascularization. The early phase of behavioral recovery was not dependent on neovascularization, but the late phase coincided with neovascularization. The same investigators have also demonstrated that human NSPCs can increase dendritic plasticity in stroke-lesioned cortex as well as on the side contralateral to the lesion in parallel to functional recovery (Andres et al., 2011). Stroke-subjected rats, which were transplanted with human NSPCs, also exhibited increased corticocortical, corticostriatal, corticothalamic, and corticospinal axonal rewiring from the intact, contralesional hemisphere and transcallosal and corticospinal axonal sprouting correlating with functional recovery. This clearly illustrates the capacity of human fetal brain-derived NSPCs to promote sprouting in the host brain.

We have shown that NSPCs derived from human fetal striatum and cortex exhibit region-specific differentiation *in vitro* and survive, migrate, and form neurons to a similar extent after intrastriatal transplantation in intact newborn rats (Kallur et al., 2006). When transplanted into the stroke-damaged striatum of adult rats, both types of cells exhibited a similar robust survival and migrated throughout the damaged striatum. However, striatal NSPCs migrated further and occupied a larger volume of striatum. The grafted cells showed very low proliferation rate without tumor formation or signs of overgrowth. Although most of the cells in the graft core exhibited features of immature neural cells, the grafted cells outside the core displayed mature neuronal morphology and expressed mature neuronal markers such as HuD, calbindin, and parvalbumin (Darsalia et al., 2007). Astrocytes were virtually absent.

For the clinical application, it is important to determine the optimal time after the stroke for stem cell transplantation and number of cells to be implanted for maximum survival and neuronal differentiation. In an attempt to address these issues, we transplanted human fetal striatum-derived NSPCs into the stroke-lesioned striatum (Darsalia et al., 2011). Our data demonstrated that transplantation of NSPCs shortly after the stroke (48 h) resulted in better cell survival than transplantation 6 weeks later. The delayed transplantation did not influence the magnitude of migration, neuronal differentiation, and cell proliferation in the grafts. Interestingly, transplanting higher number of NSPCs did not result in a greater number of surviving cells or increased neuronal differentiation. Stroke-induced brain damage causes an inflammatory response including activation of resident microglia and infiltration of blood-derived immune cells. A substantial number of activated microglia was observed at 48 h after the insult in the injured striatum and reached maximum levels at 1–6 weeks. These findings indicate that the best survival of grafted human NSPCs in stroke-damaged brain requires optimum numbers of cells to be transplanted in the early post-stroke phase, that is, before the inflammatory response has reached its maximum level.

Human fetal NSPCs have been also grafted into stroke-damaged brain of three adult cynomolgous monkeys (Roitberg et al., 2006). The magnetic resonance (MR) image-guided stereotactic intracerebral injections of NSPCs labeled with bromodeoxyuridine (BrdU) were performed in areas surrounding the ischemic lesion. Animals were killed at 45, 75, or 105 days after transplantation. Double-label immunofluorescence showed that the grafted human-derived NSPCs survived up to 105 days in the primate brain when transplanted 1 week after the ischemic event and partly underwent neuronal differentiation expressing the early neuronal marker βIII-tubulin.

Compared to systemic injection, intracerebral transplantation is the most suitable mode of cell delivery aiming for replacement of dead and damaged neurons in the stroke-damaged brain. First, based on the location and extent of the damage, which is extremely variable between stroke patients, cells can be placed directly in the desired target which can be defined on MR scan. Second, the exact number of cells can be delivered and distributed in the selected brain areas. Third, by intracerebral transplantation grafted cells avoid ending up in the ischemic core and/or peripheral organs of the body, which occurs using systemic delivery. Fourth, the relatively short time period after stroke during which the blood–brain barrier is opened significantly shortens the therapeutic time window which is available when targeting the recovery phase with systemic delivery. Fifth, unless the occluded vessel is recanalized, the access of the systemically injected cells to the damaged area is restricted and less efficient. Despite these disadvantages of systemic versus intracerebral delivery, human-derived NSPCs have been injected systemically in stroke-damaged animals at least in two studies and showed functional effects.

In the first study (Chu et al., 2004), stroke-subjected rats were injected with human fetal brain-derived NSPCs intravenously the day after surgery. Several weeks later, animals with NSPCs showed better behavioral performance and less hemispheric atrophy as compared to control animals. Human NSPCs were detected at the lesion site as long as 540 days after transplantation and expressed markers for neurons and astrocytes. Injected cells were present also in the kidneys, lungs, and spleen. Counting of cell numbers and their proliferative activity were not carried out, and no mechanisms explaining the functional effect of cell delivery were presented. It can be concluded, though, that the observed improvement of recovery was not mediated through neuronal replacement.

In the other study, rats were subjected to MCA occlusion and then received human fetal brain-derived NSPCs by injection into the carotid artery ipsilateral to the stroke (Jiang et al., 2006). MRI revealed that the grafted NSPCs selectively migrated toward the ischemic boundary. The NSPC-treated group revealed early and large increase in fractional anisotropy in the ischemic recovery region which was coincident with white matter reorganization. Transplantation of NSPCs improved the recovery in several behavioral tests, starting already at 1 week after stroke and 5 days after cell injection. Such an early effect of cell injection strongly argues against the involvement of neuronal replacement as a mechanism for improved recovery. As speculated by the authors, the NSPCs may have activated an intrinsic mechanism of white matter recovery by induction of oligodendrocyte progenitor cells or by release of neurotrophic factors.

The only human NSPC line currently undergoing clinical trial in stroke patients is CTX0E03, which was derived from human fetal brain cortex following conditional immortalization (Pollock et al., 2006). Intrastriatal transplantation of CTX0E03 in a rat model of stroke at 3–4 weeks after the insult caused normalization of gross motor asymmetry and improvements in impaired sensorimotor function at 6–12 weeks postgrafting. Most importantly, the grafted cells substantially decreased their proliferative activity. They differentiated into primitive, neurofilament-positive cells but without the elaborated dendritic arborizations characteristic of cortical or striatal interneurons or projection neurons. Taken together, it is unlikely that these cells act by neuronal replacement and become integrated into host neuronal network.

NSPCs with strong capacity for neuronal differentiation can be obtained not only from the fetal brain but also from other sources. One such source is human ES cells which have been widely used in recent years for the generation of different types of neurons (Buhnemann et al., 2006; Daadi et al., 2009; Kriks et al., 2011; Ma et al., 2012; Shi et al., 2012; Stacpoole et al., 2011). The first study in which human ES cell-derived NSPCs were implanted into the ischemic striatum in adult rats subjected to stroke demonstrated migration of grafted cells toward the ischemia-injured brain parenchyma and improvement in the independent use of the impaired forelimb in the cylinder test 2 months posttransplantation (Daadi et al., 2008). The ES cell-derived NSPCs used in this study were expanded as an adherent monolayer culture and they were fully neuralized, uniformly expressing nestin, vimentin, and radial glial markers. The grafted cells virtually ceased proliferation and about 60% of them differentiated into Tuj1+ neurons, half of which were GABAergic and only 2% glutamatergic. Further analysis in a follow-up study (Daadi et al., 2009) showed that the grafted cells exhibited synaptophysin staining as punctate clusters, providing evidence for synaptic connectivity from host neurons. Electron microscopy showed establishment of host presynaptic terminals on the grafted cells. In addition, electrophysiological recordings from grafted cells revealed the presence of voltage-gated sodium currents characteristic for neurons and displayed excitatory postsynaptic currents, suggesting the establishment of synaptic input. These data were in close agreement with those in a previous study demonstrating that about 30% of NSPCs derived from mouse ES cells and grafted into the brain of stroke-lesioned rats differentiated into mature, NeuN+ neurons, exhibiting electrophysiological neuronal properties (Buhnemann et al., 2006). It has also been shown that human ES cell-derived NSPCs implanted into a cortical lesion partly differentiated to neurons and induced some improvement of simple sensorimotor functions (Hicks et al., 2009). These studies clearly demonstrate the capacity of human ES cell-derived NSPCs to give rise to functional neurons which integrate in host neuronal circuitry after transplantation in the stroke-damaged brain. However, it is still unclear whether neuronal replacement is responsible for the functional recovery by grafted NSPCs derived from human ES cells or if the underlying mechanisms are more complex and include modulation of microglia response and enhancement of axonal sprouting as observed in neonatal hypoxic-ischemic brain injury (Daadi et al., 2010).

The possibility of reconstruction of stroke-damaged neuronal network using human ES cell-derived NSPCs is further supported by recent data (Steinbeck et al., 2012). Steinbeck et al. (2012) demonstrated widespread and long-distance axonal projections of neurons generated from grafted NSPCs derived from human ES cells. The projections of the NSPCs after intracerebral transplantation were closely mimicking the intrinsic projections in the adult rodent brain. Thus, the cells implanted into motor cortex extended their axons via the external and internal capsule into the cervical spinal cord and via the corpus callosum into the contralateral cortex, and established synaptic connections with host neurons. Recently, it was shown that it is possible to efficiently direct human ES cells to produce striatal GABAergic neurons (Ma et al., 2012). Transplantation of these cells into the striatum of quinolinic acid-lesioned mice resulted in the generation of large numbers of DARPP32+ GABAergic neurons, which projected to the substantia nigra, received glutamatergic and dopaminergic inputs, and improved impaired motor function. These findings open new possibilities to generate specific striatal projection neurons also for implantation in the stroke-damaged brain.

Human ES cells can give rise to unlimited numbers of NSPCs but are associated with a risk of tumor formation (Seminatore et al., 2010). A problem for clinical translation is that the risk of tumorigenesis is difficult to assess in the preclinical xenograft situation (Erdo et al., 2003). Human fetal NSPCs are less tumorigenic than ES cells and, importantly, in the clinical trial with human NSPCs in Batten's disease, there were no tumors in any of the five patients at 2 years after transplantation (www.stemcellsinc.com). A growing number of studies show that if human ES cells are predifferentiated and/or fated in culture toward neuronal lineage, the formation of tumors after transplantation is virtually abolished. However, more longitudinal studies in immune-deficient animals with transplantation of human ES-derived cells are needed in order to obtain solid and convincing safety data.

Different types of somatic cells including fibroblasts can be reprogrammed into pluripotent stem cells by introduction of transcription factors (Takahashi and Yamanaka, 2006). Similar to ES cells, these so-called iPS cells possess the capacity to differentiate into specific neuron types such as dopaminergic neurons (Hargus et al., 2010; Wernig et al., 2008) and motor neurons (Dimos et al., 2008; Ebert et al., 2009). The iPS cell technology allows patient-specific cells to be produced without the need for immunosuppressive treatment after transplantation, and the ethical issues associated with the use of human ES cells are avoided. However, iPS cells derived from somatic cells of patients with chronic neurodegenerative disorders may exhibit increased susceptibility to the pathological process. In contrast, stroke is an acute neurodegenerative disorder, and it is unlikely that patients who have suffered a stroke will have any specific genetic alterations which would influence the phenotype or the survival of the neurons produced from somatic cells. Therefore, autologous transplantation of neurons derived from iPS cells of stroke patients seems to be an attractive approach. Recently, human fibroblast-derived iPS cells implanted into striatum of stroke-damaged rats were found to migrate to injured brain areas and improve sensorimotor recovery at 4–16 days postgrafting. Whether any neurons were formed is unclear, and the observed effect on recovery was most likely not mediated by neuronal replacement (Jiang et al., 2011). Another study did not detect any effect of human fibroblast-derived iPS cells on stroke-induced behavioral impairments at 4 weeks after transplantation, despite the fact that surviving grafts were detected in 80% of the animals and that many cells expressed the neuronal markers MaP2 and βIII-tubulin (Jensen et al., 2011). In these studies, the observation time was limited, the tumorigenicity issue was not properly addressed, and the analysis on neuronal differentiation of grafted cell was not comprehensive. Moreover, the functional integration of grafted cells into host brain was not assessed at all.

In order to address these issues, we have transplanted long-term expandable neuroepithelial-like stem (lt-NES) cells, generated from adult human fibroblast-derived iPS cells, into the stroke-damaged mouse and rat brain (Oki et al., 2012). The grafted lt-NES cells ceased to proliferate and survived up to at least 4 months after implantation in T cell-deficient rats without forming tumors. Most importantly, we showed for the first time that grafted human iPS cell-derived lt-NES cells generated neurons with mature morphological properties and electrophysiological characteristics of functional neurons *in vivo*. These neurons sent axonal projections from the site of implantation in the striatum to the proper target area, the globus pallidus of the host brain. Providing evidence for their functional integration into host neuronal circuitries, we found that the grafted iPS cell-derived lt-NES cells received spontaneous synaptic input from surrounding host neurons, and they responded with induced potentials when the host tissue surrounding the graft was stimulated. Intrastriatal transplantation of the human iPS cell-derived lt-NES cells into stroke-damaged mice improved motor recovery as documented in the paw reaching test, which is relevant for human stroke and reflects the impairment of fine movements in upper extremities. Since the improvement of motor function occurred as early as 1 week after transplantation, it is unlikely that this effect was mediated by neuronal replacement. Our findings suggest that the early effect is induced by modulation of cellular plasticity triggered by the increased levels of VEGF which we found to be present in astrocytes and blood vessels of transplanted animals. Several other studies have also demonstrated that NSPCs can promote functional recovery in the stroke-damaged brain early after transplantation (Ramos-Cabrer et al., 2010), ruling out the contribution of neuronal replacement. However, neuronal replacement and incorporation of new neurons into the host neuronal network could still contribute to the late phases of the recovery and may be of importance for persistence of the functional improvement.

Stem cells of nonneural origin (umbilical cord blood, bone marrow-derived hematopoietic stem cells, mesenchymal stem cells) have also been shown to reduce functional impairments after stroke in rodents through mechanisms other than neuronal replacement (Lindvall and Kokaia, 2011). For example, intravenously administered mesenchymal stem cells overexpressing VEGF enhanced angiogenesis and improved functional recovery in rodents (Lee et al., 2007; Onda et al., 2008). Gene-modified human mesenchymal stem cells expressing neurotrophic factors provided neuroprotection and improved impaired neurological function (Horita et al., 2006; Liu et al., 2006; Nomura et al., 2005).

Another source for generation of functional neurons for cell replacement in stroke could be the direct reprogramming of human somatic cells (Pang et al., 2011). This conversion does not occur through a pluripotent stem cell stage and thereby eliminates the risk for tumor formation. It remains to be demonstrated, though, that this approach can efficiently generate large numbers of human-derived cells. For use in stroke, it has to be shown that different neuronal subtypes can be produced and survive transplantation.

3 ENDOGENOUS NEURAL STEM CELLS IN STROKE

In the adult mammalian brain, NSPCs are found in the area underlying the lateral ventricles, the so-called subventricular zone (SVZ). The NSPCs proliferate throughout life in rodents and give rise to neuroblasts, which migrate to the olfactory bulb where they differentiate into functional interneurons (Fig. 1). About 10 years ago, the first evidence was presented that stroke increases NSPC proliferation in the SVZ and that neuroblasts generated from these cells migrate to the stroke-damaged striatum (Arvidsson et al., 2002), a region where neurogenesis does not occur in the intact brain. After several weeks, these migrated cells differentiate into mature striatal neurons (Arvidsson et al., 2002; Parent et al., 2002; Yamashita et al., 2006). The vast majority of the newly formed neuroblasts die during the early stages of their development. Caspase-mediated apoptotic death is involved in the loss of the stroke-generated neuroblasts, and caspase inhibitors can partly rescue these neuroblasts (Thored et al., 2006). Initially, it was thought that stroke-induced neurogenesis operates only acutely after stroke, with the number of generated neuroblasts being too small to have any impact on post-stroke regenerative processes. However, later data showed that post-stroke neurogenesis continues for several months (Thored et al., 2006), even up to 1 year after stroke (Kokaia et al., 2006). Thus, the number of new neurons that potentially could replace the dead neurons is substantial. Moreover, experimental studies to explore mechanisms of neuronal replacement from endogenous NSPCs in the damaged brain, as well as various interventions to promote neurogenesis, are not restricted to the acute postischemic phase but can be applied over an extended time period.

Maximum proliferation of NSPCs in the ipsilateral SVZ occurs 1–2 weeks after the stroke (Arvidsson et al., 2002; Jin et al., 2001; Parent et al., 2002; Zhang et al., 2004). However, the number of DCX+ neuroblasts migrating toward the lesioned striatum is similar at 1, 2, 6, 8, 12, and 16 weeks after the insult, indicating that increased proliferation is not required for migration of neuroblasts. The identity of the signal(s) triggering this ectopic migration is not completely understood. Several lines of evidence indicate that SDF-1α/CXCR4 signaling is involved in the persistent migration of new neuroblasts from SVZ to the damaged area after stroke (Thored et al., 2006). First, SDF-1 expression was upregulated in reactive astrocytes of ischemic rat striatum. Second, neural progenitors from SVZ expressed the CXCR4 receptor. Third, infusion of the CXCR4 antagonist AMD 3100 caused a significant but partial suppression of the migration of the newly formed neurons in the striatal parenchyma. Other factors promoting neuroblast migration after stroke are monocyte chemoattractant protein-1 (Yan et al., 2006), matrix metalloproteinase-9 (Lee et al., 2006), and osteopontin (Yan et al., 2009).

The majority of neuroblasts migrating into the striatum after stroke express two transcription factors characteristic of striatal medium spiny neurons during embryonic development, Pbx, and Meis2 (Arvidsson et al., 2002). After several weeks of differentiation, about 50% of the new neurons express DARPP-32, a specific marker for medium spiny neurons, that is, the type of neuron mainly affected by the ischemic insult. However, the ischemic insult also causes death of striatal interneurons, although they are less vulnerable than projection neurons (Andsberg et al., 2001). Teramoto et al. (2003) reported the generation of one type of striatal interneuron, parvalbumin-expressing neurons, in the striatum of epidermal growth factor-treated mice subjected to stroke. Later, Hou and colleagues demonstrated the production of GABAergic and cholinergic neurons from SVZ in response to stroke (Hou et al., 2008). Using a different model of striatal injury, that is, intrastriatal injection of the excitotoxin quinolinic acid, we showed generation of new neurons in the striatum expressing not only DARPP-32 but also markers of two types of striatal interneurons, parvalbumin and neuropeptide Y (Collin et al., 2005). Taken together, these data indicate that the SVZ can produce both striatal projection neurons and interneurons in response to damage. During embryonic development, striatal projection neurons and interneurons originate from the lateral and medial ganglionic eminence, respectively (Marin et al., 2000). Whether the different striatal neuronal phenotypes generated in the adult brain after injury arise from the same NSPC population in the SVZ or if they have separate origins within the SVZ is currently unknown.

Several studies have demonstrated that the new striatal neuroblasts generated after stroke differentiate to cells with mature phenotype. However, for these new neurons to contribute to restoration of stroke-damaged neuronal circuitry, functional maturation and synaptic integration with neighboring neurons are required. So far, only one study has reported electrophysiological maturation of new striatal neurons generated from NSPCs in SVZ following stroke. Hou et al. (2008) demonstrated that new neurons, labeled through viral transfection of SVZ cells, gradually formed neuronal polarity and showed a time-dependent increase in dendrite length and branch formation. EGFP+ cells were double stained for NeuN and GAD67, and MaP-2 and ChAT, respectively. They were also expressing phosphorylated synapsin I and showed typical synaptic structures comprising dendrites and spines. Both the presumed GABAergic and cholinergic newborn neurons, labeled with DiI, injected in the ventricle, and BrdU, fired action potentials and received excitatory and inhibitory synaptic inputs. Although this study needs to be reproduced by other investigators, it provides the first experimental evidence that the ischemia-induced new striatal neurons can become functionally integrated into neural networks in the adult brain after stroke.

Although there is a lot of circumstantial evidence supporting that stroke-induced neurogenesis contributes to the spontaneous functional recovery, it has still not been demonstrated that there is a casual relationship between neurogenesis and post-stroke improvement (Lagace, 2012). A recent attempt to ablate DCX+ SVZ neuroblasts in order to determine the contribution of these cells to post-stroke behavioral recovery did not give clear results (Jin et al., 2010). In this study, transgenic mice expressing herpes simplex virus thymidine kinase under control of the DCX promoter were treated for 14 days with the antiviral drug ganciclovir. This treatment depleted the DCX-expressing and BrdU-labeled cells from the SVZ and abolished neurogenesis, which caused increased infarct size and exacerbated postischemic sensorimotor behavioral deficits. However, the worsening of behavior and of ischemic damage already at 24 h makes it inconceivable that lack of new mature neurons replacing the dead ones had contributed to the observed effect. It seems more likely that the newly generated neuroblasts improve functional recovery at this early time-point through release of, for example, growth factors, stimulating brain plasticity, and being neuroprotective. This does not exclude the possibility that improvement due to neuronal replacement can occur later.

Stroke is an age-related disease and the majority of stroke patients are older than 70 years. Therefore, it is important that the occurrence of stroke-induced striatal neurogenesis has been demonstrated also in aged animals (Darsalia et al., 2005). Although aging caused a dramatic reduction of cell proliferation in the SVZ under basal conditions, the number of new neuroblasts and mature neurons at 7 weeks after stroke was similarly in the ischemic striatum in young-aged animals. Thus, despite the low basal NSPC proliferation in the SVZ, this potential mechanism of self-repair operates also after stroke in the aged brain.

Recent experimental evidence suggests that new neurons can be generated following cerebral ischemia not only from SVZ but also from local NSPCs in the brain parenchyma. Shimada et al. (2010) have reported the existence of local NSPCs that may possess regenerative capacity in the peri-infarct area of the cerebral cortex. Another example of a potential self-repair mechanism following cerebral ischemia from local NSPCs was reported by Ohira et al. (2010). They found that after global ischemia, NSPCs located in the subpial region of the neocortex in rats proliferated and then migrated into deeper cortical layers where they differentiated into

GABAergic interneurons. Further experimental studies based on cell fate-tracing methods are highly warranted.

Whether neuroblasts from SVZ migrate through the rostral migratory stream to the olfactory bulb in adult humans similar to rodents (Curtis et al., 2007) or migration stops at 18 months after birth (Sanai et al., 2011) is controversial. However, it is well established that the adult human brain contains NSPCs in the SVZ (Bernier et al., 2000; Eriksson et al., 1998; Johansson et al., 1999; Sanai et al., 2004). Moreover, increased neuroblast production after stroke has been demonstrated in the human brain (Jin et al., 2006; Macas et al., 2006; Marti-Fabregas et al., 2010; Minger et al., 2007). In an 84-year-old patient, who suffered a cerebrovascular accident 1 week prior to death, cells expressing specific markers for NSPCs and new blood vessels, as well as VEGF-immunopositive cells, were detected both around and distant from the infarcted area but not in the corresponding brain areas of the healthy control. In addition, an increased number of NSPCs was observed in the neurogenic region of the lateral ventricular wall (Minger et al., 2007). In another study, analysis of more than 50 postmortem human brain tissue samples revealed that cerebral ischemia gave rise to increased number of PSA-NCAM-positive neuroblasts close to the SVZ compared to brains of similar age without neuropathologic changes (Macas et al., 2006). Moreover, this study showed that the ischemia led to elevated number of Ki-67+ proliferating cells in the ipsilateral SVZ without concomitant apoptotic cell death. Clear signs of increased proliferative activity in the SVZ after stroke was also detected in seven aged patients (mean age 82 ± 5 years) who died at 10 ± 5 days after the ischemic stroke and from whom brain samples were obtained 4 ± 2 h after death (Marti-Fabregas et al., 2010). Compared to the contralateral SVZ, there was increased cell density of the ribbon layer, enlargement of the cytoplasmic volume of astrocytes, and higher number of Ki-67+ as well as Tuj-1+ or PSA-NCAM+ cells in the ipsilateral SVZ. Signs of newborn neurons in the human brain after stroke have also been detected in areas remote from the SVZ, in the ischemic penumbra surrounding cerebral cortical infarcts, where the new cells were preferentially localized in the vicinity of blood vessels (Jin et al., 2006). Although it is very difficult to study proliferative and neurogenic activity in the human brain, the reports so far provide good evidence for the idea that neurogenesis occurs from endogenous NSPCs also in the human brain after stroke and could be part of a self-repair mechanism.

In order to become clinically valuable, the stroke-induced neurogenic response has to be markedly enhanced. This could be achieved through modulation of the different steps of neurogenesis including NSPC proliferation, neuroblast formation, survival, migration, neuronal differentiation, and functional integration. Several growth factors, neurotrophins, and neuropeptides have been identified that promote these different steps (Aberg, 2010; Calvo et al., 2011; Decressac et al., 2009; Kolb et al., 2006; Lee et al., 2006; Robin et al., 2005; Yan et al., 2007). Promotion of endogenous neurogenesis through enhancement of SVZ cell proliferation can also be achieved by NSPC transplantation in stroke-damaged brain (Nakano-Doi et al., 2010; Zhang et al., 2011). Of particular importance for the promotion of neurogenesis and its functional benefit would be to increase the survival of the stroke-induced neuroblasts and mature neurons. The majority of the new neuroblasts die soon after formation (Arvidsson et al., 2002), but their survival can be substantially increased by administration of anti-inflammatory (Kluska et al., 2005) and antiapoptotic (Doeppner et al., 2009; Kobayashi et al., 2006) substances. Before modulation of endogenous neurogenesis after stroke can be considered for clinical use, it has to be shown in animal models of stroke that increased neurogenesis is beneficial for functional recovery, and additional research will be required regarding the safety and practicality of different ways to enhance neurogenesis. A major challenge for the clinical translation of endogenous stem cell-based therapy is to develop strategies to direct the new cells to the stroke-damaged areas in the patient's brain. Studies in rodents involve ischemic damage to the striatum and cortex, which are located in close proximity to SVZ. The generated neuroblasts only need to migrate over relatively short distances (several millimeters). In contrast, the requirements for migration are much greater in the human brain. It is still unclear whether human neuroblasts derived from NSPCs possess the capacity for such migration.

4 CONCLUSIONS

Results from studies in experimental models are promising and provide supportive evidence that stem cell-based approaches can be developed into clinically useful strategies to promote recovery after stroke. It is clear that transplantation of NSPCs as well as modulation of endogenous neurogenesis is viable approaches. However, our recent clinical study (Delavaran et al., 2012) has shown that stroke patients with infarcts which predominantly include striatum are rare (less than 10%) and tend to have mild neurological deficits and a favorable short-term prognosis. These observations indicate that future cell replacement studies should primarily be aimed for patients with cortical and/or combined cortical–subcortical lesions. There is currently no proven stem cell therapy for stroke, and there is still a long way for stem cell-based neuronal replacement strategies to reach clinical application in stroke patients. Only by combining high-quality basic and clinical research will it be possible to develop stem cell-based treatments for stroke patients and offer them substantial improvements of their quality of life.

Acknowledgments

Our own work was supported by the Swedish Research Council, EU FP7 grants and Swedish Government Initiative for Strategic Research Areas (Stem Therapy).

References

Aberg, D., 2010. Role of the growth hormone/insulin-like growth factor 1 axis in neurogenesis. Endocr. Dev. 17, 63–76.

Andres, R.H., Horie, N., Slikker, W., Keren-Gill, H., Zhan, K., Sun, G., Manley, N.C., Pereira, M.P., Sheikh, L.A., McMillan, E.L., Schaar, B.T., Svendsen, C.N., Bliss, T.M., Steinberg, G.K., 2011. Human neural stem cells enhance structural plasticity and axonal transport in the ischaemic brain. Brain 134, 1777–1789.

Andsberg, G., Kokaia, Z., Lindvall, O., 2001. Upregulation of p75 neurotrophin receptor after stroke in mice does not contribute to differential vulnerability of striatal neurons. Exp. Neurol. 169, 351–363.

Arvidsson, A., Collin, T., Kirik, D., Kokaia, Z., Lindvall, O., 2002. Neuronal replacement from endogenous precursors in the adult brain after stroke. Nat. Med. 8, 963–970.

Bernier, P.J., Vinet, J., Cossette, M., Parent, A., 2000. Characterization of the subventricular zone of the adult human brain: evidence for the involvement of Bcl-2. Neurosci. Res. 37, 67–78.

Buhnemann, C., Scholz, A., Bernreuther, C., Malik, C.Y., Braun, H., Schachner, M., Reymann, K.G., Dihne, M., 2006. Neuronal differentiation of transplanted embryonic stem cell-derived precursors in stroke lesions of adult rats. Brain 129, 3238–3248.

Calvo, C.F., Fontaine, R.H., Soueid, J., Tammela, T., Makinen, T., Alfaro-Cervello, C., Bonnaud, F., Miguez, A., Benhaim, L., Xu, Y., Barallobre, M.J., Moutkine, I., Lyytikka, J., Tatlisumak, T., Pytowski, B., Zalc, B., Richardson, W., Kessaris, N., Garcia-Verdugo, J.M., Alitalo, K., Eichmann, A., Thomas, J.L., 2011. Vascular endothelial growth factor receptor 3 directly regulates murine neurogenesis. Genes Dev. 25, 831–844.

Chu, K., Kim, M., Park, K.I., Jeong, S.W., Park, H.K., Jung, K.H., Lee, S.T., Kang, L., Lee, K., Park, D.K., Kim, S.U., Roh, J.K., 2004. Human neural stem cells improve sensorimotor deficits in the adult rat brain with experimental focal ischemia. Brain Res. 1016, 145–153.

Collin, T., Arvidsson, A., Kokaia, Z., Lindvall, O., 2005. Quantitative analysis of the generation of different striatal neuronal subtypes in the adult brain following excitotoxic injury. Exp. Neurol. 195, 71–80.

Curtis, M.A., Kam, M., Nannmark, U., Anderson, M.F., Axell, M.Z., Wikkelso, C., Holtas, S., Van Roon-Mom, W.M., Bjork-Eriksson, T., Nordborg, C., Frisen, J., Dragunow, M., Faull, R.L., Eriksson, P.S., 2007. Human neuroblasts migrate to the olfactory bulb via a lateral ventricular extension. Science 315, 1243–1249.

Daadi, M.M., Maag, A.L., Steinberg, G.K., 2008. Adherent self-renewable human embryonic stem cell-derived neural stem cell line: functional engraftment in experimental stroke model. PLoS One 3, e1644.

Daadi, M.M., Li, Z., Arac, A., Grueter, B.A., Sofilos, M., Malenka, R.C., Wu, J.C., Steinberg, G.K., 2009. Molecular and magnetic resonance imaging of human embryonic stem cell-derived neural stem cell grafts in ischemic rat brain. Mol. Ther. 17, 1282–1291.

Daadi, M.M., Davis, A.S., Arac, A., Li, Z., Maag, A.L., Bhatnagar, R., Jiang, K., Sun, G., Wu, J.C., Steinberg, G.K., 2010. Human neural stem cell grafts modify microglial response and enhance axonal sprouting in neonatal hypoxic-ischemic brain injury. Stroke 41, 516–523.

Darsalia, V., Heldmann, U., Lindvall, O., Kokaia, Z., 2005. Stroke-induced neurogenesis in aged brain. Stroke 36, 1790–1795.

Darsalia, V., Kallur, T., Kokaia, Z., 2007. Survival, migration and neuronal differentiation of human fetal striatal and cortical neural stem cells grafted in stroke-damaged rat striatum. Eur. J. Neurosci. 26, 605–614.

Darsalia, V., Allison, S.J., Cusulin, C., Monni, E., Kuzdas, D., Kallur, T., Lindvall, O., Kokaia, Z., 2011. Cell number and timing of transplantation determine survival of human neural stem cell grafts in stroke-damaged rat brain. J. Cereb. Blood Flow Metab. 31, 235–242.

Decressac, M., Prestoz, L., Veran, J., Cantereau, A., Jaber, M., Gaillard, A., 2009. Neuropeptide Y stimulates proliferation, migration and differentiation of neural precursors from the subventricular zone in adult mice. Neurobiol. Dis. 34, 441–449.

Delavaran, H., Sjunnesson, H., Arvidsson, A., Lindvall, O., Norrving, B., van Westen, D., Kokaia, Z., Lindgren, A. 2012. Proximity of brain infarcts to regions of endogenous neurogenesis and involvement of striatum in ischemic stroke. Eur. J. Neurol. (in press).

Dimos, J.T., Rodolfa, K.T., Niakan, K.K., Weisenthal, L.M., Mitsumoto, H., Chung, W., Croft, G.F., Saphier, G., Leibel, R., Goland, R., Wichterle, H., Henderson, C.E., Eggan, K., 2008. Induced pluripotent stem cells generated from patients with ALS can be differentiated into motor neurons. Science 321, 1218–1221.

Doeppner, T.R., Nagel, F., Dietz, G.P., Weise, J., Tonges, L., Schwarting, S., Bahr, M., 2009. TAT-Hsp70-mediated neuroprotection and increased survival of neuronal precursor cells after focal cerebral ischemia in mice. J. Cereb. Blood Flow Metab. 29, 1187–1196.

Ebert, A.D., Yu, J., Rose Jr., F.F., Mattis, V.B., Lorson, C.L., Thomson, J.A., Svendsen, C.N., 2009. Induced pluripotent stem cells from a spinal muscular atrophy patient. Nature 457, 277–280.

Erdo, F., Buhrle, C., Blunk, J., Hoehn, M., Xia, Y., Fleischmann, B., Focking, M., Kustermann, E., Kolossov, E., Hescheler, J., Hossmann, K.A., Trapp, T., 2003. Host-dependent tumorigenesis of embryonic stem cell transplantation in experimental stroke. J. Cereb. Blood Flow Metab. 23, 780–785.

Eriksson, P.S., Perfilieva, E., Bjork-Eriksson, T., Alborn, A.M., Nordborg, C., Peterson, D.A., Gage, F.H., 1998. Neurogenesis in the adult human hippocampus. Nat. Med. 4, 1313–1317.

Hargus, G., Cooper, O., Deleidi, M., Levy, A., Lee, K., Marlow, E., Yow, A., Soldner, F., Hockemeyer, D., Hallett, P.J., Osborn, T., Jaenisch, R., Isacson, O., 2010. Differentiated Parkinson patient-derived induced pluripotent stem cells grow in the adult rodent brain and reduce motor asymmetry in Parkinsonian rats. Proc. Natl. Acad. Sci. U.S.A. 107, 15921–15926.

Hicks, A.U., Lappalainen, R.S., Narkilahti, S., Suuronen, R., Corbett, D., Sivenius, J., Hovatta, O., Jolkkonen, J., 2009. Transplantation of human embryonic stem cell-derived neural precursor cells and enriched environment after cortical stroke in rats: cell survival and functional recovery. Eur. J. Neurosci. 29, 562–574.

Horie, N., Pereira, M.P., Niizuma, K., Sun, G., Keren-Gill, H., Encarnacion, A., Shamloo, M., Hamilton, S.A., Jiang, K., Huhn, S., Palmer, T.D., Bliss, T.M., Steinberg, G.K., 2011. Transplanted stem cell-secreted vascular endothelial growth factor effects poststroke recovery, inflammation, and vascular repair. Stem Cells 29, 274–285.

Horita, Y., Honmou, O., Harada, K., Houkin, K., Hamada, H., Kocsis, J.D., 2006. Intravenous administration of glial cell line-derived neurotrophic factor gene-modified human mesenchymal stem cells protects against injury in a cerebral ischemia model in the adult rat. J. Neurosci. Res. 84, 1495–1504.

Hou, S.W., Wang, Y.Q., Xu, M., Shen, D.H., Wang, J.J., Huang, F., Yu, Z., Sun, F.Y., 2008. Functional integration of newly generated neurons into striatum after cerebral ischemia in the adult rat brain. Stroke 39, 2837–2844.

Jensen, M.B., Yan, H., Krishnaney-Davison, R., Al Sawaf, A., Zhang, S.C., 2011. Survival and Differentiation of transplanted neural stem cells derived from human induced pluripotent stem cells in a rat stroke model. J. Stroke Cerebrovasc. Dis. Nov 10. [Epub ahead of print].

Jiang, Q., Zhang, Z.G., Ding, G.L., Silver, B., Zhang, L., Meng, H., Lu, M., Pourabdillah-Nejed, D.S., Wang, L., Savant-Bhonsale, S., Li, L., Bagher-Ebadian, H.,

Hu, J., Arbab, A.S., Vanguri, P., Ewing, J.R., Ledbetter, K.A., Chopp, M., 2006. MRI detects white matter reorganization after neural progenitor cell treatment of stroke. Neuroimage 32, 1080–1089.

Jiang, M., Lv, L., Ji, H., Yang, X., Zhu, W., Cai, L., Gu, X., Chai, C., Huang, S., Sun, J., Dong, Q., 2011. Induction of pluripotent stem cells transplantation therapy for ischemic stroke. Mol. Cell. Biochem. 354, 67–75.

Jin, K., Minami, M., Lan, J.Q., Mao, X.O., Batteur, S., Simon, R.P., Greenberg, D.A., 2001. Neurogenesis in dentate subgranular zone and rostral subventricular zone after focal cerebral ischemia in the rat. Proc. Natl. Acad. Sci. U.S.A. 98, 4710–4715.

Jin, K., Wang, X., Xie, L., Mao, X.O., Zhu, W., Wang, Y., Shen, J., Mao, Y., Banwait, S., Greenberg, D.A., 2006. Evidence for stroke-induced neurogenesis in the human brain. Proc. Natl. Acad. Sci. U.S.A. 103, 13198–13202.

Jin, K., Wang, X., Xie, L., Mao, X.O., Greenberg, D.A., 2010. Transgenic ablation of doublecortin-expressing cells suppresses adult neurogenesis and worsens stroke outcome in mice. Proc. Natl. Acad. Sci. U.S.A. 107, 7993–7998.

Johansson, C.B., Svensson, M., Wallstedt, L., Janson, A.M., Frisen, J., 1999. Neural stem cells in the adult human brain. Exp. Cell Res. 253, 733–736.

Kallur, T., Darsalia, V., Lindvall, O., Kokaia, Z., 2006. Human fetal cortical and striatal neural stem cells generate region-specific neurons in vitro and differentiate extensively to neurons after intrastriatal transplantation in neonatal rats. J. Neurosci. Res. 84, 1630–1644.

Kelly, S., Bliss, T.M., Shah, A.K., Sun, G.H., Ma, M., Foo, W.C., Masel, J., Yenari, M.A., Weissman, I.L., Uchida, N., Palmer, T., Steinberg, G.K., 2004. Transplanted human fetal neural stem cells survive, migrate, and differentiate in ischemic rat cerebral cortex. Proc. Natl. Acad. Sci. U.S.A. 101, 11839–11844.

Kluska, M.M., Witte, O.W., Bolz, J., Redecker, C., 2005. Neurogenesis in the adult dentate gyrus after cortical infarcts: effects of infarct location, N-methyl-D-aspartate receptor blockade and anti-inflammatory treatment. Neuroscience 135, 723–735.

Kobayashi, T., Ahlenius, H., Thored, P., Kobayashi, R., Kokaia, Z., Lindvall, O., 2006. Intracerebral infusion of glial cell line-derived neurotrophic factor promotes striatal neurogenesis after stroke in adult rats. Stroke 37, 2361–2367.

Kokaia, Z., Thored, P., Arvidsson, A., Lindvall, O., 2006. Regulation of stroke-induced neurogenesis in adult brain - recent scientific progress. Cereb. Cortex 16 (Suppl 1), i162–i167.

Kolb, B., Morshead, C., Gonzalez, C., Kim, M., Gregg, C., Shingo, T., Weiss, S., 2006. Growth factor-stimulated generation of new cortical tissue and functional recovery after stroke damage to the motor cortex of rats. J. Cereb. Blood Flow Metab. 27, 983–997.

Kriks, S., Shim, J.W., Piao, J., Ganat, Y.M., Wakeman, D.R., Xie, Z., Carrillo-Reid, L., Auyeung, G., Antonacci, C., Buch, A., YanG, L., Beal, M.F., Surmeier, D.J., Kordower, J.H., Tabar, V., Studer, L., 2011. Dopamine neurons derived from human ES cells efficiently engraft in animal models of Parkinson's disease. Nature 480, 547–551.

Lagace, D.C., 2012. Does the endogenous neurogenic response alter behavioral recovery following stroke? Behav. Brain Res. 227, 426–432.

Lee, S.R., Kim, H.Y., Rogowska, J., Zhao, B.Q., Bhide, P., Parent, J.M., Lo, E.H., 2006. Involvement of matrix metalloproteinase in neuroblast cell migration from the subventricular zone after stroke. J. Neurosci. 26, 3491–3495.

Lee, H.J., Kim, K.S., Park, I.H., Kim, S.U., 2007. Human neural stem cells over-expressing VEGF provide neuroprotection, angiogenesis and functional recovery in mouse stroke model. PLoS One 2, e156.

Lindvall, O., Björklund, A., 2004. Cell therapy in Parkinson's disease. NeuroRx 1, 382–393.

Lindvall, O., Kokaia, Z., 2011. Stem cell research in stroke: how far from the clinic? Stroke 42, 2369–2375.

Liu, H., Honmou, O., Harada, K., Nakamura, K., Houkin, K., Hamada, H., Kocsis, J.D., 2006. Neuroprotection by PlGF gene-modified human mesenchymal stem cells after cerebral ischaemia. Brain 129, 2734–2745.

Ma, L., Hu, B., Liu, Y., Vermilyea, S.C., Liu, H., Gao, L., Sun, Y., Zhang, X., Zhang, S.C., 2012. Human embryonic stem cell-derived GABA neurons correct locomotion deficits in quinolinic acid-lesioned mice. Cell Stem Cell 10, 455–464.

Macas, J., Nern, C., Plate, K.H., Momma, S., 2006. Increased generation of neuronal progenitors after ischemic injury in the aged adult human forebrain. J. Neurosci. 26, 13114–13119.

MaRIN, O., Anderson, S.A., Rubenstein, J.L., 2000. Origin and molecular specification of striatal interneurons. J. Neurosci. 20, 6063–6076.

Marti-Fabregas, J., Romaguera-Ros, M., Gomez-Pinedo, U., Martinez-Ramirez, S., Jimenez-Xarrie, E., Marin, R., Marti-Vilalta, J.L., Garcia-Verdugo, J.M., 2010. Proliferation in the human ipsilateral subventricular zone after ischemic stroke. Neurology 74, 357–365.

Minger, S.L., Ekonomou, A., Carta, E.M., Chinoy, A., Perry, R.H., Ballard, C.G., 2007. Endogenous neurogenesis in the human brain following cerebral infarction. Regen. Med. 2, 69–74.

Nakano-Doi, A., Nakagomi, T., Fujikawa, M., Nakagomi, N., Kubo, S., Lu, S., Yoshikawa, H., Soma, T., Taguchi, A., Matsuyama, T., 2010. Bone marrow mononuclear cells promote proliferation of endogenous neural stem cells through vascular niches after cerebral infarction. Stem Cells 28, 1292–1302.

Nomura, T., Honmou, O., Harada, K., Houkin, K., Hamada, H., Kocsis, J.D., 2005. I.V. infusion of brain-derived neurotrophic factor gene-modified human mesenchymal stem cells protects against injury in a cerebral ischemia model in adult rat. Neuroscience 136, 161–169.

Ohira, K., Furuta, T., Hioki, H., Nakamura, K.C., Kuramoto, E., Tanaka, Y., Funatsu, N., Shimizu, K., Oishi, T., Hayashi, M., Miyakawa, T., Kaneko, T., Nakamura, S., 2010. Ischemia-induced neurogenesis of neocortical layer 1 progenitor cells. Nat. Neurosci. 13, 173–179.

Oki, K., Tatarishvili, J., Woods, J., Koch, P., Wattananit, S., Mine, Y., Monni, E., Prietro, D.T., Ahlenius, H., Ladewig, J., Brüstle, O., Lindvall, O., Kokaia, Z., 2012. Human induced pluripotent stem cells form functional neurons and improve recovery after grafting in stroke-damaged brain. Stem Cells 30, 1120–1133.

Onda, T., Honmou, O., Harada, K., Houkin, K., Hamada, H., Kocsis, J.D., 2008. Therapeutic benefits by human mesenchymal stem cells (hMSCs) and Ang-1 gene-modified hMSCs after cerebral ischemia. J. Cereb. Blood Flow Metab. 28, 329–340.

Pang, Z.P., Yang, N., Vierbuchen, T., Ostermeier, A., Fuentes, D.R., Yang, T.Q., Citri, A., Sebastiano, V., Marro, S., Sudhof, T.C., Wernig, M., 2011. Induction of human neuronal cells by defined transcription factors. Nature 476, 220–223.

Parent, J.M., Vexler, Z.S., Gong, C., Derugin, N., Ferriero, D.M., 2002. Rat forebrain neurogenesis and striatal neuron replacement after focal stroke. Ann. Neurol. 52, 802–813.

Pollock, K., Stroemer, P., Patel, S., Stevanato, L., Hope, A., Miljan, E., Dong, Z., Hodges, H., Price, J., Sinden, J.D., 2006. A conditionally immortal clonal stem cell line from human cortical neuroepithelium for the treatment of ischemic stroke. Exp. Neurol. 199, 143–155.

Ramos-Cabrer, P., Justicia, C., Wiedermann, D., Hoehn, M., 2010. Stem cell mediation of functional recovery after stroke in the rat. PLoS One 5, e12779.

Robin, A.M., Zhang, Z.G., Wang, L., Zhang, R.L., Katakowski, M., Zhang, L., Wang, Y., Zhang, C., Chopp, M., 2005. Stromal cell-derived factor 1alpha mediates neural progenitor cell motility after focal cerebral ischemia. J. Cereb. Blood Flow Metab. 26, 125–134.

Roitberg, B.Z., Mangubat, E., Chen, E.Y., Sugaya, K., Thulborn, K.R., Kordower, J.H., Pawar, A., Konecny, T., Emborg, M.E., 2006. Survival and early differentiation of human neural stem cells transplanted in a nonhuman primate model of stroke. J. Neurosurg. 105, 96–102.

Sanai, N., Tramontin, A.D., Quinones-Hinojosa, A., Barbaro, N.M., Gupta, N., Kunwar, S., Lawton, M.T., McDermott, M.W., Parsa, A.T., Manuel-Garcia Verdugo, J., Berger, M.S., Alvarez-Buylla, A., 2004. Unique astrocyte ribbon in adult human brain contains neural stem cells but lacks chain migration. Nature 427, 740–744.

Sanai, N., Nguyen, T., Ihrie, R.A., Mirzadeh, Z., Tsai, H.H., Wong, M., Gupta, N., Berger, M.S., Huang, E., Garcia-Verdugo, J.M., Rowitch, D.H., Alvarez-Buylla, A., 2011. Corridors of migrating neurons in the human brain and their decline during infancy. Nature 478, 382–386.

Seminatore, C., Polentes, J., Ellman, D., Kozubenko, N., Itier, V., Tine, S., Tritschler, L., Brenot, M., Guidou, E., Blondeau, J., Lhuillier, M., Bugi, A., Aubry, L., Jendelova, P., Sykova, E., Perrier, A.L., Finsen, B., Onteniente, B., 2010. The postischemic environment differentially impacts teratoma or tumor formation after transplantation of human embryonic stem cell-derived neural progenitors. Stroke 41, 153–159.

Shi, Y., Kirwan, P., Smith, J., Robinson, H.P., Livesey, F.J., 2012. Human cerebral cortex development from pluripotent stem cells to functional excitatory synapses. Nat. Neurosci. 15 (477–86), S1.

Shimada, I.S., Peterson, B.M., Spees, J.L., 2010. Isolation of locally derived stem/progenitor cells from the peri-infarct area that do not migrate from the lateral ventricle after cortical stroke. Stroke 41, 552–560.

Stacpoole, S.R., Bilican, B., Webber, D.J., Luzhynskaya, A., He, X.L., Compston, A., Karadottir, R., Franklin, R.J., Chandran, S., 2011. Efficient derivation of NPCs, spinal motor neurons and midbrain dopaminergic neurons from hESCs at 3% oxygen. Nat. Protoc. 6, 1229–1240.

Steinbeck, J.A., Koch, P., Derouiche, A., Brüstle, O., 2012. Human embryonic stem cell-derived neurons establish region-specific, long-range projections in the adult brain. Cell. Mol. Life Sci. 69, 461–470.

Takahashi, K., Yamanaka, S., 2006. Induction of pluripotent stem cells from mouse embryonic and adult fibroblast cultures by defined factors. Cell 126, 663–676.

Teramoto, T., Qiu, J., Plumier, J.C., Moskowitz, M.A., 2003. EGF amplifies the replacement of parvalbumin-expressing striatal interneurons after ischemia. J. Clin. Invest. 111, 1125–1132.

Thored, P., Arvidsson, A., Cacci, E., Ahlenius, H., Kallur, T., Darsalia, V., Ekdahl, C.T., Kokaia, Z., Lindvall, O., 2006. Persistent production of neurons from adult brain stem cells during recovery after stroke. Stem Cells 24, 739–747.

Wernig, M., Zhao, J.P., Pruszak, J., Hedlund, E., Fu, D., Soldner, F., Broccoli, V., Constantine-Paton, M., Isacson, O., Jaenisch, R., 2008. Neurons derived from reprogrammed fibroblasts functionally integrate into the fetal brain and improve symptoms of rats with Parkinson's disease. Proc. Natl. Acad. Sci. U.S.A. 105, 5856–5861.

Yamashita, T., Ninomiya, M., Hernandez Acosta, P., Garcia-Verdugo, J.M., Sunabori, T., Sakaguchi, M., Adachi, K., Kojima, T., Hirota, Y., Kawase, T., Araki, N., Abe, K., Okano, H., Sawamoto, K., 2006. Subventricular zone-derived neuroblasts migrate and differentiate into mature neurons in the post-stroke adult striatum. J. Neurosci. 26, 6627–6636.

Yan, Y.P., Sailor, K.A., Lang, B.T., Park, S.W., Vemuganti, R., Dempsey, R.J., 2006. Monocyte chemoattractant protein-1 plays a critical role in neuroblast migration after focal cerebral ischemia. J. Cereb. Blood Flow Metab. 27, 1213–1224.

Yan, J., Xu, L., Welsh, A.M., Hatfield, G., Hazel, T., Johe, K., Koliatsos, V.E., 2007. Extensive neuronal differentiation of human neural stem cell grafts in adult rat spinal cord. PLoS Med. 4, e39.

Yan, Y.P., Lang, B.T., Vemuganti, R., Dempsey, R.J., 2009. Osteopontin is a mediator of the lateral migration of neuroblasts from the subventricular zone after focal cerebral ischemia. Neurochem. Int. 55, 826–832.

Zhang, R., Zhang, Z., Zhang, C., Zhang, L., Robin, A., Wang, Y., Lu, M., Chopp, M., 2004. Stroke transiently increases subventricular zone cell division from asymmetric to symmetric and increases neuronal differentiation in the adult rat. J. Neurosci. 24, 5810–5815.

Zhang, P., Li, J., Liu, Y., Chen, X., Lu, H., Kang, Q., Li, W., Gao, M., 2011. Human embryonic neural stem cell transplantation increases subventricular zone cell proliferation and promotes peri-infarct angiogenesis after focal cerebral ischemia. Neuropathology 31, 384–391.

CHAPTER

In vivo imaging of cell transplants in experimental ischemia

5

Joanna Adamczak, Mathias Hoehn[1]

Max-Planck-Institute for Neurological Research, In vivo *NMR, Cologne, Germany*

[1]*Corresponding author. Tel.: +49-221-4726315, Fax: +49-221-4726337, e-mail address: mathias@nf.mpg.de*

Abstract

The therapeutic potential of stem cells for regeneration after cerebral lesion has become of great interest. This is particularly so for neurodegenerative diseases as well as for stroke. Contrary to more conventional, cerebroprotective treatment approaches, the focus of regeneration lies in a longer time window during the chronic phase of the lesion evolution. Thus, in order to assess the true potential of a treatment strategy and to investigate the underlying mechanisms, observation of the temporal profile of both the cell dynamics as well as the organ response to the treatment is of paramount importance. This need for intraindividual longitudinal studies can be optimally met by the application of noninvasive imaging modalities. This chapter presents in breadth the potential of noninvasive imaging modalities for cell tracking with application focus to experimental stroke. While the lion's share of discussed studies is based on MRI, we have also included the contributions of positron emission tomography and of the increasingly important optical imaging modality.

Keywords

cell migration, cell vitality, DNA of imaging reporters, magnetic resonance imaging, multimodal molecular imaging, optical imaging, positron emission tomography, stem cells

Abbreviations

^{18}F-FLT	[^{18}F]-Fluor-deoxythymidine
BLI	bioluminescence imaging
BMSC	bone marrow stromal cells
D2R	dopamine 2 receptor
ESC	embryonic stem cells
Gd-DTPA	gadolinium diethylenetriaminepentaacetate

Progress in Brain Research, Volume 201, ISSN 0079-6123, http://dx.doi.org/10.1016/B978-0-444-59544-7.00004-4

GFP	green fluorescent protein
MCAO	middle cerebral artery occlusion
MPIO	micron sized particles of iron oxide
MRI	magnetic resonance imaging
NPCs	neural progenitor cells
OI	optical imaging
PET	positron emission tomography
RMS	rostral migratory stream
SGZ	subgranular zone
SVZ	subventricular zone
USPIO	ultrasmall paramagnetic iron oxide particle

1 INTRODUCTION

Stroke remains the third-leading cause of death and the main cause of disability worldwide, leaving the majority of survivors dependent on institutional care (Donnan et al., 2008). Ischemia represents the most common kind of a cerebrovascular event accounting for 85–90% of all incidences (Donnan et al., 2008). Despite numerous studies on neuroprotective agents, the only effective and FDA approved therapy for ischemic stroke is the removal of the thrombus/embolus either pharmacologically by thrombolysis with recombinant tissue plasminogen activator, which has a very short-therapeutic time window of 4–6 h after onset of ischemia (Wahlgren et al., 2008), or mechanically with a retriever (Berlis et al., 2004; Smith et al., 2008). Up to now, no effective treatment to improve functional recovery exists in the subacute or chronic phase.

Stem cell-based approaches are potential new therapies for neuroregeneration after ischemic stroke and already have shown promising results in preclinical studies (Luo, 2011; Ramos-Cabrer et al., 2010). Transplanted into an ischemic rat brain, stem cells have been observed to migrate toward the lesion (Guzman et al., 2007; Hoehn et al., 2002; Modo et al., 2004; Modo et al., 2009; Zhang et al., 2003), differentiate into neurons (Hoehn et al., 2004; Modo et al., 2002; Modo et al., 2004), and improve functional recovery (Ramos-Cabrer et al., 2010; Zhang et al., 2003). However, the underlying mechanisms of the benefit in functional performance remain rather speculative, suggesting paracrine support (Couillard-Despres et al., 2011; Einstein and Ben-Hur, 2008; Pluchino et al., 2005; Walker et al., 2011) rather than integration into the remaining neuronal network (Chopp et al., 2009). Also, stem cell-based therapy has to deal with the possible complication of spontaneous tumor formation. It is therefore necessary to carefully monitor transplant dynamics over a long period of time. Immunohistochemical staining techniques have been commonly used to identify the location and fate of transplanted stem cells. However, the information provided by immunohistochemistry is restricted to one specific time point, and many animals are needed to assess the whole temporal dynamics of a graft. Noninvasive imaging techniques like magnetic resonance imaging (MRI), optical imaging (OI), and positron emission tomography (PET) offer the possibility to follow the dynamics of a graft within one animal over time. These techniques can be exploited to gain information about the location and migration of the transplanted cells in space and time. Furthermore, imaging can be used to indicate the cell fate and to follow the impact on functional recovery in animal models of ischemic stroke. Combination of two or more methods into a multimodal approach holds the opportunity to look at complementary aspects of a stem cell graft, provided by the different imaging modalities, within the same animal. Integration of these pieces of information will allow for a more stringent control over the cell grafts and elucidate the mechanisms of action of stem cell therapy in cerebral ischemia.

2 CELL TRACKING BY MRI

MRI offers many advantages for the field of *in vivo* cell tracking. Its noninvasive nature makes longitudinal studies possible, allowing for repetitive measurements on the same animal and thus producing the full temporal profile of mechanistic events for each individual. A further, important ethical aspect of the modality's noninvasiveness is the fact that provision of intraindividual temporal profiles drastically reduces the need for sacrifice of whole groups at different survival times, otherwise necessary to obtain comparable information with invasive, single time point methods. A high-spatial resolution down to 50 μm or even beyond can be achieved, and even single cell detection has been reported (Shapiro et al., 2006b). Several contrast mechanisms are available in MRI, which make it a diverse imaging technique for many different applications. For tracking of stem cells *in vivo*, the use of contrast agents is required in order to discriminate between the transplanted cells and the host tissue. The contrast agents make use of the contrast mechanisms T1, T2, and T2* and specifically modify these relaxation times in their surroundings. More details about the methodological aspects of designing and characterizing MR contrast agents useful for this purpose can be found in the literature (Cormode et al., 2010; Himmelreich and Dresselaers, 2009; Hoehn et al., 2007; Liu and Frank, 2009; Modo et al., 2005).

Cells are usually labeled *in vitro* before transplantation, but also *in vivo* labeling of special cell populations is possible and has been performed (Nieman et al., 2010; Saleh et al., 2004; Shapiro et al., 2006a; Sumner et al., 2009; Vande Velde et al., 2011; Vreys et al., 2010). The most commonly used contrast agents for cell tracking are based on iron oxide nanoparticles (super paramagnetic iron oxides, SPIOs), which induce a strong contrast on T2 and especially on T2* images. The presence of SPIOs disrupts the MR signal in their vicinity and thereby causes a negative contrast on the MR images.

Hoehn et al. (2002) were the first to use this contrast mechanism to track the transplanted stem cells *in vivo* over 3–4 weeks in a rat model of middle cerebral artery occlusion (MCAO). Murine embryonic stem cells (ESCs) were labeled with the ultrasmall paramagnetic iron oxide particle (USPIO) Sinerem® and transplanted at 2 weeks post stroke into the contralateral cortex and striatum. Migrating clusters of cells were observed on T2*-weighted images along the corpus callosum toward the ischemic hemisphere at 6 days post transplantation. ESCs first lined the lateral ventricle in the ischemic hemisphere and at 2 weeks post implantation were found spread across the ischemic striatum and lateral cortex. The authors confirmed the *in vivo* findings by immunohistochemical staining for green fluorescent protein (GFP), which was stably expressed by the transplanted cells. When neural precursor cells were labeled with iron oxide particles (Feridex®) and transplanted 7 days prior to stroke into the healthy brain, they started targeted migration toward the lesion after the onset of ischemia, again using the corpus callosum as travel substrate (Guzman et al., 2008). Hypointense regions along the corpus callosum were already present at 7 days post transplantation and after 3 weeks, a clear distribution along the stroke border was visible. Cells transplanted into animals that underwent sham surgery showed only little and undirected extension of the contralateral hypointense cell cluster. Migration along the corpus callosum toward the lesion was also observed from contralaterally transplanted bone marrow stromal cells (BMSC) and mouse ESCs using the SPIO particle Endorem® as label (Jendelova et al., 2003, 2004). Intravenously administered labeled BMSC resulted as well in hypointense areas along the lesion edge indicating the accumulation of BMSC, which was confirmed by Prussian blue staining (Jendelova et al., 2003). However, when Walczak et al. (2008) compared the amount of cells reaching the brain after intravenous (i.v.) or intra-arterial (i.a.) administration in a rat model of stroke, the number of iron oxide particle labeled cells after i.v. injection was not detectable with MRI and only few cells were found after histological evaluation (Walczak et al., 2008). The authors used Feridex®-labeled mesenchymal stem cells and injected 1×10^6 cells 30 min after ischemia. Injection into the internal carotid artery resulted in greater cell delivery to the brain. Iron-labeled cells were mainly found in the vascular territory of the internal carotid artery, but occasionally also on the contralateral side (Walczak et al., 2008). Intra-arterial injection resulted in a high variability in engraftment efficacy, with some animals also having no MRI-detectable engraftment. High cell delivery was confirmed by transcranial laser Doppler flowmetry, when blood flow dropped during cell injection below 50%. It must be cautioned, however, that the i.a. cell injection resulted in high animal mortality (Walczak et al., 2008), probably due to thrombosis by clotted cell aggregates.

The behavior of human neural stem cells has been investigated recently. Graft viability over time has been analyzed quantitatively for several weeks by Kallur et al. (2011) showing the slow shrinkage of the graft volume in healthy rat brain during 6 weeks observation period with T2*-weighted MRI (Fig. 1). In the context of ischemic stroke, human neural stem cells have been studied with MRI using the iron oxide particle Feridex® (Guzman et al., 2007). The label had no effect on cell viability. The cells even kept their electrophysiological characteristics. Transplantation in rats into the cortex close to the lesion was performed 6 days after distal MCAO; and in a separate group of animals, cells were transplanted further away from the lesion. MRI at 1 week post transplantation could clearly visualize the implantation site. Transparenchymal migration was only observed in the group with close

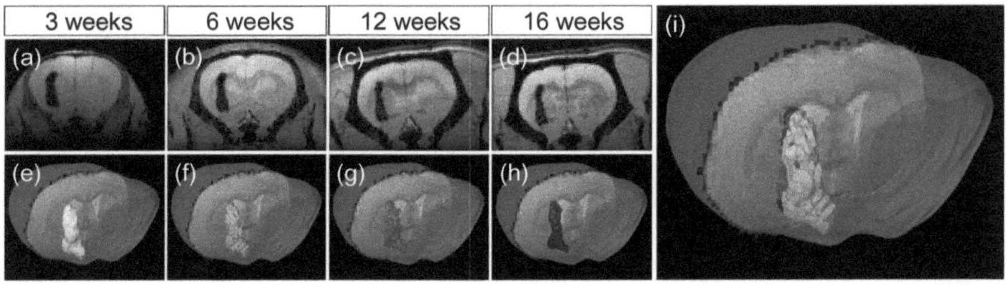

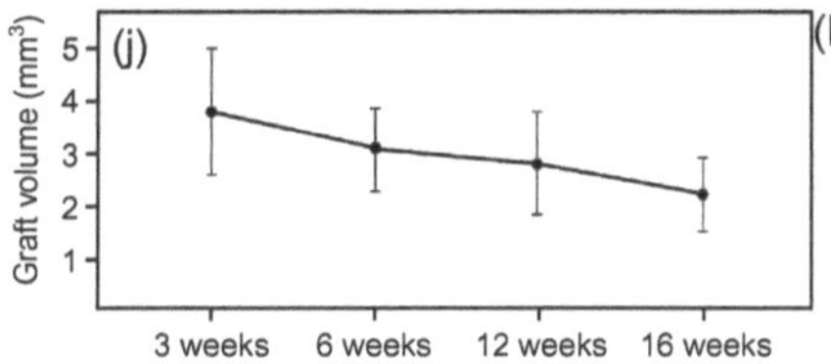

(k)

Time after transplantation	Absolute volume of graft (mm^3)	Absolute volume of brain (mm^3)	‰ Graft relative to brain volume
3 weeks	3.83 ± 1.21	659 ± 38.1	5.79 ± 1.78
6 weeks	3.07 ± 0.81	899 ± 55.6	3.40 ± 0.76
12 weeks	2.82 ± 1.01	957 ± 34.4	2.94 ± 1.02
16 weeks	2.21 ± 0.69	925 ± 75.3	2.41 ± 0.79

FIGURE 1

(A–D) Estimated volume of human cell grafts over time. Coronal FLASH 3D images of a representative animal illustrating the grafted human NSCs and corresponding three-dimensional reconstruction and surface rendering from high resolution FLASH 3D images at 3 (E; yellow), 6 (F; green), 12 (G; blue), and 16 (H; red) weeks after transplantation. (I) Compilation of the grafted cells at all four time points with respective color code maintained. Note the similar volume and shape of the grafts from 3 to 16 weeks. (J) Quantitative analysis of all animals across the four time points ($n=7$) shows a gradual decrease in the absolute graft volume over time ($p=0.014$). (K) Absolute values for both, graft and brain volume and the relative size of the graft as per mille (‰) of the brain volume at 3, 6, 12, and 16 weeks after grafting. Note the decrease of both absolute and relative graft volume over time, and the substantial increase in brain volume especially between 3 and 6 weeks after transplantation. (For interpretation of the references to color in this figure legend, the reader is referred to the online version of this chapter.)

(Reproduced with permission from Kallur et al., 2011).

proximity between graft and lesion and was verified by immunohistochemistry at 5 weeks post transplantation. Grafts further away (mean distance of 3.48 mm) only migrated toward the ischemic cortex when the graft had contact to the corpus callosum. MR signal and graft volume reduction were interpreted as clearance of dead cells (Guzman et al., 2007).

The adult brain executes a continuous generation of endogenous stem cells in at least two locations, the subgranular zone (SGZ) of the hippocampus and the subventricular zone (SVZ). Within the SVZ along the lateral ventricle, dividing stem cells give rise to neural progenitor cells (NPCs; Alvarez-Buylla and Garcia-Verdugo, 2002), which normally travel along the rostral migratory stream (RMS) to the olfactory bulb where they differentiate into inhibitory interneurons. These subventricular NPCs were labeled *in vitro* with ferromagnetic particles and transplanted into the magna cisterna of rats, which received an embolic occlusion of the middle cerebral artery 48 h prior to transplantation. Daily MRI for 1 week and weekly MRI for up to

8 weeks showed the migration of the cells along the fourth ventricle reaching the lateral ventricle at 2 day post implantation. The cells spread toward the ischemic area during the time between 3 and 7 days. From their noninvasive and longitudinal observation, the authors calculated a migratory speed of 65 ± 15 μm/h (Zhang et al., 2003). It must be noted, however, that this value of migration speed is substantially higher than other values found in several independent reports where cells had clearly migrated only through parenchyma. A series of such migration speed values are compiled in Table 1. No contrast changes were observed on MR images of animals that received transplantations of dead SVZ cells.

These residual stem cells in the adult brain can also be labeled *in vivo*. Injection of micron sized particles of iron oxide (MPIO) into the lateral ventricle results in an uptake of the particles by all the cells surrounding the ventricle. Amongst these are the endogenous stem cells and the early NPCs, which will migrate out of the unspecifically labeled territory and along the rostral migratory path towards the olfactory bulb. These cells have now been imaged with MRI (Fig. 2) and the rostral

Table 1 Cell migration speed of endogenous neurogenesis and of intracerebrally implanted (stem) cell grafts

Migration speed[a] (μm/h)	Cell type	Model	Species	Observation method	Reference
8–28	Neuroblasts	Neuroblasts	Birds	BrdU staining/ histology	Alvarez-Buylla and Nottebohm (1988)
~30	Neuroblasts	RMS	Mice	BrdU staining/ histology	Lois and Alvarez-Buylla (1994)
~10	Neurospheres	EAE model	Mice	MRI	Ben-Hur et al. (2007)
20–30	Murine ESC	Stroke	Rats	MRI	Hoehn et al. (2002)
65 ± 15	Neuroblasts	Implant/ stroke	Rats	MRI	Zhang et al. (2003)
15	Neural stem cells	Implant/ stroke	Mice	BLI	Kim et al. (2004)
30 ± 2	Neuroblasts	Living slices/ healthy	Mice	Multiphoton microscopy	Zhao and Nam (2007)
21 ± 1	Neuroblasts	Living slices/ stroke	Mice	Multiphoton microscopy	Zhao and Nam (2007)

BLI, Bioluminescence imaging; BrdU, bromodeoxyuridine; EAE, experimental allergic encephalitis; ESC, embryonic stem cells; MRI, magnetic resonance imaging; RMS, rostral migratory stream.
[a]Migration speed calculated from the published data.

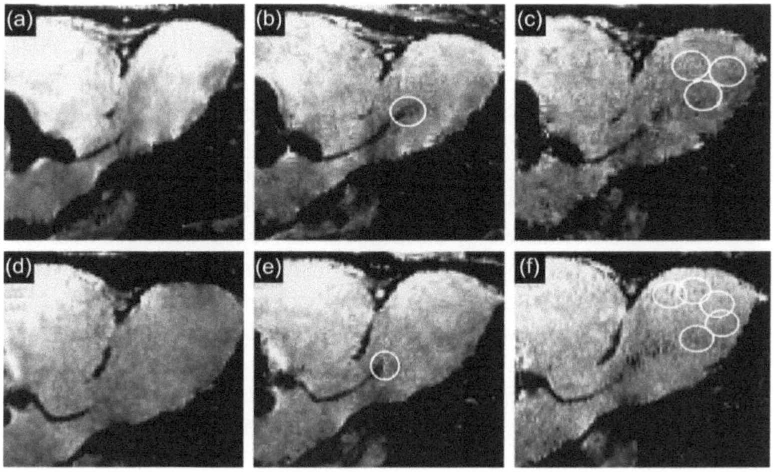

FIGURE 2

Observation of progenitor cell migration from the subventricular zone (SVZ) along the rostral migratory stream (RMS) to the olfactory bulb. MPIOs (20 μl) were injected into the anterior ventricle proximal to the SVZ in adult rats. 3D gradient echo images of two injected rat brains were acquired at 1 day (A and D), 2–3 days (B and E) and 8–9 days (C and F) post injection. Circles show the entrance to the olfactory bulb (B and E) and the spread of dark spots in the olfactory bulb (C and F).

(Reproduced with permission from Granot et al., 2011).

migratory path is visible as a hypointense track on the MR images (Granot et al., 2011; Nieman et al., 2010; Shapiro et al., 2006a; Sumner et al., 2009; Vreys et al., 2010). Using a transgenic approach, endogenous SVZ neural stem cells were transduced *in vivo* to overexpress ferritin, an iron storage protein, thus accumulating intracellular iron. These cells thus became visible for *in vivo* tracking with T2*-weighted MRI (Vande Velde et al., 2011). Furthermore, transgenic MRI reporter mice already exist in which only a specific pool of cells are overexpressing the ferritin protein, to create a contrast for MRI detection and tracking (Cohen et al., 2005).

MRI tracking of iron oxide labeled cells or ferritin overexpressing cells within the brain is based on hypointensity on T2- and T2*-weighted images. However, iron is an abundant element in the body. Upon ischemic stroke, dilated blood vessels may also present as hypointense areas on T2*-weighted images, due to the blood oxygenation level dependence effect, and may be mistaken for migrating labeled cells (Kallur et al., 2011). The inflammatory response can further confound the interpretation of hypointense regions along the ischemic area. Reactive glial cell and infiltrating macrophages are able to incorporate iron from their environment during the process of debris removal (Saleh et al., 2004; Weber et al., 2005) thereby also creating a strong T2* contrast (Hoehn et al., 2007; Vandeputte et al., 2011). This phagocytic characteristic of macrophages has been purposely used to label peripherally circulating macrophages by direct injection of contrast agent into the blood stream.

This has allowed participation of these cells in the inflammatory response to be followed after brain injury (Nighoghossian et al., 2007; Saleh et al., 2004, 2007). However, this approach has recently been questioned, on the grounds that interpretation of the results may be strongly dependent on the pathological model or experimental condition (Farr et al., 2011).

A second family of contrast agents widely used to label cells is based on gadolinium chelates, which induce a positive or white contrast, appearing as hyperintense signals on T1-weighted MR images. T1 contrast agents are less sensitive than iron oxide particles (Helm, 2010), but they are unambiguously distinguishable on MR images. Cells labeled with a contrast agent consisting of gadolinium (Gd-DTPA) and rhodamine particles linked by a dextran polymer were implanted in an animal model of global ischemia. Dispersion of cells along the corpus callosum and the CA1 field of the hippocampus were observed at 7 days post injection. At 14 days, MRI could not detect the cells any longer, and they were assumed to have migrated out of the implantation site and integrated into the hippocampus (Modo et al., 2002). The authors repeated the study in a rat model of MCAO. Transplantation of stem cells into the contralateral hemisphere was performed at 3 months post MCAO (Modo et al., 2004). After 1 day, no migration was visible on MR images, but after 7 days, the change in contrast indicated dispersion into the surrounding striatum and along the corpus callosum, in agreement with observations made using iron oxide labeled cells (Guzman et al., 2008; Hoehn et al., 2002; Jendelova et al., 2003, 2004). At 14 days, a substantial number of cells reached the ipsilateral perilesional area, but some cells still remained in the injection tract (Modo et al., 2004). The Gd-DTPA based contrast agent did not affect cell differentiation and migration, either *in vitro* or *in vivo* (Modo et al., 2002, 2004), but whether these labeled cells would still possess repair functionality was assessed in a one-year-long study using a behavioral test as readout for functional recovery (Modo et al., 2009). Cells showed the same migration patterns as previously described (Modo et al., 2004), but, interestingly, behavioral results suggest an interference of the label with stem cell induced recovery (Modo et al., 2009).

New contrast agents using fluorinated substances are emerging in the field of MRI cell tracking. Since there are only minute traces of endogenous ^{19}F in the body, cells can be tracked with ^{19}F MRI without any background or other confounding signals. The most frequently used labels are perfluorocarbon (PFC) nanoparticles (Ahrens et al., 2005; Helfer et al., 2010). This technique has already been successfully applied to T cell migration in models of diabetes (Srinivas et al., 2007), inflammation (Srinivas et al., 2009), or tumor studies (Partlow et al., 2007). Also, human neural stem cells (Kallur et al., 2011), which have been demonstrated to have high therapeutic potential for stroke (Darsalia et al., 2007, 2011), have been successfully labeled with PFC nanoparticles and detected sensitively after implantation into mouse brains (Böhm-Sturm et al., 2011). Very recently, in an elegant experiment, PFC-labeled human neural stem cells have been transplanted in combination with an extracellular matrix bioscaffold into the lesion cavity of stroke-damaged rat brains (Fig. 3; Bible et al., 2012). After 1 week of transplantation, a uniform distribution of cells throughout the cavity was seen, but the area covered by the ^{19}F signal decreased from 1 to 7 days

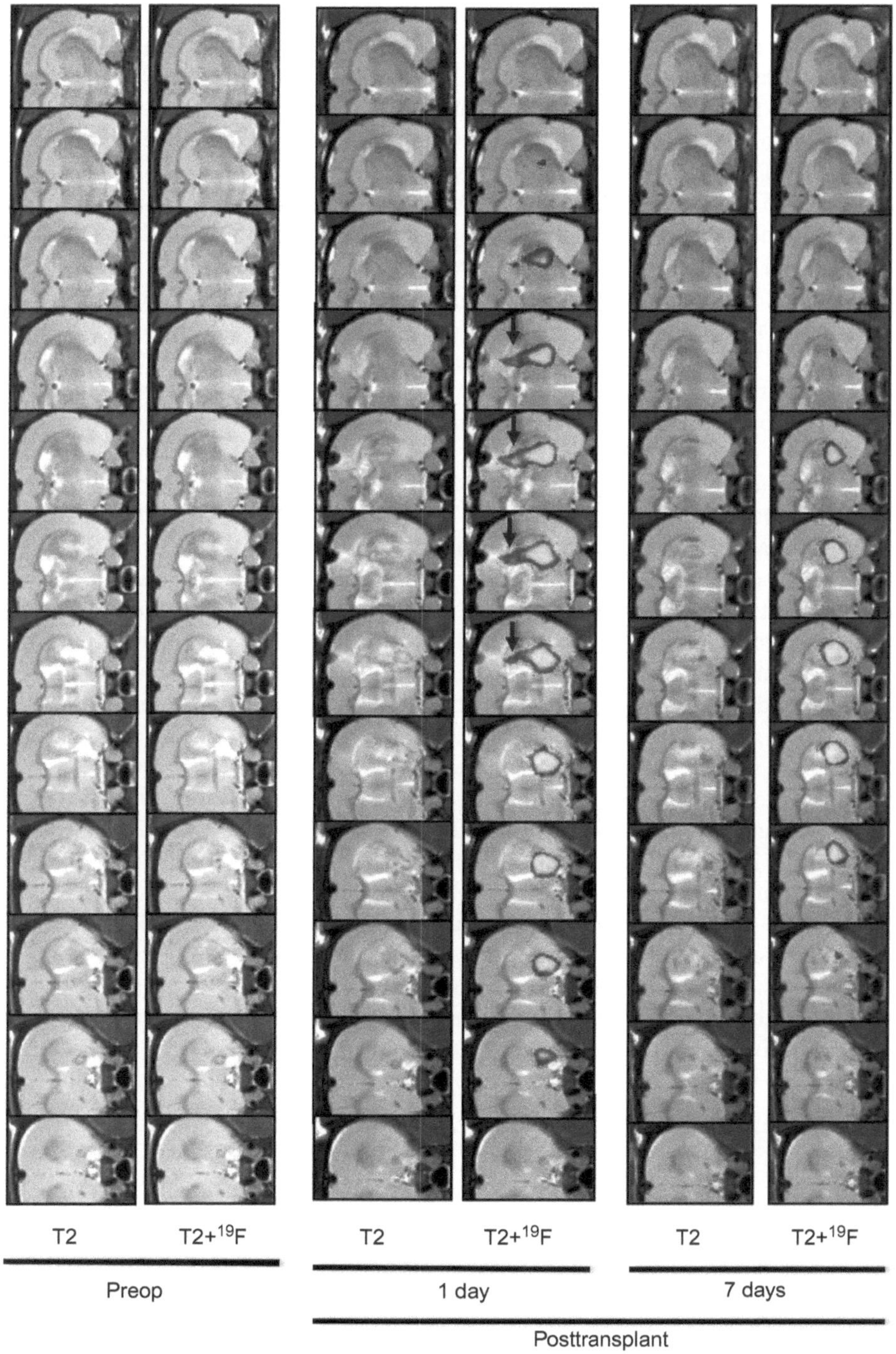
T2
T2+19F
T2
T2+19F
T2
T2+19F
Preop
1 day
7 days
Posttransplant

(Bible et al., 2012). In another experiment by Böhm-Sturm and colleagues, ischemic mouse brains received PFC-labeled mouse ESCs, and their fate has been followed for up to 5 weeks with combined ^{19}F- and ^{1}H-MRI (Böhm-Sturm et al., 2012).

In order to successfully establish a stem cell therapy for ischemic stroke, it is necessary to be able to follow the cells in space and time once they have been implanted. The issue of locating the cells with MRI within the host tissue has been addressed in the previous paragraphs. However, the final aim of a stem cell-based therapy is the recovery of function, and evidence exists that stem cell transplantations have a beneficial effect on the recovery from ischemic stroke (Lindvall and Kokaia, 2011; Shimada and Spees, 2011). Functional improvement after cell transplantation has mainly been assessed by behavioral tests in preclinical studies (Chen et al., 2001; Daadi et al., 2008; England et al., 2009; Li et al., 2002, 2004). MRI not only offers the possibility to follow the location of transplanted stem cells in the brain, it also offers the opportunity to image their functionality and the effect they have on the organism at the same time. Using a noninvasive functional MRI protocol, functional recovery after stem cell transplantation was observed in a rat model of MCAO. Cells were delivered into the periphery of the ischemic lesion at 3 weeks after ischemia, and approximately 10 weeks later, a return of the lost functional activation to somatosensory stimulation could be observed (Ramos-Cabrer et al., 2010). Nevertheless, the underlying mechanisms for the observed recovery remain under debate (Darsalia et al., 2007; Einstein and Ben-Hur, 2008; Hoehn et al., 2002, 2004; Karussis et al., 2008; Pluchino et al., 2005; Shimada and Spees, 2011). A question about the fate of the transplanted cells arises (Bulte et al., 1999, 2001; Hoehn et al., 2002; Jendelova et al., 2003; Modo et al., 2002; Zhang et al., 2003) and has been usually assessed with histological and immunohistological methods which study the differentiation into different cell types and the integration into the host tissue (Darsalia et al., 2007; Hoehn et al., 2002, 2004; Lundberg et al., 1997). To gain more insight into the therapeutic effect, the functional status of the transplanted cells should be assessable noninvasively. MRI contrast agents based on gadolinium chelates can be designed to be enzyme- or cell specific. Thus, they may report when a cell enters a specific functional status. Although no application to ischemic stroke has yet been reported, a gadolinium chelate-based responsive contrast agent has been designed to produce contrast activated by glutamate decarboxylase, which is present only in inhibitory

FIGURE 3

In vivo serial ^{19}F MRI imaging. The distribution of ^{19}F-labeled human neural stem cells within decellularized extracellular matrix encompasses the lesion cavity. ^{19}F-labeled cells can also be seen along the injection tract (green arrow) that crosses through "intact" overlying cortex on day 1. However, by 7 days post transplantation, the signal from the injection tract is no longer visible, although the cells distributed throughout the lesion cavity can still be reliably detected, although the extent of distribution is reduced. (For interpretation of the references to color in this figure legend, the reader is referred to the online version of this chapter.)

(Reproduced with permission from Bible et al., 2012).

GABAergic neurons, and thus allows the functional status of the transplanted cells to be visualized by MRI (Aswendt et al., 2012).

3 CELL TRACKING BY OI

Similar to MRI, OI is a noninvasive imaging technique and qualifies for longitudinal tracking of transplanted cells. Although it has a very low spatial resolution compared to MRI, it offers a much higher sensitivity so that even small numbers of cells can be detected (Sutton et al., 2008). Furthermore, it allows high-throughput studies due to low costs and fast acquisition times.

OI can be subdivided into two modalities: bioluminescence and fluorescence imaging. Bioluminescence imaging (BLI) is based on an enzyme-mediated production of light, which employs the oxidation of a specific substrate. The necessary enzyme is luciferase, which in most species needs the substrate luciferin to produce light. Several different luciferases from different species exist (firefly, gaussia, sea pansy, bacteria), and the spectrum of the emitted light is dependent on the luciferase used (Zhao et al., 2005). In order to employ bioluminescence for cell tracking, cells have to be transgenically modified to express luciferase. Upon systemic injection of its substrate, photon emission takes place only at the site of luciferase expression so that there is no background signal from the residual tissue. This allows even detection of signal from deep in the brain (Couillard-Despres et al., 2011; Sutton et al., 2008). For some luciferases, like the firefly luciferase, the light emitting reaction is dependent on ATP, thus allowing its detection as a viability marker of the luciferase expressing cells. In addition, the light emission is directly proportional to the number of cells, which further permits cell quantification (Kruttwig et al., 2010; Sutton et al., 2008). However, this is only true for cells at the brain surface, since light emission from deeper lying cells is scattered and absorbed by the overlying tissue. Using BLI, cell tracking can be performed over long periods of time since the luciferase gene is stably integrated into the genome, and luciferase expression is preserved during proliferation or differentiation so that all cell progeny express luciferase without dilution (Kim et al., 2004). These characteristics of BLI provide major advantages for the field of imaging stem cell therapy while several studies have successfully employed this modality for tracking tumor metastasis and tumor growth as well as for following stem cell therapy in cardiac injury (de Almeida et al., 2011).

Kim et al. (2004) used BLI to track murine C17.2 NPCs after transplantation in a murine model of stroke. Their aim was to follow migration, but also to study possible proliferation and quantification of the transplanted cells. The NPC line was modified to express firefly luciferase as well as GFP and β-galactosidase for histological validation of *in vivo* findings. Immediately after electrocoagulation of the right middle cerebral artery, the group injected 1×10^6 cells into the contralateral frontal lobe and, in a separate group, into the left ventricle. Cell location and viability was followed for 21 days (Fig. 4). Directly after injection, the strongest photon emission was located at the injection site. Thereafter, the signal was decreasing over the injection site while the signal was shifting toward the ischemic hemisphere, crossing the midline at

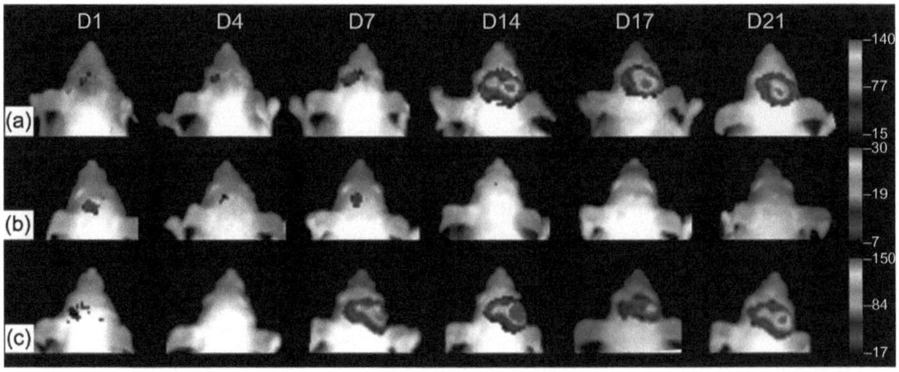

FIGURE 4

Representative bioluminescence imaging of NPC injections. (A) Intraparenchymal injection contralateral to a site of superficial infarction. Cells are initially seen at the injection site and then migrate to the infarct, crossing the midline by day 7 (D7), with subsequent increasing photon emissions at the infarct. (B) Intraparenchymal injection into the left lobe of a control animal with sham surgery. Note the decline of photon emissions over time, with eventual disappearance of signal. There is no migration to the contralateral hemisphere. (C) Intraventricular injection contralateral to a site of superficial MCA infarction. Note the initial distributed appearance, consistent with random cell distribution through the CSF. The infarct is populated by day 7 and shows increasing photon emissions thereafter. CSF, cerebrospinal fluid. (For color version of this figure, the reader is referred to the online version of this chapter.)

(Reproduced with permission from Kim et al., 2004).

7 days after transplantation. At this time point, already 40% of the injected cells were located at the ischemic hemisphere, and at 21 days, 70% had traveled to the ischemic site, which was confirmed by X-gal staining of β-galactosidase (Kim et al., 2004). In the ischemic animals, transplanted cells migrated at a speed of 15 μm/h (see Table 1) while in healthy control animals injected cells did not migrate, and bioluminescence from the injection site decreased until it was not detectable at 7 days post transplantation, probably due to clearance of the cells. A similar migration pattern of transplanted BMSC cells contralaterally to an ischemic lesion was observed by Jang et al. (2010). These authors tested different transplantation time points: 2 h, 1 day, 3 days, and 7 days after infarction. The best efficacy was achieved in the 2 h group, where highest photon emission could be observed from the ischemic side 25 days after transplantation (Jang et al., 2010).

Several groups have used BLI to follow also endogenous neurogenesis in health (Couillard-Despres et al., 2008; Reumers et al., 2008) and disease (Farr et al., 2009). Reumers et al. (2008) labeled cells in the SVZ by injecting lentiviruses which stably express firefly luciferase. Bioluminescence was recorded with high sensitivity to follow the migration of infected neuroblasts from the SVZ to the olfactory bulb

(Reumers et al., 2008). A different approach for bioluminescence of migration of endogenous neural stem cells was used by Couillard-Despres et al. (2008), who created a transgenic mouse, which expresses luciferase under the control of the doublecortin promoter, a protein expressed by migratory neuroblasts of the SVZ (Couillard-Despres et al., 2008). Using these transgenic mice, an increase of SVZ cell proliferation after cerebral ischemia could be followed with OI, and a clear shift of photon emission toward the ischemic hemisphere was observed (Farr et al., 2009). These observations are in line with earlier invasive studies where stroke results in increased proliferation of the ipsilateral SVZ and more lateral migration of these early neurons toward the lesion (Arvidsson et al., 2002; Parent et al., 2002; Thored et al., 2006).

The possibility of quantification and a readout of cell viability outline the enormous potential of BLI for cell tracking within the brain. However, the quantification has to overcome the problem of limited depth of light penetration making quantification only possible if cells stay at the same tissue level in the organ. To address this issue of light absorption through thick portions of tissue, the second modality of optical imaging, fluorescent imaging (FLI), can make use of light in the far red or near infrared spectrum, which has a higher penetration power through tissue (Kawabori et al., 2011; Lin et al., 2007). The general principle behind fluorescence imaging is the excitation of a fluorophore by an external light source and the capture of the emitted light using appropriate filters. Prominent examples are green fluorescent protein (GFP) and its derivates for shifted light emission like yellow fluorescent protein (YFP). Using GFP expressing bone marrow stromal cells, Shichinohe et al (Shichinohe et al., 2004) followed their migration towards the ischemic lesion for 12 weeks after transplantation into the ipsilateral striatum with in-vivo fluorescence optical imaging. The authors showed that thinning or removing the skull yielded similar results as imaging through the intact skull (Shichinohe et al., 2004). Recently, small nanocrystals, so-called quantum dots, have been developed, which possess the advantage of narrow emission spectra and resistance to photo bleaching (Rak-Raszewska et al., 2012). Although quantum dots have been shown to have no effect on cell viability (Rak-Raszewska et al., 2012), their usefulness for cell tracking over long periods of time is debatable, since dividing cells lost 90% of their label within 3 days (Rak-Raszewska et al., 2012). Nevertheless, quantum dots have been successfully used for stem cell tracking (Kawabori et al., 2011; Rosen et al., 2007). Especially quantum dots with emission spectra in the far red or near infrared have potential for in-vivo tracking of cells within the brain, as demonstrated by Kawabori et al. (Kawabori et al., 2011). Using a rat model of permanent MCAO, the study tested the potential of bone marrow stromal cells for functional recovery when transplantation occurred 7 days post stroke via two different delivery routes. Increasing fluorescence around the lesion was observed in the intracerebral transplantation group for 21 days postimplantation but strong decrease thereafter. Intravenous transplantation resulted in no detectable fluorescence from the brain, similar to vehicle treated animals. Improved motor performance was significant only in the group of intracerebral transplantation. The fluorescent probes mentioned above need to be excited by an external light source, which makes the excitation of deeply transplanted cells difficult. Two-photon or multiphoton excitation techniques have been proposed in the literature to overcome this issue (Fumagalli et al., 2011; Zhao and Nam, 2007). Although only on living brain slices, multiphoton microscopy was employed to study the behavior of the NPCs along the RMS (Zhao and Nam, 2007) and a migratory speed could be calculated of 30 ± 2 μm/h. Between 3 and 8 h after ischemia migration along the RMS was reduced, but recovered between 13 and 18 h after ischemia (Zhao and Nam, 2007; cf. Table 1).

There are many hundreds of fluorescent dyes available in the literature with various cell tracking applications. They can be nonspecific for exogenous cell labeling by incorporation into the cell, or they can be integrated in the genome under the expression of a reporter gene (Sutton et al., 2008; Tung, 2004). Fluorescence can be the result of an enzyme-activatable (smart) contrast agent and even specific cell type targeting can be achieved (Lu et al., 2010).

4 CELL TRACKING BY PET

Although PET has a low spatial resolution compared to MRI, it offers higher sensitivity, which allows the detection of cellular events at lower scale. Due to its high sensitivity, it has been used to monitor the fate of transplanted cells in myocardial infarction (Doyle et al., 2007; Higuchi et al., 2009; Kang et al., 2006) and in neurodegenerative diseases like Parkinson's, Huntington's, or traumatic brain injury (reviewed in Wang et al., 2011; Zhang et al., 2008). Imaging strategies include direct cell labeling with a PET radionucleotide or reporter gene-based cell labeling. Direct cell labeling employs the incubation of the cells with a radiotracer *in vitro* prior to transplantation into the living subject. Since the used radionucleotides have relatively short half-life (^{18}F: 110 min, ^{64}Cu: 12.7 h) only short-term distribution studies are possible (Wang et al., 2011). If a reporter gene-based approach is chosen, the therapeutic cells are modified to express special receptors, enzymes, or transporters, which will facilitate the specific uptake of a PET tracer. For example cells can express the herpes simplex virus thymidine kinase, which can phosphorylate exogenously administered radioactive substrates to produce a metabolite, or cells can express the sodium–iodine symporter, which can then be tracked with a tracer employing iodine (Wang et al., 2011). However, the above mentioned gene-based approaches cannot be used for tracking cells in the brain, since their tracers cannot pass the blood–brain barrier. Another reporter gene-based system uses the dopamine 2 receptor (D2R). The available tracers 3-(2′-(^{18}F)flouoroethyl)spiprone (^{18}F-FESP), ^{11}C-raclopride, or ^{11}C-*N*-methylspiperone (^{11}C-NMSP) can easily cross the BBB, so that transplanted cell distribution can be studied in the brain. However, the endogenous D2R occurrence in the striatum provides a high background signal and outlines the limitation of this strategy (Wang et al., 2011). Nevertheless, this strategy was successfully employed to track transplanted neural stem cells in a rat model of traumatic brain injury (Zhang et al., 2008). Higher binding of ^{11}C-NMSP in the cerebral lesion compared to the intact side was observed 1 day after transplantation, but gradually decreased thereafter, which was assumed to be due to the loss of NSC viability (Zhang et al., 2008).

Using ^{18}F-FLT, a radiolabelled thymidine analog, PET has been employed to track endogenous stem cell populations. ^{18}F-FLT is phosphorylated by thymidine kinase 1 during the onset of S-phase, thus it is possible to visualize cell proliferation with PET (Kong et al., 1992). In a healthy brain, proliferation of cells has ceased except for the SVZ and the SGZ, where neural stem and progenitor cells reside. Rueger et al. (2010) detected these endogenous neural stem cells in healthy rat brains. One week after stroke, proliferation in the SVZ and SGZ was increased (Fig. 5), but significance was only achieved in the SGZ (Rueger et al., 2010). Since stroke-induced proliferation in the SVZ was reported to be higher than in the SGZ, these findings suggest that ^{18}F-FLT binding is additionally influenced by other parameters (Couillard-Despres et al., 2011). Furthermore, ^{18}F-FLT is not a neurogenesis-specific PET tracer but a proliferation marker; thus, following stroke, reactive gliosis or immune cell infiltration will be undistinguishable from neural progenitor proliferation (Couillard-Despres et al., 2011).

5 MULTIMODAL CELL TRACKING APPROACHES

Much knowledge has been gained since the successful visualization of transplanted cells in the living organism. Graft location, viability, migration, and even differentiation can be assessed using noninvasive imaging techniques, as has been reviewed above. However, each technique has advantages and disadvantages, and can only provide information about specific aspects of the graft. Combination of two or more modalities offers the possibility to integrate modality-specific strengths and to gain a more complete picture of graft behavior. For example, a complementary use of OI or PET for high sensitivity and specificity together with MRI for spatial and temporal resolution has found applications of cell tracking following cerebral ischemia.

Since PET has a low spatial resolution, anatomical MRI has been used to locate the SVZ and the SGZ in PET images. It was then possible to follow the behavior of the signal of these specific neurogenic regions over time, and an increased cell proliferation following stroke was observed (Rueger et al., 2010).

BLI also has a very low spatial resolution and has therefore similarly been matched with MRI (Daadi et al., 2009). Using the firefly luciferase system, whose light production is dependent on ATP, it is possible to follow the viability status of the transplanted cells. Also a linear relationship between the number of cells and luciferase activity gives real time information about cell quantity and growth. Human neural stem cells were transduced to express firefly luciferase and subsequently labeled with SPIOs (Daadi et al., 2009) for transplantation at 1 week after stroke. Different graft sizes were followed for 8 weeks, and MRI clearly detected the graft

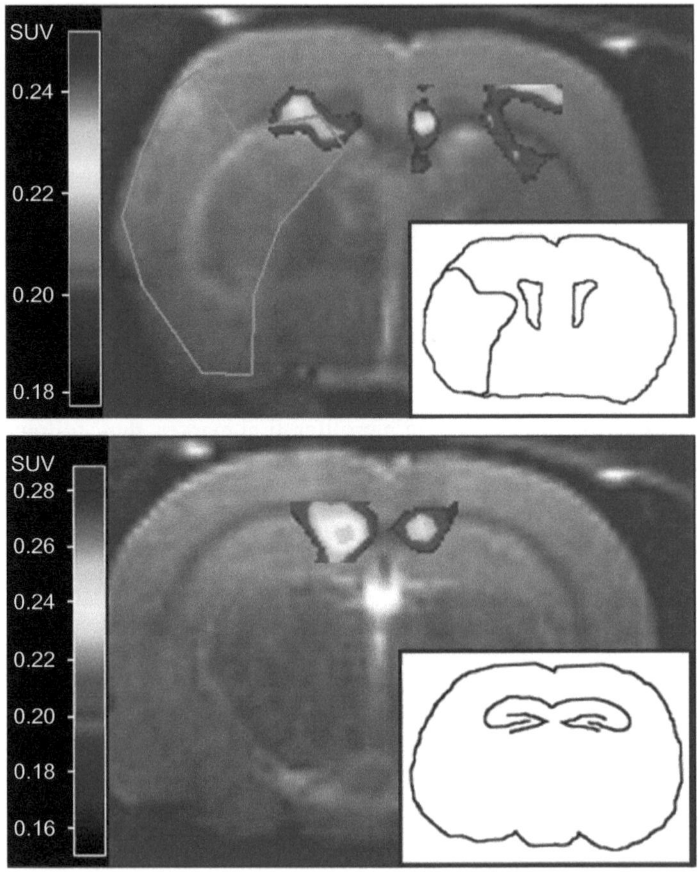

FIGURE 5

Stroke-induced expansion of the NSC niches can be distinguished with PET. (A) Coregistration of [^{18}F]FLT-PET and gadolinium-enhanced MRI 1 week after stroke (infarct outlined in purple, for schematic overview see inset). [^{18}F]FLT binding within the infarct core was repressed by image processing to emphasize comparatively weak tracer binding in the subventricular zone. (B) Coregistration of [^{18}F]FLT-PET and gadolinium-enhanced MRI 1 week after stroke shows [^{18}F]FLT binding within the hippocampus region (for schematic overview see inset). (For interpretation of the references to color in this figure legend, the reader is referred to the online version of this chapter.)

(*Reproduced with permission from Rueger et al., 2010*).

position next to the lesion. BLI showed stable graft survival without proliferation (Daadi et al., 2009).

Böhm-Sturm et al. (2012) use ^{19}F-MRI to unambiguously identify transplanted cells and locate their position on simultaneously obtained ^{1}H anatomical MRI (Fig. 6). Additional expression of click beetle luciferase by the transplanted cells reports cell viability after transplantation into the vicinity of the ischemic territory.

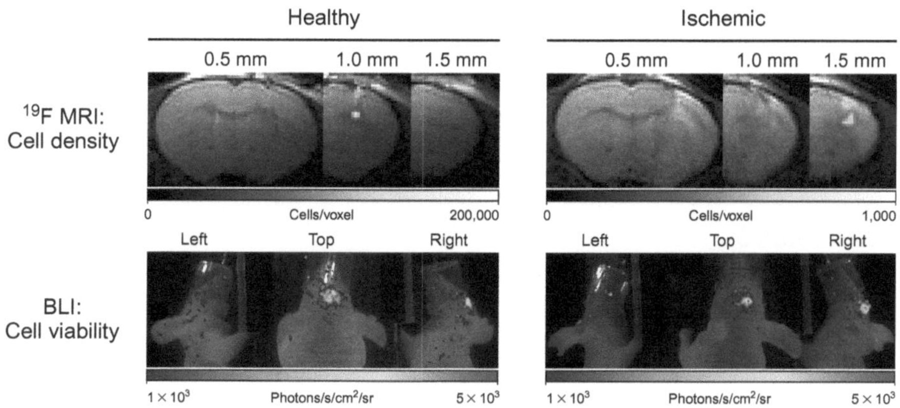

FIGURE 6

Multimodal imaging of stem cell graft location and viability. Genetically modified murine neural stem cells expressing the click beetle luciferase can be labeled with a perfluorocarbon emulsion for bimodal detection with bioluminescence imaging (BLI) and ^{19}F magnetic resonance imaging (^{19}F MRI). Quantitative ^{19}F MRI (top panel) allows to map the cell density within the graft both in the healthy (left) and the ischemic (right) mouse brain. The lesion can be depicted as the hyperintense region on a T2 weighted background ^{1}H image acquired in the same experiment (gray scale). Three coronal slices through the brain are shown (0.5, 1.0, and1.5 mm from Bregma). After administration of the substrate luciferin the transgenic cells generate a BLI signal which can be detected with an optical camera (bottom panel). Two mirrors placed next to the animal provide side views (left/right) for better localization of the signal source. The luciferase gene is expressed only in viable cells and the light reaction is ATP dependent, thus the strength of the BLI signal is a measure of cell viability. Due to the noninvasiveness of both methods, graft location and function can now be followed side by side in long-term studies of stem cell mediated repair of the ischemic brain. (For color version of this figure, the reader is referred to the online version of this chapter.)

(With permission from P. Boehm-Sturm, M. Aswendt, A. Minassian, personal communication.)

6 CONCLUSIONS

Stem cell therapy can only be taken to the clinic, if safe and beneficial behavior of the transplanted cells can be guaranteed. Therefore, it is important to refine and optimize tracking techniques, so that they can be used in a clinical setting. PET and MRI are already routinely used in patients and hold also potential for cell tracking in patients with cerebral ischemia. However, many more possibilities are available to the experimental research community and advantage should be taken of the potential of multimodal cell tracking in experimental cerebral ischemia as some groups already do. Possible beneficial treatment strategies can thus be narrowed down in the experimental environment before costly clinical trials are organized.

Acknowledgments

Support of EU grants of the 7th Framework (ENCITE, HEALTH-F5-2008-201842; TargetBrain, HEALTH-F2-2012-279017) and of the Volkswagen Foundation (I/83 443) are gratefully acknowledged.

References

Ahrens, E.T., Flores, R., Xu, H., Morel, P.A., 2005. In vivo imaging platform for tracking immunotherapeutic cells. Nat. Biotechnol. 23, 983–987.

Alvarez-Buylla, A., Garcia-Verdugo, J.M., 2002. Neurogenesis in adult subventricular zone. J. Neurosci. 22, 629–634.

Alvarez-Buylla, A., Nottebohm, F., 1988. Migration of young neurons in adult avian brain. Nature 335, 353–354.

Arvidsson, A., Collin, T., Kirik, D., Kokaia, Z., Lindvall, O., 2002. Neuronal replacement from endogenous precursors in the adult brain after stroke. Nat. Med. 8, 963–970.

Aswendt, M., Gianolio, E., Pariani, G., Napolitano, R., Fedeli, F., Himmelreich, U., Aime, S., Hoehn, M., 2012. In vivo imaging of inhibitory, GABAergic neurons by MRI. Neuroimage 62, 1685–1693.

Ben-Hur, T., van Heeswijk, R., Einstein, O., Aharonowitz, M., Xue, R., Frost, E.E., Mori, S., Reubinoff, B.E., Bulte, J.W.M., 2007. Serial *in vivo* MR tracking of magnetically labeled neural spheres transplanted in chronic EAE mice. Magn. Reson. Med. 57, 164–171.

Berlis, A., Lutsep, H., Barnwell, S., Norbash, A., Wechsler, L., Jungreis, C.A., Woolfenden, A., Redekop, G., Hartmann, M., Schumacher, M., 2004. Mechanical thrombolysis in acute ischemic stroke with endovascular photoacoustic recanalization. Stroke 35, 1112–1116.

Bible, E., Dell'Acqua, F., Solanky, B., Balducci, A., Crapo, P.M., Badylak, S.F., Ahrens, E.T., Modo, M., 2012. Non-invasive imaging of transplanted human neural stem cells and ECM scaffold remodeling in the stroke-damaged rat brain by 19F- and diffusion-MRI. Biomaterials 33, 2858–2871.

Böhm-Sturm, P., Mengler, L.M., Wecker, S.W., Hoehn, M., 2011. In vivo tracking of human neural stem cells in the mouse brain with 19F MRI. In: Proceedings of the European Molecular Imaging Meeting in Leiden, NL.

Böhm-Sturm, P., Aswendt, M., Mengler, L.M., Adamczak, J., Wecker, S.W., Hoehn, M., 2012. In-vivo tracking of cells in ischemic mice with BLI and 19F MRI. In: Proceedings of the International Society for Magnetic Resonance Imaging, Melbourne, Australia.

Bulte, J.W.M., Zhang, S.-C., van Gelderen, P., Herynek, V., Jordan, E.K., Duncan, I.D., Frank, J.A., 1999. Neurotransplantation of magnetically labeled oligodendrocyte progenitors: magnetic resonance tracking of cell migration and myelination. Proc. Natl. Acad. Sci. 96, 15256–15261.

Bulte, J.W.M., Douglas, T., Witwer, B., Zhang, S.-C., Strable, E., Lewis, B.K., Zywicke, H., Miller, B., Gelderen, P.v., Moskowitz, B.M., Duncan, I.D., Frank, J.A., 2001. Magnetodendrimers allow endosomal magnetic labeling and *in vivo* tracking of stem cells. Nat. Biotechnol. 19, 1141–1147.

Chen, J., Sanberg, P.R., Li, Y., Wang, L., Lu, M., Willing, A.E., Sanchez-Ramos, J., Chopp, M., 2001. Intravenous administration of human umbilical cord blood reduces behavioral deficits after stroke in rats. Stroke 32, 2682–2688.

Chopp, M., Li, Y., Zhang, Z.G., 2009. Mechanisms underlying improved recovery of neurological function after stroke in the rodent after treatment with neurorestorative cell-based therapies. Stroke 40, S143–S145.

Cohen, B., Dafni, H., Meir, G., Harmelin, A., Neeman, M., 2005. Ferritin as an endogenous MRI reporter for noninvasive imaging of gene expression in C6 glioma tumors. Neoplasia 7, 109–117.

Cormode, D.P., Jarzyna, P.A., Mulder, W.J., Fayad, Z.A., 2010. Modified natural nanoparticles as contrast agents for medical imaging. Adv. Drug Deliv. Rev. 62, 329–338.

Couillard-Despres, S., Finkl, R., Winner, B., Ploetz, S., Wiedermann, D., Aigner, R., Bogdahn, U., Winkler, J., Hoehn, M., Aigner, L., 2008. In vivo optical imaging of neurogenesis: watching new neurons in the intact brain. Mol. Imaging 7, 28–34.

Couillard-Despres, S., Vreys, R., Aigner, L., Van der Linden, A., 2011. In vivo monitoring of adult neurogenesis in health and disease. Front. Neurosci. 5, 67.

Daadi, M.M., Maag, A.-L., Steinberg, G.K., 2008. Adherent self-renewable human embryonic stem cell-derived neural stem cell line: functional engraftment in experimental stroke model. PLoS One 3, e1644.

Daadi, M.M., Li, Z., Arac, A., Grueter, B.A., Sofilos, M., Malenka, R.C., Wu, J.C., Steinberg, G.K., 2009. Molecular and magnetic resonance imaging of human embryonic stem cell-derived neural stem cell grafts in ischemic rat brain. Mol. Ther. 17, 1282–1291.

Darsalia, V., Kallur, T., Kokaia, Z., 2007. Survival, migration and neuronal differentiation of human fetal striatal and cortical neural stem cells grafted in stroke-damaged rat striatum. Eur. J. Neurosci. 26, 605–614.

Darsalia, V., Allison, S.J., Cusulin, C., Monni, E., Kuzdas, D., Kallur, T., Lindvall, O., Kokaia, Z., 2011. Cell number and timing of transplantation determine survival of human neural stem cell grafts in stroke-damaged rat brain. J. Cereb. Blood Flow Metab. 31, 235–242.

de Almeida, P.E., van Rappard, J.R., Wu, J.C., 2011. In vivo bioluminescence for tracking cell fate and function. Am. J. Physiol. Heart Circ. Physiol. 301, H663–H671.

Donnan, G.A., Fisher, M., Macleod, M., Davis, S.M., 2008. Stroke. Lancet 371, 1612–1623.

Doyle, B., Kemp, B.J., Chareonthaitawee, P., Reed, C., Schmeckpeper, J., Sorajja, P., Russell, S., Araoz, P., Riederer, S.J., Caplice, N.M., 2007. Dynamic tracking during intracoronary injection of 18F-FDG-labeled progenitor cell therapy for acute myocardial infarction. J. Nucl. Med. 48, 1708–1714.

Einstein, O., Ben-Hur, T., 2008. The changing face of neural stem cell therapy in neurologic diseases. Arch. Neurol. 65, 452–456.

England, T., Martin, P., Bath, P.M.W., 2009. Stem cells for enhancing recovery after stroke: a review. Int. J. Stroke 4, 101–110.

Farr, T.D., Kallur, T., Wiedermann, D., Couillard-Despres, S., Aigner, L., Hoehn, M., 2009. Live imaging of stroke induced neurogenesis in the mouse brain. In: Proceedings of the Annual Scientific Meeting of the ISMRM, Honolulu, Hawaii.

Farr, T.D., Seehafer, J.U., Nelles, M., Hoehn, M., 2011. Challenges towards MR imaging of the peripheral inflammatory response in the subacute and chronic stages of transient focal ischemia. NMR Biomed. 24, 35–45.

Fumagalli, S., Coles, J.A., Ejlerskov, P., Ortolano, F., Bushell, T.J., Brewer, J.M., De Simoni, M.G., Dever, G., Garside, P., Maffia, P., Carswell, H.V., 2011. In vivo real-time multiphoton imaging of T lymphocytes in the mouse brain after experimental stroke. Stroke 42, 1429–1436.

Granot, D., Scheinost, D., Markakis, E.A., Papademetris, X., Shapiro, E.M., 2011. Serial monitoring of endogenous neuroblast migration by cellular MRI. Neuroimage 57, 817–824.

Guzman, R., Uchida, N., Bliss, T.M., He, D., Christopherson, K.K., Stellwagen, D., Capela, A., Greve, J., Malenka, R.C., Moseley, M.E., Palmer, T.D., Steinberg, G.K., 2007. Long-term monitoring of transplanted human neural stem cells in developmental and pathological contexts with MRI. Proc. Natl. Acad. Sci. U.S.A. 104, 10211–10216.

Guzman, R., Bliss, T., De Los Angeles, A., Moseley, M., Palmer, T., Steinberg, G., 2008. Neural progenitor cells transplanted into the uninjured brain undergo targeted migration after stroke onset. J. Neurosci. Res. 86, 873–882.

Helfer, B.M., Balducci, A., Nelson, A.D., Janjic, J.M., Gil, R.R., Kalinski, P., de Vries, I.J., Ahrens, E.T., Mailliard, R.B., 2010. Functional assessment of human dendritic cells labeled for *in vivo* (19)F magnetic resonance imaging cell tracking. Cytotherapy 12, 238–250.

Helm, L., 2010. Optimization of gadolinium-based MRI contrast agents for high magnetic-field applications. Future Med. Chem. 2, 385–396.

Higuchi, T., Anton, M., Saraste, A., Dumler, K., Pelisek, J., Nekolla, S.G., Bengel, F.M., Schwaiger, M., 2009. Reporter gene PET for monitoring survival of transplanted endothelial progenitor cells in the rat heart after pretreatment with VEGF and atorvastatin. J. Nucl. Med. 50, 1881–1886.

Himmelreich, U., Dresselaers, T., 2009. Cell labeling and tracking for experimental models using magnetic resonance imaging. Methods 48, 112–124.

Hoehn, M., Kustermann, E., Blunk, J., Wiedermann, D., Trapp, T., Wecker, S., Focking, M., Arnold, H., Hescheler, J., Fleischmann, B.K., Schwindt, W., Buhrle, C., 2002. Monitoring of implanted stem cell migration *in vivo*: a highly resolved *in vivo* magnetic resonance imaging investigation of experimental stroke in rat. Proc. Natl. Acad. Sci. U.S.A. 99, 16267–16272.

Hoehn, M., Küstermann, E., Wiedermann, D., Bührle, C., Blunk, J., Weber, R., Wegener, S., Trapp, T., Hescheler, J., Fleischmann, B.K., Kandal, K., Ramos-Cabrer, P., 2004. Stem cell migration after stroke observed by *in vivo* MR imaging. In: Krieglstein, J., Klumpp, S. (Eds.), Pharmacology of Cerebral Ischemia. Medpharm Scientific Publishers, Stuttgart, pp. 503–512.

Hoehn, M., Wiedermann, D., Justicia, C., Ramos-Cabrer, P., Kruttwig, K., Farr, T., Himmelreich, U., 2007. Cell tracking using magnetic resonance imaging. J. Physiol. 584, 25–30.

Jang, K.S., Lee, K.S., Yang, S.H., Jeun, S.S., 2010. In vivo tracking of transplanted bone marrow-derived mesenchymal stem cells in a murine model of stroke by bioluminescence imaging. J. Korean Neurosurg. Soc. 48, 391–398.

Jendelova, P., Herynek, V., DeCroos, J., Glogarova, K., Andersson, B., Hajek, M., Sykova, E., 2003. Imaging the fate of implanted bone marrow stromal cells labeled with superparamagnetic nanoparticles. Magn. Reson. Med. 50, 767–776.

Jendelova, P., Herynek, V., Urdzikova, L., Glogarova, K., Kroupova, J., Andersson, B., Bryja, V., Burian, M., Hajek, M., Sykova, E., 2004. Magnetic resonance tracking of transplanted bone marrow and embryonic stem cells labeled by iron oxide nanoparticles in rat brain and spinal cord. J. Neurosci. Res. 76, 232–243.

Kallur, T., Farr, T.D., Bohm-Sturm, P., Kokaia, Z., Hoehn, M., 2011. Spatio-temporal dynamics, differentiation and viability of human neural stem cells after implantation into neonatal rat brain. Eur. J. Neurosci. 34, 382–393.

Kang, W.J., Kang, H.J., Kim, H.S., Chung, J.K., Lee, M.C., Lee, D.S., 2006. Tissue distribution of 18F-FDG-labeled peripheral hematopoietic stem cells after intracoronary administration in patients with myocardial infarction. J. Nucl. Med. 47, 1295–1301.

Karussis, D., Kassis, I., Kurkalli, B.G.S., Slavin, S., 2008. Immunomodulation and neuroprotection with mesenchymal bone marrow stem cells (MSCs): a proposed treatment for multiple sclerosis and other neuroimmunological/neurodegenerative diseases. J. Neurol. Sci. 265, 131–135.

Kawabori, M., Kuroda, S., Sugiyama, T., Ito, M., Shichinohe, H., Houkin, K., Kuge, Y., Tamaki, N., 2011. Intracerebral, but not intravenous, transplantation of bone marrow stromal cells enhances functional recovery in rat cerebral infarct: an optical imaging study. Neuropathology 32 (3), 217–226.

Kim, D.E., Schellingerhout, D., Ishii, K., Shah, K., Weissleder, R., 2004. Imaging of stem cell recruitment to ischemic infarcts in a murine model. Stroke 35, 952–957.

Kong, X.B., Zhu, Q.Y., Vidal, P.M., Watanabe, K.A., Polsky, B., Armstrong, D., Ostrander, M., Lang Jr., S.A., Muchmore, E., Chou, T.C., 1992. Comparisons of anti-human immunodeficiency virus activities, cellular transport, and plasma and intracellular pharmacokinetics of 3′-fluoro-3′-deoxythymidine and 3′-azido-3′-deoxythymidine. Antimicrob. Agents Chemother. 36, 808–818.

Kruttwig, K., Brueggemann, C., Kaijzel, E., Vorhagen, S., Hilger, T., Lowik, C., Hoehn, M., 2010. Development of a three-dimensional *in vitro* model for longitudinal observation of cell behavior: monitoring by magnetic resonance imaging and optical imaging. Mol. Imaging Biol. 12, 367–376.

Li, Y., Chen, J., Chen, X.G., Wang, L., Gautam, S.C., Xu, Y.X., Katakowski, M., Zhang, L.J., Lu, M., Janakiraman, N., Chopp, M., 2002. Human marrow stromal cell therapy for stroke in rat. Neurology 59, 514–523.

Li, X., Blizzard, K.K., Zeng, Z., DeVries, A.C., Hurn, P.D., McCullough, L.D., 2004. Chronic behavioral testing after focal ischemia in the mouse: functional recovery and the effects of gender. Exp. Neurol. 187, 94–104.

Lin, S., Xie, X., Patel, M.R., Yang, Y.H., Li, Z., Cao, F., Gheysens, O., Zhang, Y., Gambhir, S.S., Rao, J.H., Wu, J.C., 2007. Quantum dot imaging for embryonic stem cells. BMC Biotechnol. 7, 67.

Lindvall, O., Kokaia, Z., 2011. Stem cell research in stroke: how far from the clinic? Stroke 42, 2369–2375.

Liu, W., Frank, J.A., 2009. Detection and quantification of magnetically labeled cells by cellular MRI. Eur. J. Radiol. 70, 258–264.

Lois, C., Alvarez-Buylla, A., 1994. Long-distance neuronal migration in the adult mammalian brain. Science 264, 1145–1148.

Lu, S., Xu, X., Zhao, W., Wu, W., Yuan, H., Shen, H., Zhou, C., Li, L.S., Ma, L., 2010. Targeting of embryonic stem cells by peptide-conjugated quantum dots. PLoS One 5, e12075.

Lundberg, C., Martínez-Serrano, A., Cattaneo, E., McKay, R.D.G., Björklund, A., 1997. Survival, integration, and differentiation of neural stem cell lines after transplantation to the adult rat striatum. Exp. Neurol. 145, 342–360.

Luo, Y., 2011. Cell-based therapy for stroke. J. Neural Transm. 118, 61–74.

Modo, M., Cash, D., Mellodew, K., Williams, S.C., Fraser, S.E., Meade, T.J., Price, J., Hodges, H., 2002. Tracking transplanted stem cell migration using bifunctional, contrast agent-enhanced, magnetic resonance imaging. Neuroimage 17, 803–811.

Modo, M., Mellodew, K., Cash, D., Fraser, S.E., Meade, T.J., Price, J., Williams, S.C., 2004. Mapping transplanted stem cell migration after a stroke: a serial, *in vivo* magnetic resonance imaging study. Neuroimage 21, 311–317.

Modo, M., Hoehn, M., Bulte, J.W., 2005. Cellular MR imaging. Mol. Imaging 4, 143–164.

Modo, M., Beech, J.S., Meade, T.J., Williams, S.C., Price, J., 2009. A chronic 1 year assessment of MRI contrast agent-labelled neural stem cell transplants in stroke. Neuroimage 47 (Suppl 2), T133–T142.

Nieman, B.J., Shyu, J.Y., Rodriguez, J.J., Garcia, A.D., Joyner, A.L., Turnbull, D.H., 2010. In vivo MRI of neural cell migration dynamics in the mouse brain. Neuroimage 50, 456–464.

Nighoghossian, N., Wiart, M., Cakmak, S., Berthezene, Y., Derex, L., Cho, T.H., Nemoz, C., Chapuis, F., Tisserand, G.L., Pialat, J.B., Trouillas, P., Froment, J.C., Hermier, M., 2007. Inflammatory response after ischemic stroke: a USPIO-enhanced MRI study in patients. Stroke 38, 303–307.

Parent, J.M., Vexler, Z.S., Gong, C., Derugin, N., Ferriero, D.M., 2002. Rat forebrain neurogenesis and striatal neuron replacement after focal stroke. Ann. Neurol. 52, 802–813.

Partlow, K.C., Chen, J., Brant, J.A., Neubauer, A.M., Meyerrose, T.E., Creer, M.H., Nolta, J.A., Caruthers, S.D., Lanza, G.M., Wickline, S.A., 2007. 19F magnetic resonance imaging for stem/progenitor cell tracking with multiple unique perfluorocarbon nanobeacons. FASEB J. 21, 1647–1654.

Pluchino, S., Zanotti, L., Rossi, B., Brambilla, E., Ottoboni, L., Salani, G., Martinello, M., Cattalini, A., Bergami, A., Furlan, R., Comi, G., Constantin, G., Martino, G., 2005. Neurosphere-derived multipotent precursors promote neuroprotection by an immunomodulatory mechanism. Nature 436, 266–271.

Rak-Raszewska, A., Marcello, M., Kenny, S., Edgar, D., See, V., Murray, P., 2012. Quantum dots do not affect the behaviour of mouse embryonic stem cells and kidney stem cells and are suitable for short-term tracking. PLoS One 7, e32650.

Ramos-Cabrer, P., Justicia, C., Wiedermann, D., Hoehn, M., 2010. Stem cell mediation of functional recovery after stroke in the rat. PLoS One 5, e12779.

Reumers, V., Deroose, C.M., Krylyshkina, O., Nuyts, J., Geraerts, M., Mortelmans, L., Gijsbers, R., Van den Haute, C., Debyser, Z., Baekelandt, V., 2008. Noninvasive and quantitative monitoring of adult neuronal stem cell migration in mouse brain using bioluminescence imaging. Stem Cells 26, 2382–2390.

Rosen, A.B., Kelly, D.J., Schuldt, A.J., Lu, J., Potapova, I.A., Doronin, S.V., Robichaud, K.J., Robinson, R.B., Rosen, M.R., Brink, P.R., Gaudette, G.R., Cohen, I.S., 2007. Finding fluorescent needles in the cardiac haystack: tracking human mesenchymal stem cells labeled with quantum dots for quantitative *in vivo* three-dimensional fluorescence analysis. Stem Cells 25, 2128–2138.

Rueger, M.A., Backes, H., Walberer, M., Neumaier, B., Ullrich, R., Simard, M.L., Emig, B., Fink, G.R., Hoehn, M., Graf, R., Schroeter, M., 2010. Noninvasive imaging of endogenous neural stem cell mobilization *in vivo* using positron emission tomography. J. Neurosci. 30, 6454–6460.

Saleh, A., Wiedermann, D., Schroeter, M., Jonkmanns, C., Jander, S., Hoehn, M., 2004. Central nervous system inflammatory response after cerebral infarction as detected by magnetic resonance imaging. NMR Biomed. 17, 163–169.

Saleh, A., Schroeter, M., Ringelstein, A., Hartung, H.P., Siebler, M., Modder, U., Jander, S., 2007. Iron oxide particle-enhanced MRI suggests variability of brain inflammation at early stages after ischemic stroke. Stroke 38, 2733–2737.

Shapiro, E.M., Gonzalez-Perez, O., Manuel Garcia-Verdugo, J., Alvarez-Buylla, A., Koretsky, A.P., 2006a. Magnetic resonance imaging of the migration of neuronal precursors generated in the adult rodent brain. Neuroimage 32, 1150–1157.

Shapiro, E.M., Sharer, K., Skrtic, S., Koretsky, A.P., 2006b. In vivo detection of single cells by MRI. Magn. Reson. Med. 55, 242–249.

Shichinohe, H., Kuroda, S., Lee, J.B., Nishimura, G., Yano, S., Seki, T., Ikeda, J., Tamura, M., Iwasaki, Y., 2004. In vivo tracking of bone marrow stromal cells transplanted into mice cerebral infarct by fluorescence optical imaging. Brain Res. Brain Res. Protoc. 13, 166–175.

Shimada, I.S., Spees, J.L., 2011. Stem and progenitor cells for neurological repair: minor issues, major hurdles, and exciting opportunities for paracrine-based therapeutics. J. Cell. Biochem. 112, 374–380.

Smith, W.S., Sung, G., Saver, J., Budzik, R., Duckwiler, G., Liebeskind, D.S., Lutsep, H.L., Rymer, M.M., Higashida, R.T., Starkman, S., Gobin, Y.P., Frei, D., Grobelny, T., Hellinger, F., Huddle, D., Kidwell, C., Koroshetz, W., Marks, M., Nesbit, G., Silverman, I.E., 2008. Mechanical thrombectomy for acute ischemic stroke: final results of the Multi MERCI trial. Stroke 39, 1205–1212.

Srinivas, M., Morel, P.A., Ernst, L.A., Laidlaw, D.H., Ahrens, E.T., 2007. Fluorine-19 MRI for visualization and quantification of cell migration in a diabetes model. Magn. Reson. Med. 58, 725–734.

Srinivas, M., Turner, M.S., Janjic, J.M., Morel, P.A., Laidlaw, D.H., Ahrens, E.T., 2009. In vivo cytometry of antigen-specific t cells using 19F MRI. Magn. Reson. Med. 62, 747–753.

Sumner, J.P., Shapiro, E.M., Maric, D., Conroy, R., Koretsky, A.P., 2009. In vivo labeling of adult neural progenitors for MRI with micron sized particles of iron oxide: quantification of labeled cell phenotype. Neuroimage 44, 671–678.

Sutton, E.J., Henning, T.D., Pichler, B.J., Bremer, C., Daldrup-Link, H.E., 2008. Cell tracking with optical imaging. Eur. Radiol. 18, 2021–2032.

Thored, P., Arvidsson, A., Cacci, E., Ahlenius, H., Kallur, T., Darsalia, V., Ekdahl, C.T., Kokaia, Z., Lindvall, O., 2006. Persistent production of neurons from adult brain stem cells during recovery after stroke. Stem Cells 24, 739–747.

Tung, C.H., 2004. Fluorescent peptide probes for *in vivo* diagnostic imaging. Biopolymers 76, 391–403.

Vande Velde, G., Rangarajan, J.R., Toelen, J., Dresselaers, T., Ibrahimi, A., Krylychkina, O., Vreys, R., Van der Linden, A., Maes, F., Debyser, Z., Himmelreich, U., Baekelandt, V., 2011. Evaluation of the specificity and sensitivity of ferritin as an MRI reporter gene in the mouse brain using lentiviral and adeno-associated viral vectors. Gene Ther. 18, 594–605.

Vandeputte, C., Thomas, D., Dresselaers, T., Crabbe, A., Verfaillie, C., Baekelandt, V., Van Laere, K., Himmelreich, U., 2011. Characterization of the inflammatory response in a photothrombotic stroke model by MRI: implications for stem cell transplantation. Mol. Imaging Biol. 13, 663–671.

Vreys, R., Vande Velde, G., Krylychkina, O., Vellema, M., Verhoye, M., Timmermans, J.P., Baekelandt, V., Van der Linden, A., 2010. MRI visualization of endogenous neural progenitor cell migration along the RMS in the adult mouse brain: validation of various MPIO labeling strategies. Neuroimage 49, 2094–2103.

Wahlgren, N., Ahmed, N., Davalos, A., Hacke, W., Millan, M., Muir, K., Roine, R.O., Toni, D., Lees, K.R., 2008. Thrombolysis with alteplase 3–4.5 h after acute ischaemic stroke (SITS-ISTR): an observational study. Lancet 372, 1303–1309.

Walczak, P., Zhang, J., Gilad, A.A., Kedziorek, D.A., Ruiz-Cabello, J., Young, R.G., Pittenger, M.F., van Zijl, P.C., Huang, J., Bulte, J.W., 2008. Dual-modality monitoring of targeted intraarterial delivery of mesenchymal stem cells after transient ischemia. Stroke 39, 1569–1574.

Walker, P.A., Letourneau, P.A., Bedi, S., Shah, S.K., Jimenez, F., Jr, C.S., 2011. Progenitor cells as remote "bioreactors": neuroprotection via modulation of the systemic inflammatory response. World J. Stem Cells 3, 9–18.

Wang, J., Tian, M., Zhang, H., 2011. PET molecular imaging in stem cell therapy for neurological diseases. Eur. J. Nucl. Med. Mol. Imaging 38, 1926–1938.

Weber, R., Wegener, S., Ramos-Cabrer, P., Wiedermann, D., Hoehn, M., 2005. MRI detection of macrophage activity after experimental stroke in rats: new indicators for late appearance of vascular degradation? Magn. Reson. Med. 54, 59–66.

Zhang, Z.G., Jiang, Q., Zhang, R., Zhang, L., Wang, L., Zhang, L., Arniego, P., Ho, K.L., Chopp, M., 2003. Magnetic resonance imaging and neurosphere therapy of stroke in rat. Ann. Neurol. 53, 259–263.

Zhang, H., Zheng, X., Yang, X., Fang, S., Shen, G., Zhao, C., Tian, M., 2008. 11C-NMSP/18F-FDG microPET to monitor neural stem cell transplantation in a rat model of traumatic brain injury. Eur. J. Nucl. Med. Mol. Imaging 35, 1699–1708.

Zhao, L.R., Nam, S.C., 2007. Multiphoton microscope imaging: the behavior of neural progenitor cells in the rostral migratory stream. Neurosci. Lett. 425, 83–88.

Zhao, H., Doyle, T.C., Coquoz, O., Kalish, F., Rice, B.W., Contag, C.H., 2005. Emission spectra of bioluminescent reporters and interaction with mammalian tissue determine the sensitivity of detection *in vivo*. J. Biomed. Opt. 10, 41210.

CHAPTER

Bone marrow stem cells in experimental stroke

6

Jeffery D. Kocsis[*,†,1], **Osamu Honmou**[*,‡]

[*]*Department of Neurology, Yale University, School of Medicine, New Haven, CT, USA*

[†]*Center for Neuroscience and Regeneration Research, VA Connecticut Healthcare System, West Haven, CT, USA*

[‡]*Department of Neural Regenerative Medicine, Sapporo Medical University, Sapporo, Japan*

[1]*Corresponding author. Tel.: +1-203-9373802, Fax: +1-203-9373801, e-mail address: jeffery.kocsis@yale.edu*

Abstract

Adult bone marrow-derived mesenchymal stem cells (MSCs) have a remarkable spectrum of functional properties. Transplantation of MSCs improves clinical outcome in animal models of cerebral ischemia, and traumatic brain and spinal cord injury, via mechanisms that may include induction of axonal sprouting and remyelination, protection of injured neurons, neovascularization, and, potentially, replacement of damaged cells. Beneficial therapeutic effects have been reported in experimental models of stroke following intravenous delivery of MSCs. Initial clinical studies using intravenously delivered MSCs have been initiated in human subjects with stroke. This chapter reviews multiple beneficial effects of MSCs in experimental stroke models, describes initial human clinical studies on intravenous MSC delivery in stroke, and provides a perspective on further experimental and clinical studies of MSCs.

Keywords

bone marrow, stem cells, stroke, neural protection

1 INTRODUCTION

Treatment of stroke is presently primarily limited to acute thrombolysis therapy with recombinant tissue plasminogen activator inhibitor (tPA), which must be initiated within the early hours following stroke onset (Pereira et al., 2001). Hemorrhagic etiology must be ruled out and, therefore, only a limited number of patients can receive this therapy. It is estimated that as of 2009, 3.4–5.2% of acute ischemic stroke patients received thrombolytic therapy (Adeoye et al., 2011). Moreover, about half of

Progress in Brain Research, Volume 201, ISSN 0079-6123, http://dx.doi.org/10.1016/B978-0-444-59544-7.00005-6

the patients receiving thrombolytic therapy show little or no improvement in functional outcome (Wahlgren et al., 2007). Thus, there is a need for novel therapeutic approaches that will promote functional recovery.

Variable degrees of apparently spontaneous functional recovery occur in stroke patients (Cramer et al., 1999) and in animal models of stroke, even though the size of the ischemic lesion may stay the same or increase during recovery (Jiang et al., 2001). These observations suggest that compensatory neural plasticity or brain remodeling may contribute to time-dependent functional recovery. Cellular therapies have the objective of introducing new mechanisms that will promote functional recovery or enhance these endogenous repair processes (Chopp and Li, 2002). While an early assumption in stem cell therapeutic approaches for neurological diseases was that stem cells would replace injured cells, current work suggests several alternative mechanisms: stem cells may release or stimulate the release of trophic factors that may be neuroprotective and/or promote neovascularization, remyelination, and axonal sprouting.

Cell-based therapeutic approaches are currently being studied for a number of neurological diseases. Importantly, improved neurological function has been observed following intravenous infusion of mesenchymal stem cells (MSCs) in a number of disease models including experimental autoimmune encephalomyelitis and rodent models of stroke (e.g., Honmou et al., 2012). Suggested mechanisms include attenuation of inflammation and demyelination, and elevation of trophic factors that may be neuroprotective or that may stimulate axonal sprouting and oligodendrogliogenesis. These different effects are not mutually exclusive, raising the possibility that a cell-based therapy may have the advantage of exerting multiple therapeutic effects at various sites and times within the lesion, as the cells respond to an evolving pathological microenvironment.

During development, the mesodermal layer contains multipotent progenitors that can give rise to bone, cartilage, muscle, and other mesenchymal tissues. It was hypothesized that a population of MSCs in bone marrow could differentiate along multiple distinct cell lineage pathways (Caplan, 1991; Caplan and Correa, 2011). Several investigators demonstrated that MSCs are capable of self-renewal and differentiation and are multipotent for osteogenic, adipogenic, and chondrogenic lineages (Caplan, 2005; Kobune et al., 2003; Krause et al., 2001; Pittenger et al., 1999; Prockop, 1997; Sanchez-Ramos et al., 2000). It has also been suggested that MSCs derived from bone marrow can differentiate into cells of neuronal and glial lineage (Akiyama et al., 2002; Azizi et al., 1998; Brazelton et al., 2000; Kim et al., 2006; Kopen et al., 1999; Prockop, 1997; Sanchez-Ramos et al., 2000; Sasaki et al., 2001; Woodbury et al., 2000). Yet, others have challenged neural differentiation of bone marrow-derived cells that enter the brain (Massengale et al., 2005). *In vivo* identification of MSCs is problematic because there is no unambiguous marker for these cells (Caplan and Correa, 2011), and it is possible that the local microenvironment may direct different cell characteristics (Bianco et al., 2008).

MSCs, unlike hematopoietic stem cells, are $CD34^-$ and $CD45^-$, providing a basis for their isolation by flow cytometry. Molecular markers for human MSCs include

$CD73^+$ and $CD105^+$ (Honma et al., 2006; Kobune et al., 2003; Majumdar et al., 1998). Isolated cultured bone marrow-derived MSCs have been shown to secrete multiple trophic factors including brain-derived neurotrophic factor (BDNF), glial-derived neurotrophic factor (GDNF), nerve growth factor, hepatocyte growth factor (HGF), vascular endothelial growth factors (VEGF), angiopoietin-1 (Ang-1), and others (Chen et al., 2002; Hamano et al., 2000; Iihoshi et al., 2004; Kurozumi et al., 2004; Onda et al., 2008). This molecular profile of MSCs suggests that they may be able to promote neovascularization and/or provide trophic support to vulnerable cells within the penumbral region of a cerebral infarction lesion where neural tissue may be vulnerable.

Translation from animal models to clinical studies using intravenously delivered MSCs is now in progress. This chapter will describe experimental work showing therapeutic effects of MSC transplantation in experimental stroke models, and provide an update of initial clinical studies of intravenous MSC infusions in human subjects with stroke.

2 THE IMPACT OF MSC TRANSPLANTATION IN EXPERIMENTAL STROKE MODELS: STRUCTURAL AND FUNCTIONAL CHANGES

Transplantation of rodent MSCs directly into the infarcted brain within hours to days after induction of cerebral ischemia can reduce infarct size and improve functional outcome (Chen et al., 2001; Li et al., 2002). Moreover, intravenous infusion of MSCs derived from human bone marrow (hMSCs) in rodent stroke models provides functional benefit (Honma et al., 2006; Iihoshi et al., 2004; Nomura et al., 2005). While some of the beneficial effects of MSCs may result from neurogenesis, other mechanisms including neuroprotection, axonal sprouting, and angiogenesis have been suggested to be critically important (Chen et al., 2001; Honmou et al., 2012). Indeed, intravenous delivery of MSCs in stroke models leads to reduced apoptosis of cells at the ischemic lesion boundary (Liu et al., 2006) and promotes endogenous cell proliferation (Li et al., 2002). Figure 1 shows the reduction in lesion volume and improved functional outcome following intravenous infusion of human bone marrow-derived MSCs in a rodent stroke model.

2.1 Secretion or stimulation of trophic factors

The capacity of MSCs to release growth and trophic factors, or to stimulate their release from resident brain cells, has been suggested to be a key component to their beneficial effects in cerebral ischemia (Chen et al., 2002). A low-level basal secretion of multiple neurotrophic factors by MSCs has been observed in culture, and ischemic rat brain extracts can induce the production of neurotrophins and angiogenic growth factors in MSCs (Chen et al., 2002). One of these, BDNF, is constitutively expressed at low-levels in primary human MSC cultures (Kurozumi et al., 2004;

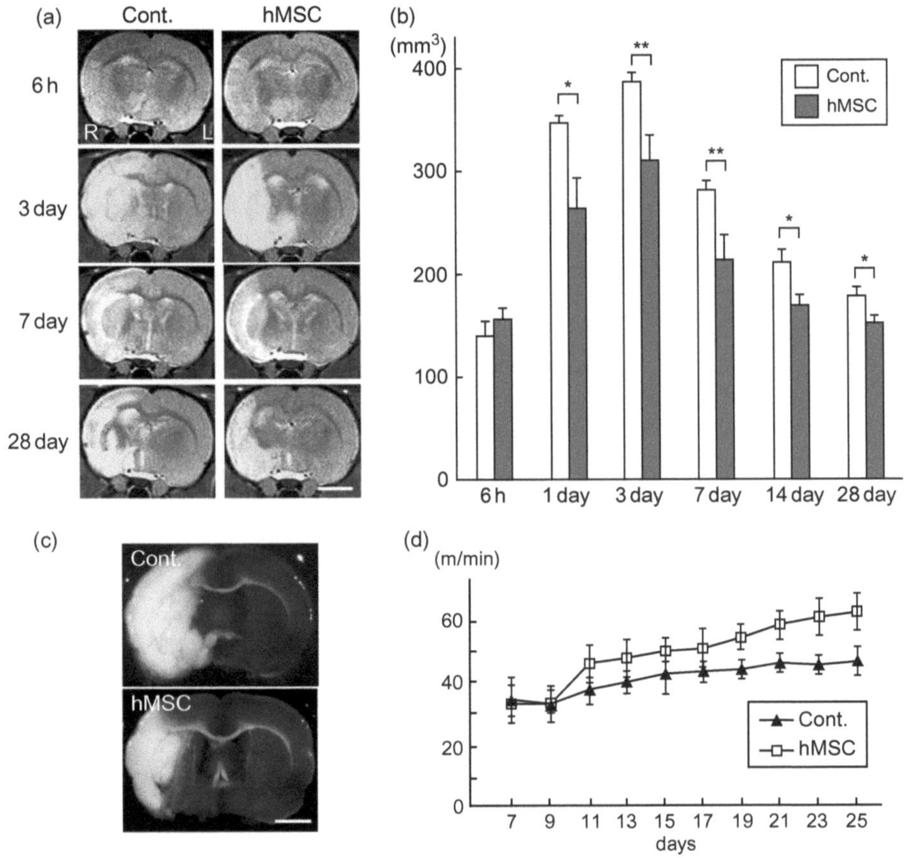

FIGURE 1

Ischemic lesion volume is reduced and functional outcome is improved following hMSC injection. MRI imaging (A) of the rat brain at various times after systemic delivery of hMSCs indicates reduced lesion volume as compared to control animals (Cont.) without cell infusions. Quantification of the reduced lesion volume is shown in (B). Reduced lesion (white area) is evident in the stained sections in (C). Behavioral testing indicates that the maximum speed on a treadmill test was greater in the cell infusion group (D). Thus lesion volume was reduced and functional outcome was improved in the cell therapy group. (For color version of this figure, the reader is referred to the online version of this chapter.)

Modified from Horita et al. (2006) (A–C) and Honma et al. (2006) (D).

Nomura et al., 2005) and is increased in the ischemic lesion following intravenous MSC treatment in the rat middle cerebral artery occlusion (MCAO) model (Kurozumi et al., 2004; Nomura et al., 2005). Compared to transplantation of MSCs which reduce lesion volume and improve functional outcome, transplantation of BDNF gene-modified human MSCs results in increased BDNF levels in the ischemic lesion and stronger therapeutic effects (Kurozumi et al., 2004; Nomura et al., 2005).

Moreover, reduction in ischemic lesion volume is enhanced and functional outcome improved in a rat cerebral ischemia model following treatment with human MSCs genetically modified to express GDNF (Horita et al., 2006). Transplantation of genetically modified MSCs that secrete BDNF into a spinal cord injury model also results in improved functional outcome and enhanced sprouting of raphé–spinal axons (Sasaki et al., 2009), suggesting that the therapeutic effects of MSCs may extend to the spinal cord. A potential advantage of a cell-based therapy that delivers trophic factors to injury sites rather than systemic pharmacological delivery is the reduction in potential adverse effects of systemic drug delivery.

2.2 Angiogenic stimulation

Culture studies indicate that MSCs secrete angiogenic cytokines including VEGF (Toyama et al., 2009), Ang-1 (Onda et al., 2008; Toyama et al., 2009), and placental growth factor (PlGF) (Kinnaird et al., 2004; Liu et al., 2006). These factors promote the growth of new and stable vessels in cardiac and limb ischemia (Autiero et al., 2003; Luttun et al., 2002). VEGF has been reported to show strong angiogenic effects in brain (Zhang and Chopp, 2002), limb ischemia (Baumgartner et al., 1998), and myocardial ischemia models (Kastrup, 2003) and is required for initiation of formation of immature vessels by vasculogenesis or angiogenesis (Carmeliet and Collen, 1997). However, VEGF also potently enhances vascular permeability to blood plasma proteins within minutes after an ischemic insult, an effect which can contribute to edema after the cerebral infarction (Bates et al., 2002; Wang et al., 1996). Direct injection of VEGF into CNS tissues results in the opening of the blood–brain barrier, which allows the entry of large molecules and cells (Sasaki et al., 2010). Consistent with these results, intravenous infusion of genetically modified MSCs that express VEGF into a rat cerebral infarction model has been shown to result in increased functional deficits (Toyama et al., 2009). On the other hand, Ang-1 is involved in maturation, stabilization, and remodeling of short vessels (Davis et al., 1996; Suri et al., 1998; Yancopoulos et al., 2000), and promotes angiogenesis in the brain (Onda et al., 2008; Ward and Lamanna, 2004). Ang-1 protects the peripheral vasculature from vascular leakage (Thurston et al., 1999), an action that may account for its antiedemic effects following cerebral ischemia.

Ang-1 is produced by pericytes (Sundberg et al., 2002) and signals through the Tie2 family of tyrosine kinase receptors on endothelial cells to promote survival and stabilization of blood vessels and reduce "leakiness" (Baffert et al., 2006; Carmeliet, 2003; Han et al., 2010; Thurston et al., 1999). Moreover, following traumatic brain injury, pericytes migrate from the vascular wall (Dore-Duffy et al., 2000), and the neurovascular unit (endothelial cell, pericyte, and astrocyte) and blood–brain barrier are compromised. If a similar disruption of the neurovascular unit occurs following stroke, it would be expected that MSCs would provide support to the microvasculature via Ang-1 signaling to vulnerable endothelial cells. Indeed, intravenous delivery of MSCs genetically modified to express Ang-1 resulted in greater neovascularization (Fig. 2) and functional recovery than vehicle control or MSC delivery alone in a

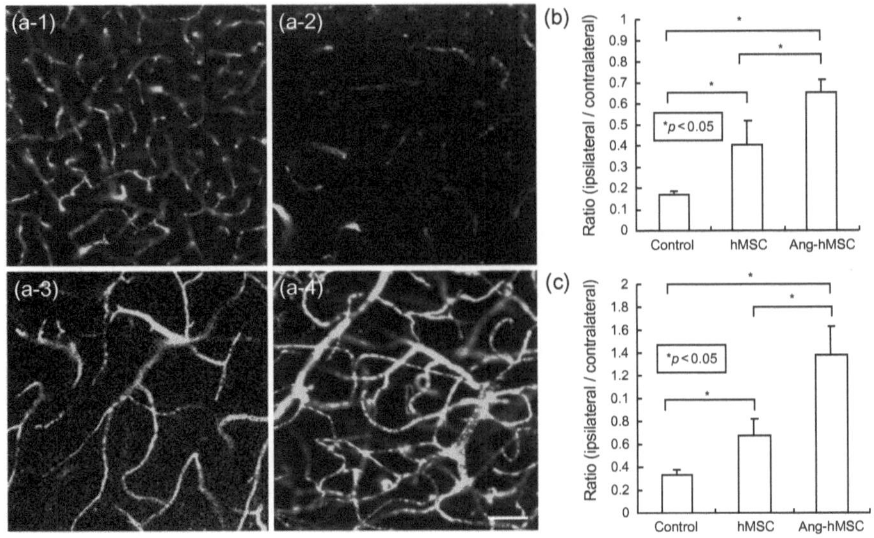

FIGURE 2

Increased cerebral vascularization following hMSC and Ang-hMSC delivery. Seven and 28 days after MCAO, the angiogenesis in boundary zone was analyzed using a three-dimensional analysis system. (A-1) Three-dimensional capillary image with systemically perfused FITC-dextran in the normal rat brain is shown. The total volume of the microvessels in the sampled lesion site decreased (A-2) 28 days after MCAO, but was greater in (A-3) the hMSC-treated group, and (A-4) the Ang-hMSC-treated group. These results are summarized in (B) (7 days after MCAO) and (C) (28 days after MCAO). The ratio (ipsilateral/contralateral) was significantly higher in both the hMSC- and the Ang-hMSC-treated groups as compared with the medium-treated group. Also the ratio of the Ang-hMSC-treated group was significantly higher than the ratio of the hMSC-treated group. Scale bar = 150 mm. (For color version of this figure, the reader is referred to the online version of this chapter.)

Modified from Onda et al. (2008).

model of cerebral ischemia (Onda et al., 2008), In contrast, intravenous infusion of genetically modified MSCs that express VEGF into a rat cerebral infarction model resulted in increased functional deficits (Toyama et al., 2009). However, Miki et al. (2007) report that marrow stromal cells genetically modified to express VEGF have a greater therapeutic effect than MSCs alone. It is possible that the level of VEGF expression may be critical in terms of potential therapeutic effects. During development, VEGF stimulates early vessel outgrowth and Ang-1 then leads to vessel stabilization; thus, both are important for neovascularization (Toyama et al., 2009). Intravenous injection of MSCs genetically modified to express both Ang-1 and VEGF resulted in the greatest neovascularization and functional recovery (Toyama et al., 2009). Thus, a tempo-spatial orchestration of VEGF and Ang-1 expression may be important for appropriate neovascularization.

PlGF, another VEGF family member (Maglione et al., 1991), has a neuroprotective effect as well as an angiogenic effect in rodent stroke models (Liu et al., 2006). Transplantation of PlGF genetically modified human MSCs leads to elevated PlGF levels in ischemic brain and a concomitant reduction in lesion volume, together with increased angiogenesis and functional improvement (Liu et al., 2006). This suggests that PlGF may contribute to the therapeutic effects of MSCs in cerebral ischemia.

It has been suggested that pericytes are a source of MSCs (Caplan and Correa, 2011). Given that pericytes are disrupted after cerebral trauma and that MSC delivery may have reparative effects on microvasculature, it will be important to determine if the microvasculature is a therapeutic target for MSCs, by either replacing pericytes or stimulating endogenous pericyte mobilization.

2.3 Stimulation of neurogenesis

New neurons can be derived from progenitor cells within the subventricular zone (SVZ) of the lateral ventricle and the dentate gyrus of the hippocampus even in the adult brain (Alvarez-Buylla and Lim, 2004). Neural precursor cells in the SVZ migrate through the rostral migratory stream (RMS) to the olfactory bulb where they differentiate into interneurons (Luskin et al., 1997). The cells that migrate from the SVZ in the RMS are neurogenic cells and have been harvested for transplantation studies. There is an increase in the number of cells within the SVZ following cerebral ischemia; several reports indicate that this increase in neural progenitor cell number is enhanced by MSC treatment (Chen et al., 2003; Chopp et al., 2007). Thus, endogenous neurogenesis from brain progenitor regions may be a response to brain injury including stroke, and MSCs may facilitate this process. Interestingly, Shen et al. (2007) demonstrated that synaptophysin expression increases in MSC-treated ischemic brains, suggesting an increase in axonal sprouting. Taken together, the evidence suggests that transplanted MSCs may have multiple therapeutic effects over protracted periods of time, including reduction of apoptotic cell death, promotion of glial and neuronal survival, and vascular and synaptic reorganization.

2.4 Possible rapid effects on neuronal excitability

As described by Honmou et al. (2011) and discussed later, beneficial effects can be observed within days following MSC delivery. These early effects occur too rapidly to be explained by plastic changes in neural tissue or neovascularization. An explanation may lie in the secretion of neurotrophins by MSCs (Horita et al., 2006; Liu et al., 2006; Nomura et al., 2005). One of these, BDNF, modulates several types of potassium channels, including the voltage-dependent potassium channel (Kv1.3) that contributes to the resting potential (Tucker and Fadool, 2002), G-protein-activated inward rectifying potassium channels (Kir3) (Ippolito et al., 2002; Rogalski et al., 2000), and calcium-activated potassium currents, which contribute to spike repolarization and afterhyperpolarization in CNS neurons (Berninger et al., 1993; Matsumoto et al., 2001; Mizoguchi et al., 2002; Numakawa et al., 2001; Weiger et al., 2002). BDNF also evokes PLC-g-dependent currents (nonselective cation: mainly sodium ion) in pontine neurons (Li et al., 1999), and affects TRPC3 channels (Li et al., 1999), which are expressed in the brain. BDNF may also affect synaptic transmission via increases in phosphorylation that increase NMDA (*N*-Methyl-D-aspartate) receptor open probability (Levine et al., 1998), upregulation of surface expression of AMPA (2-amino-3-(3-hydroxy-5-methyl-isoxazol-4-yl)propanoic acid) receptors (Narisawa-Saito et al., 2002), and/or modulation of norepinephrine uptake and release (Rodriguez Fermepin et al., 2009).

Honmou et al. (2011) have suggested that BDNF secreted by MSCs may protect against hyperexcitability of spinal reflexes, thus reducing spasticity. They suggest that downregulation of the potassium–chloride cotransporter KCC2 following injury reduces the rate-dependent depression (RDD) of the Hoffmann reflex, as is observed in spasticity. BDNF upregulates KCC2 and restores RDD. Consistent with this suggestion, Honmou et al. (2011) observed that spasticity after stroke in patients was reduced within a few days after hMSC injection. These observations raise the possibility that MSC delivery may result in neurotrophin-mediated modulation of neural excitability in the poststroke brain. These potential effects on excitability modulation by MSCs are highly speculative and clearly more detailed research addressing the early functional changes in stroke after MSC delivery will be an important area of research to distinguish between neural modulatory vs. plastic changes.

BDNF is also known to modulate microvascular permeability (Sharma, 2003), blood cerebrospinal fluid barrier and BBB breakdown, and brain edema formation (Sharma and Johanson, 2007). Intravenously injected hMSCs increase BDNF and GDNF levels in the brain and reduce edema formation in rat cerebral ischemia models (Horita et al., 2006; Nomura et al., 2005). It is possible that the early effects of MSCs (i.e., within hours or days) may result from resolution of cerebral edema. Honmou et al. (2011) observed a rapid fade of High Intensity Area in MRI (FLAIR images) in stroke patients following infusion of MSC, which may be caused by such an antiedematous effect mediated by neurotrophins.

2.5 Homing mechanisms of MSCs to injured CNS

The precise mechanisms responsible for homing of transplanted MSCs to a lesion site after intravenous infusion are not fully understood. A simple explanation for the relatively selective distribution of intravenously delivered cells to sites of cerebral edema is that trauma-induced BBB breakdown selectively allows cells to reach the infarction site. However, a number of cytokines, chemokines, and adhesion molecules are involved in the homing of immune cells (Charo and Ransohoff, 2006). Several of these have been reported to play a role in MSC migration including CXCL12 and its receptor CXCR4 (Spaeth et al., 2008), CCL2, CCL3, IL-8 (Wang et al., 2002), HGF (Forte et al., 2006), platelet-derived growth factor-AB, insulin-like growth factor-1 (IGF-1), CCL5 and CCL22 (Ponte et al., 2007), and integrin β1 (Ip et al., 2007). With regard to MSC homing to injury sites in the CNS, CCL2 (Wang et al., 2002), CXCL12/CXCR4 and CX3CL1/CX3CR1

(Ji et al., 2004), and CCR3, CCR5, CXCR3 (Song et al., 2011) have been implicated and may play important roles in both site- and phase-specific features of the MSC effect (Song et al., 2011). However, knowledge of the precise mechanism by which MSCs migrate to pathological lesions in the CNS after intravenous infusion is incomplete and further work is necessary to elucidate this mechanism.

3 INTRAVENOUS DELIVERY OF HUMAN MSCs IN A NONHUMAN PRIMATE MODEL OF STROKE AS A PRELUDE TO PHASE I HUMAN CLINICAL STUDY

While rodent studies suggest the potential of MSC-based therapies to improve outcome in clinical stroke studies, direct extrapolation from the rodent to human cannot be made given issues of species differences and scale. Demonstration of efficacy in a nonhuman primate (NHP) stroke model would provide greater confidence in this approach as the organization of the vascular, and sensory and motor systems in the brain are more similar among primates. However, a limited number of stroke studies have been carried out in the NHP using a relatively small number of animals (de Crespigny et al., 2005; Roitberg et al., 2006). We recently established a stroke model where the left middle cerebral artery (MCA) was occluded with an aneurysm clip for several hours in the African green monkey (Sasaki et al., 2011; Fig. 3A–C). Immediately after MCAO, the cerebral venous blood darkened and there was collapse of the cortical microvasculature (Fig. 3E); after removal of the clip, virtually normal circulation was observed (Fig. 3F). The animals were given regular neurological examinations during the postinfarction period (up to 12 months) and MRI was periodically performed. Two groups of animals were observed in terms of outcome: (1) severe long-lasting hemiplegia and (2) modest initial hemiplegia with near complete recovery. The severely impaired group had greater cerebral infarction from MRI (Fig. 3I), and examination of the brain showed necrosis and tissue loss in the distribution area of the MCA (Fig. 3E and G). Features of cortical surface vascular patterns were predictive of lesion severity. Animals where venous blood return was stopped during M1 occlusion had poorer outcome than those where venous return was extremely slow, but present. Thus, animals with poor collateral circulation (collateral-poor) had greater infarction volume than those with more rich (collateral-rich) collateral circulation. These results indicate that variation in collateral circulation in the NHP will lead to significant differences in clinical outcome even with identical lesion parameters. Safety and preclinical effects of intravenous infusion of human mesenchymal stem cells (hMSCs) were studied in this model system. Human MSCs were infused at a single time point in a group of NHPs and human serum alone was infused in another group. The study indicated that there was a trend for greater functional improvement that was not statistically significant (Sasaki et al., 2011). Importantly, no adverse effects were observed from the intravenous infusion of the human MSCs in the NHP, suggesting potential safety in a human clinical stroke study.

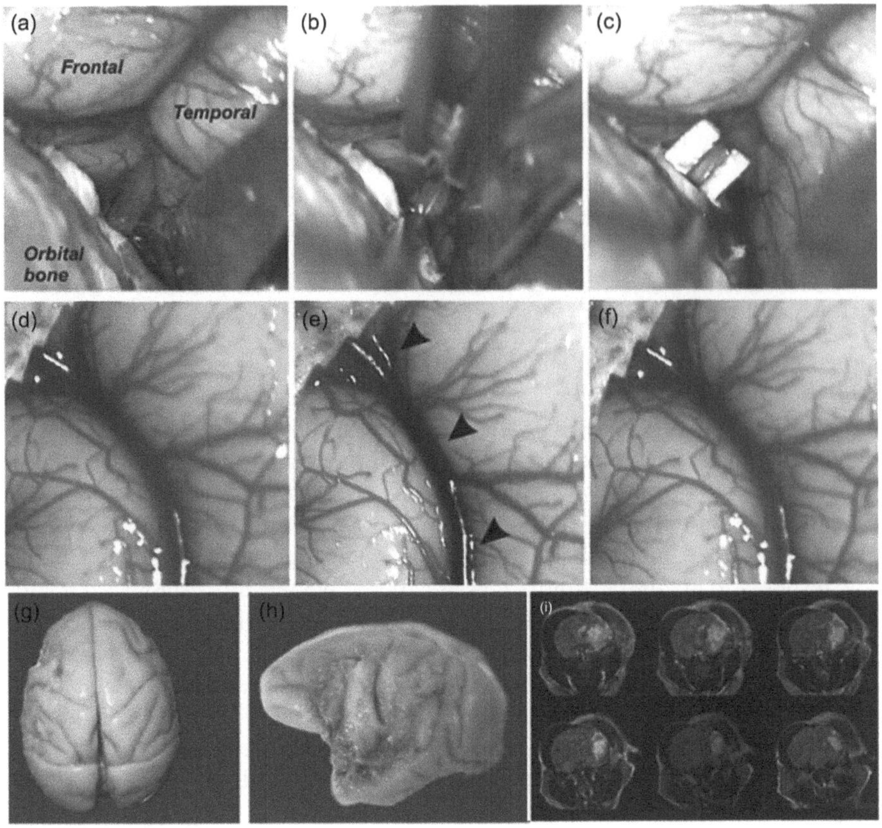

FIGURE 3

Occlusion (clipping) of M1 portion of left middle cerebral artery in the African green monkey brain (A). A small vascular clip is applied to M1 portion. (B) MCA is occluded by a clip. (C). (D–F) Cortical venous flow. Before clipping (D), the color of venous blood is normal. Complete blood pooling of cortical veins is observed during occlusion of MCA (arrowheads in (E)). After releasing the clip (F), the color of venous blood appears to be normal. Necrosis in left hemisphere several months after MCA occlusion is observed in dorsal (G) and lateral (H) view. (I) MRI (T2) series showing high intensity regions in the infarction area. Frontal is frontal lobe, Temporal is temporal lobe. M1 portion of the MCA is exposed by a retractor in the deep Sylvian fissure. (For color version of this figure, the reader is referred to the online version of this chapter.)

Modified from Sasaki et al. (2011).

4 CLINICAL STUDIES IN STROKE PATIENTS UTILIZING INTRAVENOUSLY APPLIED hMSCs

The first study to examine the feasibility and safety of a cell therapy approach in stroke patients using culture-expanded autologous MSCs was carried out by Bang et al. (2005). The investigators prospectively and randomly allocated 30 patients with large cerebral infarcts in the MCA territory; 25 subjects did not receive MSCs while 5 received intravenous infusions of 1×10^8 autologous MSCs. One week after admission, bone marrow aspirates were obtained and mononuclear cells isolated by Ficoll density centrifugation. Plastic adherent cells were expanded in culture in fetal bovine serum and cells characterized as $CD34^-$ $CD45^-$, $SH2^+$, and $SH4^+$. Neurological evaluation was carried out over the course of a year. Two infusions of MSCs (5×10^7 per infusion) were given: 4–5 weeks and 7–9 weeks after symptom onset.

The patients in the MSC group showed no deaths, stroke recurrence, or serious adverse events. While the study was not fully blinded, the MSC group showed greater functional recovery as measured on the Barthel index. While the investigators concluded that the cell therapy approach was safe, they emphasized that double-blinded studies with larger cohorts would be necessary to reach a definitive conclusion regarding MSC therapy. A 5-year follow-up confirmed that there were no adverse events related to the human MSC transplantation in these five stroke patients (Lee et al., 2010).

We carried out a Phase I study on 12 stroke patients who received intravenous infusions of autologous bone marrow-derived MSCs cultured in autoserum at the Sapporo Medical University in Japan (Honmou et al., 2011). The overall structure of the study is outlined in Fig. 4. Bone marrow aspirates were obtained within weeks after patient admission to the hospital and the bone marrow cells were grown as plastic adherent cells in culture with autoserum (i.e., serum derived from the patient) and expanded in culture for about 2–3 weeks, at which time they were tested for safety to determine that no pathogens were present. Autologous human MSCs (ahMSCs) grow to confluence in appropriate culture conditions as flattened fibroblast-like cells. Although MSCs may be present in different proportions in various species, human MSCs have a distinct cell surface antigen pattern including $CD34^-$, $CD45^-$, $CD73^+$, $CD105^+$, and we used methodologies that allowed for culture of human MSCs in very high purity (Kobune et al., 2003; Majumdar et al., 1998). After the cells were expanded and safety and antigenic phenotype analyses were completed, they were cryopreserved and stored. On the day of patient infusion, cryopreserved units were thawed at the bedside and infused intravenously into the patients over 30 min. All patients were monitored closely during and after MSC infusion and for the following 24 h. Oxygen saturation, temperature, blood pressure, pulse, and respiratory rate were carefully monitored before and after injection. Brain MRIs were obtained before and at intervals up to 1 year after MSC infusion.

Human bone marrow-derived MSCs were successfully isolated from bone marrow aspirates from all 12 stroke patients, and all were successfully culture-expanded. Importantly, serial evaluations showed no cell-related, serological, or imaging-defined adverse effects during the 12-month period of observation. MRIs following cell injection showed no tumor or abnormal cell growth in any of the 12 patients studied in this series over the time course of 1 year. Changes in NIH stroke scale (NIHSS) score and lesion volume within the first weeks after cell infusion suggested a potential therapeutic benefit of infusion of ahMSC expanded in human serum (Honmou et al., 2011). Notably, the rate of recovery measured in terms of change in NIHSS

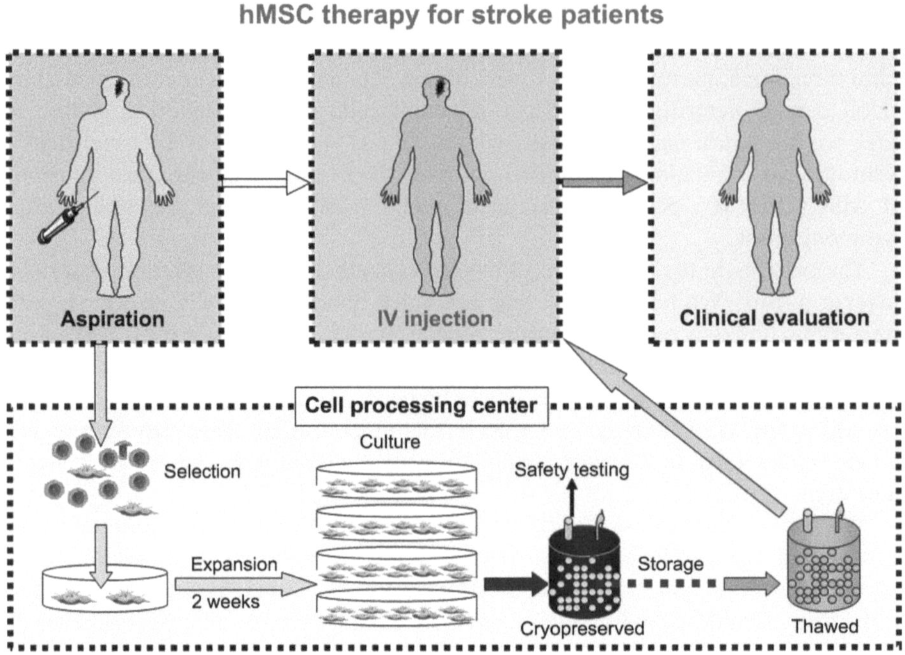

FIGURE 4

Schematic representation of the sequence of events in a clinical study infusing autologous hMSCs. After stroke and enrollment into the study, bone marrow aspirates were obtained from each subject. The cells were processed in a cell tissue processing center where they were selected, expanded, and cryopreserved. The cells were tested for safety and, after thawing, were used for intravenous infusion. Clinical evaluation was carried out over 1 year. (For color version of this figure, the reader is referred to the online version of this chapter.)

Modified from Honmou et al. 2012.

scores per day dramatically improved within the first 2 weeks after hMSC injections in the majority of patients. Enhancement of the rate of recovery was most pronounced during the initial phase (within 1 week) after injection. The daily rate of decrease of the median NIHSS score was 0.36 in the week following infusion, compared to a median daily rate of change of 0.04 in the period before infusion; 4 of the 12 patients studied displayed a reduction in NIHSS scores of 4 or more points within the first 7 days after ahMSC infusion, and 3 of the 12 patients displayed a reduction of 5 or more points within the first 7 days after ahMSC infusion. In the middle phase (2 weeks to 3 months after cell injection), the increase in the recovery rate was less pronounced. In the late phase (3 months to 1 year after cell injection), recovery essentially reached a plateau. The improvement in NIHSS scores was maintained for 1 year in all patients. There was also a reduction of at least 15% in lesion volume at 7 days post-hMSC infusion, compared to preinfusion, and there tended to be a correlation between the change in lesion volume and the change in NIHSS score. As a Phase I study, this initial series was not blinded and did not include placebo controls. Thus, we stress that these results must be interpreted with caution. We cannot exclude a contribution of spontaneous recovery to postinfusion changes in these patients and would emphasize that these unblinded observations do not prove a therapeutic effect with the current safety study design. Moreover, the degree to which changes in the NIHSS score of the magnitude that we observed translate into clinically meaningful functional improvement in any given patient remains to be studied. Although there was no control group in this study, the time-locked increase in the rate of recovery and lesion volume is suggestive of a potential therapeutic benefit of ahMSC injections in stroke patients. Future blinded studies are being designed, and they will be important to determine if this approach has efficacy.

5 PROSPECTS

Interventional approaches for stroke patients are generally limited to the very acute phase after stroke onset, for example, the therapeutic time window of fibrinolytic agents such as tissue-type plasminogen activator (tPA) is limited to hours after stroke onset. If found to be efficacious in blinded studies, MSC therapy might extend the therapeutic time window from acute to subacute and potentially to the chronic phase. A current assumption has been that neurons and axons in high intensity areas in MRI (FLAIR images) which are seen after stroke will die or are going to die. But our data suggest that these areas can be reduced after MSC therapy (Honmou et al., 2011) and that a subset of neural tissue in even more long-term stroke lesions can be rescued by cell therapy. A double-blinded crossover study may allow us to test this hypothesis and to conclusively demonstrate the therapeutic potential of MSC infusion.

MSCs display a remarkable ensemble of properties. Systemically delivered MSCs have been examined in clinical studies for a number of neurological diseases including Krabbe's disease (Escolar et al., 2005), Hurler's syndrome (Staba et al., 2004), metachromatic leukodystrophy (Koc et al., 2002), and stroke (Bang et al., 2005; Honmou et al., 2011). Their relatively benign safety profile as delivered in these studies and in our study in stroke (Honmou et al., 2011) suggests the prospect of therapeutic use of MSCs for a large number of CNS diseases. The optimal therapeutic protocol, in terms of number of cells, preparation protocols, and delivery timing require future study. Although we previously demonstrated in a rat model that a relatively large dose of hMSCs administered at an early time point is more effective than temporally dispersed (fractionated) dosing for stroke (Omori et al., 2008), the therapeutic benefits may differ between diseases and species, or depending on the phase of pathological conditions. A temporal sequence of potential mechanisms for the therapeutic benefits of MSC therapy is outlined in. Early (days) beneficial effects may be the result of excitability modulation by MSC-release of neuromodulators such as BDNF. Anti-inflammatory response with reduction of neural edema may also have relatively early therapeutic effects. Importantly, the MSCs may

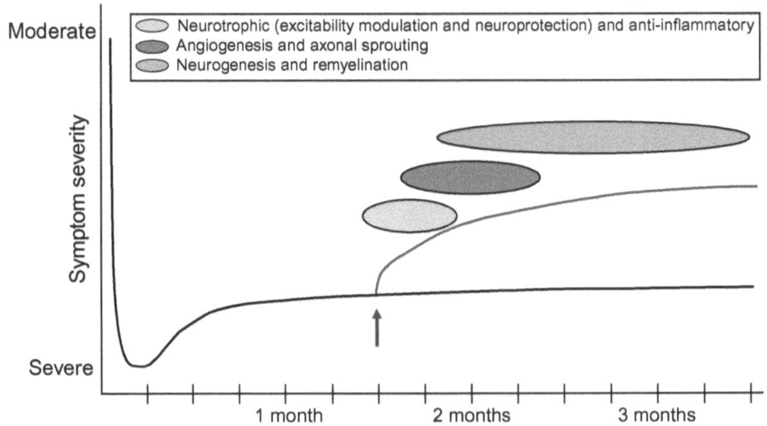

FIGURE 5

Schematic representation of potential therapeutic mechanisms of hMSCs after stroke as a function of time. The dashed line indicates an idealized spontaneous recovery curve following stroke with initial severe deficits that show some endogenous recovery that plateaus. The solid line indicates incremental recovery of function following MSC infusion (arrow). Early improvement in function may result from neurotrophic effects that may modulate excitability, confer neuroprotection and anti-inflammatory responses. An intermediate phase of recovery may result from angiogenesis, axonal sprouting. If neurogenesis and remyelination contribute to functional recovery it would likely contribute to the later phase of recovery. (For color version of this figure, the reader is referred to the online version of this chapter.)

Modified from Honmou et al., 2012.

contribute to neovascularization and vascular stabilization thereby protecting CNS tissue. MSCs could also provide trophic support for vulnerable neurons, particularly in the penumbra, thus leading to enhanced tissue sparing. With increased time, the MSCs may also stimulate local axonal sprouting with new synaptic connections and mobilization of resident progenitor cells that may contribute to neurogenesis and remyelination of axons. Each of these potential mechanisms merits careful investigation. Figure 5 is a diagram showing the potential time frame of potential changes that could occur after MSC therapy.

The therapeutic time window for pharmacological intervention in stroke is mainly the acute phase in current clinical practice. Most of the therapeutic strategies are related to recirculation (tPA, free radical scavengers, etc.) and/or prevention of recurrence (antiplatelet agents or anticoagulants). Therapeutic benefits of these agents on neurological deficits are minimal when applied in the subacute and chronic phases. Although physical rehabilitation is an important primary therapy in both the subacute and chronic phases, therapeutic benefits are important but limited. One reason why medical interventional therapy for stroke in the subacute and chronic phases is essentially absent may be based on the assumption that the infarcted lesion has been fully established in the acute phase and the necrotic tissue is untreatable.

Indeed, neither MRI findings nor clinical symptoms (neurological deficits) change much with conventional medication in the subacute and chronic phase. However, our clinical study (Honmou et al., 2011) shows changes in both MRI findings and clinical neurological symptom which may be from neurotrophic mechanisms suggesting a potential novel cellular interventional approach for treatment of stroke. At least a subset of neurons and their axons in lesion areas including the penumbra of stroke patients in the subacute phase may be compromised, but not dead, with loss of tissue continuing in the subacute and chronic phases. It will be important to determine if therapeutic intervention, either cellular or noncellular approaches, in the subacute and early chronic phase can impact cell survival and improve the clinical outcome in stroke.

References

Adeoye, O., Hornung, R., Khatri, P., Kleindorfer, D., 2011. Recombinant tissue-type plasminogen activator use for ischemic stroke in the United States: a doubling of treatment rates over the course of 5 years. Stroke 42 (7), 1952–1955.

Akiyama, Y., Radtke, C., Kocsis, J.D., 2002. Remyelination of the rat spinal cord by transplantation of identified bone marrow stromal cells. J. Neurosci. 22, 6623–6630.

Alvarez-Buylla, A., Lim, D.A., 2004. For the long run: maintaining germinal niches in the adult brain. Neuron 41, 683–686.

Autiero, M., Luttun, A., Tjwa, M., Carmeliet, P., 2003. Placental growth factor and its receptor, vascular endothelial growth factor receptor-1: novel targets for stimulation of ischemic tissue revascularization and inhibition of angiogenic and inflammatory disorders. J. Thromb. Haemost. 1, 1356–1370.

Azizi, S.A., Stokes, D., et al., 1998. Engraftment and migration of human bone marrow stromal cells implanted in the brains of albino rats—similarities to astrocyte grafts. Proc. Natl. Acad. Sci. U.S.A. 95, 3908–3913.

Baffert, F., Le, T., et al., 2006. Angiopoietin-1 decreases plasma leakage by reducing number and size of endothelial gaps in venules. Am. J. Physiol. Heart Circ. Physiol. 290, H107–H118.

Bang, O.Y., Lee, J.S., et al., 2005. Autologous mesenchymal stem cell transplantation in stroke patients. Ann. Neurol. 57, 874–882.

Bates, D.O., Hillman, N.J., et al., 2002. Regulation of microvascular permeability by vascular endothelial growth factors. J. Anat. 200, 581–597.

Baumgartner, I., Pieczek, A., Manor, O., Blair, R., Kearney, M., Walsh, K., et al., 1998. Constitutive expression of phVEGF165 after intramuscular gene transfer promotes collateral vessel development in patients with critical limb ischemia. Circulation 97, 1114–1123.

Berninger, B., Garcia, D.E., Inagaki, N., Hahnel, C., Lindholm, D., 1993. BDNF and NT-3 induce intracellular Ca2+ elevation in hippocampal neurones. Neuroreport 4, 1303–1306.

Bianco, P., Robey, P.G., et al., 2008. Mesenchymal stem cells: revisiting history, concepts, and assays. Cell Stem Cell 2, 313–319.

Brazelton, T.R., Rossi, F.M., et al., 2000. From marrow to brain: expression of neuronal phenotypes in adult mice. Science 290, 1775–1779.

Caplan, A.I., 1991. Mesenchymal stem cells. J. Orthop. Res. 9, 641–650.

Caplan, A.I., 2005. Mesenchymal stem cells: cell-based reconstructive therapy in orthopedics. Tissue Eng. 11, 1198–1211.

Caplan, A.I., Correa, D., 2011. The MSC: an injury drugstore. Cell Stem Cell 9, 11–15.

Carmeliet, P., 2003. Blood vessels and nerves: common signals, pathways and diseases. Nat. Rev. Genet. 4, 710–720.

Carmeliet, P., Collen, D., 1997. Molecular analysis of blood vessel formation and disease. Am. J. Physiol. 273, H2091–H2104.

Charo, I.F., Ransohoff, R.M., 2006. The many roles of chemokines and chemokine receptors in inflammation. N. Engl. J. Med. 354, 610–621.

Chen, J., Li, Y., et al., 2001. Therapeutic benefit of intravenous administration of bone marrow stromal cells after cerebral ischemia in rats. Stroke 32, 1005–1011.

Chen, X., Li, Y., et al., 2002. Ischemic rat brain extracts induce human marrow stromal cell growth factor production. Neuropathology 22, 275–279.

Chen, J., Li, Y., et al., 2003. Intravenous bone marrow stromal cell therapy reduces apoptosis and promotes endogenous cell proliferation after stroke in female rat. J. Neurosci. Res. 73, 778–786.

Chopp, M., Li, Y., 2002. Treatment of neural injury with marrow stromal cells. Lancet Neurol. 1, 92–100.

Chopp, M., Zhang, Z.G., et al., 2007. Neurogenesis, angiogenesis, and MRI indices of functional recovery from stroke. Stroke 38, 827–831.

Cramer, S.C., Finklestein, S.P., et al., 1999. Activation of distinct motor cortex regions during ipsilateral and contralateral finger movements. J. Neurophysiol. 81, 383–387.

Davis, S., Aldrich, T.H., Jones, P.F., Acheson, A., Compton, D.L., Jain, V., et al., 1996. Isolation of angiopoietin-1, a ligand for the TIE2 receptor, by secretion-trap expression cloning. Cell 87, 1161–1169.

de Crespigny, A.J., D'Arceuil, H.E., Maynard, K.I., He, J., McAuliffe, D., Norbash, A., et al., 2005. Acute studies of a new primate model of reversible middle cerebral artery occlusion. J. Stroke Cerebrovasc. Dis. 14, 80–87.

Dore-Duffy, P., Owen, C., et al., 2000. Pericyte migration from the vascular wall in response to traumatic brain injury. Microvasc. Res. 60, 55–69.

Escolar, M.L., Poe, M.D., Provenzale, J.M., Richards, K.C., et al., 2005. Transplantation of umbilical-cord blood in babies with infantile Krabbe's disease. N. Engl. J. Med. 352 (20), 2069–2081.

Forte, G., Minieri, M., Cross, P., Antenucci, D., et al., 2006. Hepatocyte growth factor effects on mesenchymal stem cells: proliferation, migration, and differentiation. Stem Cells 24, 23–33.

Hamano, K., Li, T.S., Kobayashi, T., Kobayashi, S., Matsuzaki, M., Esato, K., 2000. Angiogenesis induced by the implantation of self-bone marrow cells: a new material for therapeutic angiogenesis. Cell Transplant 9, 439–443.

Han, S., Arnold, S.A., Sithu, S.D., et al., 2010. Rescuing vasculature with intravenous angiopoietin-1 and alpha v beta 3 integrin peptide is protective after spinal cord injury. Brain 133(Pt4), 1026–1042. Erratum in: Brain 2011, 134(Pt5), 1575.

Honma, T., Honmou, O., et al., 2006. Intravenous infusion of immortalized human mesenchymal stem cells protects against injury in a cerebral ischemia model in adult rat. Exp. Neurol. 199, 56–66.

Honmou, O., Houkin, K., et al., 2011. Intravenous administration of auto serum-expanded autologous mesenchymal stem cells in stroke. Brain 134, 1790–1807.

Honmou, O., Sasaki, M., Rie, O., Waxman, S.G., Kocsis, J.D., 2012. Mesenchymal stem cells: therapeutic outlook for stroke. Trends Mol. Med. 18, 292–297.

Horita, Y., Honmou, O., et al., 2006. Intravenous administration of glial cell line-derived neurotrophic factor gene-modified human mesenchymal stem cells protects against injury in a cerebral ischemia model in the adult rat. J. Neurosci. Res. 84, 1495–1504.

Iihoshi, S., Honmou, O., et al., 2004. A therapeutic window for intravenous administration of autologous bone marrow after cerebral ischemia in adult rats. Brain Res. 1007, 1–9.

Ip, J.E., Wu, Y., Huang, J., Zhang, L., Pratt, R.E., Dzau, V.J., 2007. Mesenchymal stem cells use integrin beta1 not CXC chemokine receptor 4 for myocardial migration and engraftment. Mol. Biol. Cell 18, 2873–2882.

Ippolito, D.L., Temkin, P.A., Rogalski, S.L., Chavkin, C., 2002. N-terminal tyrosine residues within the potassium channel Kir3 modulate GTPase activity of Galphai. J. Biol. Chem. 277, 32692–32696.

Ji, J.F., He, B.P., Dheen, S.T., Tay, S.S., 2004. Interactions of chemokines and chemokine receptors mediate the migration of mesenchymal stem cells to the impaired site in the brain after hypoglossal nerve injury. Stem Cells 22, 415–427.

Jiang, Q., Ewin, J.R., et al., 2001. Magnetization transfer MRI: application to treatment of middle cerebral artery occlusion in rat. J. Magn. Reson. Imaging 13, 178–184.

Kastrup, J., 2003. Therapeutic angiogenesis in ischemic heart disease: gene or recombinant vascular growth factor protein therapy? Curr. Gene Ther. 3, 197–206.

Kim, S., Honmou, O., et al., 2006. Neural differentiation potential of peripheral blood- and bone-marrow-derived precursor cells. Brain Res. 1123, 27–33.

Kinnaird, T., Stabile, E., Burnett, M.S., Shou, M., Lee, C.W., Barr, S., et al., 2004. Local delivery of marrow-derived stromal cells augments collateral perfusion through paracrine mechanisms. Circulation 109, 1543–1549.

Kobune, M., Kawano, Y., et al., 2003. Telomerized human multipotent mesenchymal cells can differentiate into hematopoietic and cobblestone area-supporting cells. Exp. Hematol. 31 (8), 715–722.

Koc, O.N., Day, J., Nieder, M., Gerson, S.L., Lazarus, H.M., Krivit, W., 2002. Allogeneic mesenchymal stem cell infusion for treatment of metachromatic leukodystrophy (MLD) and Hurler syndrome (MPS-IH). Bone Marrow Transplant 30, 215–222.

Kopen, G.C., Prockop, D.J., Phinney, D.G., 1999. Marrow stromal cells migrate throughout forebrain and cerebellum, and they differentiate into astrocytes after injection into neonatal mouse brains. Proc. Natl. Acad. Sci. U.S.A 96, 10711–10716.

Krause, D.S., Theise, N.D., Collector, M.I., Henegariu, O., Hwang, S., Gardner, R., et al., 2001. Multi-Organ, Multi-Lineage Engraftment by a Single Bone Marrow-Derived Stem Cell. Cell 105, 369–377.

Kurozumi, K., Nakamura, K., et al., 2004. BDNF gene-modified mesenchymal stem cells promote functional recovery and reduce infarct size in the rat middle cerebral artery occlusion model. Mol. Ther. 9, 189–197.

Lee, J.S., Hong, J.M., et al., 2010. A long-term follow-up study of intravenous autologous mesenchymal stem cell transplantation in patients with ischemic stroke. Stem Cells 28, 1099–1106.

Levine, E.S., Crozier, R.A., Black, I.B., Plummer, M.R., 1998. Brain-derived neurotrophic factor modulates hippocampal synaptic transmission by increasing N-methyl-D-aspartic acid receptor activity. Proc. Natl. Acad. Sci. U.S.A 95, 10235–10239.

Li, H.S., Xu, X.Z., Montell, C., 1999. Activation of a TRPC3-dependent cation current through the neurotrophin BDNF. Neuron 24, 261–273.

Li, Y., Chen, J., et al., 2002. Human marrow stromal cell therapy for stroke in rat: neurotrophins and functional recovery. Neurology 59, 514–523.

Liu, H., Honmou, O., et al., 2006. Neuroprotection by PlGF gene-modified human mesenchymal stem cells after cerebral ischaemia. Brain 129, 2734–2745.

Luskin, M.B., Zigova, T., et al., 1997. Neuronal progenitor cells derived from the anterior subventricular zone of the neonatal rat forebrain continue to proliferate in vitro and express a neuronal phenotype. Mol. Cell. Neurosci. 8, 351–366.

Luttun, A., Tjwa, M., Moons, L., Wu, Y., Angelillo-Scherrer, A., Liao, F., et al., 2002. Revascularization of ischemic tissues by PlGF treatment, and inhibition of tumor angiogenesis, arthritis and atherosclerosis by anti-Flt1. Nat. Med. 8, 831–840.

Maglione, D., Guerriero, V., Viglietto, G., Delli-Bovi, P., Persico, M.G., 1991. Isolation of a human placenta cDNA coding for a protein related to the vascular permeability factor. Proc. Natl. Acad. Sci. U.S.A 88, 9267–9271.

Majumdar, M.K., Thiede, M.A., et al., 1998. Phenotypic and functional comparison of cultures of marrow-derived mesenchymal stem cells (MSCs) and stromal cells. J. Cell. Physiol. 176, 57–66.

Massengale, M., Wagers, A.J., et al., 2005. Hematopoietic cells maintain hematopoietic fates upon entering the brain. J. Exp. Med. 201, 1579–1589.

Matsumoto, T., Numakawa, T., Adachi, N., Yokomaku, D., Yamagishi, S., Takei, N., et al., 2001. Brain-derived neurotrophic factor enhances depolarization-evoked glutamate release in cultured cortical neurons. J. Neurochem. 79, 522–530.

Miki, Y., Nonoguchi, N., et al., 2007. Vascular endothelial growth facto gene-transferred bone marrow stromal cells engineered with a herpes simplex virus type 1 vector can improve neurological deficits and reduced infarction volume in rat brain ischemia. Neurosurgery 61, 586–594 discussion 594–595.

Mizoguchi, Y., Monji, A., Nabekura, J., 2002. Brain-derived neurotrophic factor induces long-lasting Ca2+-activated K+ currents in rat visual cortex neurons. Eur. J. Neurosci. 16, 1417–1424.

Narisawa-Saito, M., Iwakura, Y., Kawamura, M., Araki, K., Kozaki, S., Takei, N., et al., 2002. Brain-derived neurotrophic factor regulates surface expression of alpha-amino-3-hydroxy-5-methyl-4-isoxazoleproprionic acid receptors by enhancing the N-ethylmaleimide-sensitive factor/GluR2 interaction in developing neocortical neurons. J. Biol. Chem. 277, 40901–40910.

Nomura, T., Honmou, O., et al., 2005. I.V. infusion of brain-derived neurotrophic factor gene-modified human mesenchymal stem cells protects against injury in a cerebral ischemia model in adult rat. Neuroscience 136, 161–169.

Numakawa, T., Matsumoto, T., Adachi, N., Yokomaku, D., Kojima, M., Takei, N., et al., 2001. Brain-derived neurotrophic factor triggers a rapid glutamate release through increase of intracellular Ca(2+) and Na(+) in cultured cerebellar neurons. J. Neurosci. Res. 66, 96–108.

Omori, Y., Honmou, O., Harada, K., Suzuki, J., Houkin, K., Kocsis, J.D., 2008. Optimization of a therapeutic protocol for intravenous injection of human mesenchymal stem cells after cerebral ischemia in adult rats. Brain Res. 1236, 30–38.

Onda, T., Honmou, O., et al., 2008. Therapeutic benefits by human mesenchymal stem cells (hMSCs) and Ang-1 gene-modified hMSCs after cerebral ischemia. J. Cereb. Blood Flow Metab. 28, 329–340.

Pereira, A.C., Martin, P.J., Warburton, E.A., 2011. Thrombolysis in acute ischaemic stroke. Postgrad. Med. J. 77, 166–171.

Pittenger, M.F., Mackay, A.M., et al., 1999. Multilineage potential of adult human mesenchymal stem cells. Science 284, 143–147.

Ponte, A.L., Marais, E., Gallay, N., Langonne, A., Delorme, B., Herault, O., et al., 2007. The in vitro migration capacity of human bone marrow mesenchymal stem cells: comparison of chemokine and growth factor chemotactic activities. Stem Cells 25, 1737–1745.

Prockop, D.J., 1997. Marrow stromal cells as stem cells for nonhematopoietic tissues. Science 276, 71–74.

Rodriguez Fermepin, M., Trinchero, M., Minetto, J., Beltran, A., Fernandez, B.E., 2009. Brain derived neurotrophic factor and neurotrophin-4 employ different intracellular pathways to modulate norepinephrine uptake and release in rat hypothalamus. Neuropeptides 43, 275–282.

Rogalski, S.L., Appleyard, S.M., Pattillo, A., Terman, G.W., Chavkin, C., 2000. TrkB activation by brain-derived neurotrophic factor inhibits the G protein-gated inward rectifier Kir3 by tyrosine phosphorylation of the channel. J. Biol. Chem. 275, 25082–25088.

Roitberg, B.Z., Mangubat, E., Chen, E.Y., Sugaya, K., Thulborn, K.R., Kordower, J.H., et al., 2006. Survival and early differentiation of human neural stem cells transplanted in a non-human primate model of stroke. J. Neurosurg. 105, 96–102.

Sanchez-Ramos, J., Song, S., et al., 2000. Adult bone marrow stromal cells differentiate into neural cells in vitro. Exp. Neurol. 164, 247–256.

Sasaki, M., Honmou, O., Akiyama, Y., Uede, T., Hashi, K., Kocsis, J.D., 2001. Transplantation of an acutely isolated bone marrow fraction repairs demyelinated adult rat spinal cord axons. Glia 35, 26–34.

Sasaki, M., Radtke, C., et al., 2009. BDNF-hypersecreting human mesenchymal stem cells promote functional recovery, axonal sprouting, and protection of corticospinal neurons after spinal cord injury. J. Neurosci. 29, 14932–14941.

Sasaki, M., Lankford, K.L., et al., 2010. Focal experimental autoimmune encephalomyelitis in the Lewis rat induced by immunization with myelin oligodendrocyte glycoprotein and intraspinal injection of vascular endothelial growth factor. Glia 58, 1523–1531.

Sasaki, M., Honmou, O., et al., 2011. Development of a middle cerebral artery occlusion model in the nonhuman primate and a safety study of I.V. infusion of human mesenchymal stem cells. PLoS One 6 (10), e26577.

Sharma, H.S., 2003. Neurotrophic factors attenuate microvascular permeability disturbances and axonal injury following trauma to the rat spinal cord. Acta. Neurochir. Suppl. 86, 383–388.

Sharma, H.S., Johanson, C.E., 2007. Intracerebroventricularly administered neurotrophins attenuate blood cerebrospinal fluid barrier breakdown and brain pathology following whole-body hyperthermia: an experimental study in the rat using biochemical and morphological approaches. Ann. N. Y. Acad. Sci. 1122, 112–129.

Shen, L.H., Li, Y., et al., 2007. One-year follow-up after bone marrow stromal cell treatment in middle-aged female rats with stroke. Stroke 38, 2150–2156.

Song, C.H., Honmou, O., Furuoka, H., Horiuchi, M., 2011. Identification of chemoattractive factors involved in the migration of bone marrow-derived mesenchymal stem cells to brain lesions caused by prion. J. Virol. 85 (21), 11069–11078.

Spaeth, E., Klopp, A., Dembinski, J., Andreeff, M., Marini, F., 2008. Inflammation and tumor microenvironments: defining the migratory itinerary of mesenchymal stem cells. Gene Ther. 15, 730–738.

Staba, S.L., Escolar, M.L., Poe, M., Kim, Y., et al., 2004. Cord-blood transplants from unrelated donors in patients with Hurler's syndrome. N. Engl. J. Med. 350 (19), 1960–1969.

Sundberg, C., Kowanetz, M., et al., 2002. Stable expression of angiopoietin-1 and other markers by cultured pericytes: phenotypic similarities to a subpopulation of cells in maturing vessels during later stages of angiogenesis in vivo. Lab. Invest. 82, 387–401.

Suri, C., McClain, J., et al., 1998. Increased vascularization in mice overexpressing angiopoietin-1. Science 282, 468–471.

Thurston, G., Smith, K., et al., 1999. Leakage-resistant bloods in mice transgenically overexpressing angiopoietin-1. Science 286, 2511–2514.

Toyama, K., Honmou, O., et al., 2009. Therapeutic benefits of angiogenetic gene-modified human mesenchymal stem cells after cerebral ischemia. Exp. Neurol. 216, 47–55.

Tucker, K., Fadool, D.A., 2002. Neurotrophin modulation of voltage-gated potassium channels in rat through TrkB receptors is time and sensory experience dependent. J. Physiol. 542, 413–429.

Wahlgren, N., Ahmed, N., Davalos, A., Ford, G.A., Grond, M.I., Hacke, W., et al., 2007. Thrombolysis with alterplase for acute ischaemic stroke in the Safe Implementation of Thrombolysis in Stroke-Monitoring Study (SITS-MOST): an observational study. Lancet 369 (9558), 275–282.

Wang, L., Li, Y., Chen, X., Chen, J., Gautam, S.C., Xu, Y., et al., 2002. MCP-1, MIP-1, IL-8 and ischemic cerebral tissue enhance human bone marrow stromal cell migration in interface culture. Hematology 7, 113–117.

Wang, W., Merrill, M.J., Borchardt, R.T., 1996. Vascular endothelial growth factor affects permeability of brain microvessel endothelial cells in vitro. Am. J. Physiol. 271, C1973–C1980.

Ward, N.L., Lamanna, J.C., 2004. The neurovascular unit and its growth factors: coordinated response in the vascular and nervous systems. Neurol. Res. 26, 870–883.

Weiger, T.M., Hermann, A., Levitan, I.B., 2002. Modulation of calcium-activated potassium channels. J. Comp. Physiol. A. Neuroethol. Sens. Neural. Behav. Physiol. 188, 79–87.

Woodbury, D., Schwarz, E.J., et al., 2000. Adult rat and human bone marrow stromal cells differentiate into neurons. J. Neurosci. Res. 61, 364–370.

Yancopoulos, G.D., Davis, S., et al., 2000. Vascular-specific growth factors and blood vessel formation. Nature 407, 242–248.

Zhang, Z., Chopp, M., 2002. Vascular endothelial growth factor and angiopoietins in focal cerebral ischemia. Trends Cardiovasc. Med. 12, 62–66.

Advantages and challenges of alternative sources of adult-derived stem cells for brain repair in stroke

Paul R. Sanberg[1], **David J. Eve, Christopher Metcalf, Cesario V. Borlongan**

Center of Excellence for Aging & Brain Repair, Department Neurosurgery and Brain Repair, University of South Florida Morsani College of Medicine, Tampa, FL, USA

[1]Corresponding author. Tel.: +1-813-9743154, Fax: 1-813-9743078, e-mail address: psanberg@health.usf.edu

Abstract

Considerable promise has been demonstrated by cell therapy for the treatment of stroke. Adult-derived stem cells avoid the ethical dilemmas of using embryonic and fetal stem cells and thus are the ideal type of cell to study. There are a number of different types of stem cells that could prove to be useful, but there are potential concerns associated with each one. This review summarizes the current knowledge on the use of the different possible adult-derived stem cell types including their benefits and challenges. While the optimal conditions are still to be determined, these cells may prove to be at the forefront of stem cell research and ultimately therapy for stroke and other disorders.

Keywords

adipose tissue, breast milk, induced pluripotent cells, menstrual blood, neural tissue, teeth, transplantation, umbilical cord blood

1 INTRODUCTION

The bone marrow has historically been the primary source of adult-derived stem cells (see Chapter 6), and in this chapter, we are going to discuss the numerous alternative sources of adult stem cells other than bone marrow that could be available as possible clinical therapies. We will further discuss whether one source of cells may be better than others and highlight the challenges that must be surmounted before adult stem cells could be used in large-scale clinical trials. An overview of the stem cells

Progress in Brain Research, Volume 201, ISSN 0079-6123, http://dx.doi.org/10.1016/B978-0-444-59544-7.00006-8

currently under investigation from a neurorestoratology standpoint has previously been published by us (Huang et al., 2010).

2 ADULT STEM CELLS

As discussed in the previous chapter, the bone marrow contains a diverse population of cells which can be purified to consist of specific cell types or used as a mixture. One of the first hurdles with any alternative source would be determining what the cellular make up is and how to purify the specific cell populations if necessary. Due to the heterogeneity of stroke itself, determining the optimal cell type for possible treatment of stroke is currently under investigation and may ultimately depend on the grafted cell's proposed mode of action.

3 NEURAL STEM CELLS

While initially, the most obvious adult-derived stem cell type for brain repair would appear to be neural stem cells (NSCs) due to their propensity to develop into the required neural cells, one of the biggest challenges to overcome with this cell type is the difficulty in obtaining these cells from an adult. Major surgery is required to procure these cells from a patient and thus the likelihood of them being used as an applied autologous treatment is small. However, one alternative method in obtaining NSCs could be from harvesting cells during other surgical procedures (Chaichana et al., 2009). In all likelihood, if NSCs were to be used, they would either be from an allogeneic fetal source or derived entirely from a different cell type. Determining whether or not to stimulate endogenous NSCs or implant exogenous cells invites a new set of challenges. However, while the use of exogenous adult NSCs is likely to be rare, methods to control or maximize the activity of endogenous NSCs could prove to be more useful. Endogenous NSCs can be found in the subgranular zone of the dentate gyrus, the subventricular zone (SVZ), and also the subependymal zone of the spinal cord. The activities of these regions are upregulated following an injury such as stroke, but this alone is not sufficient to elicit full functional repair and ultimately no real degree of cell replacement is observed (Nakayama et al., 2010). A review of the current literature provides evidence suggesting that stem cell transplantation is beneficial by promoting endogenous NSC production (Bachstetter et al., 2008; Jin et al., 2011; Park et al., 2010; van Velthoven et al., 2010), although some studies suggest that endogenous neurogenesis may be reduced due to microglial activation after a stroke, while dendritic plasticity was increased (Minnerup et al., 2011). In contrast, Deierborg et al. (2010) demonstrated in rat animal models that *ex vivo* culture of endogenous NSCs from the SVZ following an experimental stroke promoted neurogenesis by the presence of microglia. This suggests that the microglial activation due to stroke in the presence of NSCs may have a different effect than the microglia activated in their absence. In earlier studies, the migration of new cells into the injured striatum was reduced, and when migration did occur, the cells were unable to differentiate into neurons, implying that the lesioned area created a microenvironment that is unfavorable for cell replacement (Deierborg et al., 2009; Nygren et al., 2006). These observations suggest that any form of potential treatment may need to address the modification of the lesioned microenvironment to favor cell replacement and/or survival of injured cells. However, a study in human brain tissue postmortem, 1 week following a stroke, demonstrated an increased level of neurogenesis and angiogenesis in the region of the infarct area when compared with a noninfarcted brain (Minger et al., 2007). A human stroke study demonstrated enhanced neurogenesis using antibody markers from collecting tissue samples obtained from individuals who suffered a subarachnoid hemorrhage during an aneurysm clipping and deep brain tumor removal (Sgubin et al., 2007). A similar observation was seen in nonhuman primates (Koketsu et al., 2006). As previously discussed, the human stroke study used antibody markers for NSCs; thus, it did not determine whether the cells would eventually become neurons. Other studies have used the synthetic nucleoside bromodeoxyuridine (BrdU) by incorporating the molecule into the proliferating cells and observing its colocalization with NeuN for determining a neuronal fate. Additional studies have suggested that the maturation of the NSCs may be slower after ischemia (Taupin, 2006; Tonchev et al., 2005). This highlights the fact that the timing of BrdU administration and cell fate determination are of paramount importance and may explain some of the discrepancies observed between studies (Deierborg et al., 2009).

A major concern amongst the current stem cell research community is the potential of stem cells to form tumors. In general, the more undifferentiated a cell is, the greater the likelihood of tumors appears to be; that is, embryonic-derived stem cells are more likely to form tumors than adult-derived stem cells. Recently, this problem was highlighted by a publication demonstrating tumor formation after transplantation of fetally derived NSCs in a child with ataxia telangiectasia (Amariglio et al., 2009). However, further investigation of this study has revealed that the transplantation involved the implantation of a heterogenous mixture of cells, and only some could be considered true NSCs. Therefore, the cause of the tumors may not arise from the NSCs themselves. Studies such as these highlight the importance of using a purified cell population for transplantations (Amariglio and Rechavi, 2010; Jandial and Snyder, 2009).

Several studies have suggested that endogenous NSCs arising from the SVZ are redirected from their customary route along the rostral migratory stream and migrate along blood vessels toward the ischemic region (Kojima et al., 2010; Thored et al., 2007). This migration pattern has led researchers to postulate that the stem cells utilize an endogenously established orientation pathway to reach the ischemic region. Parallel migration of chemokine signals such as stromal-derived factor-1, vascular endothelial growth factor, and angiopoietin that are released from ischemic tissue can aid in the migration of stem cells to the infarcted area (Barkho et al., 2008; Carbajal et al., 2010; Liu et al., 2009; Zhang et al., 2009).

The proliferation capacity of adult-derived stem cells tends to be reduced when compared to embryonic or fetal stem cells. Therefore, obtaining a sufficient number of cells for transplantation presents a potential problem depending on how abundant the cells are within their source. Researchers have devised several techniques in overcoming this potential low population of cells. One way is the long-term culturing of the cells. However, long-term culturing of NSCs may increase the likelihood of the cells to spontaneously become a non-neural cell type such as a tumor precursor cell (Wu et al., 2011). An alternative way to long-term culturing is to immortalize the cells. One of the first immortalized cell lines used in animal studies of stroke was the hNT neuron, which originated from a teratocarcinoma (Newman et al., 2005a). In 1998, this cell line was tested in a Phase I clinical trial (with a Phase II clinical trial following soon after) in stroke patients demonstrating the safety and feasibility of the cell line, but no significant improvements were observed (Kondziolka et al., 2005). An alternative method to immortalizing the cells is the insertion of an oncogene. ReNeuron Ltd, a stem cell therapeutics company based in England, chose this route to generate an immortalized neural cell line using a c-Myc regulator gene and mutated estrogen receptor transgene (Pollock et al., 2006). This cell line is currently undergoing clinical trials for stroke in the United Kingdom (Mack, 2011) and will be discussed further in Chapter 8.

An alternate method of generating NSCs and other cell types is to create them by partial differentiation of other types of stem cells, for example, embryonic stem cells or induced pluripotent stem cells. The latter could be derived from adults and will be discussed in section 11 and Chapter 9.

4 MESENCHYMAL STROMAL CELLS

Mesenchymal stromal cells (MSCs) can be obtained from almost all tissues of the body including bone marrow, placenta, teeth, and adipose tissue, but not the nervous system (though they do migrate to this region under injury conditions). The ubiquitous nature of MSCs means that they are a favorable cell type for autologous cell transplantations. However, there is growing evidence suggesting that MSCs from different sources are not entirely the same (Barlow et al., 2008; Dmitrieva et al., 2012; Jansen et al., 2010; Kim et al., 2011; Strioga et al., 2012; but, see Lin et al., 2012). The differences may relate to the various ways of extracting, isolating, and proliferation of the cells and therefore, one source of MSCs may turn out to be more suitable than another based on the required characteristics for a specific disorder. For instance, bone marrow-derived cells from horses have demonstrated senesce at significantly earlier passages than adipose- and umbilical cord-derived cells (Vidal et al., 2012), suggesting that alternative sources to bone marrow may be optimal. While there are numerous research publications implying that MSCs are able to differentiate into neuron-like cells, there is still some debate about whether they can actually form functional neurons (Breunig et al., 2007; Burns et al., 2006; White, 2011).

MSCs have been shown to induce neurogenesis after stroke, and in general, the cells do not demonstrate a long-term survival time frame after transplantation (Chen et al., 2003). This suggests that the benefit of transplantation primarily arises from the transient neurotrophic-mediated effects rather than the actual cell replacement. In addition to the neurotrophic-mediated effect, the modulation of the vasculature was observed equally with MSCs derived from four different sources (bone marrow, adipose tissue, skeletal muscle, and myocardium; Lin et al., 2012). Despite the earlier mention of the potential tumorigenic property of stem cells, a survey of current studies reveals only a few instances of tumors that have followed MSC transplantation (e.g., Tolar et al., 2007). Also, Subramanian et al. (2012) reported that umbilical cord-derived MSCs do not appear to become tumor progenitor cells in the presence of tumor cells, unlike bone marrow-derived MSCs. This observation advocates that while tumors are rare they are still a valid concern, and not all MSCs are equal in their propensity to form tumors.

5 EXTRAEMBRYONIC TISSUE STEM CELLS

While MSCs may be the most commonly studied stem cell derived from the extra-embryonic tissues such as the umbilical cord, placenta, amnion, and Wharton's jelly; additional types of stem cells can be found within these tissues including amniotic epithelial cells, amnion-derived stem cells, placental-derived stem cells, and umbilical cord matrix stem cells (Marcus and Woodbury, 2008). Amniotic epithelial cells and the amnion MSCs differ in their primary germ cell layer of origin; the amniotic epithelial cells arise from the ectoderm, whereas the amnion MSCs arise from the mesoderm (Yu et al., 2009). The main differences between these two cell populations appear to be their yield and multipotency for mesodermal cell lineages, with the amnion MSCs possessing greater mesodermal capacity (Diaz-Prado et al., 2010), though less endothelial capabilities (Konig et al., 2011). Within the scientific community, there is an increasing number of extraembryonic tissue stem cell studies, however, due to variations in the cell's exact source as well as profound species differences in the development of the source tissue; it is difficult to interpret exactly how these cells will function (Dobreva et al., 2010). Such inconsistencies require either direct comparison of the different cell types to determine how (or if) they differ or an agreement on the standard isolation and proliferation techniques for a certain cell type. For instance, the International Society for Cellular Therapy (ISCT) has announced a minimal criteria required to define a stem cell as an MSC based on their plastic adherence, cluster of differentiation (CD) expression, and differentiation ability (Dominici et al., 2006). ISCT's proposed form of standardization will make it easier to interpret stem cell studies and should not be limited to just cell definition. The Stem Cell Therapies as an Emerging Paradigm in Stroke (STEPS) consortium was designed to interpret all studies, including clinical, and emphasizes that standardization will make procedures and end points easier to understand (Borlongan, 2009; Borlongan et al., 2008; Chopp et al., 2009; Stem Cell Therapies as an Emerging Paradigm in Stroke Participants, 2009). These STEPS guidelines in adult stroke have recently been extended to neonatal hypoxic-ischemic injury or cerebral palsy (Borlongan and Weiss, 2011).

Human placental-derived MSCs have been transplanted into animal models of stroke and observed to promote endogenous neurogenesis, rather than cell replacement (Chen et al., 2012; Yarygin et al., 2009). The medium in which cells are cultured during a neurotrophic factor induction protocol exhibits an increased neuroprotective capacity when compared to medium from non-induced cells, on cultured neuronal cells subjected to oxidative stress (Yust-Katz et al., 2012).

The umbilical cord lining MSCs appear to exhibit greater immunological immaturity when compared to aged-bone marrow MSCs, and they exert a greater immunosuppressive effect on the immune response cascade (Deuse et al., 2011). This phenomenon may be partly related to a greater secretion of leukemia inhibitory factor by the umbilical cord lining MSCs when compared to bone marrow-derived cells, as demonstrated by mixed lymphocyte reactions (Najar et al., 2010).

6 UMBILICAL CORD BLOOD

A promising and recent trend is the banking of umbilical cord blood at the time of birth for future use by the patient (or an allogeneic transplant). In the majority of cases, umbilical cord blood primarily refers to the mononuclear fraction of the blood which includes hematopoietic progenitors, lymphocytes, monocytes, and MSCs. While this heterogeneous mixture of cells is generally immunologically immature, the cells are still removed from the body within approximately 30 days (Walczak et al., 2007). Human umbilical cord blood transplantation in experimental animal models of stroke have shown optimal positive outcome with the combination of intravenous administration, 48 h post-stroke, and a cell count of 1 million cells (Newcomb et al., 2006; Vendrame et al., 2004; Willing et al., 2003). Additional studies have demonstrated that the spleen organ, cytokine profile of the cells, cell migratory capacity, and the contribution of the various components of the mononuclear fraction all have an involvement in the beneficial response, monocytes, and to a lesser extent the stem cells and B cells being the most relevant cell types (Ajmo et al., 2008; Newman et al., 2005b, 2006; Womble et al., 2009). However, a recent study demonstrated that the CD34+ stem cell in the mononuclear fraction appears only to be effective during *in vitro* studies (Boltze et al., 2012). The ability of the cells to modulate the immune response and reduce proinflammatory cytokine levels was also reported (Vendrame et al., 2005).

7 ADIPOSE TISSUE

Autologous adipose tissue can be obtained from any individual or species (e.g., dog; Vieira et al., 2010), or from the discarded lipoaspirate following a liposuction procedure. Numerous names have been given to the obtained adipose tissue cells including adipose tissue MSCs, although the International Fat Applied Technology Society proposes that adipose-derived stem cells be used as the common designation.

The cells are plastic-adherent and multipotent and have also been observed to spontaneously become cancerous following extensive passaging (Rubio et al., 2005). However, this report has since been retracted due to problems with replicating the data (de la Fuente et al., 2010). Further studies suggest that long-term passaging (MacIsaac et al., 2012) and transplantation of adipose-derived stem cells (Ra et al., 2011) do not result in tumors by themselves, but they can promote cancerous cells to form tumors (Yu et al., 2008; Zhang et al., 2010b). This latter observation raises possible concerns over potential side effects of their transplantation and therefore further studies are required. The cells have been transplanted in a number of animal models of stroke where they were observed to reduce infarct size, improve neurological function, and reduce the level of cerebral inflammation and chronic degeneration in an intracerebral hemorrhage model (Kim et al., 2007; Leu et al., 2010).

8 MENSTRUAL BLOOD

Every month, women of reproductive age slough off a newly formed endometrial lining of cells during their menstrual cycle. This lining and the associated menstrual blood has been shown to contain stem cells which have been called endometrial regenerative cells and MenSCs by two different research groups, although it is unclear whether they are referring to the same population of cells as they were cultured differently (Meng et al., 2007; Patel et al., 2008). Both cell types have been shown to possess multipotency toward a number of different lineages, including neural. The MenSCs have been demonstrated to have a protective role against hypoxia in neuronal cultures *in vitro*. This observation is believed to relate to the cells' secretion of growth factors since conditioned media had a similar effect (Borlongan et al., 2010). In rat studies *in vivo*, the utilization of the occluded middle cerebral artery model of stroke has also shown a promising benefit following transplantation without suppressing the immune system (Allickson et al., 2011; Borlongan et al., 2010; Rodrigues et al., 2012). These cells can be regularly obtained relatively easily and for autologous use in females; they can also be banked for future applications, theoretically including allogeneic transplants in males. Menstrual blood has the potential to be a proven and convenient treatment for a number of disorders.

9 BREAST MILK

A further possible source of stem cells has been found in breast milk (McGregor and Rogo, 2006). Study of these cells has been limited, but a number of different stem cell types have been identified, although one study suggests that the cells do not respond to normal cell culture protocols (Fan et al., 2011). This could possibly present another potential valuable source of stem cells that could be banked for autologous transplantation into women, similar to cells from menstrual blood as previously discussed.

10 TEETH

Over the past decade, it has been shown that stem cells can also be derived from an unusual source: teeth. Teeth-derived stem cells can be obtained at various stages of their development (Miura et al., 2003). These cells appear to possess neurogenic potential (Yang et al., 2009) and one study suggests that this propensity is greater than for bone marrow-derived cells (Karaoz et al., 2011). This therefore suggests that teeth-derived cells could prove to be a useful tool in stem cell therapy and further studies are awaited.

11 INDUCED PLURIPOTENT STEM CELLS

It was originally believed that once a stem cell had matured and terminally differentiated, it could NOT regain its multipotency. However, in the past few years, scientists have successfully dedifferentiated fibroblasts back to embryonic-like stem cells by transfecting specific transcription factors into the cells and culturing them. One of the inherent concerns with this technique is the oncogenicity of the transcription factors and the use of viral vectors to integrate them. The transfection technique has been performed using a number of different species and cell types and can also be applied to precursor cells to increase their potency, for example, umbilical cord and placental MSCs (Cai et al., 2010) and NSCs or adipose-derived cells (Tat et al., 2010). One of the primary advantages in using precursor cells is that they are less likely to have inherent mutations, which would also be incorporated into the induced cells, that are present in adult cells. The success rate of incorporation and dedifferentiation is relatively low, but once the cells have been obtained, they can be proliferated many times to generate abundant quantities. The cells are very similar, but not exactly identical, to embryonic stem cells (Liu et al., 2011), and express a different gene profile (Chin et al., 2009). Induced pluripotent stem cells have the same potential to form tumors, but they have been reported to be less efficient at generating human neurons in culture compared to embryonic-derived stem cells (Hu et al., 2010). In addition, induced pluripotent stem cells were found to incorporate into chimeric mice (a standard test of pluripotency) less readily than embryonic stem cells even when the cells had identical DNA (Dolgin, 2010). Several studies also suggest that the induced cells carry residual transcriptional memory of their former selves (Lister et al., 2011; Ohi et al., 2011). The original derivation of induced pluripotent stem cells involved retroviral incorporation of four to six transcription factors, several of which are oncogenic. This has led researchers to devise different methods and reduce the number of required transcription factors. These preventative measures indicate that there are a wide range of induced pluripotent stem cell types with slight differences in their preparation, which may also result in subtle variations in their activity. In one study, human induced pluripotent stem cells were generated using the original set of four transcription factors (Oct4, Sox2, Nanog, and Lin-28) and subsequently transplanted into a rat model of stroke, resulting in subsequent improvements in sensorimotor function (Jiang et al., 2011). However, the cells were only followed for a short period of time, and it is unclear whether the animals remained tumor free. Another issue due to the short period of time following the cells is that they could also be predifferentiated prior to transplantation. Recent evidence suggests that the process of induction may alter the immunogenicity of the cells. Another study has demonstrated rejection, even though the cells were originally autologous (Zhao et al., 2011), suggesting that induced pluripotent stem cells may not be the optimal cell type for use in transplantation, but instead may work better as a source of cells to generate tissue to model a disease.

12 AUTOLOGOUS VERSUS ALLOGENEIC

One important consideration with respect to cell therapy is whether to use autologous or allogeneic cells. The limiting factors include cell yield, ease of harvesting, proliferation capacity, and cell immunogenicity. When the initial cell yield is determined to be inadequate, the cells could be proliferated. For example, with respect to stroke, there is evidence suggesting that the optimal time for delivery with umbilical cord blood cells is 48 h post-stroke (Newcomb et al., 2006). This is an insufficient amount of time to generate the necessary number of cells if harvested recently (though of course this would be unlikely for autologous), suggesting that the cells would require proliferation and storage (cell banking) before they could be used. Numerous cell banks for umbilical cord blood exist today and additional umbilical cord, placenta, menstrual blood, adipose tissue banks are appearing across the landscape. These banks could act as either autologous or allogeneic repositories depending on their operating procedures. Autologous cells are much less likely to be rejected than allogeneic types, and therefore immunosuppression is a major consideration when experimenting with allogeneic cells. A recent study suggests that the use of cyclosporine A promotes recovery of cortical injury following a stroke due to promotion of endogenous NSC activity and migration, suggesting that immunosuppression itself may influence the activity of stem cells (Erlandsson et al., 2011). These series of observations may cast doubt on the benefits of stem cell transplantation if no non-immunosuppression studies were performed in conjunction to determine the effects of immunosuppression. In addition, stroke studies in immunocompromised animals show increased endogenous neurogenesis via a CD4 + T cell, but not a CD25 + T cell dependent mechanism (Saino et al., 2010). The immunogenic status of the cells has become an important issue, and in general, the less mature or lineage-specific a cell is, the less likely it is to induce an immune response. For example, umbilical cord blood is considered to be immunologically immature and its survival following transplantation is generally not affected by immunosuppression. This means that the umbilical cord blood cells are less likely to require strict human leukocyte antigen matching compared with bone marrow transplants to prevent rejection, making them an ideal cell (Willing et al., 2007). However, placental MSCs appear to be less immune immature than bone marrow MSCs. Although placental MSCs were found to be a heterogenous population of adult and neonatal cells (Fazekasova et al., 2011), chorionic plate-derived cells may have additional immunomodulatory properties compared to bone marrow and adipose-derived cells (Lee et al., 2012). MSCs derived from the decidua have been observed to be only from adult (i.e., maternal) origin (Kanematsu et al., 2011), while maternally derived cells still exhibited plasticity (Sabapathy et al., 2012).

13 CO-TRANSPLANTATION AND COMBINATION THERAPY

An important consideration is whether or not to transplant just one cell type. For instance, in models of spinal cord injury, the co-transplantation of NSCs and adipose-derived stem cells resulted in an increased level of NSC survival (Oh et al., 2011), while bone marrow stromal cells that were co-transplanted alongside embryonic stem cells, prevented the formation of tumors (Matsuda et al., 2009). In a stroke model, co-transplantation of NSCs and epithelial cells resulted in an increased level of NSC survival and differentiation (Nakagomi et al., 2009). This demonstrates that transplantation of two different cell types modifies the microenvironment to allow the cells to survive even longer. In addition to co-transplantation, combination therapy, whereby stem cells and a second treatment are used, is also a possible avenue for therapy. Typically, combination therapies could be either a manipulation of the cells to express a factor prior to transplantation (e.g., bone marrow-derived cells and neurotrophin-3; Zhang et al., 2010a) or the stem cells are administered in conjunction with a second treatment (e.g., a nitric oxide donor and bone marrow-derived cells, Chen et al., 2004, or stem cells and a scaffold system, Jin et al., 2010; Shi et al., 2012). In this way, the use of two or more therapies can promote the beneficial effects that are desired.

14 OTHER FACTORS

There are numerous factors that need to be resolved. Previously discussed is a summary of the different adult-derived stem cell types that could be used and it is unclear which type will turn out to be the best. Several other factors have recently been reviewed by us (Sanberg et al., 2012a) and so they are only briefly summarized below. Data on the optimal dose and route of administration are still ongoing and are likely to be cell dependent. One alternative to the potential problem of an insufficient quantity of cells surviving is to perform multiple small-dose transplantation of cells over time rather than using a single large transplantation of stem cells that may be difficult to obtain. We have explored this idea for umbilical cord blood treatments for a number of disorders, including amyotrophic lateral sclerosis, Alzheimer's disease, and Sanfilippo syndrome (Sanberg et al., 2012b), but not yet for stroke, as our earlier studies suggested that there was an optimal time point for treatment (Newcomb et al., 2006). Since umbilical cord blood is known to modulate the spleen, it is therefore important to transplant these cells intravenously. The sex of the donor/recipient may also play an important role in the success of these therapies, since it has previously been shown that male and female cells respond differently to insult.

15 MODE OF ACTION

Most studies seem to suggest that stem cells do not remain within the brain and thus cell replacement is not the normal mode of action. Other possibilities include the release of growth factors to modulate cell survival and promote endogenous stem cells. The former may also modulate the inflammatory response, thus making the microenvironment more conducive for cell survival and has been reviewed by us previously (Park et al., 2009; Sanberg et al., 2011, 2012a).

16 CONCLUSIONS

There is considerable promise from a number of different adult-derived stem cell types for the treatment of stroke. This is highlighted by the ongoing limited clinical trials of stem cell therapy for stroke patients, such as ReNeuron, Sanbio, and Celgene Cellular Therapeutics, which have been discussed elsewhere (Chapter 9). However, a number of factors still need to be determined including the optimal cell type and route of administration to ensure safety and efficacy of cell therapy for large-scale clinical trials.

Acknowledgments

P. R. S. and C. V. B. hold patents in stem cell technologies for the treatment of neurodegenerative disorders.

References

Ajmo Jr., C.T., Vernon, D.O., Collier, L., Hall, A.A., Garbuzova-Davis, S., Willing, A., Pennypacker, K.R., 2008. The spleen contributes to stroke-induced neurodegeneration. J. Neurosci. Res. 86, 2227–2234.

Allickson, J.G., Sanchez, A., Yefimenko, N., Borlongan, C.V., Sanberg, P.R., 2011. Recent studies assessing the proliferative capability of a novel adult stem cell identified in menstrual blood. Open Stem Cell J. 3, 4–10.

Amariglio, N., Hirshberg, A., Scheithauer, B.W., Cohen, Y., Loewenthal, R., Trakhtenbrot, L., Paz, N., Koren-Michowitz, M., Waldman, D., Leider-Trejo, L., Toren, A., Constantini, S., Rechavi, G., 2009. Donor-derived brain tumor following neural stem cell transplantation in an ataxia telangiectasia patient. PLoS Med. 6, e1000029.

Amariglio, N., Rechavi, G., 2010. On the origin of glioneural neoplasms after neural cell transplantation. Nat. Med. 16, 157, author reply pp. 157–158.

Bachstetter, A.D., Pabon, M.M., Cole, M.J., Hudson, C.E., Sanberg, P.R., Willing, A.E., Bickford, P.C., Gemma, C., 2008. Peripheral injection of human umbilical cord blood stimulates neurogenesis in the aged rat brain. BMC Neurosci. 9, 22.

Barkho, B.Z., Munoz, A.E., Li, X., Li, L., Cunningham, L.A., Zhao, X., 2008. Endogenous matrix metalloproteinase (MMP)-3 and MMP-9 promote the differentiation and migration of adult neural progenitor cells in response to chemokines. Stem Cells 26, 3139–3149.

Barlow, S., Brooke, G., Chatterjee, K., Price, G., Pelekanos, R., Rossetti, T., Doody, M., Venter, D., Pain, S., Gilshenan, K., Atkinson, K., 2008. Comparison of human placenta- and bone marrow-derived multipotent mesenchymal stem cells. Stem Cells Dev. 17, 1095–1107.

Boltze, J., Reich, D.M., Hau, S., Reymann, K.G., Strassburger, M., Lobsien, D., Wagner, D.C., Kamprad, M., Stahl, T., 2012. Assessment of neuroprotective effects of human umbilical cord blood mononuclear cell subpopulations in vitro and in vivo. Cell Transplant. 21, 723–737.

Borlongan, C.V., 2009. Cell therapy for stroke: remaining issues to address before embarking on clinical trials. Stroke 40, S146–S148.

Borlongan, C.V., Chopp, M., Steinberg, G.K., Bliss, T.M., Li, Y., Lu, M., Hess, D.C., Kondziolka, D., 2008. Potential of stem/progenitor cells in treating stroke: the missing steps in translating cell therapy from laboratory to clinic. Regen. Med. 3, 249–250.

Borlongan, C.V., Kaneko, Y., Maki, M., Yu, S.J., Ali, M., Allickson, J.G., Sanberg, C.D., Kuzmin-Nichols, N., Sanberg, P.R., 2010. Menstrual blood cells display stem cell-like phenotypic markers and exert neuroprotection following transplantation in experimental stroke. Stem Cells Dev. 19, 439–452.

Borlongan, C.V., Weiss, M.D., 2011. Baby STEPS: a giant leap for cell therapy in neonatal brain injury. Pediatr. Res. 70, 3–9.

Breunig, J.J., Arellano, J.I., Macklis, J.D., Rakic, P., 2007. Everything that glitters isn't gold: a critical review of postnatal neural precursor analyses. Cell Stem Cell 1, 612–627.

Burns, T.C., Ortiz-Gonzalez, X.R., Gutierrez-Perez, M., Keene, C.D., Sharda, R., Demorest, Z.L., Jiang, Y., Nelson-Holte, M., Soriano, M., Nakagawa, Y., Luquin, M.R., Garcia-Verdugo, J.M., Prosper, F., Low, W.C., Verfaillie, C.M., 2006. Thymidine analogs are transferred from prelabeled donor to host cells in the central nervous system after transplantation: a word of caution. Stem Cells 24, 1121–1127.

Cai, J., Li, W., Su, H., Qin, D., Yang, J., Zhu, F., Xu, J., He, W., Guo, X., Labuda, K., Peterbauer, A., Wolbank, S., Zhong, M., Li, Z., Wu, W., So, K.F., Redl, H., Zeng, L., Esteban, M.A., Pei, D., 2010. Generation of human induced pluripotent stem cells from umbilical cord matrix and amniotic membrane mesenchymal cells. J. Biol. Chem. 285, 11227–11234.

Carbajal, K.S., Schaumburg, C., Strieter, R., Kane, J., Lane, T.E., 2010. Migration of engrafted neural stem cells is mediated by CXCL12 signaling through CXCR4 in a viral model of multiple sclerosis. Proc. Natl. Acad. Sci. U.S.A. 107, 11068–11073.

Chaichana, K.L., Guerrero-Cazares, H., Capilla-Gonzalez, V., Zamora-Berridi, G., Achanta, P., Gonzalez-Perez, O., Jallo, G.I., Garcia-Verdugo, J.M., Quinones-Hinojosa, A., 2009. Intraoperatively obtained human tissue: protocols and techniques for the study of neural stem cells. J. Neurosci. Methods 180, 116–125.

Chen, J., Li, Y., Katakowski, M., Chen, X., Wang, L., Lu, D., Lu, M., Gautam, S.C., Chopp, M., 2003. Intravenous bone marrow stromal cell therapy reduces apoptosis and promotes endogenous cell proliferation after stroke in female rat. J. Neurosci. Res. 73, 778–786.

Chen, J., Li, Y., Zhang, R., Katakowski, M., Gautam, S.C., Xu, Y., Lu, M., Zhang, Z., Chopp, M., 2004. Combination therapy of stroke in rats with a nitric oxide donor and human bone marrow stromal cells enhances angiogenesis and neurogenesis. Brain Res. 1005, 21–28.

Chen, J., Shehadah, A., Pal, A., Zacharek, A., Cui, X., Cui, Y., Roberts, C., Lu, M., Zeitlin, A., Hariri, R., Chopp, M., 2012. Neuroprotective effect of human placenta-derived cell treatment of stroke in rats. Cell Transplant. (in press).

Chin, M.H., Mason, M.J., Xie, W., Volinia, S., Singer, M., Peterson, C., Ambartsumyan, G., Aimiuwu, O., Richter, L., Zhang, J., Khvorostov, I., Ott, V., Grunstein, M., Lavon, N., Benvenisty, N., Croce, C.M., Clark, A.T., Baxter, T., Pyle, A.D., Teitell, M.A., Pelegrini, M., Plath, K., Lowry, W.E., 2009. Induced pluripotent stem cells and embryonic stem cells are distinguished by gene expression signatures. Cell Stem Cell 5, 111–123.

Chopp, M., Steinberg, G.K., Kondziolka, D., Lu, M., Bliss, T.M., Li, Y., Hess, D.C., Borlongan, C.V., 2009. Who's in favor of translational cell therapy for stroke: STEPS forward please? Cell Transplant. 18, 691–693.

de la Fuente, R., Bernad, A., Garcia-Castro, J., Martin, M.C., Cigudosa, J.C., 2010. Retraction: spontaneous human adult stem cell transformation. Cancer Res. 70, 6682.

Deierborg, T., Roybon, L., Inacio, A.R., Pesic, J., Brundin, P., 2010. Brain injury activates microglia that induce neural stem cell proliferation ex vivo and promote differentiation of neurosphere-derived cells into neurons and oligodendrocytes. Neuroscience 171, 1386–1396.

Deierborg, T., Staflin, K., Pesic, J., Roybon, L., Brundin, P., Lundberg, C., 2009. Absence of striatal newborn neurons with mature phenotype following defined striatal and cortical excitotoxic brain injuries. Exp. Neurol. 219, 363–367.

Deuse, T., Stubbendorff, M., Tang-Quan, K., Phillips, N., Kay, M.A., Eiermann, T., Phan, T.T., Volk, H.D., Reichenspurner, H., Robbins, R.C., Schrepfer, S., 2011. Immunogenicity and immunomodulatory properties of umbilical cord lining mesenchymal stem cells. Cell Transplant. 20, 655–667.

Diaz-Prado, S., Muinos-Lopez, E., Hermida-Gomez, T., Rendal-Vazquez, M.E., Fuentes-Boquete, I., de Toro, F.J., Blanco, F.J., 2010. Multilineage differentiation potential of cells isolated from the human amniotic membrane. J. Cell. Biochem. 111, 846–857.

Dmitrieva, R.I., Minullina, I.R., Bilibina, A.A., Tarasova, O.V., Anisimov, S.V., Zaritskey, A.Y., 2012. Bone marrow- and subcutaneous adipose tissue-derived mesenchymal stem cells: differences and similarities. Cell Cycle 11, 377–383.

Dobreva, M.P., Pereira, P.N., Deprest, J., Zwijsen, A., 2010. On the origin of amniotic stem cells: of mice and men. Int. J. Dev. Biol. 54, 761–777.

Dolgin, E., 2010. Gene flaw found in induced stem cells. Nature 464, 663.

Dominici, M., Le Blanc, K., Mueller, I., Slaper-Cortenbach, I., Marini, F., Krause, D., Deans, R., Keating, A., Prockop, D., Horwitz, E., 2006. Minimal criteria for defining multipotent mesenchymal stromal cells. The International Society for Cellular Therapy position statement. Cytotherapy 8, 315–317.

Erlandsson, A., Lin, C.H., Yu, F., Morshead, C.M., 2011. Immunosuppression promotes endogenous neural stem and progenitor cell migration and tissue regeneration after ischemic injury. Exp. Neurol. 230, 48–57.

Fan, Y., Chong, Y.S., Choolani, M.A., Cregan, M.D., Chan, J.K., 2011. Unravelling the mystery of stem/progenitor cells in human breast milk. PLoS One 5, e14421.

Fazekasova, H., Lechler, R., Langford, K., Lombardi, G., 2011. Placenta-derived MSCs are partially immunogenic and less immunomodulatory than bone marrow-derived MSCs. J. Tissue Eng. Regen. Med. 5, 684–694.

Hu, B.Y., Weick, J.P., Yu, J., Ma, L.X., Zhang, X.Q., Thomson, J.A., Zhang, S.C., 2010. Neural differentiation of human induced pluripotent stem cells follows developmental principles but with variable potency. Proc. Natl. Acad. Sci. U.S.A. 107, 4335–4340.

Huang, H., Chen, L., Sanberg, P., 2010. Cell therapy from bench to bedside translation in CNS neurorestoratology era. Cell Med. 1, 15–46.

Jandial, R., Snyder, E.Y., 2009. A safer stem cell: on guard against cancer. Nat. Med. 15, 999–1001.

Jansen, B.J., Gilissen, C., Roelofs, H., Schaap-Oziemlak, A., Veltman, J.A., Raymakers, R.A., Jansen, J.H., Kogler, G., Figdor, C.G., Torensma, R., Adema, G.J., 2010. Functional differences between mesenchymal stem cell populations are reflected by their transcriptome. Stem Cells Dev. 19, 481–490.

Jiang, M., Lv, L., Ji, H., Yang, X., Zhu, W., Cai, L., Gu, X., Chai, C., Huang, S., Sun, J., Dong, Q., 2011. Induction of pluripotent stem cells transplantation therapy for ischemic stroke. Mol. Cell. Biochem. 354, 67–75.

Jin, K., Mao, X., Xie, L., Galvan, V., Lai, B., Wang, Y., Gorostiza, O., Wang, X., Greenberg, D.A., 2010. Transplantation of human neural precursor cells in Matrigel scaffolding improves outcome from focal cerebral ischemia after delayed postischemic treatment in rats. J. Cereb. Blood Flow Metab. 30, 534–544.

Jin, K., Xie, L., Mao, X., Greenberg, M.B., Moore, A., Peng, B., Greenberg, R.B., Greenberg, D.A., 2011. Effect of human neural precursor cell transplantation on endogenous neurogenesis after focal cerebral ischemia in the rat. Brain Res. 1374, 56–62.

Kanematsu, D., Shofuda, T., Yamamoto, A., Ban, C., Ueda, T., Yamasaki, M., Kanemura, Y., 2011. Isolation and cellular properties of mesenchymal cells derived from the decidua of human term placenta. Differentiation 82, 77–88.

Karaoz, E., Demircan, P.C., Saglam, O., Aksoy, A., Kaymaz, F., Duruksu, G., 2011. Human dental pulp stem cells demonstrate better neural and epithelial stem cell properties than bone marrow-derived mesenchymal stem cells. Histochem. Cell Biol. 136, 455–473.

Kim, J.M., Lee, S.T., Chu, K., Jung, K.H., Song, E.C., Kim, S.J., Sinn, D.I., Kim, J.H., Park, D.K., Kang, K.M., Hyung Hong, N., Park, H.K., Won, C.H., Kim, K.H., Kim, M., Kun Lee, S., Roh, J.K., 2007. Systemic transplantation of human adipose stem cells attenuated cerebral inflammation and degeneration in a hemorrhagic stroke model. Brain Res. 1183, 43–50.

Kim, S.H., Kim, Y.S., Lee, S.Y., Kim, K.H., Lee, Y.M., Kim, W.K., Lee, Y.K., 2011. Gene expression profile in mesenchymal stem cells derived from dental tissues and bone marrow. J. Periodontal Implant Sci. 41, 192–200.

Kojima, T., Hirota, Y., Ema, M., Takahashi, S., Miyoshi, I., Okano, H., Sawamoto, K., 2010. Subventricular zone-derived neural progenitor cells migrate along a blood vessel scaffold toward the post-stroke striatum. Stem Cells 28, 545–554.

Koketsu, D., Furuichi, Y., Maeda, M., Matsuoka, N., Miyamoto, Y., Hisatsune, T., 2006. Increased number of new neurons in the olfactory bulb and hippocampus of adult non-human primates after focal ischemia. Exp. Neurol. 199, 92–102.

Kondziolka, D., Steinberg, G.K., Wechsler, L., Meltzer, C.C., Elder, E., Gebel, J., Decesare, S., Jovin, T., Zafonte, R., Lebowitz, J., Flickinger, J.C., Tong, D., Marks, M.P., Jamieson, C., Luu, D., Bell-Stephens, T., Teraoka, J., 2005. Neurotransplantation for patients with subcortical motor stroke: a phase 2 randomized trial. J. Neurosurg. 103, 38–45.

Konig, J., Huppertz, B., Desoye, G., Parolini, O., Frohlich, J.D., Weiss, G., Dohr, G., Sedlmayr, P., Lang, I., 2011. Amnion-derived mesenchymal stromal cells show angiogenic properties but resist differentiation into mature endothelial cells. Stem Cells Dev. 21, 1309–1320.

Lee, J.M., Jung, J., Lee, H.J., Cho, K.J., Hwang, S.G., Kim, G.J., 2012. Comparison of immunomodulatory effects of placenta mesenchymal stem cells with bone marrow and adipose mesenchymal stem cells. Int. Immunopharmacol. 13, 219–224.

Leu, S., Lin, Y.C., Yuen, C.M., Yen, C.H., Kao, Y.H., Sun, C.K., Yip, H.K., 2010. Adipose-derived mesenchymal stem cells markedly attenuate brain infarct size and improve neurological function in rats. J. Transl. Med. 8, 63.

Lin, R.Z., Moreno-Luna, R., Zhou, B., Pu, W.T., Melero-Martin, J.M., 2012. Equal modulation of endothelial cell function by four distinct tissue-specific mesenchymal stem cells. Angiogenesis 15, 443–455.

Lister, R., Pelizzola, M., Kida, Y.S., Hawkins, R.D., Nery, J.R., Hon, G., Antosiewicz-Bourget, J., O'Malley, R., Castanon, R., Klugman, S., Downes, M., Yu, R., Stewart, R., Ren, B., Thomson, J.A., Evans, R.M., Ecker, J.R., 2011. Hotspots of aberrant epigenomic reprogramming in human induced pluripotent stem cells. Nature 471, 68–73.

Liu, X.S., Chopp, M., Zhang, R.L., Hozeska-Solgot, A., Gregg, S.C., Buller, B., Lu, M., Zhang, Z.G., 2009. Angiopoietin 2 mediates the differentiation and migration of neural progenitor cells in the subventricular zone after stroke. J. Biol. Chem. 284, 22680–22689.

Liu, X.S., Chopp, M., Zhang, R.L., Tao, T., Wang, X.L., Kassis, H., Hozeska-Solgot, A., Zhang, L., Chen, C., Zhang, Z.G., 2011. MicroRNA profiling in subventricular zone after stroke: MiR-124a regulates proliferation of neural progenitor cells through Notch signaling pathway. PLoS One 6, e23461.

MacIsaac, Z.M., Shang, H., Agrawal, H., Yang, N., Parker, A., Katz, A.J., 2012. Long-term in-vivo tumorigenic assessment of human culture-expanded adipose stromal/stem cells. Exp. Cell Res. 318, 416–423.

Mack, G.S., 2011. ReNeuron and StemCells get green light for neural stem cell trials. Nat. Biotechnol. 29, 95–97.

Marcus, A.J., Woodbury, D., 2008. Fetal stem cells from extra-embryonic tissues: do not discard. J. Cell. Mol. Med. 12, 730–742.

Matsuda, R., Yoshikawa, M., Kimura, H., Ouji, Y., Nakase, H., Nishimura, F., Nonaka, J., Toriumi, H., Yamada, S., Nishiofuku, M., Moriya, K., Ishizaka, S., Nakamura, M., Sakaki, T., 2009. Cotransplantation of mouse embryonic stem cells and bone marrow stromal cells following spinal cord injury suppresses tumor development. Cell Transplant. 18, 39–54.

McGregor, J.A., Rogo, L.J., 2006. Breast milk: an unappreciated source of stem cells. J. Hum. Lact. 22, 270–271.

Meng, X., Ichim, T.E., Zhong, J., Rogers, A., Yin, Z., Jackson, J., Wang, H., Ge, W., Bogin, V., Chan, K.W., Thebaud, B., Riordan, N.H., 2007. Endometrial regenerative cells: a novel stem cell population. J. Transl. Med. 5, 57.

Minger, S.L., Ekonomou, A., Carta, E.M., Chinoy, A., Perry, R.H., Ballard, C.G., 2007. Endogenous neurogenesis in the human brain following cerebral infarction. Regen. Med. 2, 69–74.

Minnerup, J., Kim, J.B., Schmidt, A., Diederich, K., Bauer, H., Schilling, M., Strecker, J.K., Ringelstein, E.B., Sommer, C., Scholer, H.R., Schabitz, W.R., 2011. Effects of neural progenitor cells on sensorimotor recovery and endogenous repair mechanisms after photothrombotic stroke. Stroke 42, 1757–1763.

Miura, M., Gronthos, S., Zhao, M., Lu, B., Fisher, L.W., Robey, P.G., Shi, S., 2003. SHED: stem cells from human exfoliated deciduous teeth. Proc. Natl. Acad. Sci. U.S.A. 100, 5807–5812.

Najar, M., Raicevic, G., Boufker, H.I., Fayyad-Kazan, H., De Bruyn, C., Meuleman, N., Bron, D., Toungouz, M., Lagneaux, L., 2010. Adipose-tissue-derived and Wharton's jelly-derived mesenchymal stromal cells suppress lymphocyte responses by secreting leukemia inhibitory factor. Tissue Eng. Part A 16, 3537–3546.

Nakagomi, N., Nakagomi, T., Kubo, S., Nakano-Doi, A., Saino, O., Takata, M., Yoshikawa, H., Stern, D.M., Matsuyama, T., Taguchi, A., 2009. Endothelial cells support survival, proliferation, and neuronal differentiation of transplanted adult ischemia-induced neural stem/progenitor cells after cerebral infarction. Stem Cells 27, 2185–2195.

Nakayama, D., Matsuyama, T., Ishibashi-Ueda, H., Nakagomi, T., Kasahara, Y., Hirose, H., Kikuchi-Taura, A., Stern, D.M., Mori, H., Taguchi, A., 2010. Injury-induced neural stem/progenitor cells in post-stroke human cerebral cortex. Eur. J. Neurosci. 31, 90–98.

Newcomb, J.D., Ajmo Jr., C.T., Sanberg, C.D., Sanberg, P.R., Pennypacker, K.R., Willing, A.E., 2006. Timing of cord blood treatment after experimental stroke determines therapeutic efficacy. Cell Transplant. 15, 213–223.

Newman, M.B., Misiuta, I., Willing, A.E., Zigova, T., Karl, R.C., Borlongan, C.V., Sanberg, P.R., 2005a. Tumorigenicity issues of embryonic carcinoma-derived stem cells: relevance to surgical trials using NT2 and hNT neural cells. Stem Cells Dev. 14, 29–43.

Newman, M.B., Willing, A.E., Manresa, J.J., Davis-Sanberg, C., Sanberg, P.R., 2005b. Stroke-induced migration of human umbilical cord blood cells: time course and cytokines. Stem Cells Dev. 14, 576–586.

Newman, M.B., Willing, A.E., Manresa, J.J., Sanberg, C.D., Sanberg, P.R., 2006. Cytokines produced by cultured human umbilical cord blood (HUCB) cells: implications for brain repair. Exp. Neurol. 199, 201–208.

Nygren, J., Wieloch, T., Pesic, J., Brundin, P., Deierborg, T., 2006. Enriched environment attenuates cell genesis in subventricular zone after focal ischemia in mice and decreases migration of newborn cells to the striatum. Stroke 37, 2824–2829.

Oh, J.S., Kim, K.N., An, S.S., Pennant, W.A., Kim, H.J., Gwak, S.J., Yoon do, H., Lim, M.H., Choi, B.H., Ha, Y., 2011. Cotransplantation of mouse neural stem cells (mNSCs) with adipose tissue-derived mesenchymal stem cells improves mNSC survival in a rat spinal cord injury model. Cell Transplant. 20, 837–849.

Ohi, Y., Qin, H., Hong, C., Blouin, L., Polo, J.M., Guo, T., Qi, Z., Downey, S.L., Manos, P.D., Rossi, D.J., Yu, J., Hebrok, M., Hochedlinger, K., Costello, J.F., Song, J.S., Ramalho-Santos, M., 2011. Incomplete DNA methylation underlies a transcriptional memory of somatic cells in human iPS cells. Nat. Cell Biol. 13, 541–549.

Park, D.H., Eve, D.J., Musso 3rd, J., Klasko, S.K., Cruz, E., Borlongan, C.V., Sanberg, P.R., 2009. Inflammation and stem cell migration to the injured brain in higher organisms. Stem Cells Dev. 18, 693–702.

Park, D.H., Eve, D.J., Sanberg, P.R., Musso 3rd, J., Bachstetter, A.D., Wolfson, A., Schlunk, A., Baradez, M.O., Sinden, J.D., Gemma, C., 2010. Increased neuronal proliferation in the dentate gyrus of aged rats following neural stem cell implantation. Stem Cells Dev. 19, 175–180.

Patel, A.N., Park, E., Kuzman, M., Benetti, F., Silva, F.J., Allickson, J.G., 2008. Multipotent menstrual blood stromal stem cells: isolation, characterization, and differentiation. Cell Transplant. 17, 303–311.

Pollock, K., Stroemer, P., Patel, S., Stevanato, L., Hope, A., Miljan, E., Dong, Z., Hodges, H., Price, J., Sinden, J.D., 2006. A conditionally immortal clonal stem cell line from human cortical neuroepithelium for the treatment of ischemic stroke. Exp. Neurol. 199, 143–155.

Ra, J.C., Shin, I.S., Kim, S.H., Kang, S.K., Kang, B.C., Lee, H.Y., Kim, Y.J., Jo, J.Y., Yoon, E.J., Choi, H.J., Kwon, E., 2011. Safety of intravenous infusion of human adipose tissue-derived mesenchymal stem cells in animals and humans. Stem Cells Dev. 20, 1297–1308.

Rodrigues, M.C., Voltarelli, J., Sanberg, P.R., Allickson, J.G., Kuzmin-Nichols, N., Garbuzova-Davis, S., Borlongan, C.V., 2012. Recent progress in cell therapy for basal ganglia disorders with emphasis on menstrual blood transplantation in stroke. Neurosci. Biobehav. Rev. 36, 177–190.

Rubio, D., Garcia-Castro, J., Martin, M.C., de la Fuente, R., Cigudosa, J.C., Lloyd, A.C., Bernad, A., 2005. Spontaneous human adult stem cell transformation. Cancer Res. 65, 3035–3039.

Sabapathy, V., Ravi, S., Srivastava, V., Srivastava, A., Kumar, S., 2012. Long-term cultured human term placenta-derived mesenchymal stem cells of maternal origin displays plasticity. Stem Cells Int. 2012, 174328.

Saino, O., Taguchi, A., Nakagomi, T., Nakano-Doi, A., Kashiwamura, S., Doe, N., Nakagomi, N., Soma, T., Yoshikawa, H., Stern, D.M., Okamura, H., Matsuyama, T., 2010. Immunodeficiency reduces neural stem/progenitor cell apoptosis and enhances neurogenesis in the cerebral cortex after stroke. J. Neurosci. Res. 88, 2385–2397.

Sanberg, P.R., Eve, D.J., Cruz, L.E., Borlongan, C.V., 2012a. Neurological disorders and the potential role for stem cells as a therapy. Br. Med. Bull. 101, 163–181.

Sanberg, P.R., Eve, D.J., Willing, A.E., Garbuzova-Davis, S., Tan, J., Sanberg, C.D., Allickson, J.G., Cruz, L.E., Borlongan, C.V., 2011. The treatment of neurodegenerative disorders using umbilical cord blood and menstrual blood-derived stem cells. Cell Transplant. 20, 85–94.

Sanberg, P.R., Garbuzova-Davis, S., Tan, J., Eve, D.J., Borlongan, C.V., Cruz, L.E., Kuzmin-Nichols, N., Sanberg, C.D., 2012b. Human umbilical cord blood cells for neurodegenerative diseases: importance of lower dose multiple injections. Cell Transplant. 21, 788.

Sgubin, D., Aztiria, E., Perin, A., Longatti, P., Leanza, G., 2007. Activation of endogenous neural stem cells in the adult human brain following subarachnoid hemorrhage. J. Neurosci. Res. 85, 1647–1655.

Shi, W., Nie, D., Jin, G., Chen, W., Xia, L., Wu, X., Su, X., Xu, X., Ni, L., Zhang, X., Chen, J., 2012. BDNF blended chitosan scaffolds for human umbilical cord MSC transplants in traumatic brain injury therapy. Biomaterials 33, 3119–3126.

Stem Cell Therapies as an Emerging Paradigm in Stroke Participants, 2009. Stem Cell Therapies as an Emerging Paradigm in Stroke (STEPS): bridging basic and clinical science for cellular and neurogenic factor therapy in treating stroke. Stroke 40, 510–515.

Strioga, M.M., Viswanathan, S., Darinskas, A., Slaby, O., Michalek, J., 2012. Same or not the same? Comparison of adipose tissue-derived versus bone marrow-derived mesenchymal stem and stromal cells. Stem Cells Dev. 21, 2724–2752.

Subramanian, A., Shu-Uin, G., Kae-Siang, N., Gauthaman, K., Biswas, A., Choolani, M., Bongso, A., Chui-Yee, F., 2012. Human umbilical cord Wharton's jelly mesenchymal stem cells do not transform to tumor-associated fibroblasts in the presence of breast and ovarian cancer cells unlike bone marrow mesenchymal stem cells. J. Cell. Biochem. 113, 1886–1895.

Tat, P.A., Sumer, H., Jones, K.L., Upton, K., Verma, P.J., 2010. The efficient generation of induced pluripotent stem (iPS) cells from adult mouse adipose tissue-derived and neural stem cells. Cell Transplant. 19, 525–536.

Taupin, P., 2006. Stroke-induced neurogenesis: physiopathology and mechanisms. Curr. Neurovasc. Res. 3, 67–72.

Thored, P., Wood, J., Arvidsson, A., Cammenga, J., Kokaia, Z., Lindvall, O., 2007. Long-term neuroblast migration along blood vessels in an area with transient angiogenesis and increased vascularization after stroke. Stroke 38, 3032–3039.

Tolar, J., Nauta, A.J., Osborn, M.J., Panoskaltsis Mortari, A., McElmurry, R.T., Bell, S., Xia, L., Zhou, N., Riddle, M., Schroeder, T.M., Westendorf, J.J., McIvor, R.S., Hogendoorn, P.C., Szuhai, K., Oseth, L., Hirsch, B., Yant, S.R., Kay, M.A., Peister, A., Prockop, D.J., Fibbe, W.E., Blazar, B.R., 2007. Sarcoma derived from cultured mesenchymal stem cells. Stem Cells 25, 371–379.

Tonchev, A.B., Yamashima, T., Sawamoto, K., Okano, H., 2005. Enhanced proliferation of progenitor cells in the subventricular zone and limited neuronal production in the striatum and neocortex of adult macaque monkeys after global cerebral ischemia. J. Neurosci. Res. 81, 776–788.

van Velthoven, C.T., Kavelaars, A., van Bel, F., Heijnen, C.J., 2010. Mesenchymal stem cell treatment after neonatal hypoxic-ischemic brain injury improves behavioral outcome and induces neuronal and oligodendrocyte regeneration. Brain Behav. Immun. 24, 387–393.

Vendrame, M., Cassady, J., Newcomb, J., Butler, T., Pennypacker, K.R., Zigova, T., Sanberg, C.D., Sanberg, P.R., Willing, A.E., 2004. Infusion of human umbilical cord blood cells in a rat model of stroke dose-dependently rescues behavioral deficits and reduces infarct volume. Stroke 35, 2390–2395.

Vendrame, M., Gemma, C., de Mesquita, D., Collier, L., Bickford, P.C., Sanberg, C.D., Sanberg, P.R., Pennypacker, K.R., Willing, A.E., 2005. Anti-inflammatory effects of human cord blood cells in a rat model of stroke. Stem Cells Dev. 14, 595–604.

Vidal, M.A., Walker, N.J., Napoli, E., Borjesson, D.L., 2012. Evaluation of senescence in mesenchymal stem cells isolated from equine bone marrow, adipose tissue, and umbilical cord tissue. Stem Cells Dev. 21, 273–283.

Vieira, N.M., Brandalise, V., Zucconi, E., Secco, M., Strauss, B.E., Zatz, M., 2010. Isolation, characterization, and differentiation potential of canine adipose-derived stem cells. Cell Transplant. 19, 279–289.

Walczak, P., Chen, N., Eve, D., Hudson, J., Zigova, T., Sanchez-Ramos, J., Sanberg, P.R., Sanberg, C.D., Willing, A.E., 2007. Long-term cultured human umbilical cord neural-like cells transplanted into the striatum of NOD SCID mice. Brain Res. Bull. 74, 155–163.

White, R.B., 2011. Looks like a duck, quacks like a duck ... still behaves like a bone marrow stromal cell. Transl. Res. 157, 53–55.

Willing, A.E., Eve, D.J., Sanberg, P.R., 2007. Umbilical cord blood transfusions for prevention of progressive brain injury and induction of neural recovery: an immunological perspective. Regen. Med. 2, 457–464.

Willing, A.E., Lixian, J., Milliken, M., Poulos, S., Zigova, T., Song, S., Hart, C., Sanchez-Ramos, J., Sanberg, P.R., 2003. Intravenous versus intrastriatal cord blood administration in a rodent model of stroke. J. Neurosci. Res. 73, 296–307.

Womble, T.A., Green, S.M., Nelson, A.P., Shahaduzzaman, M.D., Golden, J.E., Sanberg, P.R., Pennypacker, K.R., Willing, A.E., 2009. CD14+ and CD133+ human umbilical cord blood cells are essential for neurological recovery following MCAO. Cell Transplant. 18, 240.

Wu, W., He, Q., Li, X., Zhang, X., Lu, A., Ge, R., Zhen, H., Chang, A.E., Li, Q., Shen, L., 2011. Long-term cultured human neural stem cells undergo spontaneous transformation to tumor-initiating cells. Int. J. Biol. Sci. 7, 892–901.

Yang, K.L., Chen, M.F., Liao, C.H., Pang, C.Y., Lin, P.Y., 2009. A simple and efficient method for generating Nurr1-positive neuronal stem cells from human wisdom teeth (tNSC) and the potential of tNSC for stroke therapy. Cytotherapy 11, 606–617.

Yarygin, K.N., Kholodenko, I.V., Konieva, A.A., Burunova, V.V., Tairova, R.T., Gubsky, L.V., Cheglakov, I.B., Pirogov, Y.A., Yarygin, V.N., Skvortsova, V.I., 2009. Mechanisms of positive effects of transplantation of human placental mesenchymal stem cells on recovery of rats after experimental ischemic stroke. Bull. Exp. Biol. Med. 148, 862–868.

Yu, J.M., Jun, E.S., Bae, Y.C., Jung, J.S., 2008. Mesenchymal stem cells derived from human adipose tissues favor tumor cell growth in vivo. Stem Cells Dev. 17, 463–473.

Yu, S.J., Soncini, M., Kaneko, Y., Hess, D.C., Parolini, O., Borlongan, C.V., 2009. Amnion: a potent graft source for cell therapy in stroke. Cell Transplant. 18, 111–118.

Yust-Katz, S., Fisher-Shoval, Y., Barhum, Y., Ben-Zur, T., Barzilay, R., Lev, N., Hod, M., Melamed, E., Offen, D., 2012. Placental mesenchymal stromal cells induced into neurotrophic factor-producing cells protect neuronal cells from hypoxia and oxidative stress. Cytotherapy 14, 45–55.

Zhang, R.L., Chopp, M., Gregg, S.R., Toh, Y., Roberts, C., Letourneau, Y., Buller, B., Jia, L., Nejed Davarani, S.P., Zhang, Z.G., 2009. Patterns and dynamics of subventricular zone neuroblast migration in the ischemic striatum of the adult mouse. J. Cereb. Blood Flow Metab. 29, 1240–1250.

Zhang, W., Yan, Q., Zeng, Y.S., Zhang, X.B., Xiong, Y., Wang, J.M., Chen, S.J., Li, Y., Bruce, I.C., Wu, W., 2010a. Implantation of adult bone marrow-derived mesenchymal stem cells transfected with the neurotrophin-3 gene and pretreated with retinoic acid in completely transected spinal cord. Brain Res. 1359, 256–271.

Zhang, Y., Bellows, C.F., Kolonin, M.G., 2010b. Adipose tissue-derived progenitor cells and cancer. World J. Stem Cells 2, 103–113.

Zhao, T., Zhang, Z.N., Rong, Z., Xu, Y., 2011. Immunogenicity of induced pluripotent stem cells. Nature 474, 212–215.

CHAPTER

Prospects for stem cell-derived therapy in stroke

8

John D. Sinden*,[1], Indira Vishnubhatla*, Keith W. Muir[†]

[*]ReNeuron Limited, Surrey Research Park, Guildford, Surrey, UK

[†]Institute of Neurosciences and Psychology, University of Glasgow, Southern General Hospital, Glasgow, Scotland, UK

[1]Corresponding author. Tel.: +44-1483-269119, Fax: +44-1483-534864, e-mail address: john-sinden@reneuron.com

Abstract

The prospects for stem cell-derived therapy in stroke look promising, with a myriad of cell therapy products developed from brain, blood, bone marrow, and adipose tissue in early clinical development. Eight clinical trials have now reported final results, and several are currently registered recruiting patients or pending to start. Products passing the safety hurdle are recruiting patients for large efficacy studies. Besides identifying the most appropriate cell type, other issues to resolve include optimal timing for intervention, optimal delivery route, cell dose, patient selection, relevant clinical endpoints, and monitoring for effectiveness, to advance cell therapy through the hurdles of clinical research. In this chapter, we present the products and strategies used in the current cell therapy trials in ischemic stroke, provide an update on relevant preclinical research, and discuss the vital developments still needed to advance their clinical application as a future therapeutic option.

Keywords

stem cells, cell therapy, ischemic stroke, functional recovery, clinical trials

Abbreviations

4-OHT	4-hydoxytamoxifen
ADSCs	adipose tissue-derived stromal cells
AEs	adverse events
ALDH	aldehyde dehydrogenase
AMI	acute myocardial infarction
BBB	blood–brain barrier

Progress in Brain Research, Volume 201, ISSN 0079-6123, http://dx.doi.org/10.1016/B978-0-444-59544-7.00007-X

BI	Barthel Index
BOLD	blood oxygen level dependent
CCA	common carotid artery
CNS	central nervous system
EBST	elevated body swing test
EPO	erythropoietin
ESS	European Stroke Scale
G-CSF	granulocyte-colony-stimulating factor
GMP	good manufacturing practice
hNT	human teratocarcinoma (cell line)
HSC	hematopoietic stem cells
IA	intra-arterial
IC	intracerebral
ICH	intracerebral hemorrhage
IL	interleukin
IV	intravenous
MAPCs	multipotent adult progenitor cells
MCA	middle cerebral artery
MCAo	middle cerebral artery occlusion
MI	myocardial infarction
MMSE	Mini-Mental State Examination
MNCs	mononuclear cells (from bone marrow)
mNSS	modified Neurological Severity Scores
MRI	magnetic resonance imaging
mRS	modified Rankin Scale
MSCs	mesenchymal/marrow stromal cells
NICD	Notch-1 intracellular domain
NIHSS	National Institutes of Health Stroke Scale
PET	positron emission tomography
SAEs	serious adverse events
SDF-1α	stromal-derived factor-1α
SPECT	single-photon emission computed tomography
SVZ	subventricular zone
TMS	transcranial magnetic stimulation
t-PA	serine protease tissue-type plasminogen activator

1 INTRODUCTION

For many decades, the complex pathology of ischemic stroke has posed an insurmountable clinical challenge. Approximately half of all stroke survivors sustain permanent neurological deficits, resulting in many cases in the inability to sustain an independent quality of everyday life. The need to develop therapies to reduce neurological deficits or to improve diminished function remains a major scientific and clinical imperative.

Ischemic stroke starts with the interruption of arterial blood flow to a focal area of the brain, causing neuronal death in the affected core region due to loss of oxygen and glucose. The necrotic core is surrounded briefly by an ischemic penumbra consisting of neurons that are functionally impaired but not yet necrotic. The ischemic penumbra is the primary target for early therapeutic intervention, since prompt restoration of blood flow can salvage a large part of this tissue, and neuroprotectant strategies (at least in experimental models) may also prevent penumbral tissue loss. A framework derived from experimental work suggests three broad time windows for different therapeutic approaches (Fig. 1; Dirnagl et al., 1999; Zaleska et al., 2009). The first, and to date only, window to be targeted successfully, covers the minutes to hours following the ischemic insult. Breaking up or lysing the occluding thrombus can restore cerebral blood flow and limit further damage. The second opportunity targets the hours to days after ischemia, when secondary injury may result from the biochemical cascade triggered by ischemia, which leads to neuronal death in the penumbra, predominantly within a few hours. However, secondary processes that may exacerbate injury and extend damage beyond the conventionally defined penumbra include tissue inflammation, brain swelling, oxidative stress, and apoptosis. Clinically useful therapies that target these processes are lacking. The third opportunity targets the chronic phase, aiming either to repair or to restore the core ischemic zone or to promote brain remodeling whereby different substrates and circuits can adapt to restore function. To date, the only approved therapeutic stroke product is a serine protease tissue-type plasminogen activator (t-PA), which is effective when given within 4.5 h after stroke (Hacke et al., 2008). However, beyond the thrombolytic window, decades of research have shown that there are no compounds in the existing chemical pharmacopeia able to prevent or repair the neurological damage. Population benefits from t-PA have been limited by its

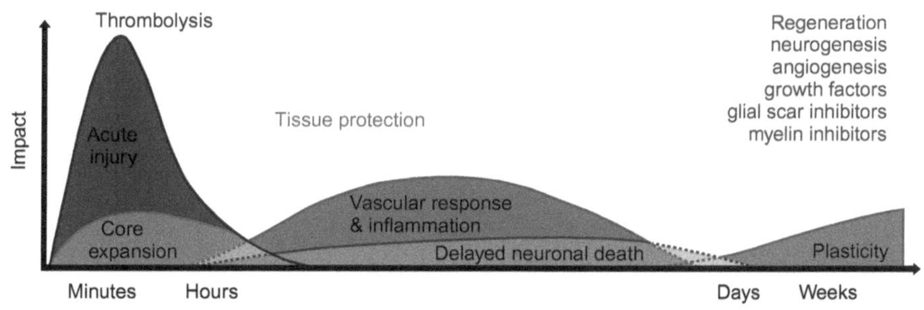

FIGURE 1

Potential windows for cell therapy interventions in stroke (adapted from Dirnagl et al., 1999; Zaleska et al., 2009). (For color version of this figure, the reader is referred to the online version of this chapter.)

Reproduced with permission from Elsevier Limited, 2012.

accessibility and short therapeutic window, with fewer than 4% of ischemic stroke patients in the United Kingdom and about 2% in the United States reported to receive treatment (Kleindorfer et al., 2008; Macrae, 2011). Pharmacological attempts at neuroprotection, even when deployed within 3–6 h, were ineffective. Reasons include administration of many drugs well beyond the time windows that had been tested in the laboratory, narrow therapeutic dose ranges that prevented doses that were effective in preclinical studies being attained in clinical studies because of adverse effects, model-specific effects (e.g., many drugs are effective in transient middle cerebral artery occlusion, MCAo, models but are ineffective in permanent occlusion models), and trials designed with overoptimistic effect sizes (Muir and Teal, 2005). Furthermore, without sufficient tissue perfusion—the blood–brain barrier (BBB) is intact up to 3 h after onset of ischemia (Hatashita and Hoff, 1990) - many neuroprotectant drugs cannot reach the damaged tissue.

Stem cells have recently taken center stage in the search for new therapies to provide benefit beyond thrombolysis, with different stem cell types showing potential in acute, subacute, and even chronic phases. A variety of stem cells can be identified that differ in their potency or the diversity of cell types they can generate. Pluripotent stem cells, such as embryonic stem cells or induced pluripotent stem cells, have the potential to differentiate into all cell lineages within the developing embryo. Multipotent stem cells, derived from human tissues (adult or fetal), have more limited potency and typically differentiate only into mature cell types from the tissues of origin. For clinical application, pluripotent cells must be at least partially differentiated prior to delivery, because of their inherent potent teratoma forming ability. In contrast, most clinical applications of multipotent stem cells are of undifferentiated cells as they do not generally form tumors after implantation. It is likely that many therapeutic properties of multipotent stem cells arise from the properties of the stem cell itself rather than its differentiated progeny. Stem cells have a natural tropism to areas of inflammation and injury. Systemically administered cells have been found to migrate to a damaged region and survive from 1 week to several months (Iihoshi et al., 2004; Li et al., 2001a,b; Shen et al., 2007). Migrated or transplanted cells, unlike pharmacological treatments, have the capacity to interact with local cells depending on their microenvironment within the injured brain, secreting cytokines, and other trophic factors to modulate the immune response, inhibit apoptosis, and improve or regenerate the vasculature (Ourednik et al., 2002; Shen et al., 2007; Smith et al., 2012; Stroemer et al., 2009).

Further, stem cells are produced endogenously in response to damage and recruitment may persist for at least several months after ischemic injury (Kokaia et al., 2006; Leker et al., 2007; Thored et al., 2006). The capacity for self-repair, where differentiation of cells into neurons phenotypically similar to those lost in the ischemic lesion, has been reported in animal models (Arvidsson et al., 2002; Chu et al., 2012). There is some evidence that stroke-induced neurogenesis occurs in humans, including the elderly (Jin et al., 2006; Minger et al., 2007). Cells expressing immature neuronal markers such as doublecortin and βIII-tubulin as well as ki67, a marker of proliferation, were observed near blood vessels in the penumbra of cortical infarcts (Jin et al., 2006). In contrast, neurogenesis does not occur in the normal human cortex in the absence of injury (Bhardwaj et al., 2006). Spontaneous neurogenesis in response to brain injury is inadequate to allow full functional recovery, and strategies are being sought to amplify this endogenous regenerative response (Androutsellis-Theotokis et al., 2006). Exogenous cell therapy may act synergistically or by enhancing the endogenous neurogenic response to promote brain repair (Burns and Steinberg, 2011; Lindvall and Kokaia, 2011).

The multifunctional capacity of stem cells (Gray et al., 2000) has numerous positive therapeutic consequences and because of their inherent complexity, cell therapies offer the opportunity to intervene at several points in the pathological process. Proposed mechanisms for exogenous cell therapy involve neuroprotection, immunomodulation, stimulation of angiogenesis (vascular repair), neurogenesis, synaptogenesis, and recruitment of endogenous neural progenitors. These mechanisms have been extensively reviewed recently (Burns and Steinberg, 2011; Lindvall and Kokaia, 2011; Miljan and Sinden, 2009; Zhang and Chopp, 2009).

The optimal timing for administering cell therapies after a stroke is not known. The brain environment not only changes dramatically over time after ischemia but also differs in different anatomical regions at any given time point. The cascade of injury peaks within hours of the initial insult and for many processes is complete within 24–48 h (Hess and Borlongan, 2008). At the least, the volume of potentially salvageable tissue declines rapidly over a period of hours, and therapeutic gain from neuroprotectant strategies is likely to be marginal unless therapy can be started within 3–6 h of injury.

Later treatment within the subacute period post-infarct targets restorative processes. These include angiogenesis, neurogenesis, and synaptogenesis rather than prevention of initial tissue injury. The optimal timing for this intervention is not known but may be limited to 1 week or 1 month (Hess and Borlongan, 2008). Neurorestoration can start later than neuroprotective treatment and therefore could potentially benefit more stroke patients (Zhang and Chopp, 2009). However, the volume of tissue amenable to this intervention and the magnitude of therapeutic gain remain unknown clinically. Therapies aiming to treat the later chronic period would potentially extend the benefits to a larger group of patients.

2 EARLY CLINICAL EXPERIENCE

Eight cell therapy clinical trials in stroke patients have been conducted as pioneering studies addressing the feasibility of a range of therapeutic cell candidates ranging from primary neural cells to bone marrow-derived cells and manufactured teratoma-derived neural-like cells. These trials are summarized in Table 1. A representative selection of published preclinical studies which support current clinical trials in stroke is presented in Table 2.

Table 1 Completed cell therapy clinical trials

Reference	Phase	Cells	Route	Number enrolled	Location of infarct/ timing of treatment	Key outcome measures	Preclinical studies published (Y/N reference)	Results
Kondziolka et al. (2000)	I	hNT	IC	12	Basal ganglia (not excluding cortical involvement)/6 months to 6 years	Safety, ESS, NIHSS, serologic and imaging tests	Y (Borlongan et al., 1998)	Safe ESS improved at 6 months ($p = -0.046$)
Kondziolka et al. (2005)	II	hNT	IC	14 treated + 4 controls	1–5 years	ESS, Fugl-Meyer	Y (Borlongan et al., 1998)	Primary outcome measure (ESS) not met
Savitz et al. (2005)	II	Fetal porcine	IC	5	1.5–10 years	Safety, NIHSS, BI, mRS, MRI	Y (Dinsmore et al., 2002)	FDA terminated due to possible side effects
Suarez-Monteagudo et al. (2009)	I	Autologous MNCs	IC	5	1–10 years	Safety, tolerance	N	Safe
Barbosa da Fonseca et al. (2010)	I	Autologous MNCs	IA	6	8–12 weeks	Neurologic deficits	N	Safe No neurologic Worsening
Lee et al. (2010)/Bang et al. (2005)	I/II	Autologous MSCs	IV	16 treated + 36 controls	5–7 weeks	Safety, mRS	Y (Shen et al., 2006)	Safe mRS improved ($p = 0.046$) Best, if SVZ is not affected by infarct Enrolment stopped due to concerns regarding culture media
Honmou et al. (2011)	I	Autologous MSCs	IV	12	Lesions affecting gray and white matter/ 36–133 days	NIHSS, lesion volume by MRI	Y (Chen et al., 2001a; Li et al., 2002)	Safe Human serum used for culturing cells Mean lesion volumes (MRI) were reduced by 20% at 1-week posttreatment Changes in NIHSS assessed case by case
Bhasin et al. (2011)	I/II	Autologous MSCs (+8 weeks physiotherapy)	IV	6 treated + 6 controls	Varied severity and lesion size/3–12 months	Fugl-Meyer, Modified BI, MRC Ashworth fMRI	Y (reference not known) Dose used based on two clinical trials (NCT00473057; NCT00395200)	24 weeks follow-up: No mortality or cell-related adverse reactions Modest increases in FM and modified BI scores fMRI evidence of neural plasticity

BI, Barthel Index; ESS, European Stroke Score; hNT, human teratocarcinoma-derived neural cell line; IC, intracerebral; IV, intravenous; IA, intra-arterial; MCAo, middle cerebral artery occlusion; MNCs, bone marrow mononuclear cells; MSCs, mesenchymal or stromal cells; MRC, Medical Research Council; mRS, modified Rankin Scale; NIHSS, National Institutes of Health Stroke Scale. (modified from Burns and Steinberg, 2011)

Table 2 Summary table of behavioral test results reported for current clinical stem cell therapy products

Cell type/ products	Animal model	Source, autologous, or donor if relevant (dose of cells)/route of administration/timing of intervention/ other details as appropriate	Time of final assessment after stroke	Behavioral outcome(s) (as compared to vehicle or no treatment) [a]: (+) significant effect ($p \leq 0.05$) or (−) no effect or marginal	References
Bone marrow mononuclear cells (MNCs)	Rat tMCAo	Autologous (1×10^7) collected 1.5 h before occlusion/IV/12 h after occlusion	31–32 days	(+) Morris water maze, (+) treadmill	Iihoshi et al. (2004)
	Rat tMCAo	Autologous (2×10^7) collected 72 h before occlusion/IA/6 h after occlusion; alone or in combination with tissue inhibitor of matrix metalloproteinase 1 and 2 and/or L-arginine and vitamin E	1 day	(+) Rotarod coordination in all treated groups	Baker et al. (2007)
	Rat tCCAo/ MCAo	Autologous (1×10^7) collected 1 h before occlusion/IA or IV/90 min after occlusion	7 days	(+) Rotarod IA group, (−) rotarod IV group	Kamiya et al. (2008)
	Rat tMCAo	Rat donor (3×10^7)/IV/1 day after occlusion	28 days	(−) Adhesion tape test, (+) cylinder test, (+) beam walking	Giraldi-Guimaraes et al. (2009)
	Rat tCCAo/ MCAo	Autologous (4×10^6 or 10×10^6) collected 22 h after occlusion/IA/24 h after occlusion	30 days	(+) Cylinder test, (+) corners test	Brenneman et al. (2010)
	Rat pMCAo	Rat donor (3×10^7)/IV/1, 7, 14, 30 days after occlusion	77 days	(−) Adhesive tape test, (+) cylinder test only for 1- and 7-day treatment	DeVasconcelos Dos Santos et al. (2010)
	Rat pMCAo	Rat donor (3×10^6)/IV/1 or 30 days after occlusion	77 days	(−) Adhesive tape test, (+) cylinder test only for treatment given 1 day after occlusion	DeVasconcelos Dos Santos et al. (2010)
	Rat tCCAo/ MCAo	Autologous (1×10^6, 10×10^6, or 30×10^6) collected 24 h after occlusion/ IV/3 or 7 days after occlusion	28 days	(+) Cylinder test*, (+) corner test*, *intervention given within 3 days post-occlusion, and best were 10×10^6 and 30×10^6 cells doses	Yang et al. (2011a)

Continued

Table 2 Summary table of behavioral test results reported for current clinical stem cell therapy products—cont'd

Cell type/ products	Animal model	Source, autologous, or donor if relevant (dose of cells)/route of administration/timing of intervention/ other details as appropriate	Time of final assessment after stroke	Behavioral outcome(s) (as compared to vehicle or no treatment) [a]: (+) significant effect ($p \leq 0.05$) or (−) no effect or marginal	References
Bone marrow-derived mesenchymal or stromal cells (MSCs)	Rat tMCAo	Rat donor (1×10^6)/IV/1 day after occlusion	35 days	(+) Adhesive tape test, (+) NSS, (−) rotarod	Chen et al. (2001a)
	Rat tMCAo	Rat donor (3×10^6)/IV/1 day after occlusion	35 days	(+) Adhesive tape test, (+) NSS, (−) rotarod	Chen et al. (2001a)
	Rat tMCAo	Rat donor (3×10^6)/IV/7 days after occlusion	14 days	(+) Adhesive tape test, (−) NSS, (+) rotarod	Chen et al. (2001a)
	Rat tMCAo	Rat donor (2×10^6)/IA/1 day after occlusion	14 days	(+) mNSS, (+) adhesive tape test	Li et al. (2001b)
	Rat right pMCAo	Rat donor (3×10^6)/IV/1 day after occlusion	28 days	(+) Adhesive tape test	Liu et al. (2007)
	Rat tMCAo	Rat donor (3×10^6)/IV/28 days after occlusion	12 weeks and 12 months	(+) mNSS, (+) adhesive tape test	Shen et al. (2006) and Shen et al. (2007)
	Rat tMCAo	Rat donor (5×10^5)/IV/1 day or 7 days after occlusion	28 days	(+) mNSS, administration 1 day after occlusion showed greater effect on functional recovery than administration 7 days after occlusion	Yang et al. (2010)
	Rat tMCAo	Human donor (5×10^5)/IV/1 day or 7 days after occlusion. Immunosuppression	28 days	(+) mNSS, administration 1 day after occlusion showed greater effect on functional recovery than administration 7 days after occlusion	Yang et al. (2010)
Selected MSC cell therapy products/	Rat hypoxia	Rat MAPCs (2.0×10^5)/IC or IV/7 or 14 days postinjury	21–28 days	(+) EBST, (+) rotarod treadmill test	Yasuhara et al. (2008)

MultiStem® (human multipotent adult progenitor cells, MAPCs)	injured neonatal				
	Rat MCAo	Rat MAPCs (4×10^5)/IC/7 days after occlusion. With and without immunosuppression	9 weeks	(+) EBST, (+) Bederson test	Mays et al. (2010)
	Rat MCAo	Human MAPCs (4×10^5)/IC/at 7 days after occlusion. With and without immunosuppression	9 weeks	(+) EBST, (+) Bederson test	Mays et al. (2010)
		Human MAPCs (4×10^5 or 1×10^6)/IV/1, 2, or 7 days after occlusion. With and without immunosuppression	9 weeks	(+,−) EBST (*1×10^6 cells only), (+) Bederson test	Mays et al. (2010)
	Rat MCAo	Human MAPCs (4×10^5, 1×10^6, 2×10^6, 2×10^6, 4×10^6, 10×10^6, or 20×10^6)/IC/2 days after occlusion. No immunosuppression	12 weeks	(+) EBST, (+) Bederson test	Mays et al. (2010)
Human Notch-1-activated MSCs (SB623)/(San Bio)	Rat tMCAo	Rat Notch-1-activated MSCs (4×10^4, 1.0×10^5, or 2.0×10^5)/IC/1 month after occlusion. Immunosuppression for the duration of study	13 weeks	(+) EBST, (+) Bederson test	Yasuhara et al. (2009)
	Rat tMCAo	Human Notch-1-activated MSCs (9×10^4 or 1.8×10^5)/IC/1 month after occlusion. Immunosuppression for the duration of study	8 weeks	(+) EBST, (+) Bederson test	Yasuhara et al. (2009)
Human adipose tissue-derived stromal cells (hADSC)	Rat tMCAo	Human ADSCs (1×10^6)/IC/1 day after occlusion. With and without brain-derived neurotrophic factor (BDNF)	14 days	(+) mNSS, (+) adhesive tape test. IC of human ADSCs and BDNF showed significantly higher recovery than human ADSCs alone	Kang et al. (2003)

Continued

Table 2 Summary table of behavioral test results reported for current clinical stem cell therapy products—cont'd

Cell type/ products	Animal model	Source, autologous, or donor if relevant (dose of cells)/route of administration/timing of intervention/ other details as appropriate	Time of final assessment after stroke	Behavioral outcome(s) (as compared to vehicle or no treatment) [a]: (+) significant effect ($p \leq 0.05$) or (−) no effect or marginal	References
	Rat tMCAo	Human ADSCs (1×10^5)/IC/1 day after occlusion	14 days	(+) mNSS, (+) rotarod test	Lee and Yoon (2008)
Human neural stem cells/ CTX0E03 DP (ReNeuron)	Rat tMCAO	Human CTX0E03 (8.0×10^5)/IC implanted both ipsilaterally and contralaterally into striatum/3–4 weeks after occlusion. Immunosuppression initiated day before implantation to end of study	4 months	(+) Bilateral asymmetry test, (+) rotarod	Pollock et al. (2006)
	Rat tMCAo	Human CTX0E03 (4.5×10^3, 4.5×10^4, 4.5×10^5)/IC/4 weeks after occlusion. Immunosuppression for 2 weeks after implantation	4 months	(+) Bilateral asymmetry test*, (+) rotameter*, (−) Whiskers, *significant in the high dose groups	Stroemer et al. (2009)
	Rat tMCAo	Human CTX0E03 (4.5×10^5 either intracerebroventricular or intraparenchymal)/IC/14 days after occlusion. Immunosuppression for 2 weeks post-transplantation	4 months	(+) Bilateral asymmetry test, (+) foot-fault test, (+) rotameter, (−) water maze	Smith et al. (2012)

CCAo, common carotid artery occlusion; EBST, elevated body swing test; IA, intra-arterial; IC, intracerebral; IV, intravenous; MCAo, middle cerebral artery occlusion; tCCAo/MCAo, transient tandem CCAo and MCAo; tMCAo, transient MCAo; pMCAo, permanent MCAo; NSS, Neurological Severity Score; mNSS, modified NSS.

[a]Preclinical test (neurological relevance): adhesive tape test/removal (fore limb asymmetry); beam walking (hind limb functions); Bederson test (neurologic deficit score consisting of spontaneous ipsilateral circling; contralateral hind limb retraction; beam walking ability; bilateral forepaw grasp); bilateral asymmetry test (tactile extinction probing sensory neglect); corners (asymmetry); cylinder (asymmetry in the use of fore limbs for postural support); EBST (asymmetric motor behavior); foot-fault test (ability to integrate motor responses); NSS, mNSS (modified Neurological Severity Scores, grading of neurologic function based on a composite score of motor and sensory functions, balance, and reflex tests); rotameter (cholinergic agonist-induced rotation used an index of striatal damage); rotarod (motor function stepping test); stepping test (learning behavior retention time); treadmill stress test (motor functions and gait); (Morris) water maze (cognitive functions); Whiskers (striatal dysfunction and loss of reflexive lifting).

3 TRIALS USING NEURAL CELLS

The use of primary human neural cells from donated fetal brain tissue is an experimental approach that becomes impractical as a therapeutic strategy. Procurement of tissues is limited by supply, and quality is variable. To provide renewable sources for these early neural cell therapy trials, two approaches were taken: in one, a manufactured human teratocarcinoma (hNT) cell line was developed (Kondziolka et al., 2000). In the second, commercially scaled and quality-assured porcine primary fetal neural cells were isolated. The trials investigated the feasibility and safety of the cells for neurotransplantation, using stereotaxic methods to deliver the cells intraparenchymally. Graft rejection was a major consideration in both studies. In the hNT trial, the tumor origin of the transplanted cells was a concern, although preclinical studies in mice showed no toxicity or tumorigenicity (Kleppner et al., 1995).

3.1 Human teratocarcinoma cells

The hNT cell line generates pure populations of postmitotic neural-like cells when exposed to retinoic acid. Rat NT cells transplanted into the rat brain produced neuron-like cells, survived, matured, and integrated into the host brain for 1-year post-transplantation without reverting to a neoplastic state (Kleppner et al., 1995; Trojanowski et al., 1993). Further, preclinical studies demonstrated functional improvements in gross motor function on the elevated body swing test (EBST) and in the learning and memory tasks on the passive avoidance test following implantation of hNT cells in the rat transient MCAo model. These studies formed the basis for an approved investigational new drug application in chronic stroke patients (Borlongan et al., 1998). A single dose of 2 or 6 million cells was injected stereotaxically by the intracerebral (IC) route into patients with stable neurological deficits after a basal ganglia infarct 6 months to 6 years previously (Table 1). Patients received immunosuppression for 2 months following implantation (Kondziolka et al., 2000). Twelve patients (males and females aged 45–75 years) were treated. Overall, no adverse effects were noted, and surviving cells in one case were observed at *postmortem* with no evidence of tumor at 27 months (Nelson et al., 2002). Improvements in the European Stroke Scale (ESS) motor score at 6 months were reported.

A Phase II randomized controlled trial followed, which recruited across two sites 14 patients and four controls aged 40–70 years who had experienced a basal ganglia infarct 1–5 years previously and had a motor deficit that was stable for at least 2 months (Kondziolka et al., 2005; Table 1). Five or ten million hNT were implanted IC and patients were followed for 6 months. Immunosuppression was used as in the Phase I study. The study reported functional improvements in some patients; but due to the very small group sizes, improvements based on a primary outcome measure of ESS at 6 months did not reach statistical significance (Kondziolka et al., 2005).

Adverse events (AEs) reported included one seizure and one subdural hematoma requiring evacuation (Kondziolka et al., 2005).

3.2 Fetal porcine cells

Transplantation of fetal cells from the pig primordial striatum was first shown to promote graft integration and to improve deficits in an animal model of Huntington's disease (Deacon et al., 1994; Isacson et al., 1995). Studies in rats showed that fetal porcine cells transplanted to the ischemic striatum 3–28 days after MCAo survived up to 7 months. Solid grafts formed within infarct cavities, which were found to have differentiated into glia and neurons, showing evidence of elaborated processes to host brain and for synaptogenesis. Animals transplanted 14 days after stroke showed significant functional improvement compared with controls (Dinsmore et al., 2002).

The open-label clinical trial used stereotaxic delivery of up to 50 million fetal porcine cells to treat patients with stable neurological deficits after a basal ganglia stroke 1.5–10 years previously (Savitz et al., 2005; Table 1). To prevent rejection, the porcine cells were pretreated with anti-major histocompatibility complex 1 antibody to avoid the use of immunosuppressive therapy. Only 5 of the 12 planned patients were treated. The FDA terminated the trial after two patients suffered serious adverse events (SAEs): one had worsening left-sided weakness and another had a seizure. The investigators attributed one of these events to a cortical vein thrombosis, which would more likely be a complication of the implantation procedure rather being related to the cells. Two of the five patients showed improvements in symptoms (speech and language and motor impairments) that persisted to 4 years (Savitz et al., 2005).

4 TRIALS USING BONE MARROW-DERIVED CELLS

Bone marrow serves as a useful and abundant source for stem cells. Bone marrow mononuclear cells (MNCs) may be rapidly isolated and permit autologous application, minimizing risk of immune rejection. The MNC fraction includes subpopulations of hematopoietic and endothelial progenitor (CD34+) cells as well as mesenchymal/marrow stromal cells (CD34−) (MSCs), all of which have been demonstrated to provide beneficial effects in animal studies. MSCs are multipotent and able to give rise to bone, cartilage, endothelial cells, neurons, and glia (Chen et al., 2008; Uccelli et al., 2008). Hematopoietic progenitor cells give rise to microglia which serve roles of immune surveillance in the brain, analogous to macrophages in other organs. These cells can rapidly respond to injury in the brain and can demonstrate both pro- and anti-inflammatory properties. Appropriate regulation of the inflammatory environment following stroke is important to minimize injury and promote recovery.

4.1 Supporting preclinical studies using MNCs

MNCs have been demonstrated to cross the BBB (Eglitis and Mezey, 1997) and are swiftly recruited to regions of ischemic injury (Shyu et al., 2006a). Administration of exogenous MNCs has been studied in animal models of stroke since 2000 (Chen et al., 2000). Current preclinical evidence indicates that the most likely mechanism of action for MNCs involves trophic, anti-inflammatory, and immunomodulatory actions that have not only acute but also persistent effects on the brain (Barnabe et al., 2009; Bliss et al., 2007; Hess and Borlongan, 2008; Mendez-Otero et al., 2007). Long-term cell survival and engraftment in the brain may not be necessary for a therapeutic effect (Bliss et al., 2007; Hess and Borlongan, 2008; Mendez-Otero et al., 2007). MNCs may not even need to enter the brain to cause an effect, but rather act in the periphery to increase trophic factor expression in the brain (Borlongan et al., 2004; Sarnowska et al., 2009).

Functional improvement and reduction in cell death in the ischemic regions have been reported after both IC and intravascular MNC delivery (Li and Chopp, 2009; Table 2). Multiple mechanisms are reported to participate in the neuroprotective effects of MNC treatment after MCAo in the rat, including induction of angiogenesis and cytokine secretion (Chen et al., 2001a,b; Iihoshi et al., 2004; Li et al., 2002). As Table 2 shows, positive effects have been reported when cell injection takes place from 12 h to 7 days after the infarct.

Both autologous and allogeneic MNCs have been shown to reduce lesion size and improve functional recovery in the MCAo rat model of cerebral ischemia (Table 2). Preclinical models included transient or permanent MCAo or tandem reversible occlusions of common carotid artery (CCA) and MCA. Each model differs in its effects on BBB permeability, the location, and type of damage and neurological deficits produced (Zea Longa et al., 1989).

The autologous MNC studies also differed in the timing of the bone marrow collection: in the first studies, bone marrow was collected 1–72 h before occlusion, processed, and reinjected 1.5–12 h after occlusion (Baker et al., 2007; Iihoshi et al., 2004; Kamiya et al., 2008). Later studies reflected the more valid clinical scenario where the bone marrow cells would be extracted after ischemia (Brenneman et al., 2010; Kamiya et al., 2008; Yang et al., 2011a).

The feasibility of acute autologous bone marrow harvesting and readministration within 1 day of occlusion was recently established in a preclinical study. Intra-arterial (IA; carotid) infusion of MNCs in young and middle-aged rats was reported to reduce infarct volume in treated rats compared to a vehicle-treated group (Brenneman et al., 2010). There were significant reductions in tumor necrosis factor-α, interleukin (IL)-1α, IL-1β, IL-6, and a significant increase in IL-10. Labeled MNCs were found in the peri-infarct area at 1 h after administration but rapidly decreased over hours and were nearly undetectable by 7 days post-infusion. Functional testing using the cylinder test and corners test showed significant improvement 1-week post-infarct in treated animals compared to untreated animals (Brenneman et al., 2010).

Acute intravenous (IV) administration of autologous MNCs, extracted from the bone marrow 24–72 h after occlusion, reduced neurological deficits in rats compared with saline-treated controls (Yang et al., 2010). The improvement in neurological outcome depended on the number of MNCs injected. At 4 weeks after stroke, animals that had received 10 million cells/kg or more showed significant reductions in neurological impairment; however, those receiving the highest dose (30 million cells/kg) or the lower dose (1 million cells/kg) did not show significant improvement compared with saline controls. The doses used in the Yang et al. study (2010) were chosen based on the expected cell yield from a limited bone marrow aspiration in patients.

4.2 Autologous MNCs clinical experience

Bone marrow-derived MNCs have already been used in human trials of end stage heart failure and acute myocardial infarction (AMI), with reports of improvement in cardiac function in end stage heart failure (Perin et al., 2003) and variable results in patients with AMI (Kamihata et al., 2001; Strauer et al., 2002).

Two small clinical studies using autologous MNCs in stroke patients have been reported (Table 1). In the Suarez-Monteagudo study (2009), autologous MNCs (extracted and isolated the day before the transplant) were stereotaxically implanted IC into five stroke patients (three ischemic and two hemorrhagic) who had suffered stroke 3–8 years previously. Patients were followed for 1 year. The study demonstrated the feasibility to extract and reimplant autologous MNCs within 24 h in chronic stroke patients. The total number of cells extracted and reimplanted varied from 14 to 55 million cells. No AEs were reported in the short-term follow-up period of 90 days; however, an asymptomatic epileptic-like activity ipsilateral to the lesion detected by electroencephalography was reported for all five patients in the longer 1-year follow-up (Suarez-Monteagudo et al., 2009).

In the second trial, six male patients (ages 24–65 years) with chronic cerebral infarcts affecting the MCA territory were included in a non-randomized, open-label Phase I clinical trial to investigate the feasibility of noninvasive monitoring of autologous MNCs labeled with 99mTc using single-photon emission computed tomography (SPECT; Barbosa da Fonseca et al., 2010). This study demonstrated that the presence of labeled MNCs in the infarct area in all patients by 2 h after IA injection. The technique allowed the monitoring of cells for at least 24 h. No AEs were reported after 4 months (Barbosa da Fonseca et al., 2010).

A third trial, involving IA infusion of MNCs in five patients with ischemic stroke 7–9 days after onset, has been presented but not yet published (Bannerjee et al., 2012). No safety issues were described in follow-up to 180 days.

5 TRIALS USING AUTOLOGOUS MESENCHYMAL OR MARROW STROMAL CELLS

Bone marrow stromal cells (MSCs) form part of the bone marrow mononuclear fraction and are a useful and abundant source of cells for cell therapy products. The main criteria for defining MSCs (according to the International Society for Cellular

Therapy) are adherence to plastic, specific surface antigen expression, and multipotent differentiation potential (Dominici et al., 2006). Human MSCs express CD105, CD73, and CD90 and lack expression of CD45, CD34, CD14, CD11b, CD79, CD19, or HLA-DR surface markers. They are able to differentiate into osteoblasts, adipocytes, and chondroblasts *in vitro* (Dominici et al., 2006) and have also been reported to develop neural phenotypes, including expression of glial and neuronal progenitor cell antigens under *in vitro* and *in vivo* conditions (Dezawa et al., 2004; Woodbury et al., 2000).

MSCs may provide benefit in the subacute/chronic window by providing micro-environment milieu alteration through growth factor production, cell-to-cell interaction, or inflammatory modulation (LeBlanc and Ringdén, 2007). They act as supportive cells, altering the fate of damaged or diseased cells. They improve cell survival or increase cellular proliferation via direct contact or by altering the local milieu through growth factor or chemokine secretion. Acute application in the autologous setting is not possible as MSCs require 3–4 weeks of growth in culture to achieve the dose needed (Bang et al., 2005; Bhasin et al., 2011; Honmou et al., 2011; Li et al., 2008).

5.1 Supporting preclinical studies

The Chopp group has shown in a number of studies that MSCs reduce neurological deficits in rodent models of stroke (Chen et al., 2001a,b). Rodents injected with MSCs show no evidence of AEs up to 1 year after injury (Shen et al., 2007). In this study, IA injection into rats 1 day after MCAo demonstrated functional improvements in the modified Neurological Severity Scores (mNSS) test, which is a composite of motor, sensory, balance, and reflex tests, and the adhesive-removal somatosensory test noted 14 days after therapy ($p < 0.01$). MSCs were reported to have survived and exerted effects on brain up to 1 year (Shen et al., 2006, 2007).

Many studies have shown beneficial effects of MSCs on functional outcomes in rodent models of stroke when delivered by IC, IV, and IA routes (Chen et al., 2001a,b; Li et al., 2001a,b; Liu et al., 2007; Yang et al., 2010; Table 2). The IV delivery route maintained efficacy even when treatment was delayed for 1 month in rodents (Shen et al., 2006, 2007). Trophic effects have been observed including angiogenesis, neurogenesis, and synaptogenesis with favorable brain remodeling (Zhang and Chopp, 2009).

5.2 Autologous MSC clinical experience

The safety and efficacy of IV MSC administration was investigated in patients who had significant persistent neurological deficits (defined as a National Institutes of Health Stroke Scale, NIHSS, score of at least 6 at day 7 post-infarct). Autologous MSCs were harvested within 7 days after stroke from the patient's bone marrow and cultured *ex vivo* for 30 days to a pure population of 1×10^8 cells, the quantity estimated to be required as the cell dose for IV administration. A total of 16 patients received the autologous MSC preparation in two equal doses given within 5 weeks of initial infarct and a further 2 weeks later (Lee et al., 2010).

The study procedure was reported to be safe and feasible at 1-year follow-up after treatment (Bang et al., 2005). Functional outcomes improved in the five MSC-treated patients as compared with 25 untreated control patients by one measure, the Barthel Index (BI), a scale used to measure performance of activities of daily living ($p=0.011$, 0.017, and 0.115 at 3, 6, and 12 months, respectively). There were no adverse cell-related, serological, or imaging-defined effects. The clinical significance of the reported improvements is difficult to interpret because of the small group sizes, lack of a placebo group, and lack of observer blinding. A long-term follow-up reported further on the safety and efficacy of the now enlarged group of 16 treated patients and the 36 control patients after 5 years following acute stroke MSC administration (Lee et al., 2010; Table 1). Functional outcome showed that, compared with the control group, the MSC-treated group had significantly decreased modified Rankin Scale (mRS), a measure of degree of dependence, at the end of study follow-up. Moreover, the improvement in mRS was found only when the subventricular zone (SVZ) was not damaged in patients. Furthermore, the study also indicated that clinical improvement was correlated to plasma stromal-derived factor-1α (SDF-1α) levels at the time of treatment. The authors suggested this finding may be an important consideration in future selection of patients for trials using MSCs. Patient recruitment was stopped because of concerns of risk of infections from the use of xenogeneic serum during *ex vivo* culturing and risk of immunologic rejection or autoimmune reactions from repeated administration of xenoprotein contaminants (Spees et al., 2004). There were no reports of serum-related adverse effects in the MSC-treated patients at the end of 5-year follow-up (Lee et al., 2010).

Subsequently, two Phase I trials have reported on the use of autologous MSCs cultured without the use of fetal bovine serum. In the Honmou study (2011), autologous human MSCs have been manufactured to good manufacturing practice (GMP) standards using autologous human serum. In this trial, 12 patients, aged from 41 to 73 years, who had lesions affecting gray matter, white matter, or both, and stable hemiparesis (some also with aphasia) received IV infusions of human MSCs between 36 and 136 days after their strokes. Patients were monitored for 1 year. There were no safety-related issues. Magnetic resonance imaging (MRI) and NIHSS results were presented as a series of case reports. Improvements in the lesion volume (reduction by 20%) and an increased daily rate of change in the NIHSS as compared to pretreatment rate were found to have occurred early in the first week post-infusion, but with no randomization or control group, and bias toward inclusion of patients at a time defined by attainment of an acceptable level of neurological function, conclusions about efficacy are impossible.

In the second Phase I trial, autologous MSCs were prepared using a serum-free culture (Bhasin et al., 2011). Twelve chronic stable patients were recruited within 3 months to 1 year of stroke and were matched for age, disease characteristics including stroke severity and neurological deficits. Six patients received treatment administered intravenously. Both control and treated groups received 8 weeks of physiotherapy. After 6 months follow-up, there were no reports of deaths or cell-related adverse reactions. Increases in the mean Fugl-Meyer and modified BI scores in the MSC-treated group, compared to the control group, showed progressive improvements, from baseline to end of physiotherapy (8 weeks) and from the 8 weeks to end of follow-up. Of particular interest is the evidence for early neuronal plasticity in the ipsilateral premotor cortex of treated patients from functional MRI, using blood oxygen level dependent (BOLD) and diffusion tensor imaging techniques, evaluating affected hand movement (Bhasin et al., 2011).

6 INCREASING TRANSLATIONAL SUCCESS FOR FUTURE CLINICAL APPLICATION

Collectively, the early clinical experience of cell therapy trials in stroke patients summarized above provides limited evidence for the safety and feasibility of the application of cell therapies in stroke. However, the small numbers, and highly selected nature of the patients included to date, mean that only major toxicity issues could have been identified, and no conclusions about potential effects on clinical change can be drawn given the limitations of the study designs. The trial experience also highlights difficulties that may impede the likelihood of translational success as clinical development proceeds.

6.1 Autologous therapies

The primary advantage of using autologous cells, namely, the lack of immune reaction, is counterbalanced by the main disadvantage, namely, the individual variability in cell quality and number, related to patient age, comorbidities, concomitant medications, etc. While it is feasible to extract and reimplant cells within 24 h, the patient may need anesthesia, which may be contraindicated in many stroke patients, and any cell processing to isolate particular cell types will delay treatment significantly. Further, the numbers of cells collected have varied substantially. These issues make clinical trial design and data interpretation very difficult. Bone marrow extraction carries some risks in the stroke population, who are commonly prescribed antiplatelet or anticoagulant drugs, or in the acute phase may have received thrombolytic agents.

6.2 Allogeneic therapies

The primary advantage is the provision of pure, well-characterized cell populations, or lines which can be fully manufactured to GMP standards. Allogeneic stem cells have to be highly scalable, so require considerable control over isolation, manufacturing, cell banking, and product preparation. Reagents need to be carefully sourced and any risks fully assessed and documented. Somatic stem cells with limited source availability may need new technologies for scaling, particularly as trials start to demand high patient numbers.

In addition, allogeneic therapies, because of their potential wide application, require extensive nonclinical safety evaluation. The key safety issues are acute immune reaction and longer term safety and toxicology. In particular, there is need to examine any potential tumor formation or complications arising from late graft rejection prior to early stage trials.

Although widely regarded as safe, some reports of MSC-derived tumors have appeared in preclinical models, suggesting that limits on passage numbers and stringent standards of cytogenetic quality control will be required for clinical applications (Tolar et al., 2007). Recently, even genetically normal autologous MSCs have been found to be tumorigenic when intraventricularly administered in a mouse model of experimental allergic encephalomyelitis (Grigoriadis et al., 2011), indicating that there is a risk that needs to be evaluated in long-term tumor studies for every cell therapy.

Moreover, in spite of small numbers of patients, the early clinical experience of cell therapies in stroke provides an indication of potential AEs and SAEs that may occur as the field develops. For IC administration, hematomas, cortical vein occlusion, and seizures have been observed (Kondziolka et al., 2000, 2005; Savitz et al., 2005; Suarez-Monteagudo et al., 2009). However, the data monitoring boards for these trials considered that these SAEs were related to the procedure rather than the cells. The intravascular route of administration represents a direct and relatively noninvasive route. For IV administration, one universal risk is pulmonary trapping. For example, luciferase positive mouse MSCs (diameter: 15–19 μm) were visualized within seconds after IV infusion trapped in the pulmonary microvasculature (Schrepfer et al., 2007). Therefore, risks associated with IV delivery of MSC cell therapy products need careful consideration. While MNCs have a cell size (5–8 μm) which will allow them passage through the pulmonary circulation without sequestration (Fischer et al., 2009), cell clumping cannot be reliably avoided, and an increased risk of strokes, presumed due to arterial occlusion by cell clumps, has been observed in animal models with IA delivery. More generally, increased plasma viscosity may not be tolerated by elderly patients with a recent thrombotic event, and acute vascular events including recurrent ischemic stroke, myocardial infarction (MI), and deep vein thrombosis, are all potentially increased by cell therapies delivered intravascularly. Distinguishing such events from the background rate in the stroke population is difficult without large trials, and specific monitoring protocols are likely to be necessary to detect subclinical events.

Publication bias, where only positive results are reported and negative or insignificant results, or data from trials terminated early remain unreported (Sena et al., 2010), is a problem that needs to be avoided in this field and in the stem cell field generally. We note that there has been a substantial increase in the number of clinical trials using cell therapies in stroke, registered at sites such as www.clinicaltrials.gov. However, it has not been possible to trace the development of many stem cell products or therapies now undergoing clinical studies because of gaps in the peer-reviewed published data. Thus, in this review, it is not possible to give any useful information about many registered cell therapy stroke trials, so we have not included these trials in our summaries. It is notable that negative studies are prominently lacking in both preclinical and clinical literature to date, which suggests publication bias.

7 CURRENT CLINICAL PRODUCTS AND APPROACHES

Table 3 summarizes the current clinical trials now known to be in progress.

8 CURRENT CELL THERAPIES TARGETING THE ACUTE PHASE

8.1 Adipose tissue-derived stromal cells

Adipose tissue has been identified as an alternative source of pluripotent stromal cells (Zuk et al., 2001). Adipose tissue, like bone marrow, is derived from the embryonic mesoderm and contains a heterogeneous stromal cell population, including MSCs, providing the rationale for developing this source as a cell therapy (Lee and Yoon, 2008). Unlike MSCs, during the first rounds of proliferation, adipose tissue-derived stromal cells (ADSCs) express the CD34 antigen (Lin et al., 2008). Human ADSCs express the MSC cell markers CD29 and CD44 and are negative for CD114 (monocyte and macrophage marker) and CD34 (a hematopoietic/endothelial stem cell marker; Kang et al., 2003; Lee and Yoon, 2008; Yang et al., 2011b). The frequency of ADSCs is much higher in adipose tissue (100- to 500-fold higher) than that of MSCs in bone marrow. The advantages offered by ADSCs for clinical application include abundance, an autologous source, and an anesthetic-free procedure for procuring the tissue and delivering it back into patients within 1 h, thus the potential to administer cells within the timescale of the acute ischemic period (Yang et al., 2011b).

ADSCs have been reported to be induced into neurospheres and neuronal-like cells *in vitro*, and thus have been touted as an alternative cell source for neuronally related cells (Safford et al., 2002; Yang et al., 2011b). Survival and migration ability of ADSCs after transplantation into rats has been reported, as has homing of adipose-derived stem cells to sites of tumor (glioma) or adenovirus infection (Lamfers et al., 2009). After implantation into the lateral ventricle of the rat brain, ADSCs were able to survive and migrate to multiple areas of the brain (Kang et al., 2003). When ADSCs were transplanted into rats after MCAo stroke, they migrated to the injured cortex, and behavioral testing demonstrated improved functional recovery in the treated animals (Table 2; Lee and Yoon, 2008). Further, ADSCs were induced to initiate neural differentiation. Following neural induction, ADSCs developed neural morphology and displayed molecular expression of Nestin, microtubule-associated protein-2, and glial fibrillary acidic protein. Better functional recovery and more reduction in hemispheric atrophy after treatment were observed in groups treated with ADSCs compared to controls (Yang et al., 2011b).

A small study of ADSCs (NCT01453829) in acute stroke patients (hemorrhagic or ischemic) is currently underway, aiming to evaluate safety, and effectiveness over

Table 3 Cell therapy clinical trials currently underway (information provided on the www.clinicaltrials.gov Website on 15 February 2012),

Sponsor/ clinicaltrials.gov registration #	Phase	Cells	Route	*N* (intend to enroll)	Location of infarct or type/timing of treatment	Key outcome measures	Preclinical functional recovery (FR), clinical design (CD), or short-term follow-up clinical trial report (CTR) published
Ageless Regenerative Institute, Mexico/ NCT01453829	I/II	Autologous ADSCs	IA/IV	10	Hemorrhagic and ischemic/ within hours	Safety NIHSS, BI	Kang et al. (2003) (FR) Lee and Yoon (2008) (FR)
University of Texas, USA/ NCT00859014	I	Autologous MNCs	IV	30	Ischemic/24–72 h (with t-PA)	Safety Not specified functional outcome	Brenneman et al. (2010) (FR) Yang et al. (2011a) (CD) Savitz et al. (2011b) (CTR)
Athersys Inc., MultiStem®, USA/ NCT01436487	II	Allogeneic MSCs	IV	140	Nonlacunar or brainstem/1–2 days after stroke	Safety, mRS, NIHSS, BI, secondary infections	Yasuhara et al. (2008) (FR) Mays et al. (2010) (CD)
Aldagen Inc., USA/ NCT01273337	II	Autologous MNCs ($ALDH^+$)	IA	100	Nonlacunar, MCA/13–19 days	Safety MRI NIHSS, mRS, BI, ESS	No
ReNeuron Ltd., UK/ NCT01151124	I	NSC/ CTX0E03	IC	12	Subcortical and/or basal ganglia MCA/ 6–60 months	Safety MRI, NIHSS, mRS, BI	Pollock et al. (2006) (FR) Stroemer et al. (2009) (CD) Smith et al. (2012) (FR)
San Bio Inc., USA/ NCT01287936	I/II	Notch-1-activated MSC (SB623)	IC	18	Subcortical MCA/6–36 months	Safety MRIs, ESS, NIHSS, mRS, Fugl-Meyer, humoral, immunological changes	Yasuhara et al. (2009) (FR and CD)
Stemedica Cell Tech, USA/ NCT01297413	I/II	Allogeneic MSCs	IV	35	Ischemic/>6 months	Safety NIHSS, MMSE BI Geriatric Depression Score	Not known

ADSCs, adipose tissue-derived stromal cells; BI, Barthel Index; ESS, European Stroke Score; hNT, human teratocarcinoma-derived neural cell line; IC, intracerebral; IV, intravenous; IA, intra-arterial; MCAo, middle cerebral artery occlusion; MMSE, Mini-Mental State Examination; MNCs, bone marrow mononuclear cells; MSCs, mesenchymal or stromal cells; mRS, modified Rankin Scale; NIHSS, National Institutes of Health Stroke Scale.
modified from Sinden and Muir (2012)

a 6-month follow-up (Table 3). To our knowledge, there have been no preclinical studies published on the intravascular delivery of this product. ADSCs will be administered via IA or IV routes in the trial.

8.2 Allogeneic MSC products

8.2.1 Multipotent adult progenitor cells in Phase II trial (Athersys)

Multipotent adult progenitor cells (MAPCs) are a subtype of the bone marrow-derived MSC population which is being developed as MultiStem® by Athersys Inc. (Mays et al., 2010). Preclinically, IV administration of MAPCs provided a similar degree of functional recovery as that seen after direct injection into the brain of animals having undergone surgically induced neonatal hypoxic-ischemic injury, as measured by behavioral improvements and reduction in ischemic cell loss (Table 2; Yasuhara et al., 2008). In a further study, human MAPCs were injected IV in varying doses into 7-day-old neonatal rats with ischemic injuries to investigate the effect when administered in a window from 1 to 7 days following the injury. The requirement for immunosuppression, route of cell administration, window for therapeutic benefit, and optimal cell dosage were also examined in the rat MCAo model of stroke (Mays et al., 2010). MCAo rats demonstrated statistically significant improvements in locomotor and neurological functions at time points up to 2 months post-occlusion for two doses of cells (4×10^5 cells or 1×10^6 cells) administered IV within 1, 2, or 7 days after occlusion, both with and without immunosuppression (Table 2; Mays et al., 2010). These functional improvements were also found at time points up to 4 months post-occlusion when the human MAPCs product was delivered IC for all doses (dose ranging from 4×10^5 to 2×10^6 cells) with no immunosuppression. An earlier study had compared the effect of IC delivery of a single dose human MAPCs product (4×10^5 cells) with and without immunosuppression and concluded that immunosuppression was not required for the therapeutic effect of these cells in this animal model (Mays et al., 2010).

Reports of efficacy of MAPCs in preclinical models of AMI and a completed Phase I clinical trial conducted in patients with first ST-elevation-myocardial infarction (STEMI) have been published (Mays et al., 2010).

The Athersys Phase II (NCT01436487) study is a large double-blind randomized placebo-controlled trial of MultiStem®, aiming to recruit 140 patients within 1–2 days of a moderate to severe, nonlacunar ischemic stroke which does not affect the brain stem. The primary outcome is safety, monitoring of AEs within 7 days of treatment and of functional deterioration by evaluating the numbers of patients with mRS scores ≤ 2 over 90 days. Secondary outcomes will evaluate efficacy using clinical scores including mRS, NIHSS, and BI to 1 year after treatment (www.clinicaltrials.gov, 15 February 2012).

8.3 Autologous MNCs

Preclinical and clinical studies that establish functional recovery and inform clinical study designs for the use of autologous MNCs in the acute period following stroke have been discussed above (Table 2; Brenneman et al., 2010; Yang et al., 2011a).

The feasibility and safety of IV administration of autologous MNCs (7–10 million cells/kg) in the 24- to 72-h window after ischemic stroke was presented as an interim study report recently (Savitz et al., 2011b). Ten patients who had severe left or right hemisphere stroke (defined by NIHSS scores at study entry, right: 6–15 or left: 6–18) had a single IV infusion and were monitored for 6 months. There were no reports of SAEs related to study procedures. Patients showed improvements in the modified Rankin, NIHSS, and BI scores at 6 months (Savitz et al., 2011b). This Phase I trial (NCT00859014) is scheduled to be completed by 2014.

9 CELL THERAPIES TARGETING THE SUBACUTE PHASE

9.1 Selected autologous ALDH$^+$ MNCs (Aldagen)

Human hematopoietic progenitors can be isolated from the bone marrow MNC fraction by using a fluorescently labeled dye specific for aldehyde dehydrogenase (ALDH) activity (Colvin et al., 1988; Russo et al., 1989). The primary function of ALDH1 in hematopoietic stem cells (HSC) relates to the production of retinoids. HSC differentiation is dependent on retinoid signaling (production of all-*trans*-retinoic acid) and inhibition of ALDH promotes HSC self-renewal via reduction of retinoic acid activity and all-*trans*-retinoic acid. Thus, modulation of ALDH activity and retinoid signaling, aiming to enhance the amplification of HSCs *in vivo*, is the basis for selecting ALDH$^+$ cells.

The ALDH$^+$ population consists of approximately 1% of nucleated cells and is enriched in cells expressing CD34, CD117, CD105, CD127, CD133, and CD166 and in primitive CD34+ CD38 and CD34+ CD133+ progenitors. The ALDH$^+$ population also includes neural progenitor cells (Keller, 2009). To our knowledge, no preclinical studies of the autologous ALDH$^+$ cells in the treatment of ischemic stroke have been published. A review of the preclinical and clinical studies of this product in ischemic heart failure and critical limb ischemia is provided in Keller (2009).

A large Phase II randomized double-blind placebo-controlled trial (NCT0127337) with a Phase I safety cohort testing is currently underway, recruiting patients with persistent neurological deficit for bone marrow harvest within 11–17 days following ischaemic stroke and dosing IA (carotid) of autologous product within 2 days of harvest. All patients will be followed for 1 year to evaluate safety and mental and physical function.

10 CELL THERAPIES TARGETING CHRONIC STROKE

10.1 Genetically modified bone marrow stromal cell product (San Bio)

SB623 cells are human bone marrow-derived neural progenitor-like cells under development by San Bio Inc., as an allogeneic cell therapy for chronic stable stroke. SB623 cells are generated by the transfection of MSCs with a plasmid encoding human Notch-1 intracellular domain (NICD), under GMP conditions (Kondziolka and Wechsler, 2008). The transfection is considered as transient because the products of the plasmid rapidly disappear with further expansion/passaging of the cells, so that the gene is only present at low levels in the cells in the final product. The result of the transfection is that the final cells, unlike other marrow stromal cells, no longer have the potential to differentiate into bone or adipose tissue (Kondziolka and Wechsler, 2008).

Notch-1 is involved in the regulation of neuronal development (Wu et al., 2001). Once released, the NICD migrates to the nucleus where it activates expression of a number of genes. Overexpression of Notch in MSCs and subsequent exposure to various growth factors led to the development of cells that resemble neurons morphologically, stain positively for neuronal markers, display electrophysiological characteristics of neurons, and reverse deficits in a rat model of stroke (Dezawa et al., 2004; Mimura et al., 2005).

SB623 cells have been evaluated in several rodent and primate studies (Kondziolka and Wechsler, 2008), including preclinical studies which showed improvements in behavioral tests related to motor functional recovery when cells were transplanted into MCAo stroked rats 1 month after occlusion (Table 2; Yasuhara et al., 2009).

A Phase I/II trial (NCT01287936) is currently recruiting chronic stroke patients with motor deficits, within 6 months to 3 years of infarct (www.clinicaltrials.gov, 15 February 2012).

10.2 Conditionally immortalized neural stem cell drug product (ReNeuron)

CTX0E03 is currently the only neural stem cell therapy product in a clinical trial for stroke. CTX0E03 is a clonal human cell line conditionally immortalized using a retroviral vector encoding the c-mycER^{TAM} fusion gene (Pollock et al., 2006). The recombinant protein product comprises human c-Myc fused with a mutant estradiol receptor (ER). The recombinant c-MycER protein is present as an inactive monomer in the cytosol of the cell and is activated by 4-hydoxytamoxifen (4-OHT), a minor tamoxifen metabolite. Dimerization of the fusion protein occurs only in the presence of 4-OHT, whereupon the fusion protein translocates into the nucleus and allows continuous and stable expansion of the CTX0E03 cells. Removing 4-OHT from *in vitro* culture conditions and prior to transplantation *in vivo*, switches off the c-Myc-dependent growth promotion. Moreover, the c-mycER^{TAM} gene has been shown to be inactivated irreversibly at the transcriptional level (Stevanato et al., 2009). The c-mycER^{TAM} technology has been central in the development of the CTX0E03 cell line and the manufacturing and banking of CTX0E03 cells to current GMP quality standards for use in clinical applications (Pollock and Sinden, 2008).

The first characterization of the CTX0E03 cell therapy product and the results of preclinical testing in animal models, including the proof of concept that intraparenchymal implantation of CTX0E03 cells could improve functional recovery after stroke, were reported in Pollock et al. (2006). A subsequent study found a significant dose effect of delayed CTX0E03 cell implantation on behavioral recovery in chronic stable rat model of stroke (Stroemer et al., 2009). CTX0E03-implanted animals compared to vehicle-treated animals showed significant improvements in recovery of sensorimotor function tests at 6–12 weeks postimplantation. These results for the bilateral asymmetry test and rotameter were reproduced in a further study, with recovery of function also observed in a foot-fault test (Smith et al., 2012).

Postmortem histological analysis found surviving cells, located in and near the infarct damage. Where differentiation of the cells is seen in the stroke model, the majority were directed to astrocytes. Comparing cells implanted to an intraventricular site with those implanted near the infarct site, the cells near the infarct improved functional recovery. In terms of lesion topology, animals with infarcts confined to the striatum showed a more significant improvement than those with infarcts affecting the cortex (Smith et al., 2012).

It has been consistently found that although there is cell survival, no expansion of implanted cells or cell replacement has been noted. *In vivo* studies have shown a significant increase in microvessels at the site of CTX0E03 implantation, suggesting that CTX0E03 cells may provide their benefit in stroke recovery via upregulation of angiogenesis in the brain (Hicks et al., 2012; Stroemer et al., 2009; Smith et al., 2012).

The PISCES trial (NCT01151124) is a Phase I trial investigating the safety of CTX0E03 intracranial implantation in chronic stable male stroke patients 60 years or over who remain moderately to severely disabled. Treatment involves a single dose of one of four doses. At the date of writing, seven of the total 12 patients have been treated.

11 FACTORS TO ENHANCE ENDOGENOUS STEM CELL THERAPY

The opportunity for interventions aiming to enhance ischemia-induced neurogenesis has also received vast attention in the past decade (Angels Font et al., 2010; Burns and Steinberg, 2011; Burns et al., 2009; Zaleska et al., 2009; Zhang and Chopp, 2009). Of the growing number of physiological stimuli, endogenous molecules and exogenous agents that have been shown to play a role in the regulation of adult neurogenesis, two molecules—erythropoietin (EPO) and granulocyte-colony-stimulating factor (G-CSF)—have reached the level of Phase II trials for use in the early acute to subacute period after a stroke. These trials are summarized in Table 4.

11.1 Erythropoietins

EPO stimulates hematopoietic progenitor cell maturation, differentiation, and survival. The level of expression of EPO and its receptor in the brain is greatly increased in neurons, neuronal progenitor cells, and cerebrovascular endothelial cells in

Table 4 Stimulating endogenous stem progenitor/cell trials

Sponsor/ reference # or reference for publication	Phase	Cells	Route	*N*	Location of infarct/ timing of treatment	Clinical outcome measures	Completed/results if published
Stem cell therapeutics/ NCT00362414	IIb	9-day regimen, β-hCG (IM) and EPO (IV) (NTx-265)	IM/IV	12	Ischemic supratentoria/ acute (within 2 days of onset)	Safety	Failed to show efficacy due to unexplained high level response in both placebo and treated groups (Aboody et al., 2011)
Ehrenreich et al. (2009)/German Multicenter EPO Stroke Trial	II/III	EPO	IV	460	MCAo/within 6, 24, or 48 h	Safety, efficacy	Negative outcome for primary outcome BI on day 90 and for other parameters for EPO treatment. An overall death rate of 16.4% ($n=42$ of 256) in the EPO and 9.0% ($n=24$ of 266) in the placebo group (OR, 1.98; 95% CI, 1.16–3.38; $p=0.01$). Also raises safety concerns, particularly in patients receiving systemic thrombolysis (63%)
Sprigg et al. (2006)	IIa	G-CSF	s.c.	36	Ischemic stroke and motor weakness (arm and/or leg, MRC grade $<5/5$)/within 7–30 days	Dose ranging	Increased CD34+ count in a dose-dependent manner, from 2.5 to 37.7 at day 5. A dose-dependent increase in white cell count ($p<0.001$). No difference in number of SAEs or dependence (mRS) between treatment groups
Schabitz et al. (2010) (AXIS)	I	G-CSF	IV	44	MCAo/within 12 h of stroke/NIHSS 4–22 with a MRI DWI/PWI mismatch	Safety	Well tolerated even at high doses. Noted substantial increase in leukocytes appears not problematic in stroke patients. Dose dependent effect of treating patients with DWI lesions >17–19
AXIS-2/ NCT00927836	II	G-CSF	IV	328	Acute/within 9 h	Mortality; mRS (90 days); NIHSS (90 days); BI; infarct size; cytokine profile	Reported Feb 2012. No effect on day 90 mRS or NIHSS endpoints, mortality, or infarct volumes on MRI in 328 randomised patients. Expected peripheral cellular responses seen and no safety or tolerability issues
Boy et al. (2011)	I/IIa	G-CSF may include standard thrombolysis	s.c.	20	Moderate MCAo/within 12 h of onset	Safety	Safe when started within 12 h of onset
England et al. (2012)	IIb	G-CSF	s.c.	60	Ischemic or hemorrhagic/ 3–30 days	Safety, dose ranging	Safe, treatment start mean of 8 ± 5 days of stroke onset. A trend toward reduction in lesion volume from baseline in treated patients ($p<0.06$) with DWI lesions 14–17 cm^3

BI, Barthel Index; Beta-hCG, beta-human chorionic gonadotropin; G-CSF, granulocyte-colony-stimulating factor; EPO, erythropoietin; IV, intravenous; IM, intramuscular; MCAo, middle cerebral artery occlusion; MRI, magnetic resonance imaging; mRS, modified Rankin Scale; NIHSS, National Institutes of Health Stroke Scale; s.c., subcutaneous.

response to injury (Xiong et al., 2010). EPO crosses the BBB and treatment improves survival after a stroke in animals (Banks et al., 2004; Brines et al., 2000; Liu et al., 2008; Siren et al., 2009). Also, studies have shown that treatment with EPO is associated with vascular remodeling, leading to improved neurobehavioral outcomes in animal models of stroke (Iwai et al., 2007; Wang et al., 2004). A Phase I clinical trial investigated the safety of recombinant human EPO given IV within 8 h of onset in 40 patients who had an ischemic stroke affecting the MCA (Ehrenreich et al., 2002) and identified the expected polycythemic response. However, the subsequent Phase II multicenter trial, involving 522 patients in whom treatment was initiated within 6 h of stroke onset, was neutral with respect to neurological change or disability measures, with no evidence of differences in BI, mRS, and NIHSS. In addition, a significantly higher mortality rate was seen in EPO-treated patients (16% vs. 9% in controls), mainly occurring in the first week after treatment and related to higher rates of intracerebral hemorrhage (ICH) and brain swelling. The causes of this finding remain unexplained but a potential interaction with IV recombinant t-PA (given to >60% of patients in the larger trial) was suggested, although statistical interaction was not formally reported. Further trials of EPO are unlikely to be initiated until the mechanism of this potentially negative interaction of recombinant t-PA and EPO is clarified (Ehrenreich et al., 2009). One proposed mechanism identified on laser Doppler flowmetry in rat MCAo is of impaired reflow following recanalization in EPO-treated animals (Zechariah et al., 2010).

A further trial (NCT00362414), a Phase II investigation of EPO with the human chorionic gonadotropin, β-hCG, given within 2 days of a stroke, has also failed to demonstrate a functional difference with treatment. The study found a high level of response in both placebo and treated groups (Aboody et al., 2011).

The carbamyolated derivative, a nonhematopoietic analogue of EPO, is under investigation in acute stroke with safety and pharmacokinetic studies currently underway (NCT00870884).

11.2 Granulocyte-colony-stimulating factor

G-CSF is a glycoprotein that regulates the production, proliferation, survival, and maturation of neutrophils in neutrophilic granulocytes. It acts through a receptor expressed on hematopoietic and neuronal cell types. G-CSF is used in clinical practice to treat neutropenia and for the production of CD34+ HSC for bone marrow transplantation. It has been shown to be safe with well described side effects (Boy et al., 2011). G-CSF crosses the BBB. In preclinical studies of stroke, G-CSF has been shown to be both neuroprotective, by reducing apoptosis in the penumbra and attenuating inflammation, and regenerative, by promoting angiogenesis and neurogenesis, by activating brain endothelial cells, and by mobilizing CD34+ cell migration to the ischemic site (Bratane et al., 2009; Gibson et al., 2005; Kawada et al., 2006; Minnerup et al., 2008; Schabitz et al., 2003; Shyu et al., 2004). Importantly, G-CSF is not only acutely protective but also improved functional recovery after stroke, even after treatment is delayed beyond 6 h post-MCAo (Minnerup et al., 2008).

Initial Phase II trials on the use of systemically administered G-CSF in patients have shown safety and confirmed aspects of the expected biological activity, namely a substantial release of CD34+ cells (Shyu et al., 2006b; Sprigg et al., 2006) and increased peripheral white blood cell counts. Trials have initiated treatment within different time windows (up to 30 days in STEMS-1 and STEMS-2, 12 h in AXIS). Subsequent small Phase I/II trials involving just over 100 patients showed no major safety issues associated with G-CSF within 12 h of stroke (including with standard IV thrombolysis in patients for whom this was indicated) (Boy et al., 2011; England et al., 2012; Schabitz et al., 2010). The AXIS-2 trial (Table 4) reported no clinical benefit on NIHSS or mRS measures at day 90 in 328 patients treated approximately 7 hours post-stroke, with a non-significantly less favourable mRS distribution in the treatment group. The investigators concluded that G-CSF was ineffective in this study (Ringelstein EB, AXIS-2 Investigators, presented at International Stroke Conference 2012).

12 CELL DELIVERY APPROACHES

12.1 Timing after stroke

As has been discussed, the timing of cell delivery depends in part on the therapeutic target, although the evolution of ideas regarding therapeutic targets has often followed the animal model experience rather than being defined *a priori*. Early administration (hours to 1–2 days) is focused on neuroprotection and shares mechanisms that have been targeted by pharmaceutical interventions for acute brain injury such as stroke and traumatic brain injury (Harting et al., 2008). The advantages of early stem cell therapy include: (i) a leaky BBB, allowing for peripheral delivery of cells to potentially deliver brain engraftment, at least after the first 24 h; (ii) acutely produced inflammatory chemokines arising from stroke damage may act as attractants to cells; and (iii) infused cells may reduce stroke-induced inflammation, locally and systemically.

As noted above, the majority of animal studies using bone MNC types delivered intravascularly find evidence of functional benefits at early time points, but find that the effect is reduced or absent at poststroke intervals of 7 days or more (Table 2). Cell replacement or regeneration may not be applicable paradigms for this early administration, at least by intravascular delivery, which is the most feasible route in the acute period. The great majority of cells delivered by intravascular routes in animal models of stroke are sequestered in the lungs and reticuloendothelial system soon after administration, and few if any are engrafted in the injured brain. Intra-arterial delivery results in higher cell migration than IV delivery (Pendharkar et al., 2010), but may be difficult to achieve in clinical use. At early time points after stroke in humans, the BBB remains intact, and many of the earliest neurochemical mechanisms targeted by pharmaceutical neuroprotectants, such as excitotoxicity, are likely not amenable to cell therapies even if delivered at these acute stages.

Later delivery of cells in the subacute stages may modulate inflammatory and apoptotic processes that contribute to secondary neuronal loss, but this is likely to represent a modest volume of tissue in comparison with the acute necrotic processes, and stimulation of early regenerative processes may be more therapeutically relevant.

Whether there is a viable clinical role for cell therapy in the chronic stages of stroke remains unclear. It is uncertain whether the "chronic" stages typically tested in rodent models ("chronic" being approximately 4 weeks after the infarct) are representative of human stroke many months or years after the event. Mechanisms that may remain relevant within weeks of stroke, such as inflammation, are unlikely to continue years after the event, so therapeutic effects in the clinic are probably dependent upon stimulation of host regenerative processes such as angiogenesis or neurogenesis, or integration of implanted cells, or some combination of these. Brain plasticity and repair potential may differ fundamentally between young rats and aged humans, with no systematic study having been undertaken, and only anecdotal clinical reports. The presence of a glial scar that effectively seals off injured tissue is a significant barrier to cell migration in spinal cord injury, and may also be relevant in chronic stroke.

12.2 Route of administration

Delivery routes at present are defined by a combination of preclinical evidence, tolerability and safety in relation to timing after stroke, constraints of cell production, and to an extent the perceived mechanism of action.

Intravenous administration offers easy access and the potential for broad distribution or for repeated doses but a potential disadvantage in that cells need to cross the BBB in order to have a therapeutic effect, particularly if the goal is cell replacement: neuroprotectant effects, on the other hand, may not be restricted by the need for a permeable BBB. Intravenous delivery of bone marrow cells has been shown to deposit cells in liver, spleen, and lungs (Chen et al., 2001a); and, as noted above, MSCs administered IV have been found trapped in the lung (Fischer et al., 2009; Harting et al., 2008). Cell numbers ending up in the central nervous system (CNS) are small.

The IA approach appears to have the advantage of greater cell delivery to the brain compared to IV infusion (Pendharkar et al., 2010) and is less invasive than IC injection. However, IA delivery carries procedural risk (including stroke, vascular injury, and ICH) and usually requires general anesthesia in the acute stroke population. Intra-arterial procedures are also relatively contraindicated within the first hours after thrombolytic drug therapy, and access is limited to centers with specialist interventional neuroradiology services. In animal models, the potential for IA delivery to cause stroke downstream of the infusion site, likely to reflect cell clumping, is a concern that has already been noted. A Phase I clinical safety study of CD34+ cell delivery via IA catheter in the affected MCA delivered 7–9 days after stroke recruited only 5 of 82 screened subjects, predominantly because of persistent occlusion of the target vessel or vascular access problems, which is likely to be a very significant barrier to recruitment in the early subacute phase (Bannerjee et al., 2012).

Intracerebral injection of cells has the advantage of maximizing cell dose at a location close to the site of tissue injury, and circumventing the BBB and perhaps also host immune responses. Procedural risks for stereotaxic delivery, at least for implantation in the basal ganglia, appear to be acceptable and consistent with those reported for stereotaxic procedures generally (Muir et al., 2011). However, the need for general anesthesia, careful imaging-based planning of trajectories, and discontinuation of antithrombotic or anticoagulant drugs to mitigate intracranial bleeding risk, may dictate use of this route in chronic or at best subacute stroke patients.

Intrathecal or intracerebroventricular routes of administration have not to date been undertaken in clinical trials but share several disadvantages common to other delivery routes.

Evidence of cell migration to the site of ischemic injury after administration by various routes has been seen, including IV, IA, and IC injection. Migration potential may differ according to cell type and route, but this has not been systematically explored for most cell lines (Kalladka and Muir, 2011), nor has the key question of whether different routes have different potential efficacy, or dose requirements.

A major potential limitation of systemic administration of cell therapy, in particular MSCs, is assuring that optimal numbers of cells reach the target tissue. The cell numbers delivered by IC injection can be defined readily, but those delivered to the CNS after either IV or IA administration cannot. Cell yields for any bone marrow-derived autologous cells are also variable and unpredictable. Clinical studies have noted more than twofold variation in cell numbers for autologous cell delivery (e.g., Lee et al., 2010: 2×50 million; Bhasin et al., 2011: 50–60 million; Honmou et al., 2011: 60–160 million cells; Bannerjee et al., 2012: CD34+ cells from 1.2 to 2.8 million). Current developments have focused on improving the homing properties of MSC cells to increase the number of cells reaching the target area and thereby enhancing functional recovery.

Products using an intravascular delivery approach may need a "homing signal" to the brain and demonstration that cells do not cause systemic organ toxicity or interfere with organ physiology (Savitz et al., 2011a). Recent developments to enhance the homing abilities of MSC cell product include genetic or chemical modifications.

12.3 Use of immunosuppression

Preclinical models generally use immunosuppression in studies of human (or other allogeneic sourced) cells implanted into rodents. However, whether this is required for clinical use is unknown and there have been few rodent studies that have compared therapeutic actions with and without immunosuppressants. Some immunosuppressant agents have independent neuroprotectant effects in stroke (e.g., tacrolimus) and there are major concerns about infection risk, drug side effects, and drug interactions in the stroke population, all issues that either cannot be addressed or are unlikely to be observed with the controlled environment and short follow-up in animal studies. Immunosuppressant use in clinical trials cannot ethically be undertaken in control populations, presenting potential difficulties in attribution of either beneficial or harmful effects that might be observed. In different clinical trials of fetal tissue transplantation in Parkinson's disease, immunosuppression has been either included or omitted, with no evidence of differences in therapeutic indices or adverse effect profile, but based on very small patient numbers (Freed et al., 2001; Olanow et al., 2003). However, the absence of any immunological reaction in a study that omitted immunosuppression offers some reassurance that this may be unnecessary as a routine procedure in allogeneic cell therapy studies.

12.4 Cell tracking

Cell tracking may be important for studies that seek to draw mechanistic inferences, but may not be a prerequisite step in clinical development if the mode of action is likely to be neuroprotection. It also presents technical challenges for autologous cells and additional developmental hurdles for allogeneic nonpatient-specific cells that are manufactured to GMP standard, whose properties may be altered by labeling procedures required for tracking. The presence of the label does not necessarily mean that viable cells are present; nor does it offer information on the differentiation or integration of viable cells. It is also possible only to gain qualitative information, since the uptake of a label by the cells to be implanted is inconsistent and influenced by a number of factors. Quantitative comparisons between patients are unlikely to be possible, and certainly are impossible for autologous labeled cells. Several techniques do not lend themselves to repeated imaging, for example, radiolabeling with short-half-life isotopes that gives a signal for a matter of a few hours only.

In a systematic review and meta-regression study (McColgan et al., 2011) to describe the range of methods employed in human cell-tracking studies and to investigate factors that predicted target-organ cell uptake efficiency, 19 relevant trials were identified, involving 145 patients with a variety of underlying diseases, including MI, Chagasic cardiomyopathy, ischemic stroke, traumatic injury of brain or spinal cord, diabetes, and cirrhosis. Cell-labeling strategies included radio-isotope labels (18-fluorodeoxyglucose-positron emission tomography, FDG-PET, 111-indium-SPECT, and 99-technetium-SPECT), and use of intracellular iron oxide particles for MRI. The meta-regression identified that labeling efficiency was proportionate to cell dose, while cell viability was lowest with indium labeling and long label incubation times. In ischemic stroke, iron-labeled cells that allow MRI tracking have been investigated both in preclinical models and in clinical use (Daadi et al., 2009). Unlike radio-isotope labels, iron-labeled cells allow repeated imaging without radiation exposure and over long time periods. The major disadvantage is that the detection of labeled cells relies on the susceptibility effect of heavy metals on appropriate MRI sequences, which leads to reduced signal. Indistinguishable signal reduction may occur with unrelated processes, such as deposition of calcium in the basal ganglia, a common finding in the elderly, and more importantly as a result of hemoglobin degradation products that indicate petechial hemorrhage following ischemic damage, a common finding that changes dynamically following brain infarction and is radiologically indistinguishable from iron-labeled cells.

There are concerns that intracellular labels may interfere with cell viability or with normal cell function. Using a gadolinium chelate as a label for human neural stem cells, Modo et al. (2009) demonstrated that cells could be visualized with MRI and exhibited evidence of migration to the site of stroke injury in rats. However, longer term follow-up found that the animals failed to exhibit behavioral recovery that had been seen consistently with equivalent doses of unlabelled cells in a similar model.

Non-iron-based MRI techniques offer possible alternatives. Fluorine-labeled cells can be visualized with MR spectroscopy and have been used in an animal study to investigate the possible additional value of extracellular matrix scaffolds to supplement cell therapies (Bible et al., 2012).

13 PATIENT SELECTION

The preponderance of ischemic, as opposed to hemorrhagic, stroke and the availability of the rat MCAo model of stroke means that the vast bulk of early phase research into novel stroke therapeutics targets ischemic stroke: however, this ignores the 10–15% of incident cases that are caused by ICH. Many of the mechanisms relevant to secondary brain injury and chronic recovery are shared between ICH and ischemic stroke, and cell therapies may therefore have similar therapeutic effects in ICH. The paucity of good animal models for ICH means that there have been few studies in this population, but positive effects have been reported. The following discussion concentrates on ischemic stroke.

13.1 Location of infarct

Ischemic stroke is a highly heterogeneous clinical entity, with a number of subtypes that are of potential relevance in considering cell therapies. Broad clinical syndromic classifications such as the Oxfordshire Community Stroke Project (Bamford et al., 1991) have clinical utility but are essentially descriptive and primarily of value in defining short-term prognosis. The mechanism of stroke differs across different clinical presentations, but varies by age and risk factor profile. While, broadly speaking, stroke related to atherothrombosis of a large artery such as the carotid is predominantly seen in late middle age onwards, with increasing incidence of cardioembolic stroke consequent to atrial fibrillation as the population ages, many risk factors are common to all stroke types and etiology can be difficult or impossible to define in individuals even with comprehensive investigation. Different etiologies may have different patterns of complications, recurrence risk, and recovery. Lacunar stroke syndromes, in particular, constitute about 20% of all incident strokes and are related to occlusion of a small penetrating artery, conventionally thought to relate to underlying hypertension and a pathological process in the vessel wall characterized pathologically as lipohyalinosis. This process tends to produce small, deeply placed infarcts (defined as diameter < 1.5 cm) in white matter, thalamus, or basal ganglia, The pathophysiological processes are poorly defined and recovery patterns generally follow a more favorable trajectory compared to the larger territorial infarcts seen with embolic vessel occlusion. However, the distinction of these stroke subtypes is often poor, with overlap of clinical syndromes, imaging findings, and risk factor history.

Brain imaging, particularly with the increasing use of MRI, is able to define the anatomical extent of ischemic stroke and the structures that are involved. However, imaging appearances are a dynamic phenomenon, and characterization of stroke may differ depending on the modality and timing after onset of ischemia. Early brain swelling, resolving over 2–5 days after ischemic stroke, later gives way to a process of inflammatory cell infiltration, gliosis, and involutional change as necrotic tissue is phagocytosed, typically leaving a shrunken cystic area of variable volume depending on the size of the occluded vessel. In chronic stroke patients, the established infarct is typically a large cavity with surrounding gliotic rim. The natural history presents inherent difficulties in characterizing stroke recovery using imaging and identifies a significant challenge to cell, and other regenerative therapies, for chronic stroke. Many clinical reports present brain imaging data to complement the clinical evaluation of individual cases (e.g., Bhasin et al., 2011; Honmou et al., 2011; Lee et al., 2010) but these have relied overwhelmingly on structural imaging. As noted, the volume of lesions will inevitably change between subacute and chronic phases, as will signal intensity of lesions on some MRI sequences. Investigators have sometimes drawn inferences about cell effects from such changes, which are largely uninterpretable without a large and well-characterized control group.

Most animal studies of IC implantation have targeted cells to the striatum since this is accessible and, in MCAo models, partially intact. The human striatum is also accessible for stereotaxic injection, with considerable surgical experience from other clinical situations. Some degree of preservation of putamen or lentiform nucleus therefore may be a practical requirement for patient selection on imaging. Stereotaxic implantation also requires sufficient intact brain along the needle trajectories to ensure accurate placement, which may place individual limitations dependent on brain anatomy and the size of infarct cavities.

There may be specific endogenous substrates required for therapeutic effects in chronic stroke: at least partial integrity of the SVZ, the site of endogenous neurogenesis, may be required, since some studies have suggested that clinical benefit is greatest when this structure is preserved (Lee et al., 2010). Other studies have correlated the likelihood of motor function recovery to the extent of damage to the descending corticospinal tract using lesion mapping (Lindenberg et al., 2010; Riley et al., 2011).

In general, cell replacement or regeneration in chronic stroke seems unlikely to have a significant impact on very large infarcts where the lesion is essentially a cystic "hole" in the brain: this type of lesion may intrinsically require a different approach with some matrix material. However, whether this speculation is correct or not remains to be seen.

In the immediate subacute phase of stroke, patient selection by infarct anatomy (location or extent) is probably of less importance than in the chronic phase, since different therapeutic paradigms are likely to underpin cell therapy mechanisms. For clinical trials, patient selection must also consider the natural recovery trajectory that is anticipated. Pure motor hemiparesis, a clinically defined lacunar syndrome, may be associated with severe early deficit and dependence, and the typically small anatomical extent of the causal lesion may appear a credible target for cell therapy in contrast to very extensive cortical infarcts. However, the recovery of these patients is generally very favorable and most are independent within a few weeks or months. Identifying a therapeutic effect in this situation is likely to be difficult without very large sample sizes.

13.2 Effects of age

Stroke incidence increases with age, and the majority of strokes in developed countries occur in people no longer of working age. However, stroke occurs in all age groups and the impact of loss of function and independence is arguably much greater when younger people, particularly those of working age, are affected. Recovery potential differs by age group: while some 57% of t-PA-treated patients in acute stroke thrombolysis studies achieved independence by 3 months, with a mean age of around 68 years, this figure was only 35% for those over the age of 80. Mortality is also much higher, 12% compared to 30% (Ford et al., 2010). This reflects increasing prevalence of comorbidities that impact functional abilities with advancing age, as well as reduced social support, and the incidence of other diseases that limit life expectancy. It may also signify reduced capacity for regeneration and plasticity in the CNS, although this has barely been studied in humans. Rat studies suggest that a number of factors, including greater apoptosis, enhanced glial reactions, and reduced neurogenesis, may represent biological barriers to recovery with increasing age (Popa-Wagner et al., 2011). At *postmortem* in a single case of recent ischemic stroke in an 84-year-old subject, endogenous neural stem cells and other reparative processes were observed in the peri-infarct region (Minger et al., 2007). Whether there are age-related differences in capacity for endogenous regeneration, or in the pathophysiological responses to stroke in general, is not known.

Older age may increase safety concerns with respect to cell delivery. General anesthesia and surgery are less well tolerated, and the aging cardiovascular system may be less tolerant of large intravascular cell loads. On the other hand, initial safety testing of allogeneic cell implants has been proposed to be restricted to older patients due to concern about lifelong tumor risk.

Old age is also commonly associated with reduced ability to participate in rehabilitation programs, due to associated cardiovascular, respiratory, or rheumatological disease, and these activities may be an important enhancement of cell therapy, or indeed, a requirement for functional recovery.

13.3 Comorbidities

Stroke most commonly occurs as a consequence of vascular risk factors and their chronic effects on vascular integrity or cardiac function, so these conditions are expected in a stroke population. Whether any specific comorbid condition, or combination of these, represents a necessary selection criterion for clinical trials is unknown. Animal models for several of these are not available (e.g., cigarette smoking) or are not readily amenable to MCAo (e.g., heart failure). In general, the presence of comorbidities in rats is associated with more severe damage after equivalent ischemic insults, with higher mortality and reduced capacity for recovery, but to what extent this is explained by other factors is unclear. The spontaneously hypertensive stroke-prone rat is an example, with this strain having significantly different collateral vessels and larger infarcts, as well as hypertension. Which of these interlinked confounding factors is responsible for poorer recovery is not readily determined. Diabetes mellitus may impair the therapeutic response to cell therapy, with a rat model of type I diabetes showing lack of behavioral recovery after MSC administration and a variety of histological differences including BBB leakage and accelerated vascular damage (Chen et al., 2011). However, MCAo in diabetic rats is associated with much more extensive infarction than in nondiabetic controls (MacDougall and Muir, 2011) and the failure to find a therapeutic effect may simply reflect greater severity of initial injury, with a reduced potential for recovery using conventional sample sizes, rather than an intrinsic inability to respond. Since approximately 20% of patients in stroke trials have a diagnosis of diabetes and another 20–30% may have insulin-resistance manifesting as impaired glucose tolerance or impaired fasting glucose, exclusion of such patients will restrict numbers of patients available for clinical trials and is unlikely to be practical.

14 CLINICAL TRIAL DEVELOPMENT

14.1 Phase I trials

Safety trials are necessary in cells that are allogeneic or have undergone modification in some way. There is sufficient experience with autologous bone marrow-derived cells from other clinical conditions that initial safety trials may be unnecessary. However, other autologous cells such as adipose tissue-derived cells lack clinical data sufficient to bypass this stage.

As outlined above, several trials with safety as the primary endpoint have been reported or are ongoing. These trials may be slow to recruit since patient populations are typically restricted and the need for safety review at each dose level—or indeed, each individual case—dictates the pace. Allogeneic cell studies have adopted a "patient volunteer" paradigm that includes patients many months to years after stroke, who have significant disability or neurological deficit and no expectation of spontaneous recovery. Data on efficacy in this population, the secondary goal for these studies, may be of limited value since clinical improvement is unlikely to occur, but even if it is observed, is very hard to interpret since few if any studies have followed patients at this time point after stroke.

Safety data are limited since patient numbers are small and trial duration short. Realistically, only common and early toxicity issues will be identifiable. The only likely short-term AE directly related to the implanted cells themselves is likely to be immunological. Longer term concerns such as tumor formation could not readily be detected, even with extended follow-up times, and given the comorbidities prevalent in the stroke population, may be impossible to distinguish from background rates. Most early safety issues are likely to relate either to procedural complications or to immunosuppression, if used.

The addition of cell tracking is a common desire for Phase I trials but may add little information. As discussed above, labeling for MRI tracking may alter cell properties, and signal change is likely to be indistinguishable from natural evolution of stroke in the subacute phase. Radiolabeling provides only short-term information on cell distribution. Neither method can reliably identify whether cells integrate into the injury site, or what their functional status might be.

Structural brain imaging is of limited value and primarily supplements clinical safety information by identifying ICH or other complications. Otherwise, needle tracks are typically visible on MRI in stereotaxically implanted patients, allowing confirmation of cell delivery at the planned location.

14.2 Phase II clinical development

Early Phase II trials will need to narrow the dose range, explore clinical and other markers of efficacy, and ideally should move toward clarifying issues around patient selection. However, it is likely that these trials will be possible with only a relatively small sample size given the technical and logistical issues with cell production, and the cost involved, which represents a significant challenge and therefore requires targeting a strong efficacy signal.

Patient selection to optimize the chances of finding an efficacy signal is a major goal for early Phase II trials. Strategies for doing this in acute stroke, particularly with reperfusion therapies, and involving a mixture of clinical criteria and advanced brain imaging, have evolved over the past 15 years, so that a credible treatment response can be detected with modest numbers of subjects (Parsons et al., 2012). The strategy should maximize the chances of enrolling patients with an appropriate biological substrate for the therapy under investigation, while minimizing the number of inevitable nonresponders (those who lack any biological substrate) and also the number of spontaneous responders (those who will recover anyway). However, the more specific the population sought, the greater the number of patients that needs to be screened for the trial (Muir, 2002). In addition, the appropriate clinical and imaging tools for maximizing the chance of finding a treatment effect in rehabilitation or regenerative therapy trials in stroke are not well developed. Clinical and imaging selection criteria will also dictate the outcome measures that are appropriate—or vice versa. For example, if the selected outcome evaluation is a scale of motor function, then only patients with a motor deficit can be recruited. Motor function was, unsurprisingly, a stronger predictor of independence in activities of daily living assessed by the BI, compared to visual field or sensory deficits (Patel et al., 2000): the BI is heavily weighted toward motor functions such as walking, climbing stairs, dressing, bathing, and feeding.

More complex interactions with common clinical deficits such as language function or visuospatial neglect are important, but harder to capture in most outcome scales.

Infarct location and size may be one imaging-based selection criterion. Specific measures such as integrity of the SVZ, or of the corticospinal tract, may be important, since patients with substantial damage to these structures may be incapable of responding to treatment. Some imaging measures, including anatomical involvement of the corticospinal tract (DeVetten et al., 2010) or functional MRI motor activation patterns (Marshal et al., 2009) may be valid predictors from very early after stroke, within hours or days of the event.

Clinical deficits need to be sufficiently great that change over time can be detected reliably. Clinical impairment early after stroke predicts the probability of subsequent recovery to independence. Since motor system damage is the most common feature of stroke, and most outcome assessments depend heavily on motor function, motor deficit is likely to constitute a selection criterion. More specific motor deficits, such as distal upper limb weakness, can be assessed with specific scales but restrict patient numbers, as seen in trials of upper limb rehabilitation (Wolf et al., 2005). In addition, there is a very wide variability in clinical motor assessments and final outcome, making individual patient prediction unlikely (Stinear, 2010). A proposed scheme for identifying potential responders that incorporates clinical assessment of motor function (shoulder abduction and finger extension strength), evidence of tract integrity using transcranial magnetic stimulation (TMS), and fractional anisotropy of the posterior limb of the internal capsule on brain MRI (Stinear, 2010) has recently undergone (Stinear et al., 2012) prospective validation. It has the considerable attraction of utilizing functional and structural information to inform investigators about the capacity of the motor system for reorganization and recovery and target recruitment to a more homogeneous patient group with reduced likelihood of recruiting patients with no likelihood of either spontaneous recovery or absent biological substrate. However, motor function is only one of many domains affected by stroke, and eligible patient numbers may be very small.

The possibility of selecting patients based on additional factors that define intrinsic stem cell mobilization such as SDF-1α levels (Bogoslovsky et al., 2011; Lee et al., 2010) has also been raised but not yet validated.

Timing of cell delivery after stroke will depend in part on biological factors related to the likely mechanism of action, but will also be constrained by the time necessary for preparation of cells (e.g., culture time for autologous MSCs), patient fitness, and delivery method. The probable biological targets for action of cells suggest that earlier administration after stroke would be more likely to yield the greatest benefit than late administration, but whether this should be within weeks, or months, is not known, and early Phase II trials may present an opportunity to test this issue by intervening at different time points. Patient numbers and natural recovery trajectories also must be considered. The number of patients with very severe stroke available for trial recruitment declines over time, as many succumb to other illnesses; mildly affected patients in contrast recover function spontaneously. The optimal point at which patients should be recruited will therefore require a balance of clinical factors that determine feasibility of patient selection, imaging, or neurophysiological criteria, logistical factors around cell preparation and delivery, and judgment about the probably targets for cell therapy.

14.3 Clinical endpoints

Clinical scales are generally used in trials and in stroke cover different levels of function, from assessment of neurological function through to assessment of participation, dependence, and quality of life. Ultimately, a gain in function sufficient to shift patients favorably across disability grades is the goal, but the necessarily broad terms of scales such as the mRS, the most commonly used scale in acute stroke research, render them of limited value in small early Phase II trials. It is likely that scales of impairment will offer a more sensitive index of biological activity. Acute neurological impairment scales such as the NIHSS may be useful as a screening tool, or to permit safety monitoring, since the clinical relevance of deterioration is generally agreed from acute stroke trials. However, these scales were designed for acute use and accordingly include some test items that are of little relevance at later time points (such as conscious level and conjugate gaze abnormalities), and some that will not be likely to constitute part of the outcome criteria (such as visual field assessment). On the other hand, the NIHSS includes only fairly crude assessments of limb function (eight points are scored for upper and lower limbs, concentrating on proximal strength), and the pattern of weakness that is sought is relevant for acute stroke but relatively uninformative for later stages, when distal upper limb and hand function is likely to have a major impact on daily functional abilities, and walking has for most patients, recovered to some extent. More detailed upper limb scales such as the Fugl-Meyer or Action Research Arm Test offer advantages at later stages of stroke. As noted above, the trade-off between patient numbers and specificity of the selected scale needs to be considered in study design and planning.

Phase II trials should primarily include measures of impairment since multiple confounding factors influence the eventual levels of disability or participation of an individual, and these are likely to vary across cultures and health care systems as well individual circumstances. This might also allow a distinction to be made between recovery of function and compensatory adaptation.

In Phase III trials, more general patient populations and broader scales of disability or handicap will likely be required to establish the generalizability of cell therapy.

14.4 Potential biomarkers

Lacking biochemical or neurophysiological markers of recovery, brain imaging currently offers the most promising prospect for a biomarker of therapeutic effect. However, the range of techniques described in the literature is wide, lacks standardization of either acquisition or analysis, and has predominantly been cross-sectional rather than longitudinal. MRI is best suited to follow-up since it offers the best level of anatomical resolution and is repeatable since ionizing radiation is not involved. However, at least 10% of patients are not compatible with MRI due to metallic implants (including pacemakers), and many do not tolerate scanning due to claustrophobia or physical discomfort in the enclosed environment, a particular issue in physically disabled patients. Radio-isotope methods such as PET may be helpful if tracers could discriminate specific processes, but can be repeated on only a limited number of occasions, which limits their usefulness.

Structural brain imaging is unlikely to be useful for reasons discussed above. Given the modes of action proposed for cell therapies, it is improbable that changes in brain structure will be observed with cell therapy even when delivered in the chronic phase. In the subacute stages, the natural evolution of brain infarction will be the dominant change.

Various newer techniques offer the possibility in future of investigating specific processes. Angiogenesis and microvascular density can be imaged with MRI methods (Seevinck et al., 2010), although many of these depend heavily upon mathematical models that may not apply in disease states and have yet to translate successfully to clinical use.

MRI techniques with which there is greater experience may offer indices of change over time. These include functional MRI to investigate changes in cortical motor activation patterns, MR spectroscopy to investigate changes in biochemical composition of relevant structures such as the internal capsule, diffusion tensor imaging to investigate the integrity of white matter tracts, and resting state connectivity analysis using the BOLD contrast technique to model brain networks. However, experience with any of these methods in stroke patients is limited, and there are significant technical challenges in analyzing data from subjects in whom there is substantial anatomical distortion as a result of stroke. Methods for acquisition or analysis of large and complex datasets are not standardized. It is possible that specific details of analysis will be unimportant, if consistent evidence of change over time is observed (Hodics et al., 2006), but systematic longitudinal data are presently scant.

14.5 Concomitant therapies

There is evidence that environmental factors are important in regenerative processes, at least in animal models. The general sensory stimulation of an "enriched environment" may be a requirement for pharmacological interventions to improve recovery (Feeney et al., 1982; Garcia-Alias et al., 2009). Physical therapy and other therapy inputs might have a complementary role to optimize recovery, but may also be a required substrate for any biological therapy intended to stimulate regenerative processes. Standardizing therapy inputs in terms of modality, intensity, and duration is a major challenge for rehabilitation trials that will apply to cell therapy trials also.

15 FUTURE PROSPECTS

Stroke is a major worldwide health problem for which therapeutic options are limited, and cell therapy offers the potential to improve neurological function and enhances recovery through a wide range of mechanisms. In addition, investigating the mechanisms of action underpinning cell therapies will, in itself, identify novel targets for therapy and enhance understanding of the natural capacity for repair and regeneration. These discoveries will therefore likely stimulate a new generation of clinical trials across a wider range of interventions that may include small molecules, growth factors, and physical interventions such as TMS. However, the significant challenges in moving beyond the current generation of small Phase I safety-focused trials to demanding Phase II and III trials cannot be underestimated (Savitz, 2012). Many issues are of generic importance to the field of regeneration and rehabilitation, including optimal patient selection, timing of intervention, and end-point selection: at present there are insufficient data to provide answers to these issues. The challenge of the next 5 years will be to utilize the forthcoming early Phase II studies to address these methodological issues and ensure that credible biological evidence of cell activity is confirmed (or indeed refuted) with sufficient confidence to progress to larger trials. It is likely that a combination of clinical assessments and advanced brain imaging will be needed to optimize the chance of achieving this goal with credible patient numbers. Detailed mechanistic understanding, while desirable, is not necessarily required before undertaking clinical studies, provided that there are sufficiently robust data from animal models of stroke to support a therapeutic effect.

Partnership between government funding bodies and small private companies developing allogeneic cell products will almost certainly be necessary to fulfill these goals. Large-scale pharmaceutical investment in stroke is unlikely after the failure of pharmacological neuroprotection trials, and equally, investigators are unlikely to be persuaded of the advisability of complex clinical trials without convincing evidence of biological activity of cell therapy. Larger scale trials of intravenously delivered autologous cells are potentially feasible as an academic undertaking, but will be very expensive, complex to standardize and quality-assure across centers, and share the common methodological issues highlighted above.

References

Aboody, K., Capela, A., Niazi, N., Stern, J.H., Temple, S., 2011. Translating stem cell studies to the clinic for CNS repair: current state of the art and the need for a Rosetta Stone. Neuron 70, 597–613.

Androutsellis-Theotokis, A., Leker, R.R., Soldner, F., Hoeppner, D.J., Ravin, R., Poser, S.W., Rueger, M.A., Bae, S.K., Kittappa, R., Mckay, R.D., 2006. Notch signalling regulates stem cell numbers in vitro and in vivo. Nature 442, 823–826.

Angels Font, M., Arboix, A., Krupinski, J., 2010. Angiogenesis, neurogenesis and neoplasticity in ischemic stroke. Curr. Cardiol. Rev. 6, 238–244.

Arvidsson, A., Collin, T., Kirik, D., Kokaia, Z., Lindvall, O., 2002. Neuronal replacement from endogenous precursors in the adult brain after stroke. Nat. Med. 8, 963–970.

Baker, A.H., Sica, V., Work, L.M., Williams-Ignarro, S., De Nigris, F., Lerman, L.O., Casamassimi, A., Lanza, A., Schiano, C., Rienzo, M., Ignarro, L.J., Napoli, C., 2007. Brain protection using autologous bone marrow cell, metalloproteinase inhibitors, and metabolic treatment in cerebral ischemia. Proc. Natl. Acad. Sci. U.S.A. 104, 3597–3602.

Bamford, J., Sandercock, P., Dennis, M., Burn, J., Warlow, C., 1991. Classification and natural history of clinically identifiable subtypes of cerebral infarction. Lancet 337, 1521–1526.

Bang, O.Y., Lee, J.S., Lee, P.H., Lee, G., 2005. Autologous mesenchymal stem cell transplantation in stroke patients. Ann. Neurol. 57, 874–882.

Banks, W.A., Jumbe, N.L., Farrell, C.L., Niehoff, M.L., Heatherington, A.C., 2004. Passage of erythropoietic agents across the blood-brain barrier: a comparison of human and murine erythropoietin and the analog darbepoetin alfa. Eur. J. Pharmacol. 505, 93–101.

Bannerjee, S., Bentley, P., Hamady, M., Davis, J., Shlebek, A., Gordon, M., Habib, J., Chataway, J., 2012. Phase I trial of autologous CD34+ haematopoietic stem cells in acute ischemic stroke. Cerebrovasc. Dis. Suppl. 2 published online.

Barbosa da Fonseca, L.M., Gutfilen, B., Rosado de Castro, P.H., Battisella, V., Goldenberg, R.C.S., Kasai-Brunswick, T., Chagas, C.L.R., Wajnberg, E., Maiolino, A., Xavier, S.S., Andre, C., Mendez-Otero, R., de Freitas, G.R., 2010. Migration and homing of bone-marrow mononuclear cells in chronic ischemic stroke after intra-arterial injection. Exp. Neurol. 221, 122–128.

Barnabe, C.C., Leclercq, S.A., Fitzgerald, A.A., 2009. The development of inflammatory arthritis and other rheumatic diseases following stem cell transplantation. Semin. Arthritis Rheum. 39, 55–60.

Bhardwaj, R.D., Curtis, M.A., Spalding, K.L., Buchholz, B.A., Fink, D., Bjork-Eriksson, T., Nordborg, C., Gage, F.H., Druid, H., Eriksson, P.S., Frisen, J., 2006. Neocortical neurogenesis in humans is restricted to development. Proc. Natl. Acad. Sci. U.S.A. 103, 12564–12568.

Bhasin, A., Srivastava, M.V., Kumaran, S.S., Mohanty, S., Bhatia, R., Bose, S., Gaikwad, S., Garg, A., Airan, B., 2011. Autologous mesenchymal stem cells in chronic stroke. Cerebrovasc. Dis. Extra 1, 93–104.

Bible, E., Dell'acqua, F., Solanky, B., Balducci, A., Crapo, P., Badylak, S.F., Ahrens, E.T., Modo, M., 2012. Non-invasive imaging of transplanted human neural stem cells and ECM scaffold remodelling in the stroke-damaged rat brain by (19)F- and diffusion-MRI. Biomaterials 33, 2858–2871.

Bliss, T., Guzman, R., Daadi, M., Steinberg, G.K., 2007. Cell transplantation therapy for stroke. Stroke 38, 817–826.

Bogoslovsky, T., Spatz, M., Chaudhry, A., Maric, D., Luby, M., Frank, J., Warach, S., the NINDS Natural History of Stroke Investigators, 2011. Stromal-derived factor-1[alpha] correlates with circulating endothelial progenitor cells and with acute lesion volume in stroke patients. Stroke 42, 618–625.

Borlongan, C.V., Tajima, Y., Trojanowski, J.Q., Lee, V.M., Sanberg, P.R., 1998. Transplantation of cryopreserved human embryonal carcinoma-derived neurons (NT2N cells) promotes functional recovery in ischemic rats. Exp. Neurol. 149, 310–321.

Borlongan, C.V., Lind, J.G., Dillon-Carter, O., Yu, G., Hadman, M., Cheng, C., Carroll, J., Hess, D.C., 2004. Bone marrow grafts restore cerebral blood flow and blood brain barrier in stroke rats. Brain Res. 1010, 108–116.

Boy, S., Sauerbruch, S., Kraemer, M., Schormann, T., Schlachetzki, F., Schuierer, G., Luerding, R., Hennemann, B., Orso, E., Dabringhaus, A., Winkler, J., Bogdahn, U., R.S. Group, 2011. Mobilisation of hematopoietic CD34+ precursor cells in patients with acute stroke is safe—results of an open-labeled non-randomized phase I/II trial. PLoS One 6, e23099.

Bratane, B.T., Bouley, J., Schneider, A., Bastan, B., Henninger, N., Fisher, M., 2009. Granulocyte-colony stimulating factor delays PWI/DWI mismatch evolution and reduces final infarct volume in permanent-suture and embolic focal cerebral ischemia models in the rat. Stroke 40, 3102–3106.

Brenneman, M., Sharma, S., Harting, M., Strong, R., Cox Jr., C.S., Aronowski, J., Grotta, J.C., Savitz, S.I., 2010. Autologous bone marrow mononuclear cells enhance recovery after acute ischemic stroke in young and middle-aged rats. J. Cereb. Blood Flow Metab. 30, 140–149.

Brines, M.L., Ghezzi, P., Keenan, S., Agnello, D., De Lanerolle, N.C., Cerami, C., Itri, L.M., Cerami, A., 2000. Erythropoietin crosses the blood-brain barrier to protect against experimental brain injury. Proc. Natl. Acad. Sci. U.S.A. 97, 10526–10531.

Burns, T.C., Verfaillie, C.M., Low, W.C., 2009. Stem cells for ischemic brain injury: a critical review. J. Comp. Neurol. 515, 25–44.

Burns, T.C., Steinberg, G.K., 2011. Stem cells and stroke: opportunities, challenges and strategies. Expert Opin. Biol. Ther. 11, 447–461.

Chen, J., Li, Y., Chopp, M., 2000. Intracerebral transplantation of bone marrow with BDNF after MCAo in rat. Neuropharmacology 39, 711–716.

Chen, J., Li, Y., Wang, L., Zhang, Z., Lu, D., Lu, M., Chopp, M., 2001a. Therapeutic benefit of intravenous administration of bone marrow stromal cells after cerebral ischemia in rats. Stroke 32, 1005–1011.

Chen, J., Li, Y., Wang, L., Lu, M., Chopp, M., 2001b. Therapeutic benefit of intracerebral transplantation of bone marrow stromal cells after cerebral ischemia in rats. J. Neurol. Sci. 189, 49–57.

Chen, J.R., Cheng, G.Y., Sheu, C.C., Tseng, G.F., Wang, T.J., Huang, Y.S., 2008. Transplanted bone marrow stromal cells migrate, differentiate and improve motor function in rats with experimentally induced cerebral stroke. J. Anat. 213, 249–258.

Chen, J., Ye, X., Yan, T., Zhang, C., Ynag, X.P., Cui, Y., Zacharek, A., Roberts, C., Liu, X., Dai, X., Lu, M., Chopp, M., 2011. Adverse effects of bone marrow stromal cell treatment of stroke in diabetic rats. Stroke 42, 3551–3558.

Chu, M., Hu, X., Lu, S., Gan, Y., Li, P., Guo, Y., Zhang, J., Chen, J., Gao, Y., 2012. Focal cerebral ischemia activates neurovascular restorative dynamics in mouse brain. Front. Biosci. (Elite Ed.) 4, 1926–1936.

Clinical Trials Gov Website: www.clinicaltrials.gov.

Colvin, M., Russo, J.E., Hilton, J., Dulik, D.M., Fenselau, C., 1988. Enzymatic mechanisms of resistance to alkylating agents in tumor cells and normal tissues. Adv. Enzyme Regul. 27, 211–221.

Daadi, M.M., Li, Z., Arac, A., Grueter, B.A., Sofilos, M., Malenka, R.C., Wu, J.C., Steinberg, G.K., 2009. Molecular and magnetic resonance imaging of human embryonic stem cell-derived neural stem cell grafts in ischemic rat brain. Mol. Ther. 17, 1282–1291.

Deacon, T.W., Pakzaban, P., Burns, L.H., Dinsmore, J., Isacson, O., 1994. Cytoarchitectonic development, axon-glia relationships, and long distance axon growth of porcine striatal xenografts in rats. Exp. Neurol. 130, 151–167.

DeVasconcelos Dos Santos, A., Da Costa Reis, J., Diaz Paredes, B., Moraes, L., Jasmin, Giraldi-Guimaraes, A., Mendez-Otero, R., 2010. Therapeutic window for treatment of cortical ischemia with bone marrow-derived cells in rats. Brain Res. 1306, 149–158.

DeVetten, G., Coutts, S.B., Hill, M.D., Goyal, M., Eesa, M., O'Brien, B., Demchuk, A.M., Kirton, A., Monitor & Vision study groups, 2010. Acute corticospinal tract Wallerian degeneration is associated with stroke outcome. Stroke 41, 751–756.

Dezawa, M., Kanno, H., Hoshino, M., Cho, H., Matsumoto, N., Itokazu, Y., Tajima, N., Yamada, H., Sawada, H., Ishikawa, H., Mimura, T., Kitada, M., Suzuki, Y., Ide, C., 2004.

Specific induction of neuronal cells from bone marrow stromal cells and application for autologous transplantation. J. Clin. Invest. 113, 1701–1710.

Dinsmore, J.H., Martin, J., Siegan, J., Morrison, J.P., Lindberg, C., Ratliff, J., Jacoby, D.I., 2002. CNS grafts for treatment of neurologic disorders. In: Lanza, R.P., Atala, A. (Eds.), Methods of Tissue Engineering. Academic Press, San Diego, pp. 1127–1134.

Dirnagl, U., Iadecola, C., Moskowitz, M.A., 1999. Pathobiology of ischemic stroke: an integrated view. Trends Neurosci. 22, 391–397.

Dominici, M., Le Blanc, K., Mueller, I., Slaper-Cortenbach, I., Marini, F., Krause, D., Deans, R., Keating, A., Prockop, D., Horwitz, E., 2006. Minimal criteria for defining multipotent mesenchymal stromal cells. The International Society for Cellular Therapy position statement. Cytotherapy 8, 315–317.

Eglitis, M.A., Mezey, E., 1997. Hematopoietic cells differentiate into both microglia and macroglia in the brains of adult mice. Proc. Natl. Acad. Sci. U.S.A. 94, 4080–4085.

Ehrenreich, H., Hasselblatt, M., Dembowski, C., Cepek, L., Lewczuk, P., Stiefel, M., Rustenbeck, H.H., Breiter, N., Jacob, S., Knerlich, F., Bohn, M., Poser, W., Ruther, E., Kochen, M., Gefeller, O., Gleiter, C., Wessel, T.C., De Ryck, M., Itri, L., Prange, H., Cerami, A., Brines, M., Siren, A.L., 2002. Erythropoietin therapy for acute stroke is both safe and beneficial. Mol. Med. 8, 495–505.

Ehrenreich, H., Weissenborn, K., Prange, H., Schneider, D., Weimar, C., Wartenberg, K., Schellinger, P.D., Bohn, M., Becker, H., Wegrzyn, M., Jahnig, P., Herrmann, M., Knauth, M., Bahr, M., Heide, W., Wagner, A., Schwab, S., Reichmann, H., Schwendemann, G., Dengler, R., Kastrup, A., Bartels, C., E.P.O.S.T. Group, 2009. Recombinant human erythropoietin in the treatment of acute ischemic stroke. Stroke 40, e647–e656.

England, T.J., Abaei, M., Auer, D.P., Lowe, J., Jones, D.R., Sare, G., Walker, M., Bath, P.M., 2012. Granulocyte-colony stimulating factor for mobilizing bone marrow stem cells in subacute stroke: the stem cell trial of recovery enhancement after stroke 2 randomized controlled trial. Stroke 43, 405–411.

Feeney, D.M., Gonzalez, A., Law, W.A., 1982. Amphetamine, haloperidol, and experience interact to affect rate of recovery after motor cortex injury. Science 217, 855–887.

Fischer, U.M., Harting, M.T., Jimenez, F., Monzon-Posadas, W.O., Xue, H., Savitz, S.I., Laine, G.A., Cox Jr., C.S., 2009. Pulmonary passage is a major obstacle for intravenous stem cell delivery: the pulmonary first-pass effect. Stem Cells Dev. 18, 683–692.

Ford, G.A., Ahmed, N., Azevedo, E., Grond, M., Larrue, V., Lindsberg, P.J., Toni, D., Wahlgren, N., 2010. Intravenous alteplase for stroke in those older than 80 years old. Stroke 41, 2568–2574.

Freed, C.R., Greene, P.E., Breeze, R.E., Tsai, W.Y., DuMouchel, W., Kao, R., Dillon, S., Winfield, H., Culver, S., Trojanowski, J.Q., Eidelberg, D., Fahn, S., 2001. Transplantation of embryonic dopamine neurons for severe Parkinson's disease. N. Engl. J. Med. 344, 710–719.

Garcia-Alias, G., Barkhuysen, S., Buckle, M., Fawcett, J.W., 2009. Chondroitinase ABC treatment opens a window of opportunity for task-specific rehabilitation. Nat. Neurosci. 12, 1145–1151.

Gibson, C.L., Bath, P.M., Murphy, S.P., 2005. G-CSF reduces infarct volume and improves functional outcome after transient focal cerebral ischemia in mice. J. Cereb. Blood Flow Metab. 25, 431–439.

Giraldi-Guimaraes, A., Rezende-Lima, M., Bruno, F.P., Mendez-Otero, R., 2009. Treatment with bone marrow mononuclear cells induces functional recovery and decreases neurodegeneration after sensorimotor cortical ischemia in rats. Brain Res. 1266, 108–120.

Gray, J.A., Grigoryan, G., Virley, D., Patel, S., Sinden, J.D., Hodges, H., 2000. Conditionally immortalized, multipotential and multifunctional neural stem cell lines as an approach to clinical transplantation. Cell Transplant. 9, 153–168.

Grigoriadis, N., Lourbopoulos, A., Lagoudaki, R., Frischer, J.-M., Polyzoidou, E., Touloumi, O., Simeonidou, C., Deretzi, G., Kountouras, J., Spandou, E., Kotta, K., Karkavelas, G., Tascos, N., Lassmann, H., 2011. Variable behavior and complications of autologous bone marrow mesenchymal stem cells transplanted in experimental autoimmune encephalomyelitis. Exp. Neurol. 230, 78–89.

Hacke, W., Kaste, M., Bluhmki, E., Brozman, M., Davalos, A., Guidetti, D., Larrue, V., Lees, K.R., Medeghri, Z., Machnig, T., Schneider, D., Von Kummer, R., Wahlgren, N., Toni, D., Investigators, E., 2008. Thrombolysis with alteplase 3 to 4.5 hours after acute ischemic stroke. N. Engl. J. Med. 359, 1317–1329.

Harting, M., Jimenez, F., Pati, S., Baumgartner, J., Cox Jr., C., 2008. Immunophenotype characterization of rat mesenchymal stromal cells. Cytotherapy 10, 243–253.

Hatashita, S., Hoff, J.T., 1990. Brain edema and cerebrovascular permeability during cerebral ischemia in rats. Stroke 21, 582–5888.

Hess, D.C., Borlongan, C.V., 2008. Stem cells and neurological diseases. Cell Prolif. 41 (Suppl. 1), 94–114.

Hicks, C., Stevanato, L., Stroemer, R.P., Tang, E., Richardson, S., Sinden, J.D., 2012. In vivo and in vitro characterization of the angiogenic effect of CTX0E03 human neural stem cells. Cell Transplant. Oct 12. [Epub ahead of print].

Hodics, T., Cohen, L.G., Cramer, S.C., 2006. Functional imaging of intervention effects in stroke motor rehabilitation. Arch. Phys. Med. Rehabil. 87, S36–S42.

Honmou, O., Houkin, K., Matsunaga, T., Niitsu, Y., Ishiai, S., Onodera, R., Waxman, S.G., Kocsis, J.D., 2011. Intravenous administration of auto serum-expanded autologous mesenchymal stem cells in stroke. Brain 134, 1790–1807.

Iihoshi, S., Honmou, O., Houkin, K., Hashi, K., Kocsis, J.D., 2004. A therapeutic window for intravenous administration of autologous bone marrow after cerebral ischemia in adult rats. Brain Res. 1007, 1–9.

Isacson, O., Deacon, T.W., Pakzaban, P., Galpern, W.R., Dinsmore, J., Burns, L.H., 1995. Transplanted xenogeneic neural cells in neurodegenerative disease models exhibit remarkable axonal target specificity and distinct growth patterns of glial and axonal fibres. Nat. Med. 1, 1189–1194.

Iwai, M., Cao, G., Yin, W., Stetler, R.A., Liu, J., Chen, J., 2007. Erythropoietin promotes neuronal replacement through revascularization and neurogenesis after neonatal hypoxia/ischemia in rats. Stroke 38, 2795–2803.

Jin, K., Wang, X., Xie, L., Mao, X.O., Zhu, W., Wang, Y., Shen, J., Mao, Y., Banwait, S., Greenberg, D.A., 2006. Evidence for stroke-induced neurogenesis in the human brain. Proc. Natl. Acad. Sci. U.S.A. 103, 13198–13202.

Kalladka, D., Muir, K.W., 2011. Stem cell therapy in stroke: designing clinical trials. Neurochem. Int. 59, 367–370.

Kamihata, H., Matsubara, H., Nishiue, T., Fujiyama, S., Tsutsumi, Y., Ozono, R., Masaki, H., Mori, Y., Iba, O., Tateishi, E., Kosaki, A., Shintani, S., Murohara, T., Imaizumi, T., Iwasaka, T., 2001. Implantation of bone marrow mononuclear cells into ischemic myocardium enhances collateral perfusion and regional function via side supply of angioblasts, angiogenic ligands, and cytokines. Circulation 104, 1046–1052.

Kamiya, N., Ueda, M., Igarashi, H., Nishiyama, Y., Suda, S., Inaba, T., Katayama, Y., 2008. Intra-arterial transplantation of bone marrow mononuclear cells immediately after reperfusion decreases brain injury after focal ischemia in rats. Life Sci. 83, 433–437.

Kang, S.K., Lee, D.H., Bae, Y.C., Kim, H.K., Baik, S.Y., Jung, J.S., 2003. Improvement of neurological deficits by intracerebral transplantation of human adipose tissue-derived stromal cells after cerebral ischemia in rats. Exp. Neurol. 183, 355–366.

Kawada, H., Takizawa, S., Takanashi, T., Morita, Y., Fujita, J., Fukuda, K., Takagi, S., Okano, H., Ando, K., Hotta, T., 2006. Administration of hematopoietic cytokines in the subacute phase after cerebral infarction is effective for functional recovery facilitating proliferation of intrinsic neural stem/progenitor cells and transition of bone marrow-derived neuronal cells. Circulation 113, 701–710.

Keller, L.H., 2009. Bone marrow-derived aldehyde dehydrogenase-bright stem and progenitor cells for ischemic repair. Congest. Heart Fail. 15, 202–206.

Kleindorfer, D., Lindsell, C.J., Brass, L., Koroshetz, W., Broderick, J.P., 2008. National US estimates of recombinant tissue plasminogen activator use: ICD-9 codes substantially underestimate. Stroke 39, 924–928.

Kleppner, S.R., Robinson, K.A., Trojanowski, J.Q., Lee, V.M., 1995. Transplanted human neurons derived from a teratocarcinoma cell line (NTera-2) mature, integrate, and survive for over 1 year in the nude mouse brain. J. Comp. Neurol. 357, 618–632.

Kokaia, Z., Thored, P., Arvidsson, A., Lindvall, O., 2006. Regulation of stroke-induced neurogenesis in adult brain—recent scientific progress. Cereb. Cortex 16 (Suppl. 1), i162–i167.

Kondziolka, D., Wechsler, L., 2008. Stroke repair with cell transplantation: neuronal cells, neuroprogenitor cells, and stem cells. Neurosurg. Focus 24, E12.

Kondziolka, D., Wechsler, L., Goldstein, S., Meltzer, C., Thulborn, K.R., Gebel, J., Jannetta, P., DeCesare, S., Elder, E.M., McGrogan, M., Reitman, M.A., Bynum, L., 2000. Transplantation of cultured human neuronal cells for patients with stroke. Neurology 55, 565–569.

Kondziolka, D., Steinberg, G.K., Wechsler, L., Meltzer, C.C., Elder, E., Gebel, J., Decesare, S., Jovin, T., Zafonte, R., Lebowitz, J., Flickinger, J.C., Tong, D., Marks, M.P., Jamieson, C., Luu, D., Bell-Stephens, T., Teraoka, J., 2005. Neurotransplantation for patients with subcortical motor stroke: a Phase 2 randomized trial. J. Neurosurg. 103, 38–45.

Lamfers, M., Idema, S., Van Milligen, F., Schouten, T., Van Der Valk, P., Vandertop, P., Dirven, C., Noske, D., 2009. Homing properties of adipose-derived stem cells to intracerebral glioma and the effects of adenovirus infection. Cancer Lett. 274, 78–87.

LeBlanc, K., Ringdén, O., 2007. Immunomodulation by mesenchymal stem cells and clinical experience. J. Intern. Med. 262, 509–525.

Lee, J.S., Hong, J.M., Moon, G.J., Lee, P.H., Ahn, Y.H., Bang, O.Y., S. Collaborators, 2010. A long-term follow-up study of intravenous autologous mesenchymal stem cell transplantation in patients with ischemic stroke. Stem Cells 28, 1099–1106.

Lee, T.-H., Yoon, J.-G., 2008. Intracerebral transplantation of human adipose tissue stromal cells after middle cerebral artery occlusion in rats. J. Clin. Neurosci. 15, 907–912.

Leker, R.R., Soldner, F., Velasco, I., Gavin, D.K., Androutsellis-Theotokis, A., Mckay, R.D., 2007. Long-lasting regeneration after ischemia in the cerebral cortex. Stroke 38, 153–161.

Li, Y., Chen, J., Chopp, M., 2001a. Adult bone marrow transplantation after stroke in adult rats. Cell Transplant. 10, 31–40.

Li, Y., Chen, J., Wang, L., Lu, M., Chopp, M., 2001b. Treatment of stroke in rat with intracarotid administration of marrow stromal cells. Neurology 56, 1666–1672.

Li, Y., Chen, J., Chen, X.G., Wang, L., Gautam, S.C., Xu, Y.X., Katakowski, M., Zhang, L.J., Lu, M., Janakiraman, N., Chopp, M., 2002. Human marrow stromal cell therapy for stroke in rat: neurotrophins and functional recovery. Neurology 59, 514–523.

Li, Y., Chopp, M., 2009. Marrow stromal cell transplantation in stroke and traumatic brain injury. Neurosci. Lett. 456, 120–123.

Li, W.Y., Choi, Y.J., Lee, P.H., Huh, K., Kang, Y.M., Kim, H.S., Ahn, Y.H., Lee, G., Bang, O.Y., 2008. Mesenchymal stem cells for ischemic stroke: changes in effects after ex vivo culturing. Cell Transplant. 17, 1045–1059.

Lin, G., Garcia, M., Ning, H., Guo, Y.-L., Lue, T.F., Lin, C.-S., 2008. Defining stem and progenitor cells within adipose tissue. Stem Cells Dev. 17, 1053–1064.

Lindenberg, R., Renga, V., Zhu, L.L., Betzler, F., Alsop, D., Schlaug, G., 2010. Structural integrity of corticospinal motor fibers predicts motor impairment in chronic stroke. Neurology 74, 280–287.

Lindvall, O., Kokaia, Z., 2011. Stem cell research in stroke: how far from the clinic? Stroke 42, 2369–2375.

Liu, X.B., Wang, J.A., Yu, S.P., Keogh, C.L., Wei, L., 2008. Therapeutic strategy of erythropoietin in neurological disorders. CNS Neurol. Disord. Drug Targets 7, 227–234.

Liu, Z., Li, Y., Qu, R., Shen, L., Gao, Q., Zhang, X., Lu, M., Savant-Bhonsale, S., Borneman, J., Chopp, M., 2007. Axonal sprouting into denervated spinal cord and synaptic and postsynaptic protein expression in the spinal cord after transplantation of bone marrow stromal cell in stroke rats. Brain Res. 148, 172–180.

MacDougall, N.J., Muir, K.W., 2011. Hyperglycaemia and infarct size in animal models of middle cerebral artery occlusion: systematic review and meta-analysis. J. Cereb. Blood Flow Metab. 31 (3), 807–818.

Macrae, I.M., 2011. Preclinical stroke research—advantages and disadvantages of the most common rodent models of focal ischaemia. Br. J. Pharmacol. 164, 1062–1078.

Marshal, R.S., Zarahn, E., Alon, L., Minzer, B., Lazar, R.M., Krakauer, J.W., 2009. Early imaging correlates of subsequent motor recovery after stroke. Ann. Neurol. 65, 596–602.

Mays, R.W., Borlongan, C.V., Yasuhara, T., Hara, K., Maki, M., Carroll, J.E., Deans, R.J., Hess, D.C., 2010. Development of an allogeneic adherent stem cell therapy for treatment of ischemic stroke. J. Exp. Stroke Transl. Med. 3, 34–46.

McColgan, P., Sharma, P., Bentley, P., 2011. Stem cell tracking in human trials: a meta-regression. Stem Cell Rev. 7 (4), 1031–1040.

Mendez-Otero, R., De Freitas, G.R., Andre, C., De Mendonca, M.L., Friedrich, M., Oliveira-Filho, J., 2007. Potential roles of bone marrow stem cells in stroke therapy. Regen. Med. 2, 417–423.

Miljan, E.A., Sinden, J.D., 2009. Stem cell treatment of ischemic brain injury. Curr. Opin. Mol. Ther. 11, 394–403.

Mimura, T., Dezawa, M., Kanno, H., Yamamoto, I., 2005. Behavioral and histological evaluation of a focal cerebral infarction rat model transplanted with neurons induced from bone marrow stromal cells. J. Neuropathol. Exp. Neurol. 64, 1108–1117.

Minger, S.L., Ekonomou, A., Carta, E.M., Chinoy, A., Perry, R.H., Ballard, C.G., 2007. Endogenous neurogenesis in the human brain following cerebral infarction. Regen. Med. 2, 69–74.

Minnerup, J., Heidrich, J., Wellmann, J., Rogalewski, A., Schneider, A., Schabitz, W.R., 2008. Meta-analysis of the efficacy of granulocyte-colony stimulating factor in animal models of focal cerebral ischemia. Stroke 39, 1855–1861.

Modo, M., Beech, J.S., Meade, T.J., Williams, S.C., Price, J., 2009. A chronic 1 year assessment of MRI contrast agent-labelled neural stem cell transplants in stroke. Neuroimage 47, T133–T142.

Muir, K.W., 2002. Heterogeneity of stroke pathophysiology and neuroprotective clinical trial design. Stroke 33, 1545–1550.

Muir, K.W., Teal, P.A., 2005. Why have neuro-protectants failed?: lessons learned from stroke trials. J. Neurol. 252, 1011–1020.

Muir, K.W., Sinden, J., Miljan, E., Dunn, L., 2011. Intracranial delivery of stem cells. Transl. Stroke Res. 2, 266–271.

Nelson, P.T., Kondziolka, D., Wechsler, L., Goldstein, S., Gebel, J., DeCesare, S., Elder, E.M., Zhang, P.J., Jacobs, A., McGrogan, M., Lee, V.M.-Y., Trojanowski, J.Q., 2002. Clonal human (hNT) neuron grafts for stroke therapy: neuropathology in a patient 27 months after implantation. Am. J. Pathol. 160, 1201–1206.

Olanow, C.W., Goetz, C.G., Kordower, J.H., Stoesse, J., Sossi, V., Brin, M.F., Shannon, K.M., Nauert, M., Perl, D.P., 2003. A double-blind controlled trial of bilateral fetal nigral transplantation in Parkinson's disease. Ann. Neurol. 54, 403–414.

Ourednik, J., Ourednik, V., Lynch, W.P., Schachner, M., Synder, E.Y., 2002. Neural stem cells display an inherent mechanism for rescuing dysfunctional neurons. Nat. Biotechnol. 20, 1103–1110.

Parsons, M., Spratt, N., Bivard, A., Campbell, B., Chung, K., Miteff, F., O'Brien, B., Bladin, C., McElduff, P., Allen, C., Bateman, G., Donnan, G., Davis, S., Levi, C., 2012. A randomized trial of tenecteplase versus alteplase for acute ischemic stroke. N. Engl. J. Med. 366, 1099–1107.

Patel, A.T., Duncan, P.W., Lai, S.-M., Studenski, S., 2000. The relation between impairments and functional outcomes poststroke. Arch. Phys. Med. Rehabil. 81, 1357–1363.

Pendharkar, A.V., Chua, J.Y., Andres, R.H., Wang, N., Gaeta, X., Wang, H., De, A., Choi, R., Chen, S., Rutt, B.K., Ganbhir, S.S., Guzman, R., 2010. Biodistribution of neural stem cells after intravascular therapy for hypoxic-ischemia. Stroke 41, 2064–2070.

Perin, E.C., Dohmann, H.F., Borojevic, R., Silva, S.A., Sousa, A.L., Mesquita, C.T., Rossi, M.I., Carvalho, A.C., Dutra, H.S., Dohmann, H.J., Silva, G.V., Belem, L., Vivacqua, R., Rangel, F.O., Esporcatte, R., Geng, Y.J., Vaughn, W.K., Assad, J.A., Mesquita, E.T., Willerson, J.T., 2003. Transendocardial, autologous bone marrow cell transplantation for severe, chronic ischemic heart failure. Circulation 107, 2294–2302.

Pollock, K., Sinden, J., 2008. Progressing neural stem cell lines to the clinic. In: Shi, Y., Clegg, D.O. (Eds.), Stem Cell Research and Therapeutics. Springer, The Netherlands, pp. 105–122.

Pollock, K., Stroemer, P., Patel, S., Stevanato, L., Hope, A., Miljan, E., Dong, Z., Hodges, H., Price, J., Sinden, J.D., 2006. A conditionally immortal clonal stem cell line from human cortical neuroepithelium for the treatment of ischemic stroke. Exp. Neurol. 199, 143–155.

Popa-Wagner, A., Buga, A.M., Kokaia, Z., 2011. Perturbed cellular response to brain injury during aging. Ageing Res. Rev. 10, 71–79.

Riley, J.D., Le, V., Der-Yeghiaian, L., See, J., Newton, J.M., Ward, N.S., Cramer, S.C., 2011. Anatomy of stroke injury predicts gains from therapy. Stroke 42, 421–426.

Ringelstein, E.B., 2012. AXIS-2 Investigators, presented at International Stroke Conference.

Russo, J.E., Hilton, J., Colvin, O.M., 1989. The role of aldehyde dehydrogenase isozymes in cellular resistance to the alkylating agent cyclophosphamide. Prog. Clin. Biol. Res. 290, 65–79.

Safford, K.M., Hicok, K.C., Safford, S.D., Halvorsen, Y.D., Wilkison, W.O., Gimble, J.M., Rice, H.E., 2002. Neurogenic differentiation of murine and human adipose-derived stromal cells. Biochem. Biophys. Res. Commun. 294, 371–379.

Sarnowska, A., Braun, H., Sauerzweig, S., Reymann, K.G., 2009. The neuroprotective effect of bone marrow stem cells is not dependent on direct cell contact with hypoxic injured tissue. Exp. Neurol. 215, 317–327.

Savitz, S.I., 2012. Stem cells and stroke: are we further away than anyone is willing to admit? Int. J. Stroke 7, 34–35.

Savitz, S.I., Dinsmore, J., Wu, J., Henderson, G.V., Stieg, P., Caplan, L.R., 2005. Neurotransplantation of fetal porcine cells in patients with basal ganglia infarcts: a preliminary safety and feasibility study. Cerebrovasc. Dis. 20, 101–107.

Savitz, S.I., Chopp, M., Deans, R., Carmichael, S.T., Phinney, D., Wechsler, L., 2011a. Stem Cell Therapy as an Emerging Paradigm for Stroke (STEPS) II. Stroke 42, 825–829.

Savitz, S.I., Misra, V., Kasam, M., Juneja, H., Cox Jr., C.S., Alderman, S., Aisiku, I., Kar, S., Gee, A., Grotta, J.C., 2011b. Intravenous autologous bone marrow mononuclear cells for ischemic stroke. Ann. Neurol. 70, 59–69.

Schabitz, W.R., Kollmar, R., Schwaninger, M., Juettler, E., Bardutzky, J., Scholzke, M.N., Sommer, C., Schwab, S., 2003. Neuroprotective effect of granulocyte colony-stimulating factor after focal cerebral ischemia. Stroke 34, 745–751.

Schabitz, W.R., Laage, R., Vogt, G., Koch, W., Kollmar, R., Schwab, S., Schneider, D., Hamann, G.F., Rosenkranz, M., Veltkamp, R., Fiebach, J.B., Hacke, W., Grotta, J.C., Fisher, M., Schneider, A., 2010. AXIS: a trial of intravenous granulocyte colony-stimulating factor in acute ischemic stroke. Stroke 41, 2545–2551.

Schrepfer, S., Deuse, T., Reichenspurner, H., Fischbein, M.P., Robbins, R.C., Pelletier, M.P., 2007. Stem cell transplantation: the lung barrier. Transplant. Proc. 39, 573–576.

Seevinck, P.R., Deddens, L.H., Dijkhuizen, R.M., 2010. Magnetic resonance imaging of brain angiogenesis after stroke. Angiogenesis 13, 101–111.

Sena, E.S., Van Der Worp, H.B., Bath, P.M., Howells, D.W., Macleod, M.R., 2010. Publication bias in reports of animal stroke studies leads to major overstatement of efficacy. PLoS Biol. 8, e1000344.

Shen, L.H., Li, Y., Chen, J., Zhang, J., Vanguri, P., Borneman, J., Chopp, M., 2006. Intracarotid transplantation of bone marrow stromal cells increases axon-myelin remodeling after stroke. Neuroscience 137, 393–399.

Shen, L.H., Li, Y., Chen, J., Cui, Y., Zhang, C., Kapke, A., Lu, M., Savant-Bhonsale, S., Chopp, M., 2007. One-year follow-up after bone marrow stromal cell treatment in middle-aged female rats with stroke. Stroke 38, 2150–2156.

Shyu, W.C., Lin, S.Z., Yang, H.I., Tzeng, Y.S., Pang, C.Y., Yen, P.S., Li, H., 2004. Functional recovery of stroke rats induced by granulocyte colony-stimulating factor-stimulated stem cells. Circulation 110, 1847–1854.

Shyu, W.C., Lee, Y.J., Liu, D.D., Lin, S.Z., Li, H., 2006a. Homing genes, cell therapy and stroke. Front. Biosci. 11, 899–907.

Shyu, W.C., Lin, S.Z., Lee, C.C., Liu, D.D., Li, H., 2006b. Granulocyte colony-stimulating factor for acute ischemic stroke: a randomized controlled trial. CMAJ 174, 927–933.

Sinden, J.D., Muir, K.W., 2012. Stem cells in stroke treatment: the promise and the challenges. Int. J. Stroke 7, 426–434.

Siren, A.L., Fasshauer, T., Bartels, C., Ehrenreich, H., 2009. Therapeutic potential of erythropoietin and its structural or functional variants in the nervous system. Neurotherapeutics 6, 108–127.

Smith, E.J., Stroemer, R.P., Gorenkova, N., Nakajima, M., Crum, W.R., Tang, E., Stevanato, L., Sinden, J.D., Modo, M., 2012. Implantation site and lesion topology determine efficacy of a human neural stem cell line in a rat model of chronic stroke. Stem Cells 30, 785–796.

Spees, J.L., Gregory, C.A., Singh, H., Tucker, H.A., Peister, A., Lynch, P.J., Hsu, S.C., Smith, J., Prockop, D.J., 2004. Internalized antigens must be removed to prepare hypoimmunogenic mesenchymal stem cells for cell and gene therapy. Mol. Ther. 9, 747–756.

Sprigg, N., Bath, P.M., Zhao, L., Willmot, M.R., Gray, L.J., Walker, M.F., Dennis, M.S., Russell, N., 2006. Granulocyte-colony-stimulating factor mobilizes bone marrow stem cells in patients with subacute ischemic stroke: the Stem cell Trial of recovery EnhanceMent after Stroke (STEMS) pilot randomized, controlled trial (ISRCTN 16784092). Stroke 37, 2979–2983.

Stevanato, L., Corteling, R.L., Stroemer, P., Hope, A., Heward, J., Miljan, E.A., Sinden, J.D., 2009. c-MycER^{TAM} transgene silencing in a genetically modified human neural stem cell line implanted into MCAo rodent brain. BMC Neurosci. 10, 86.

Stinear, C.M., 2010. Prediction of recovery of motor function after stroke. Lancet Neurol. 9, 1228–1232.

Stinear, C.M., Barber, P.A., Petoe, M., Anwar, S., Byblow, W.D., 2012. The PREP algorithm predicts potential for upper limb recovery after stroke. Brain 135 (Pt 8), 2527–2535.

Strauer, B.E., Brehm, M., Zeus, T., Kostering, M., Hernandez, A., Sorg, R.V., Kogler, G., Wernet, P., 2002. Repair of infarcted myocardium by autologous intracoronary mononuclear bone marrow cell transplantation in humans. Circulation 106, 1913–1918.

Stroemer, P., Patel, S., Hope, A., Oliveira, C., Pollock, K., Sinden, J., 2009. The neural stem cell line CTX0E03 promotes behavioral recovery and endogenous neurogenesis after experimental stroke in a dose-dependent fashion. Neurorehabil. Neural Repair 23, 895–909.

Suarez-Monteagudo, C., Hernandez-Ramirez, P., Alvarez-Gonzalez, L., Garcia-Maeso, I., De La Cuetara-Bernal, K., Castillo-Diaz, L., Bringas-Vega, M.L., Martinez-Aching, G., Morales-Chacon, L.M., Baez-Martin, M.M., Sanchez-Catasus, C., Carballo-Barreda, M., Rodriguez-Rojas, R., Gomez-Fernandez, L., Alberti-Amador, E., Macias-Abraham, C., Balea, E.D., Rosales, L.C., Del Valle Perez, L., Ferrer, B.B., Gonzalez, R.M., Bergado, J.A., 2009. Autologous bone marrow stem cell neurotransplantation in stroke patients. An open study. Restor. Neurol. Neurosci. 27, 151–161.

Thored, P., Arvidsson, A., Cacci, E., Ahlenius, H., Kallur, T., Darsalia, V., Ekdahl, C.T., Kokaia, Z., Lindvall, O., 2006. Persistent production of neurons from adult brain stem cells during recovery after stroke. Stem Cells 24, 739–747.

Tolar, J., Nauta, A.J., Osborn, M.J., Panoskaltsis Mortari, A., Mcelmurry, R.T., Bell, S., Xia, L., Zhou, N., Riddle, M., Schroeder, T.M., Westendorf, J.J., Mcivor, R.S., Hogendoorn, P.C., Szuhai, K., Oseth, L., Hirsch, B., Yant, S.R., Kay, M.A., Peister, A., Prockop, D.J., Fibbe, W.E., Blazar, B.R., 2007. Sarcoma derived from cultured mesenchymal stem cells. Stem Cells 25, 371–379.

Trojanowski, J.Q., Mantione, J.R., Lee, J.H., Seid, D.P., You, T., Inge, L.J., Lee, V.M., 1993. Neurons derived from a human teratocarcinoma cell line establish molecular and structural polarity following transplantation into the rodent brain. Exp. Neurol. 122, 283–294.

Uccelli, A., Moretta, L., Pistoia, V., 2008. Mesenchymal stem cells in health and disease. Nat. Rev. Immunol. 8, 726–736.

Wang, L., Zhang, Z., Wang, Y., Zhang, R., Chopp, M., 2004. Treatment of stroke with erythropoietin enhances neurogenesis and angiogenesis and improves neurological function in rats. Stroke 35, 1732–1737.

Wolf, S.L., Thompson, P.A., Morris, D.M., Rose, D.K., Winstein, C.J., Taub, E., Guiliani, C., Pearson, S.L., 2005. The EXCITE trial: attributes of the Wolf motor function test in patients with subacute stroke. Neurorehabil. Neural Repair 19, 194–205.

Woodbury, D., Schwarx, E.J., Prockop, D.J., Black, I.B., 2000. Adult rat and human bone marrow stromal cells differentiate into neurons. J. Neurosci. Res. 61, 364–370.

Wu, G., Lyapina, S., Das, I., Li, J., Gurney, M., Pauley, A., et al., 2001. SEL-10 is an inhibitor of notch signaling that targets notch for ubiquitin-mediated protein degradation. Mol. Cell. Biol. 21, 7403–7415.

Xiong, Y., Mahmood, A., Meng, Y., Zhang, Y., Qu, C., Schallert, T., Chopp, M., 2010. Delayed administration of erythropoietin reducing hippocampal cell loss, enhancing angiogenesis and neurogenesis, and improving functional outcome following traumatic brain injury in rats: comparison of treatment with single and triple dose. J. Neurosurg. 113, 598–608.

Yang, M., Wei, X., Li, J., Heine, L.A., Rosenwasser, R., Iacovitti, L., 2010. Changes in host blood factors and brain glia accompanying the functional recovery after systemic administration of bone marrow stem cells in ischemic stroke rats. Cell Transplant. 19, 1073–1084.

Yang, B., Strong, R., Sharma, S., Brenneman, M., Mallikarjunarao, K., Xi, X., Grotta, J.C., Aronowski, J., Savitz, S.I., 2011a. Therapeutic time window and dose response of autologous bone marrow mononuclear cells for ischemic stroke. J. Neurosci. Res. 89, 833–839.

Yang, Y.C., Liu, B.S., Shen, C.C., Lin, C.H., Chiao, M.T., Cheng, H.C., 2011b. Transplantation of adipose tissue-derived stem cells for treatment of focal cerebral ischemia. Curr. Neurovasc. Res. 8, 1–13.

Yasuhara, T., Hara, K., Maki, M., Mays, R.W., Deans, R.J., Hess, D.C., Carroll, J.E., Borlongan, C.V., 2008. Intravenous grafts recapitulate the neurorestoration afforded by intracerebrally delivered multipotent adult progenitor cells in neonatal hypoxic-ischemic rats. J. Cereb. Blood Flow Metab. 28, 1804–1810.

Yasuhara, T., Matsukawa, N., Hara, K., Maki, M., Ali, M.M., Yu, S.J., Bae, E., Yu, G., Xu, L., Mcgrogan, M., Bankiewicz, K., Case, C., Borlongan, C.V., 2009. Notch-induced rat and human bone marrow stromal cell grafts reduce ischemic cell loss and ameliorate behavioral deficits in chronic stroke animals. Stem Cells Dev. 18, 1501–1514.

Zaleska, M.M., Mercado, M.L., Chavez, J., Feuerstein, G.Z., Pangalos, M.N., Wood, A., 2009. The development of stroke therapeutics: promising mechanisms and translational challenges. Neuropharmacology 56, 329–341.

Zea Longa, E., Weinstein, P.R., Carlson, S., Cummins, R., 1989. Reversible middle cerebral artery occlusion without craniectomy in rats. Stroke 20, 84–91.

Zechariah, A., ElAli, A., Hermann, D.M., 2010. Combination of tissue-plasminogen activator with erythropoietin induces blood-brain barrier permeability, extracellular matrix disaggregation, and DNA fragmentation after focal cerebral ischemia in mice. Stroke 41, 1008–1012.

Zhang, Z.G., Chopp, M., 2009. Neurorestorative therapies for stroke: underlying mechanisms and translation to the clinic. Lancet Neurol. 8, 491–500.

Zuk, P.A., Zhu, M., Mizuno, H., Huang, J., Futrell, J.W., Katz, A.J., Benhaim, P., Lorenz, H.P., Hedrick, M.H., 2001. Multilineage cells from human adipose tissue: implications for cell-based therapies. Tissue Eng. 7, 211–228.

SECTION

3

Retinal and cortical repair

CHAPTER 9

Generation of retinal cells from pluripotent stem cells

Zi-Bing Jin*, Masayo Takahashi†,‡,1

*Laboratory for Stem Cell & Retinal Regeneration, Center for Vision Research, The Eye Hospital, School of Ophthalmology & Optometry, Wenzhou Medical College, Wenzhou, China

†Laboratory for Retinal Regeneration, RIKEN Center for Developmental Biology, Kobe, Japan

‡Center for iPS Research and Application, Kyoto University, Kyoto, Japan

[1]Corresponding author. Tel.: +81-78-306-3305, Fax: +81-78-306-3303, e-mail address: mretina@cdb.riken.jp

Abstract

Retinal degeneration is a leading cause of incurable low vision and blindness worldwide. Most retinal degenerative diseases are caused by irreversible apoptosis of retinal neural cells or adjacent supporting tissue. Because there is no radical treatment for retinal degeneration, most therapies are aimed at specific situations, such as drug or surgical intervention for late complications. Retinal cell replacement would be valuable for regenerating functional retinas, and therefore it is being examined as a next-generation treatment for retinal degeneration. With advances in stem cell biology, considerable progress has been made in recent years on generation of retinal cells. Both sensory retinal neural cells and retinal pigment epithelial cells can be induced *in vitro* from pluripotent stem cells, including embryonic stem cells and induced pluripotent stem cells. Here, we review the stepwise differentiation of retinal cells from pluripotent stem cells, with emphases on the methodology and application potential.

Keywords

differentiation, photoreceptor, pluripotent stem cell, RPE, retina

1 INTRODUCTION

The retina is part of the central nervous system and functions as the "first brain" in handling and processing visual signals. The human retina comprises two major parts (Fig. 1). The neural retina is a high cell-density tissue with high metabolic activity and is essential for phototransduction and visual signal transmission. The rod and cone photoreceptors of the neural retina are responsible for initiating the

Progress in Brain Research, Volume 201, ISSN 0079-6123, http://dx.doi.org/10.1016/B978-0-444-59544-7.00008-1

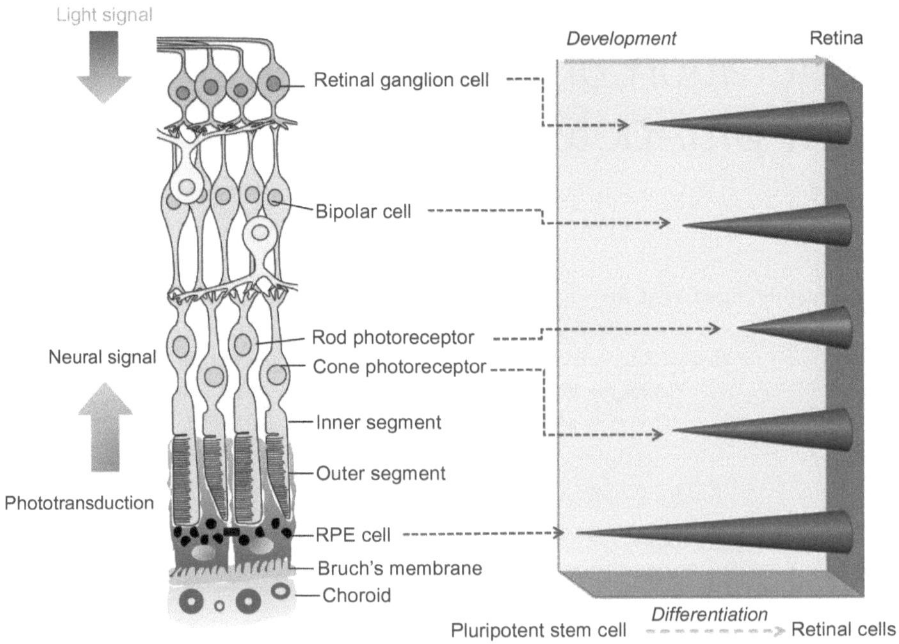

FIGURE 1

Schema of retinal cell function and developmental sequence. Different types of retinal neural cells and adjacent RPE cells (shown at the left) that develop sequentially during embryogenesis (right). (For color version of this figure, the reader is referred to the online version of this chapter.)

phototransduction cascade, a unique biological conversion of light signals to electronic signals that are transmitted to the next order neurons—the bipolar cells and retinal ganglion cells (RGCs)—and finally to the visual cortex to produce fundamental vision. The other major component of the retina is the retinal pigment epithelial (RPE) cells, a monolayered epithelial layer that functions in light absorption; transport of ions, nutrients, metabolites, and fluid; and secretion of neurotrophic factors, which are critical for photoreceptor cell survival and maintenance of function. Cell death of any of these cells leads to retinal degenerative diseases, which are the major cause of inevitable blindness throughout the world (Pizzarello et al., 2004). According to a World Health Organization report 2002 (www.who.int), age-related macular degeneration (AMD) caused by death of RPE and photoreceptor cells is a leading and increasingly significant cause of incurable blindness. In addition, glaucoma in which RGCs undergo apoptosis, diabetic retinopathy in which retinal capillary abnormalities result in neural degeneration, and genetic retinal degeneration in which RPE or retinal neural cells undergo apoptosis are other leading causes of inevitable blindness worldwide. Like other neural degenerative diseases of the central nervous system, these diseases are irreversible once the specific retinal neural cells have degenerated.

Pluripotent stem cells (PSCs) are a group of cells that can maintain an undifferentiated state indefinitely and can differentiate into most, if not all, cells of the body. Embryonic stem cells (ESCs) have been the best-studied PSCs since their first establishment from the inner cell mass of a mouse blastocyst by Evans and Kaufman (1981). Twenty-five years later, another type of PSC was established when Takahashi and Yamanaka reprogrammed somatic cells into induced PSCs (iPSCs), which are almost the same as ESCs (Takahashi and Yamanaka, 2006). At present, these two types of PSCs are the most important in regenerative medicine. In this review, we attempt to retrospectively summarize retinal differentiation from PSCs, with emphases on the methodology and translational potential.

2 EMBRYOGENESIS AND RETINAL DEVELOPMENT

Our knowledge of mammalian developmental biology is the basis of regenerative medicine. During early embryogenesis, the neural tube is folded from the neural plate, and its rostral end gives rise to the rhombencephalon, mesencephalon, and prosencephalon, which further develop into the neural retina and RPE cells. After the neural plate stage, lateral aspects of the anterior neural tube evaginate to form optic vesicles (OVs). The optic cup then forms with two layers. The inner layer, containing neural retinal progenitor cells, gives rise to different types of neural retinal cells in a temporal sequence (Fig. 1). The outer layer contains RPE progenitor cells that form RPE cells.

Differentiation and maturation of retinal cells are extremely complex processes, which are determined/regulated by a number of intrinsic and extrinsic factors (Hennig et al., 2008). It is essential to identify marker proteins at each stage of retinal development. As early as the neural plate stage, a set of eye field transcription factors (EFTFs), Pax6, Rx, Six3, Six6, and Lhx2, determine eye field specifications. Loss of EFTF function results in anophthalmos, nanophthalmos, or significant eye abnormality, whereas EFTF overexpression leads to ectopic eyes or retinal hyperproliferation (Chow et al., 1999; Furukawa et al., 1997; Lagutin et al., 2003; Mathers et al., 1997). However, it remains to be elucidated how EFTFs function in coordination to specify retinal differentiation and how the network is regulated during development. In *Xenopus*, overexpression of Frizzled-3, a Wnt receptor, leads to ectopic eyes (Cavodeassi et al., 2005). In contrast, loss of Dickkopf-1 (Dkk-1), which inhibits canonical Wnt signaling, produces anophthalmos (Mukhopadhyay et al., 2001). These results suggest that Wnt signaling is essential for eye specification. Other studies have shown that BMP and Notch signaling are important for regulating the EFTF network. For instance, the BMP inhibitor, Noggin, induces Pax6, Rx, Six3, and Lhx2 expression, indicating a positive role for Noggin in eye development (Zuber et al., 2003). The homeodomain transcription factors, Crx and Otx2, are essential for photoreceptor cell specification (Swaroop et al., 2010). Neural retinal progenitors give rise to postmitotic photoreceptor precursors with Crx expression and thereafter differentiate into rod or cone photoreceptor cells. Extrinsically, retinoic acid (RA) promotes photoreceptor maturation, and taurine enhances rod photoreceptor differentiation (Hyatt et al., 1996). In mice, neural retinal progenitors in the inner layer of the optic cup co-express Pax and Rx at early stage (E10.5), whereas the outer layer that differentiates into RPE cells co-expresses Pax6 and Mitf (Ikeda et al., 2005).

3 INDUCTION OF RETINAL PROGENITORS FROM PLURIPOTENT STEM CELLS

Retinal progenitor cells are multipotent throughout retinal development and can differentiate into different types of retinal cells. Recapitulation of neural development is essential to generate neural progenitor cells from PSCs *in vitro*. Developmentally, inactivation of BMP signaling is sufficient for induction of neural ectoderm.

There have been several milestone studies on the differentiation of retinal progenitors from PSCs (Fig. 2). Kawasaki et al. (2002) first identified a PA6 stromal cell with inducing activity (stromal cell-derived inducing activity, SDIA), which efficiently induces midbrain tyrosine hydroxylase-positive dopaminergic neurons and Pax6+ RPE cells. Ikeda et al. (2005) established a serum-free embryoid body

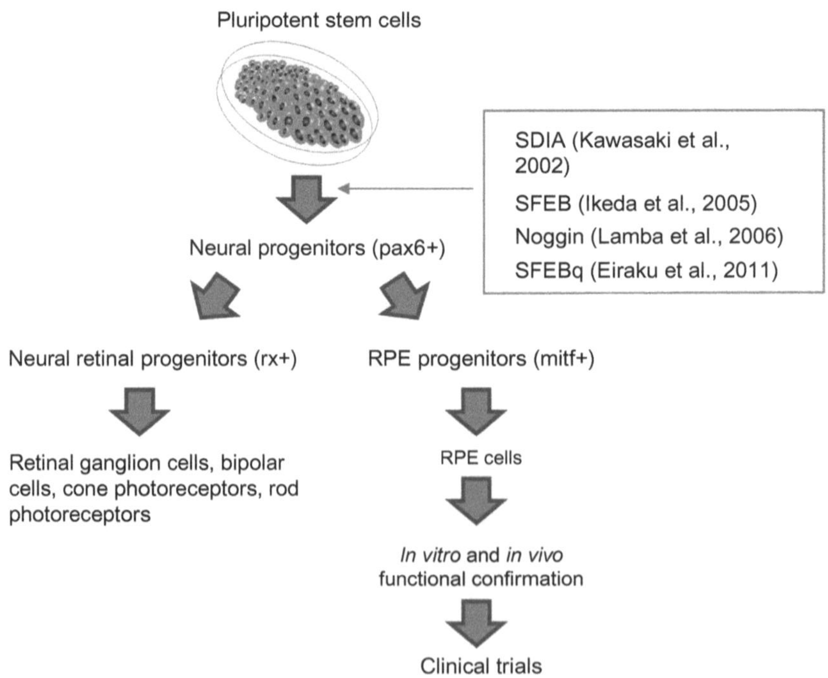

FIGURE 2

Stepwise differentiation of retinal progenitor cells, RPE cells, and neural retinal cells. (For color version of this figure, the reader is referred to the online version of this chapter.)

(SFEB) method for inducing retinal progenitors, including Pax6+Rx+ neural retinal progenitor cells and Pax6+Mitf+ RPE progenitor cells, by blocking Wnt and Nodal signaling using Dkk-1 and Lefty-A, respectively. Since insulin-like growth factor-1 (IGF-1) can promote eye development in *Xenopus* embryos, Lamba et al. (2006) combined IGF-1, Noggin, and Dkk-1 for short-term differentiation of retinal progenitor cells from human ESCs (hESCs). Recently, Meyer et al. (2011) isolated OV-like embryoid bodies at the late stage of a floating culture and found the preferential differentiation of neural progenitors, which could give rise to retinal cells. Li et al. (2011) recently reported induction of up to 95.4% of Pax6+ neural stem cells using a mixture of factors, which synergistically inhibited glycogen synthase kinase-3 and transforming growth factor-β, increasing the possibility of highly efficient retinal differentiation.

4 DIRECTED DIFFERENTIATION OF MATURE RETINAL PHOTORECEPTOR CELLS

The retina contains RPE cells, glia cells, and a number of distinct neurons. The neural retinal cells derive from common progenitors and appear in the following temporal sequence during development: RGCs, cone photoreceptors, amacrine cells, horizontal cells, bipolar cells, and rod photoreceptor cells (Fig. 1). Among these retinal neural cells, photoreceptor cells are of particular interest because of selective photoreceptor loss in many retinal diseases. To generate retinal rod and cone photoreceptor cells, Osakada et al. (2008) treated ESCs with Wnt and Nodal antagonists, fetal calf serum, and activin based on the SFEB method (SFEB/DLFA). After efficient induction of retinal progenitor cells (Pax+Rx+), the γ-secretase inhibitor DAPT, which inhibits Notch signaling, significantly promoted photoreceptor precursor cell (Crx+) differentiation. Similar to its cell fate in retinal development, the photoreceptor precursor cell is more prone to becoming a cone photoreceptor. Acidic FGF, basic FGF, taurine, Shh, and RA significantly increased the efficiency of rod differentiation, thus enhancing the production of rod photoreceptors (Osakada et al., 2008). This strategy was applied in subsequent studies to induce mouse, monkey, human, and pig iPSCs to differentiate into retinal cells (Hirami et al., 2009).

However, photoreceptor induction for cell replacement therapy still faces a series of challenges. The induced differentiated cells are not similar to photoreceptors in cell structure and morphology, for example, the disks and outer segment. A recent study successfully molded outer segment-like structures from hESC-derived retinal cells on soft lithography-patterned microchannel PLGA scaffolds (McUsic et al., 2012). Although cells express the key phototransduction-initiating protein, rhodopsin, it has been difficult to induce electrophysiological responses under light stimuli using the patch-clamp technique, indicating a functional defect in the photo response. Additionally, functional integration of induced cells into the host retina remains an essential issue to be addressed.

5 GENERATION OF RETINAL PIGMENT EPITHELIAL CELLS

Since RPE cells are essential for photoreceptor shedding, maintenance, and survival, numerous studies have focused on their differentiation from PSCs (Fig. 2). In 2002, Kawasaki et al. (2002) first described the differentiation of RPE-like cells from monkey ESCs that were cocultured with the stromal cell line PA6. These cells had RPE cell characteristics of hexagonal shape, pigmentation, and Pax6 expression. Subsequently, Haruta et al. (2004) demonstrated that the cells expressed conventional RPE markers and possessed the biological function of phagocytosis. Furthermore, transplantation of the differentiated RPE cells into a classic animal model, the Royal College of Surgeons rat, showed positive effects on the alleviation of retinal degeneration. Another group reported on hESCs spontaneously differentiating into RPE cells in the absence of FGF, although the efficiency was extremely low (Klimanskaya et al., 2004). Osakada et al. (2008, 2009) successfully generated functional RPE cells from mouse, monkey, and human ESCs based on the SFEB methods. Okamoto and colleagues demonstrated that RPE cells differentiated from monkey iPSCs had a robust capacity for phagocytosis (Okamoto and Takahashi, 2011). Furthermore, hESC-derived RPE cells were found to secrete pigment epithelium-derived factor in a polarized mode (Zhu et al., 2011). Since RPE functions as an important barrier between blood and the neural retina, Juuti-Uusitalo et al. (2012) demonstrated that hESC-derived RPE cells expressed ATP-dependent efflux transporters. Importantly, it is well known that differentiated RPE cells can be easily purified from the total cell population based on their pigmentation.

These studies together have shown robust RPE differentiation and strong evidence of physiological function in induced RPE cells and have provided a perspective for RPE cell replacement therapy in patients. With permission from the US Food and Drug Administration, Advanced Cell Technology initiated a clinical trial of ESC-derived RPE cell transplantation in two patients with advanced-stage, dry-type AMD, or Stargardt's macular dystrophy, with a 12-week immunosuppression regimen (Schwartz et al., 2012). Most importantly, the patients showed no signs of tumorigenesis, hyperproliferation, or transplant rejection 4 months after surgery, demonstrating the safety and tolerability of subretinal transplantation of hESC-RPE cells. In their preliminary report, the authors further reported a significant improvement in visual acuity in the patients. Although these results require careful interpretation and further objective examinations, for example, multifocal electroretinography and microperimetry, this study represents the first clinical trial of PSC-derived RPE cell replacement therapy and provides an encouraging result.

Despite the successful report of hESC-RPE cell transplantation, there are still several important issues to be addressed. The possibility of immunorejection at a later stage after surgery remains an open question. Previously, Tezel et al. (2007) transplanted allogeneic adult RPE cells into the eyes of 12 wet-type AMD patients with systemic immune suppression for approximately 6 months, but the patients' visual acuity, contrast sensitivity, and reading speed failed to improve 1 year after surgery, increasing the possibility of long-term immune rejection. In this sense, it would be expected to the "autogenous" iPSC-RPE cell transplantation in patients with retinal degenerative diseases (Jin et al., 2009). Since RPE cells have to adhere to survive and expand under *in vitro* culture conditions, suspended cells transplanted into the subretinal space have difficulty adhering and becoming functional. Therefore, it is possible for the injected cells to leak out into the vitreous through the injection site (Schwartz et al., 2012). In the future, engineered RPE cell sheets may be an alternative means of overcoming these problems.

6 DIFFERENTIATION OF OTHER TYPES OF RETINAL NEURAL CELLS

RGCs are the first retinal neural cells to be generated during retinal development. A few studies have attempted to differentiate RGCs from PSCs. Jagatha et al. (2009) generated RGC-like cells from neural progenitors derived from FGF-2-treated ESCs. Recently, Chen et al. (2010) used this differentiation strategy to induce neural progenitors, which were then induced to differentiate into RGCs by forced expression of Math5, which is essential for the determination of RGC fate. These results, however, remain preliminary, and regenerative medicine will require further cell characterization.

Other types of retinal cells, including bipolar, amacrine, and horizontal cells, appear during the middle stage of eye development. Using the SFEB method of stepwise induction of retinal cells from both ESCs and iPSCs, we have confirmed successful differentiation of PKC$\alpha+$ bipolar-like cells, and Islet1$+$ cells immunocytochemically represent amacrine, bipolar, or horizontal cells. However, these differentiated cells have not yet been extensively authenticated.

7 PRODUCING THREE-DIMENSIONAL RETINAL TISSUE

Although different retinal cell types can be generated via *in vitro* differentiation from PSCs, an interesting phenomenon has been observed using the SFEB method. RPE cells usually develop near the photoreceptor cell cluster, and either RPE cells or the photoreceptor cell cluster may surround the other in the culture dish (Jin et al., 2011, 2012; Osakada et al., 2008, 2009). Until Eiraku et al. (2011) reported the successful generation of compound retinal or cortical tissue from ESCs, it was not expected that a structured retinal organ could be produced in a dish. Using a low cell adhesion, round-bottomed, 96-well plate as a three-dimensional culture system (referred to as the SFEBq method), trypsinized mouse ESCs were able to form aggregates and then autonomously generate retinal primordium structure. Finally, stratified neural retinal tissue, as seen *in vivo*, was generated for the first time, demonstrating ESC-based retinal morphogenesis *in vitro*. The same group also reported the successful generation of complex pituitary-like tissue from mouse ESCs (Suga et al., 2011).

The ESC-derived pituitary was able to secrete adrenocorticotropic hormone *in vivo*, indicating at least partially functional tissue. Furthermore, when the self-organized tissue was transplanted into pituitary gland-ablated mice, it markedly compensated for the absence of pituitary function in this model system. A similar transplantation of three-dimensional retinal tissue into an animal model to investigate functional recovery from overall retinal degeneration is greatly anticipated.

8 PERSPECTIVE ON PLURIPOTENT STEM CELL-DERIVED RETINAL CELLS

PSCs have great potential for the treatment of incurable retinal degenerative diseases, in which the retina, the accessible part of the central nervous system, has been irreversibly damaged. As outlined in this review, RPE cells and retinal neural cells, including rods, cones, bipolar cells, and RGCs, can be differentiated from both ESCs and iPSCs. Among these differentiated cells, PSC-derived RPE cells have been the best characterized *in vivo*. According to the most recent clinical trial results, RPE cell transplantation is in the process of "bench-to-bedside" translation, making the possibility of a curative treatment a reality. In most retinal diseases, cone or rod photoreceptor loss is fundamental to visual disability, and photoreceptor cell replacement may be a radical treatment. However, one study has shown that donor photoreceptor precursor cells transplanted into a mouse model of retinitis pigmentosa can integrate into the degenerating retina, differentiate into rod photoreceptor cells, form synaptic connections, and eventually improve visual function (MacLaren et al., 2006). Most recently, the same group transplanted the retinal precursor cells into another animal model of congenital stationary night blindness, and demonstrated convinsingly synaptic connections with recipient retinal cells and recovery of rod-mediated visual function (Pearson et al., 2012). These studies demonstrated the feasibility of retinal cell transplantation for restoring vision in diseased retina. With advances in robust differentiation and extensive characterization of PSC-derived cone and rod photoreceptors, functional regeneration of diseased human retinas is absolutely achievable. In conclusion, the use of PSCs for retinal regeneration is moving rapidly toward clinical translation. We believe that retinal cell replacement therapy is a powerful sight-saving approach.

Acknowledgments

We are grateful to all laboratory members for critical discussion of this review. This work was supported by National Natural Science Foundation of China (#81170879) and the Ministry of Health, Labor, and Welfare, Japan (#H21-Nanchi-Ippan-216). We apologize that not all related articles were cited in the review due to space limitations.

References

Cavodeassi, F., Carreira-Barbosa, F., Young, R.M., Concha, M.L., Allende, M.L., Houart, C., Tada, M., Wilson, S.W., 2005. Early stages of zebrafish eye formation require the coordinated activity of Wnt11, Fz5, and the Wnt/beta-catenin pathway. Neuron 47, 43–56.

Chen, M., Chen, Q., Sun, X., Shen, W., Liu, B., Zhong, X., Leng, Y., Li, C., Zhang, W., Chai, F., Huang, B., Gao, Q., Xiang, A.P., Zhuo, Y., Ge, J., 2010. Generation of retinal ganglion-like cells from reprogrammed mouse fibroblasts. Invest. Ophthalmol. Vis. Sci. 51, 5970–5978.

Chow, R.L., Altmann, C.R., Lang, R.A., Hemmati-Brivanlou, A., 1999. Pax6 induces ectopic eyes in a vertebrate. Development 126, 4213–4222.

Eiraku, M., Takata, N., Ishibashi, H., Kawada, M., Sakakura, E., Okuda, S., Sekiguchi, K., Adachi, T., Sasai, Y., 2011. Self-organizing optic-cup morphogenesis in three-dimensional culture. Nature 472, 51–56.

Evans, M.J., Kaufman, M.H., 1981. Establishment in culture of pluripotential cells from mouse embryos. Nature 292, 154–156.

Furukawa, T., Kozak, C.A., Cepko, C.L., 1997. Rax, a novel paired-type homeobox gene, shows expression in the anterior neural fold and developing retina. Proc. Natl. Acad. Sci. U.S.A. 94, 3088–3093.

Haruta, M., Sasai, Y., Kawasaki, H., Amemiya, K., Ooto, S., Kitada, M., Suemori, H., Nakatsuji, N., Ide, C., Honda, Y., Takahashi, M., 2004. In vitro and in vivo characterization of pigment epithelial cells differentiated from primate embryonic stem cells. Invest. Ophthalmol. Vis. Sci. 45, 1020–1025.

Hennig, A.K., Peng, G.H., Chen, S., 2008. Regulation of photoreceptor gene expression by Crx-associated transcription factor network. Brain Res. 1192, 114–133.

Hirami, Y., Osakada, F., Takahashi, K., Okita, K., Yamanaka, S., Ikeda, H., Yoshimura, N., Takahashi, M., 2009. Generation of retinal cells from mouse and human induced pluripotent stem cells. Neurosci. Lett. 458, 126–131.

Hyatt, G.A., Schmitt, E.A., Fadool, J.M., Dowling, J.E., 1996. Retinoic acid alters photoreceptor development in vivo. Proc. Natl. Acad. Sci. U.S.A. 93, 13298–13303.

Ikeda, H., Osakada, F., Watanabe, K., Mizuseki, K., Haraguchi, T., Miyoshi, H., Kamiya, D., Honda, Y., Sasai, N., Yoshimura, N., Takahashi, M., Sasai, Y., 2005. Generation of Rx+/Pax6+ neural retinal precursors from embryonic stem cells. Proc. Natl. Acad. Sci. U.S.A. 102, 11331–11336.

Jagatha, B., Divya, M.S., Sanalkumar, R., Indulekha, C.L., Vidyanand, S., Divya, T.S., Das, A.V., James, J., 2009. In vitro differentiation of retinal ganglion-like cells from embryonic stem cell derived neural progenitors. Biochem. Biophys. Res. Commun. 380, 230–235.

Jin, Z.B., Okamoto, S., Mandai, M., Takahashi, M., 2009. Induced pluripotent stem cells for retinal degenerative diseases: a new perspective on the challenges. J. Genet. 88, 417–424.

Jin, Z.B., Okamoto, S., Osakada, F., Homma, K., Assawachananont, J., Hirami, Y., Iwata, T., Takahashi, M., 2011. Modeling retinal degeneration using patient-specific induced pluripotent stem cells. PLoS One 6, e17084.

Jin, Z.B., Okamoto, S., Xiang, P., Takahashi, M., 2012. Integration-free iPS cells derived from retinitis pigmentosa patient for disease modeling. Stem Cells Trans. Med. 1, 503–509.

Juuti-Uusitalo, K., Vaajasaari, H., Ryhanen, T., Narkilahti, S., Suuronen, R., Mannermaa, E., Kaarniranta, K., Skottman, H., 2012. Efflux protein expression in human stem cell-derived retinal pigment epithelial cells. PLoS One 7, e30089.

Kawasaki, H., Suemori, H., Mizuseki, K., Watanabe, K., Urano, F., Ichinose, H., Haruta, M., Takahashi, M., Yoshikawa, K., Nishikawa, S., Nakatsuji, N., Sasai, Y., 2002. Generation of dopaminergic neurons and pigmented epithelia from primate ES cells by stromal cell-derived inducing activity. Proc. Natl. Acad. Sci. U.S.A. 99, 1580–1585.

Klimanskaya, I., Hipp, J., Rezai, K.A., West, M., Atala, A., Lanza, R., 2004. Derivation and comparative assessment of retinal pigment epithelium from human embryonic stem cells using transcriptomics. Cloning Stem Cells 6, 217–245.

Lagutin, O.V., Zhu, C.C., Kobayashi, D., Topczewski, J., Shimamura, K., Puelles, L., Russell, H.R., McKinnon, P.J., Solnica-Krezel, L., Oliver, G., 2003. Six3 repression of Wnt signaling in the anterior neuroectoderm is essential for vertebrate forebrain development. Genes Dev. 17, 368–379.

Lamba, D.A., Karl, M.O., Ware, C.B., Reh, T.A., 2006. Efficient generation of retinal progenitor cells from human embryonic stem cells. Proc. Natl. Acad. Sci. U.S.A. 103, 12769–12774.

Li, W., Sun, W., Zhang, Y., Wei, W., Ambasudhan, R., Xia, P., Talantova, M., Lin, T., Kim, J., Wang, X., Kim, W.R., Lipton, S.A., Zhang, K., Ding, S., 2011. Rapid induction and long-term self-renewal of primitive neural precursors from human embryonic stem cells by small molecule inhibitors. Proc. Natl. Acad. Sci. U.S.A. 108, 8299–8304.

MacLaren, R.E., Pearson, R.A., MacNeil, A., Douglas, R.H., Salt, T.E., Akimoto, M., Swaroop, A., Sowden, J.C., Ali, R.R., 2006. Retinal repair by transplantation of photoreceptor precursors. Nature 444, 203–207.

Mathers, P.H., Grinberg, A., Mahon, K.A., Jamrich, M., 1997. The Rx homeobox gene is essential for vertebrate eye development. Nature 387, 603–607.

McUsic, A.C., Lamba, D.A., Reh, T.A., 2012. Guiding the morphogenesis of dissociated newborn mouse retinal cells and hES cell-derived retinal cells by soft lithography-patterned microchannel PLGA scaffolds. Biomaterials 33, 1396–1405.

Meyer, J.S., Howden, S.E., Wallace, K.A., Verhoeven, A.D., Wright, L.S., Capowski, E.E., Pinilla, I., Martin, J.M., Tian, S., Stewart, R., Pattnaik, B., Thomson, J.A., Gamm, D.M., 2011. Optic vesicle-like structures derived from human pluripotent stem cells facilitate a customized approach to retinal disease treatment. Stem Cells 29, 1206–1218.

Mukhopadhyay, M., Shtrom, S., Rodriguez-Esteban, C., Chen, L., Tsukui, T., Gomer, L., Dorward, D.W., Glinka, A., Grinberg, A., Huang, S.P., Niehrs, C., Izpisua Belmonte, J.C., Westphal, H., 2001. Dickkopf1 is required for embryonic head induction and limb morphogenesis in the mouse. Dev. Cell 1, 423–434.

Okamoto, S., Takahashi, M., 2011. Induction of retinal pigment epithelial cells from monkey iPS cells. Invest. Ophthalmol. Vis. Sci. 52, 8785–8790.

Osakada, F., Ikeda, H., Mandai, M., Wataya, T., Watanabe, K., Yoshimura, N., Akaike, A., Sasai, Y., Takahashi, M., 2008. Toward the generation of rod and cone photoreceptors from mouse, monkey and human embryonic stem cells. Nat. Biotechnol. 26, 215–224.

Osakada, F., Jin, Z.B., Hirami, Y., Ikeda, H., Danjyo, T., Watanabe, K., Sasai, Y., Takahashi, M., 2009. In vitro differentiation of retinal cells from human pluripotent stem cells by small-molecule induction. J. Cell Sci. 122, 3169–3179.

Pearson, R.A., Barber, A.C., Rizzi, M., Hippert, C., Xue, T., West, E.L., Duran, Y., Smith, A.J., Chuang, J.Z., Azam, S.A., Luhmann, U.F., Benucci, A., Sung, C.H., Bainbridge, J.W., Carandini, M., Yau, K.W., Sowden, J.C., Ali, R.R., 2012. Restoration of vision after transplantation of photoreceptors. Nature 485, 99–103.

Pizzarello, L., Abiose, A., Ffytche, T., Duerksen, R., Thulasiraj, R., Taylor, H., Faal, H., Rao, G., Kocur, I., Resnikoff, S., 2004. VISION 2020: The Right to Sight: a global initiative to eliminate avoidable blindness. Arch. Ophthalmol. 122, 615–620.

Schwartz, S.D., Hubschman, J.P., Heilwell, G., Franco-Cardenas, V., Pan, C.K., Ostrick, R.M., Mickunas, E., Gay, R., Klimanskaya, I., Lanza, R., 2012. Embryonic stem cell trials for macular degeneration: a preliminary report. Lancet 379, 713–720.

Suga, H., Kadoshima, T., Minaguchi, M., Ohgushi, M., Soen, M., Nakano, T., Takata, N., Wataya, T., Muguruma, K., Miyoshi, H., Yonemura, S., Oiso, Y., Sasai, Y., 2011. Self-formation of functional adenohypophysis in three-dimensional culture. Nature 480, 57–62.

Swaroop, A., Kim, D., Forrest, D., 2010. Transcriptional regulation of photoreceptor development and homeostasis in the mammalian retina. Nat. Rev. Neurosci. 11, 563–576.

Takahashi, K., Yamanaka, S., 2006. Induction of pluripotent stem cells from mouse embryonic and adult fibroblast cultures by defined factors. Cell 126, 663–676.

Tezel, T.H., Del Priore, L.V., Berger, A.S., Kaplan, H.J., 2007. Adult retinal pigment epithelial transplantation in exudative age-related macular degeneration. Am. J. Ophthalmol. 143, 584–595.

Zhu, D., Deng, X., Spee, C., Sonoda, S., Hsieh, C.L., Barron, E., Pera, M., Hinton, D.R., 2011. Polarized secretion of PEDF from human embryonic stem cell-derived RPE promotes retinal progenitor cell survival. Invest. Ophthalmol. Vis. Sci. 52, 1573–1585.

Zuber, M.E., Gestri, G., Viczian, A.S., Barsacchi, G., Harris, W.A., 2003. Specification of the vertebrate eye by a network of eye field transcription factors. Development 130, 5155–5167.

CHAPTER

Generation of cortical neurons from pluripotent stem cells

10

Pierre Vanderhaeghen[*,†,1]

[*]*IRIBHM (Institute for Interdisciplinary Research), Université Libre de Bruxelles (U.L.B.), Brussels, Belgium*

[†]*Welbio, Université Libre de Bruxelles (U.L.B.), Brussels, Belgium*

[1]*Corresponding author. Tel.: +32-2-5554186/4185, Fax: +32-2-5554655, e-mail address: pierre.vanderhaeghen@ulb.ac.be*

Abstract

The cerebral cortex is the most complex structure in the mammalian brain, which develops through the coordinated generation of dozens of neuronal subtypes. Here, we review recent studies that show how embryonic stem cell (ESC)-based differentiation can recapitulate the major milestones of cortical development observed *in vivo*. ESC-derived cortical-like progenitors are able to generate a diverse repertoire of neurons that display most salient features of *bona fide* cortical pyramidal neurons of different layers. Importantly, when transplanted into the mouse cerebral cortex *in vivo*, these neuronal populations develop patterns of axonal projections highly similar to those of native cortical neurons. ESC-based corticogenesis, which can be applied to cells of mouse or human origin and to induced pluripotent stem cells, opens new perspectives for the modeling of cortical development and diseases and in the long run for the rational design of brain repair strategies striking the cortex.

Keywords

cerebral cortex, cortical area, cortical layer, pluripotent stem cell, pyramidal neuron

1 INTRODUCTION

The cerebral cortex is one of the most complex structure in the brain, composed of a wide diversity of neuronal subtypes that populate specific cortical layers and areas (Gaspard and Vanderhaeghen, 2011b; Hevner, 2006; Molnar and Cheung, 2006; Molyneaux et al., 2007). The mechanisms of specification of diverse types of cortical neurons remain a fascinating question in basic neurobiology, which also has potentially major implications for the modeling and treatment of the many neurological

Progress in Brain Research, Volume 201, ISSN 0079-6123, http://dx.doi.org/10.1016/B978-0-444-59544-7.00009-3

conditions that strike specific neuronal populations of the cortex, from mental retardation to stroke, epilepsy, or various neurodegenerative diseases.

A first subdivision among cortical neurons distinguishes two main cell classes. Pyramidal neurons constitute >85% of cortical neurons, they are glutamatergic and send long range projections to other cortical or subcortical targets. The remaining 15% of cortical neurons are GABA-ergic interneurons that display only local connectivity. Pyramidal neurons and interneurons can be further subdivided into dozens of subtypes, characterized by specific molecular, cellular, and functional properties (Hevner, 2006; Hevner et al., 2003; Molnar and Cheung, 2006; Molyneaux et al., 2007; Nelson et al., 2006; Sugino et al., 2006; Wonders and Anderson, 2006).

Over the past few years, pluripotent embryonic stem cells (ESCs; Smith, 2001), whether of mouse or human origin, have emerged as a promising tool for neurobiology, allowing the *in vitro* recreation of events of organogenesis in controlled and reproducible conditions, as well as the directed differentiation of specific neuronal populations (Gaspard and Vanderhaeghen, 2010). In addition, the recent advent of induced pluripotent stem cells (iPSCs; Takahashi and Yamanaka, 2006; Yamanaka, 2007) has opened the possibility to use ESC/iPSC-based neural differentiation as tools for the modeling of human brain diseases and for the design of novel brain repair strategies (Dolmetsch and Geschwind, 2011).

Here we review recent progress on the generation of cortical neurons from pluripotent stem cells, illustrating that much of the complexity of cortical development can be recapitulated with surprisingly simple *in vitro* settings. We will also discuss results obtained with ESC-derived cortical neurons following *in vivo* transplantation in the mouse brain, and opportunities as well as challenges associated with ESC/iPSC-corticogenesis.

2 A PRIMITIVE PATHWAY OF SPECIFICATION OF THE FOREBRAIN/TELENCEPHALON

The cerebral cortex is formed in the telencephalon, the anterior-most part of the forebrain (Fig. 1A). Forebrain identity is thought to constitute a primitive pattern of neural identity, which is acquired and retained through the local inhibition of caudalizing morphogen signals (Wilson and Houart, 2004). The telencephalon then undergoes patterning along the dorsoventral axis, through the induction of ventral identities by the morphogen Sonic Hedgehog (Shh) and by complementary effects of Wnt and Bone Morphogenic Protein (BMP) signals secreted from the dorsal midline (Hebert and Fishell, 2008; Sur and Rubenstein, 2005). This regionalization process leads to the generation of the two main neuronal populations residing in the cortex: pyramidal projection neurons and inhibitory interneurons, which are generated from the dorsal and ventral part of the telencephalon, respectively (Gotz and Sommer, 2005; Hebert and Fishell, 2008; Sur and Rubenstein, 2005).

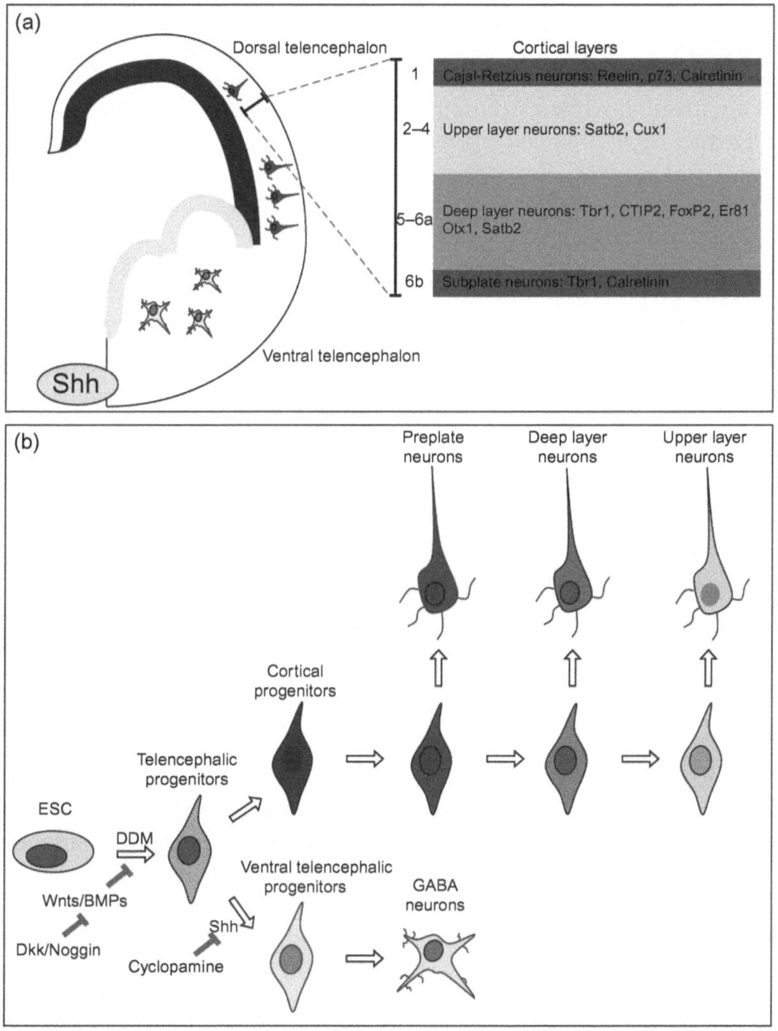

FIGURE 1

(A) Schematic representation of the developing telencephalon. Within the dorsal telencephalon, pyramidal neurons that populate distinct cortical layers express a specific combination of layer-specific markers. (B) Schematic representation of *in vitro* generation of cortical neurons. ESC cultured in a defined default medium DDM devoid of morphogens differentiate into neural progenitors that display a telencephalon-like identity. This induction can be further enhanced with Dickkopf (Dkk) or Noggin, inhibitors of Wnt and BMP signaling, respectively. Dorsal identity can be induced further with the addition of the Shh inhibitor cyclopamine, thereby allowing the generation of cortical progenitors. These cortical progenitors then spontaneously generate a wide array of different cortical neuronal types, through a progressive shift in competence similar to *in vivo* corticogenesis. (For color version of this figure, the reader is referred to the online version of this chapter.)

ESC-based models of forebrain differentiation are largely inspired by these *in vivo* findings applied to *in vitro* settings (Fig. 1B). Indeed, in most neural differentiation paradigms, neural progenitors derived from ESC initially express markers of forebrain identity, while extrinsic cues such as retinoic acid, Wnts, or Fibroblast Growth Factors (FGF) can convert them to more caudal fates, such as midbrain or spinal cord (Li et al., 2005; Tropepe et al., 2001; Watanabe et al., 2005; Wataya et al., 2008; Wichterle et al., 2002). Moreover, adding soluble inhibitors of Wnt or BMP signals during early ESC-derived neural induction can further enhance the proportion of forebrain/telencephalon progenitors (Chambers et al., 2009; Pera et al., 2004; Watanabe et al., 2005).

Based on these premises, the efficient and directed conversion of ESC into telencephalic progenitors, including cortical progenitors, was first reported through two main ways (Eiraku et al., 2008; Gaspard et al., 2008). ESCs can be cultured at low density as adherent monocultures (Ying et al., 2003) in a medium devoid of serum or any morphogen (chemically Defined Default Medium or DDM), but allowing cell survival by insulin, which will lead to a cell population comprising mostly telencephalic progenitors (Gaspard et al., 2008, 2009). Alternatively, ESCs can be cultured in nonadherent conditions (Watanabe et al., 2005) in the presence of morphogen inhibitors, leading to the growth and patterning of neural cell bodies displaying telencephalic/cortical identity (Eiraku et al., 2008).

3 MODULATING DORSOVENTRAL IDENTITY AND NEURONAL SPECIFICATION

The identity of ESC-derived telencephalic progenitors can be refined further by morphogens involved in dorsoventral patterning of the telencephalon, in particular Shh. When ESC are cultured at early stages of neural induction in DDM and cyclopamine, a natural alkaloid known to inhibit potently and specifically the Shh pathways, most of the telencephalon progenitors can adopt dorsal identity, leading to a population of neural progenitors that express the vast majority of typical markers of the cerebral cortex (Fig. 1). Most importantly, cortical-like progenitors then differentiate into neurons that display most salient generic features of cortical pyramidal neurons, including glutamatergic neurotransmission, pyramidal morphology, and expression of specific molecular markers (Gaspard et al., 2008). Interestingly, this requirement for Shh inhibition to maximize the proportion of dorsal cortical identity is mostly observed for mouse ESCs in adherent conditions, while human ESCs were shown to display mostly a dorsal identity even in the absence of Shh inhibition, which could be linked to species-specific differences in Shh signaling or responsiveness during ESC-derived neural differentiation (Shi et al., 2012; Zeng et al., 2010).

Conversely, the addition of Shh or other cues involved in ventral telencephalon specification can convert ESCs to ventral forebrain/telencephalon progenitors, which can subsequently differentiate into various populations of GABA-ergic neurons, including GABA-ergic interneurons (Danjo et al., 2011; Maroof et al., 2010;

Zeng et al., 2010), thus mimicking the *in vivo* development of the telencephalon, where neuronal specification if tightly linked to dorsoventral patterning (Hebert and Fishell, 2008; Sur and Rubenstein, 2005).

Taking advantage of the primitive pathway leading to forebrain/telencephalic identity, ESCs can thus be efficiently converted to telencephalic identity, which can be further refined by cues involved in dorsoventral patterning, thus leading to the generation of distinct types of forebrain neurons, including cortical pyramidal and interneurons.

4 GENERATION OF A DIVERSE ARRAY OF PYRAMIDAL NEURONS *IN VITRO*

Pyramidal projection neurons are diversified into subtypes that will populate distinct cortical layers (Hevner, 2006; Molnar and Cheung, 2006; Molyneaux et al., 2007) and establish specific patterns of axonal output and dendritic input, which provides the essential substrate of complex cortical circuitry (Fig. 2A). Neurons residing in distinct cortical layers display specific patterns of axonal input and output and this neuronal topology constitutes a major anatomical basis of cortical function. For instance, neurons residing in the deep layers project mainly to subcortical targets like the thalamus (layer VI), the midbrain or the spinal cord (layer V), whereas neurons of the upper layers II and III mainly connect with other cortical areas. Thanks to recent efforts aimed at identifying the molecular factors responsible for the specification of layer-specific patterns of projection (Gaspard and Vanderhaeghen, 2011b; Leone et al., 2008; Molyneaux et al., 2007), cortical neurons from different layers can be distinguished also according to the expression of specific molecular markers, the combination of which can define in a unique fashion the main populations of pyramidal neurons populating the various cortical layers (Fig. 1A).

Using these markers alone and in combination, the repertoire of pyramidal neurons generated from ESCs can be assessed despite the fact that no laminar cytoarchitecture is present in purely *in vitro* conditions. Remarkably, this revealed that ESC-derived pyramidal neurons display a high level of diversity, with every cortical layer identity being represented (Eiraku et al., 2008; Gaspard et al., 2008). The majority of them express markers of layer V and VI neurons, like CTIP2, Tbr1, or FoxP2 and a smaller fraction of them express upper layer markers like Satb2 and Cux1 (Fig. 1).

Importantly, the layer diversity of pyramidal neurons derived from mouse ESCs was further confirmed by *in vivo* grafting experiments, in which cortical-like neurons differentiated from ESCs were subsequently transplanted in the cortex of mouse neonates (Eiraku et al., 2008; Gaspard et al., 2008). Several weeks following grafting, the axonal projections of the grafted neurons were examined, which revealed that ESC-derived neurons send axonal projections to a whole array of typical brain targets of most cortical layers, including other cortical areas (targets of layers 2/3/5), thalamus (target of layer 6) and midbrain (target of layer 5) (Fig. 2A).

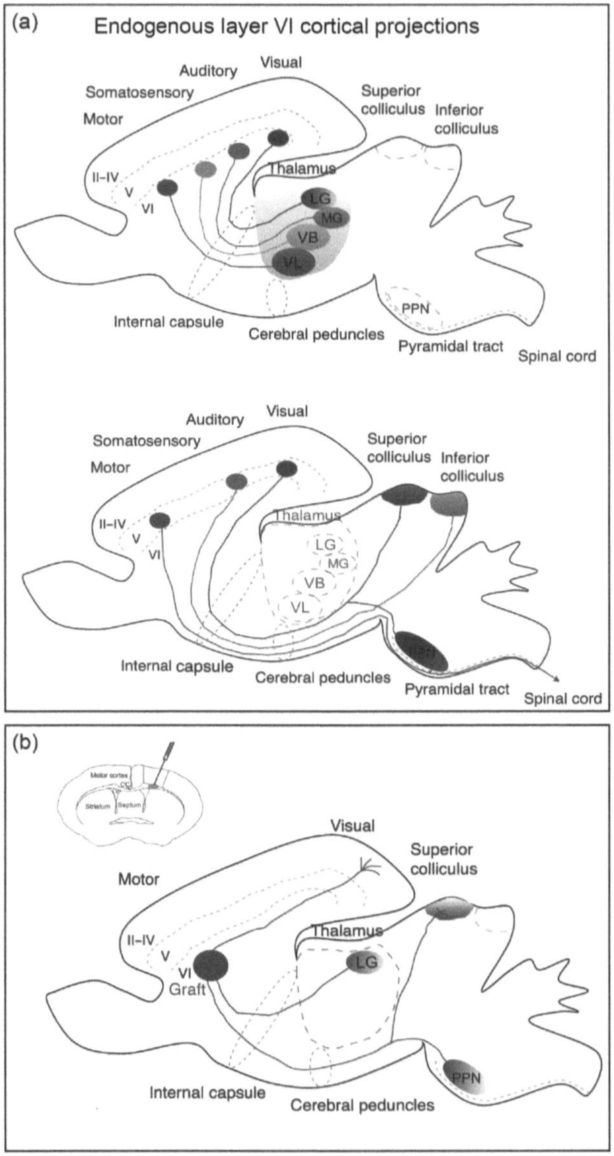

FIGURE 2

A Schematic representation of the layer and area-specific patterns of connectivity of layer VI and layer V neurons, from motor (red), somatosensory (orange), auditory (green), and visual (blue) areas. (A) Layer VI neurons in all areas project mainly to the thalamic nuclei but show area-dependent intrathalamic specificity of connectivity: motor to the ventrolateral (VL) nucleus, somatosensory to the ventrobasal (VB) nucleus, auditory to the medial geniculate (MG) nucleus, and visual to the lateral geniculate (LG) nucleus. (B) Layer V neurons in all

These findings raise the question of the mechanisms underlying the generation of pyramidal neuron diversity from ESCs. *In vivo*, the generation of layer-specific neurons follows a highly coordinated temporal pattern: pioneer neurons of the cortical plate are generated first followed by deep layer and finally upper layer neurons (Gaspard and Vanderhaeghen, 2011b; Leone et al., 2008; Molyneaux et al., 2007; Okano and Temple, 2009). Cortical progenitors are thought to be at least in part multipotent and to undergo a sequential shift in their competence to generate different subtypes of neurons, and such properties are conserved when cortical progenitors are cultured *in vitro* (Shen et al., 2006).

Remarkably, mouse or human ESC-derived cortical progenitors display the same pattern of temporal specification: they first generate pioneer neurons like Cajal-Retzius and subplate neurons, followed by deep layer neurons, and finally upper layer neurons (Eiraku et al., 2008; Gaspard et al., 2008; Shi et al., 2012). Furthermore, the different neuronal subtypes are also generated in the appropriate from single progenitor level, as revealed by clonal analyses (Gaspard et al., 2008): when cultured at clonal density, the potential of progenitors evolves with time, as early progenitors mainly generate deep layer neurons and the progeny of late progenitors is shifted to upper layer neurons (Fig. 1B). Altogether, these data represent the first demonstration that the complex events leading to the generation of neurons displaying different layer-specific patterns of identity can take place outside of the developing brain and rely mainly on a cell population-intrinsic pathway.

In addition, some other aspects of the patterned organization of the cortex can also be recreated *in vitro* (Eiraku et al., 2008): when mouse or human ESCs are cultured as bowls of cells differentiating into cortical-like progenitors, this leads to a striking polarized cellular organization, with neural progenitors occupying deeper layers of the bowls, and neurons accumulating at their periphery, following an organization highly reminiscent of a nascent cortical primordium. Altogether, these data constitute a first proof of principle that a brain-like structure can emerge as a self-organizing cytoarchitecture *in vitro*, which constitutes a promising system to decipher some of the underlying mechanisms of cortical patterning, and to optimize further *in vitro* corticogenesis.

While *in vitro* systems of corticogenesis thus display remarkable similarities with *in vivo* developmental processes, it should be noted however that the differentiation taking place *in vitro* differs from *in vivo* corticogenesis in several important aspects. First, despite the fact that neurons with all six layers' identities seem to be generated,

---

areas project to more caudal structures and also show area-dependent specificity of connectivity: motor to the caudal pedunculopontine nuclei (PPN) and the spinal cord, auditory to the inferior colliculus, and visual to the superior colliculus and the rostral PPN. (C) When grafted in mouse neonatal cortex, ESC-derived cortical neurons display a pattern of projections corresponding to most cortical layers, but to highly specific areas, in particular visual (in blue) as well as limbic (not shown) identity. (For interpretation of the references to color in this figure legend, the reader is referred to the online version of this chapter.)

one never sees the formation of a six-layered organization *in vitro*. Second, whereas *in vivo* deep and upper layer neurons each represent about half of the cortex, ESC-derived pyramidal neurons are strongly skewed toward a deep layer identity (Eiraku et al., 2008; Gaspard et al., 2008). Interestingly, this is also the case when native cortical progenitors are grown *in vitro* (Shen et al., 2006), suggesting that some extrinsic cues, coming from axons from the meninges, cerebrospinal fluid, or blood vessels (likely to be largely missing in the *in vitro* system), are required for the proper generation of the upper layer neurons (Tiberi et al., 2012). In line with this hypothesis, a higher proportion of upper layer neurons was obtained when human ESC-derived corticogenesis includes additional cues such as retinoic acid (Shi et al., 2012). In addition, the cues instructive for particular neuronal types may be present but require a specific pattern of cytoarchitecture or polarity that may be less developed in a purely *in vitro* system.

5 SPECIFICATION OF CORTICAL AREAL IDENTITY FROM ESCs: SURPRISING INSIGHTS FROM *IN VIVO* TRANSPLANTATION

In addition to layer-specific identity, neurons from different cortical areas develop selective patterns of gene expression and connectivity. For instance, the motor, somatosensory, auditory, visual, and limbic areas are composed of neurons sending modality-specific axonal projections that constitute a major landmark of areal identity (Fig. 2A). The patterning of cortical areas is a complex process resulting from the interplay between factors intrinsic to the cortex, as well as extrinsic factors from the brain and body (O'Leary et al., 2007; Rash and Grove, 2006; Sur and Rubenstein, 2005; Vanderhaeghen and Polleux, 2004).

Surprisingly, *in vivo* grafting experiments followed by detailed analyses of the patterns of axonal projections, revealed that mouse ESC-derived cortical neurons seem to acquire mainly a limbic or visual area identity, corresponding essentially to the occipital (posterior) pole of the cortex (Fig. 2B; Gaspard et al., 2008). Indeed, ESC-derived neurons grafted in the frontal cortex nonetheless send axonal projections to specific visual and limbic targets, like the lateral geniculate nucleus of the thalamus, the superior colliculus in the midbrain, or the visual and limbic cortex itself. Importantly, these results were all obtained with grafts in the frontal cortex, suggesting that the patterns observed were not due to the respecification of the grafted neurons through the influence of the host (Gaillard and Roger, 2000; Gaillard et al., 2003; Pinaudeau et al., 2000). Confirming this hypothesis, examination of the molecular identity of ES-derived cortical progenitors and neurons before grafting revealed that most of them expressed typical markers of the occipital cortex, in particular CoupTFI/II transcription factors (Armentano et al., 2007). These data thus indicate that progenitors and neurons generated *in vitro* in the ESC system can undergo an area-specific differentiation process that results in a surprisingly specific identity, corresponding mainly to occipital/visual cortex.

However, cortical-like progenitors obtained with completely different culture conditions (cocultures with stromal feeder cells) displayed areal patterns of axonal projections corresponding to their site of transplantation, that is, motor projections following grafting in motor cortex, visual-like projections following grafting in visual cortex (Ideguchi et al., 2010). Since the cells were grafted at seemingly early stages of differentiation, these results may reflect respecification of the transplanted cells by the host environment, which can indeed occur for native cortical progenitors when they are grafted at stage that is early enough to allow this respecification (Pinaudeau et al., 2000; Vanderhaeghen and Polleux, 2004).

Although the precise cellular and molecular mechanisms underlying these observations remain to be determined, they suggest that the acquisition of cortical area identity can take place *in vitro* in the absence of external cues. While the development of cortical areas is thought to occur through the interplay of intrinsic and extrinsic mechanisms (O'Leary et al., 2007; Rash and Grove, 2006; Sur and Rubenstein, 2005; Vanderhaeghen and Polleux, 2004), these data constitute the first and surprising demonstration of the acquisition of cortical areal identity without any influence from the rest of the brain.

These findings also raise the question of whether and how specific patterns of areal identity could be obtained in a directed fashion, using cues such as FGF8 known to influence cortical areal patterning (Rash and Grove, 2006). This may turn out to be a crucial issue to translate these findings into rational brain repair therapies, where in principle it would be a great advantage to transplant cortical cells of a specific areal identity (motor or visual for instance), depending on the lesioned cortical area to be repaired.

6 CORTICOGENESIS FROM PLURIPOTENT STEM CELLS: PERSPECTIVES AND CHALLENGES FOR MODELS OF DISEASE AND BRAIN REPAIR

Beyond its implications for our understanding of normal cortical development, *in vitro* corticogenesis opens new venues to apply ESC/iPSC-based technology to cortical pathology, in particular for the modeling of cortical diseases, and the rational design of cortical neuron replacement therapies. Although using such a system for brain repair seems like a relatively distant perspective, future work should aim to test whether *in vitro* corticogenesis could be used to reconstruct lesioned cortical circuits in the adult brain, as suggested recently by the remarkable ability of native cortical progenitors to engraft and connect in the adult brain (Gaillard et al., 2007). For this purpose, it will be essential to determine to which extent ESC-derived cortical neurons can indeed contribute to the specific repair of lesioned cortical areas.

Another remaining crucial issue to be solved in this respect will be to test how human ESC/iPSC-derived cortical cells can be effectively used for transplantation purposes, which could be tested using xenografting in the mouse, in order to test their ability to integrate into a native cortical network like genuine cortical neurons, as was done with mouse cells.

Another exciting and more immediate possibility is to apply such systems to human iPSCs derived from patients presenting cortical diseases of genetic origin. This will provide a unique opportunity to model diseases of the cortex, such as specific forms of mental retardation, epilepsy, neurodegeneration, or even some neuropsychiatric diseases, for which existing animal models will always remain partially inadequate given the specific features of the normal and pathological human brain (Dolmetsch and Geschwind, 2011; Gaspard and Vanderhaeghen, 2011a). It will be exciting to use iPSC models of corticogenesis to systematically characterize the cellular and molecular properties of cortical cells derived from patients suffering from neurodevelopmental or neurodegenerative diseases of genetic origin. The validity of this approach was recently illustrated for an autism-spectrum disorder, the Timothy syndrome, which revealed specific defects in neuronal specification and function, using cortical pyramidal neurons differentiated from iPSCs derived from affected patients (Pasca et al., 2011).

In conclusion, the merger of pluripotent stem cell technology and developmental neurobiology reveals unprecedented opportunities to study one of the most complex structures in our brain, the neocortex, and provides an expanding set of tools to decipher the mechanisms underlying some human brain diseases and in the long run to design brain repair strategies.

Acknowledgments

Work from the author described in this review was funded by grants from the Belgian FNRS, FRIA and FRSM, the Belgian Queen Elizabeth Medical Foundation, the Interuniversity Attraction Poles Program (IUAP), Belgian State, Federal Office for Scientific, Technical and Cultural Affairs, the Welbio and Programme d'Excellence CIBLES of the Walloon Region, the Fondations Pierre Clerdent and Roger de Spoelberch. P. V. is Research Director of the FNRS.

References

Armentano, M., Chou, S.J., Tomassy, G.S., Leingartner, A., O'Leary, D.D., Studer, M., 2007. COUP-TFI regulates the balance of cortical patterning between frontal/motor and sensory areas. Nat. Neurosci. 10, 1277–1286.

Chambers, S.M., Fasano, C.A., Papapetrou, E.P., Tomishima, M., Sadelain, M., Studer, L., 2009. Highly efficient neural conversion of human ES and iPS cells by dual inhibition of SMAD signaling. Nat. Biotechnol. 27, 275–280.

Danjo, T., Eiraku, M., Muguruma, K., Watanabe, K., Kawada, M., Yanagawa, Y., Rubenstein, J.L., Sasai, Y., 2011. Subregional specification of embryonic stem cell-derived ventral telencephalic tissues by timed and combinatory treatment with extrinsic signals. J. Neurosci. 31, 1919–1933.

Dolmetsch, R., Geschwind, D.H., 2011. The human brain in a dish: the promise of iPSC-derived neurons. Cell 145, 831–834.

Eiraku, M., Watanabe, K., Matsuo-Takasaki, M., Kawada, M., Yonemura, S., Matsumura, M., Wataya, T., Nishiyama, A., Muguruma, K., Sasai, Y., 2008. Self-organized formation of polarized cortical tissues from ESCs and its active manipulation by extrinsic signals. Cell Stem Cell 3, 519–532.

Gaillard, A., Roger, M., 2000. Early commitment of embryonic neocortical cells to develop area-specific thalamic connections. Cereb. Cortex 10, 443–453.

Gaillard, A., Nasarre, C., Roger, M., 2003. Early (E12) cortical progenitors can change their fate upon heterotopic transplantation. Eur. J. Neurosci. 17, 1375–1383.

Gaillard, A., Prestoz, L., Dumartin, B., Cantereau, A., Morel, F., Roger, M., Jaber, M., 2007. Reestablishment of damaged adult motor pathways by grafted embryonic cortical neurons. Nat. Neurosci. 10, 1294–1299.

Gaspard, N., Vanderhaeghen, P., 2010. Mechanisms of neural specification from embryonic stem cells. Curr. Opin. Neurobiol. 20, 37–43.

Gaspard, N., Vanderhaeghen, P., 2011a. From stem cells to neural networks: recent advances and perspectives for neurodevelopmental disorders. Dev. Med. Child Neurol. 53, 13–17.

Gaspard, N., Vanderhaeghen, P., 2011b. Laminar fate specification in the cerebral cortex. F1000 Biol. Rep. 3, 6.

Gaspard, N., Bouschet, T., Hourez, R., Dimidschstein, J., Naeije, G., van den Ameele, J., Espuny-Camacho, I., Herpoel, A., Passante, L., Schiffmann, S.N., Gaillard, A., Vanderhaeghen, P., 2008. An intrinsic mechanism of corticogenesis from embryonic stem cells. Nature 455, 351–357.

Gaspard, N., Gaillard, A., Vanderhaeghen, P., 2009. Making cortex in a dish: in vitro corticopoiesis from embryonic stem cells. Cell Cycle 8, 2491–2496.

Gotz, M., Sommer, L., 2005. Cortical development: the art of generating cell diversity. Development 132, 3327–3332.

Hebert, J.M., Fishell, G., 2008. The genetics of early telencephalon patterning: some assembly required. Nat. Rev. Neurosci. 9, 678–685.

Hevner, R.F., Daza, R.A., Rubenstein, J.L., Stunnenberg, H., Olavarria, J.F., Englund, C., 2003. Beyond laminar fate: toward a molecular classification of cortical projection/pyramidal neurons. Dev. Neurosci. 25, 139–151.

Hevner, R.F., 2006. From radial glia to pyramidal-projection neuron: transcription factor cascades in cerebral cortex development. Mol. Neurobiol. 33, 33–50.

Ideguchi, M., Palmer, T.D., Recht, L.D., Weimann, J.M., 2010. Murine embryonic stem cell-derived pyramidal neurons integrate into the cerebral cortex and appropriately project axons to subcortical targets. J. Neurosci. 30, 894–904.

Leone, D.P., Srinivasan, K., Chen, B., Alcamo, E., McConnell, S.K., 2008. The determination of projection neuron identity in the developing cerebral cortex. Curr. Opin. Neurobiol. 18, 28–35.

Li, X.J., Du, Z.W., Zarnowska, E.D., Pankratz, M., Hansen, L.O., Pearce, R.A., Zhang, S.C., 2005. Specification of motoneurons from human embryonic stem cells. Nat. Biotechnol. 23, 215–221.

Maroof, A.M., Brown, K., Shi, S.H., Studer, L., Anderson, S.A., 2010. Prospective isolation of cortical interneuron precursors from mouse embryonic stem cells. J. Neurosci. 30, 4667–4675.

Molnar, Z., Cheung, A.F., 2006. Towards the classification of subpopulations of layer V pyramidal projection neurons. Neurosci. Res. 55, 105–115.

Molyneaux, B.J., Arlotta, P., Menezes, J.R., Macklis, J.D., 2007. Neuronal subtype specification in the cerebral cortex. Nat. Rev. Neurosci. 8, 427–437.

Nelson, S.B., Hempel, C., Sugino, K., 2006. Probing the transcriptome of neuronal cell types. Curr. Opin. Neurobiol. 16, 571–576.

Okano, H., Temple, S., 2009. Cell types to order: temporal specification of CNS stem cells. Curr. Opin. Neurobiol. 19, 112–119.

O'Leary, D.D., Chou, S.J., Sahara, S., 2007. Area patterning of the mammalian cortex. Neuron 56, 252–269.

Pasca, S.P., Portmann, T., Voineagu, I., Yazawa, M., Shcheglovitov, A., Pasca, A.M., Cord, B., Palmer, T.D., Chikahisa, S., Nishino, S., Bernstein, J.A., Hallmayer, J., Geschwind, D.H., Dolmetsch, R.E., 2011. Using iPSC-derived neurons to uncover cellular phenotypes associated with Timothy syndrome. Nat. Med. 17, 1657–1662.

Pera, M.F., Andrade, J., Houssami, S., Reubinoff, B., Trounson, A., Stanley, E.G., Ward-van Oostwaard, D., Mummery, C., 2004. Regulation of human embryonic stem cell differentiation by BMP-2 and its antagonist noggin. J. Cell Sci. 117, 1269–1280.

Pinaudeau, C., Gaillard, A., Roger, M., 2000. Stage of specification of the spinal cord and tectal projections from cortical grafts. Eur. J. Neurosci. 12, 2486–2496.

Rash, B.G., Grove, E.A., 2006. Area and layer patterning in the developing cerebral cortex. Curr. Opin. Neurobiol. 16, 25–34.

Shen, Q., Wang, Y., Dimos, J.T., Fasano, C.A., Phoenix, T.N., Lemischka, I.R., Ivanova, N.B., Stifani, S., Morrisey, E.E., Temple, S., 2006. The timing of cortical neurogenesis is encoded within lineages of individual progenitor cells. Nat. Neurosci. 9, 743–751.

Shi, Y., Kirwan, P., Smith, J., Robinson, H.P., Livesey, F.J., 2012. Human cerebral cortex development from pluripotent stem cells to functional excitatory synapses. Nat. Neurosci. 15, 477–486.

Smith, A.G., 2001. Embryo-derived stem cells: of mice and men. Annu. Rev. Cell Dev. Biol. 17, 435–462.

Sugino, K., Hempel, C.M., Miller, M.N., Hattox, A.M., Shapiro, P., Wu, C., Huang, Z.J., Nelson, S.B., 2006. Molecular taxonomy of major neuronal classes in the adult mouse forebrain. Nat. Neurosci. 9, 99–107.

Sur, M., Rubenstein, J.L., 2005. Patterning and plasticity of the cerebral cortex. Science 310, 805–810.

Takahashi, K., Yamanaka, S., 2006. Induction of pluripotent stem cells from mouse embryonic and adult fibroblast cultures by defined factors. Cell 126, 663–676.

Tiberi, L., Vanderhaeghen, P., van den Ameele, J., 2012. Cortical neurogenesis and morphogens: diversity of cues, sources and functions. Curr. Opin. Cell Biol. 24, 269–276.

Tropepe, V., Hitoshi, S., Sirard, C., Mak, T.W., Rossant, J., van der Kooy, D., 2001. Direct neural fate specification from embryonic stem cells: a primitive mammalian neural stem cell stage acquired through a default mechanism. Neuron 30, 65–78.

Vanderhaeghen, P., Polleux, F., 2004. Developmental mechanisms patterning thalamocortical projections: intrinsic, extrinsic and in between. Trends Neurosci. 27, 384–391.

Watanabe, K., Kamiya, D., Nishiyama, A., Katayama, T., Nozaki, S., Kawasaki, H., Watanabe, Y., Mizuseki, K., Sasai, Y., 2005. Directed differentiation of telencephalic precursors from embryonic stem cells. Nat. Neurosci. 8, 288–296.

Wataya, T., Ando, S., Muguruma, K., Ikeda, H., Watanabe, K., Eiraku, M., Kawada, M., Takahashi, J., Hashimoto, N., Sasai, Y., 2008. Minimization of exogenous signals in ES cell culture induces rostral hypothalamic differentiation. Proc. Natl. Acad. Sci. U.S.A. 105, 11796–11801.

Wichterle, H., Lieberam, I., Porter, J.A., Jessell, T.M., 2002. Directed differentiation of embryonic stem cells into motor neurons. Cell 110, 385–397.

Wilson, S.W., Houart, C., 2004. Early steps in the development of the forebrain. Dev. Cell 6, 167–181.

Wonders, C.P., Anderson, S.A., 2006. The origin and specification of cortical interneurons. Nat. Rev. Neurosci. 7, 687–696.

Yamanaka, S., 2007. Strategies and new developments in the generation of patient-specific pluripotent stem cells. Cell Stem Cell 1, 39–49.

Ying, Q.L., Stavridis, M., Griffiths, D., Li, M., Smith, A., 2003. Conversion of embryonic stem cells into neuroectodermal precursors in adherent monoculture. Nat. Biotechnol. 21, 183–186.

Zeng, H., Guo, M., Martins-Taylor, K., Wang, X., Zhang, Z., Park, J.W., Zhan, S., Kronenberg, M.S., Lichtler, A., Liu, H.X., Chen, F.P., Yue, L., Li, X.J., Xu, R.H., 2010. Specification of region-specific neurons including forebrain glutamatergic neurons from human induced pluripotent stem cells. PLoS One 5, e11853.

SECTION 4

Spinal cord repair

CHAPTER

Repair involves all three surfaces of the glial cell

11

Ying Li, Daqing Li, Ahmed Ibrahim, Geoffrey Raisman[1]

Institute of Neurology, University College London, London, UK

[1]Corresponding author. Tel.: +44-207-6762172, Fax: +44-207-76762174, e-mail address: g.raisman@ion.ucl.ac.uk

Abstract

We propose that severed adult CNS axons are intrinsically capable of regeneration and reestablishing lost functions and that the key to repair lies in reconfiguring the scarring response of the astrocytic network. Astrocytes are multifunctional cells with three distinct surfaces: a *glia to glial surface*, providing the junctions needed to incorporate the astrocytes into the network; *a glia to mesodermal surface*, at which astrocytes collaborate with the meningeal fibroblasts to maintain the protective covering of the CNS; and a *glia to neuronal surface*, which provides the routes along which axons travel. After injury, the astrocytes collaborate with the meningeal fibroblasts to form a scar, which provides the necessary defensive sealing of the opened surface of the CNS, but which also has the detrimental effect of closing off the pathways along which axons could regenerate. Incorporation of glial cells transplanted from the olfactory system into a CNS injury causes a re-arrangement of the scarred astrocyte/fibroblast complex so as to produce the alignment of the *glia to neuronal surfaces* needed to provide a pathway for the regeneration of severed axons. Olfactory ensheathing cells certainly have a direct stimulatory effect on axons, but without concomitant reorganization of the glial scar, this could not in itself lead to regeneration of severed axons to their targets.

Keywords

adult, astrocytes, glial scar, fiber tracts, meninges, olfactory ensheathing cells, transplantation

1 THREE SURFACES OF GLIA

The CNS develops from an invagination of the layer of ectodermal cells of the neural plate. If we compare adult astrocytes with their cousins, the epithelial cells formed from the remainder of the somatic ectoderm, we find that astrocytes share the characteristic of having three distinct surfaces, each expressing different functions (Fig. 1). In the case of epithelial cells, these are lateral (linking adjacent epithelial

Progress in Brain Research, Volume 201, ISSN 0079-6123, http://dx.doi.org/10.1016/B978-0-444-59544-7.00010-X

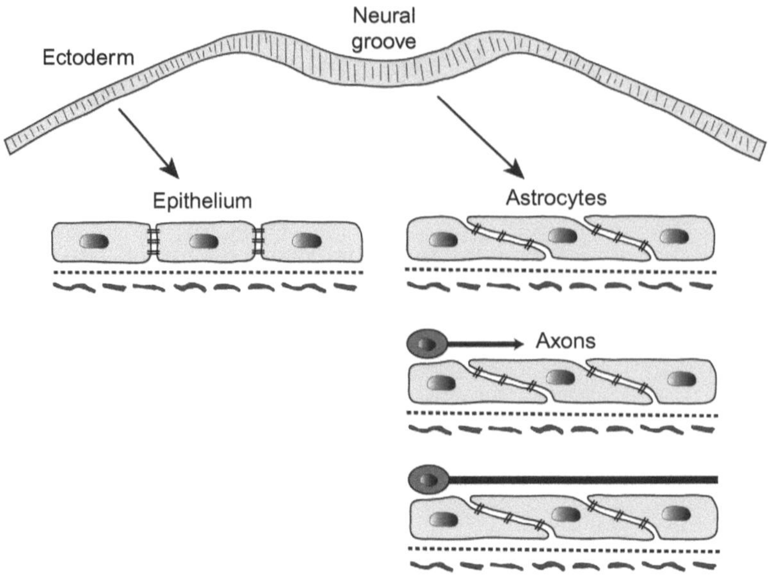

FIGURE 1

The concept of the three astrocytic surfaces illustrated by comparing the origin of the epithelial cells from the embryonic ectoderm and the astrocytes from the neural groove. Both cell types are connected into a syncytial array by desmosome/gap junction complexes (=) and have a deep, basal-lamina-lined surface (dashed line) facing the underlying mesodermal elements (fibroblast processes; broken wavy lines). Neurons develop later, and their axons come to lie on what is morphologically the former external surface of the astrocytes.

cells into a continuous sheet), basal (facing the underlying mesoderm), and apical (facing the exterior of the body). The unique feature of astrocytes is the presence of an additional element: neurons, which come to lie on the apical surface during development. Thus the comparable surfaces of astrocytes are

- A *glia to glial surface* with specialized interastrocytic junctions (Bennett et al., 2003) which link the individual astrocytes into what is effectively a syncytial network. Across these junctions, pass the signals which enable the global glial network to integrate its functions (Verkhratsky and Kettenmann, 1996). We propose that it is this surface which is involved in the incorporation of exogenous glia transplanted into the astrocytic network.
- A *glia to mesodermal surface* coated with basal lamina. This is the surface that faces the meningeal fibroblasts of the glia–pial outer covering of the CNS and lines the astrocytic end feet facing the endothelial cells of the CNS vasculature (Giaume et al., 2010; Iadecola and Nedergaard, 2007). Astrocytes are in a dynamic, two-way interaction with the surrounding meningeal fibroblasts. After injury, it is the cooperation with fibroblasts that enables astrocytes to restore the continuity of the protective boundary surface of the CNS.

- A *glia to neuronal surface*. This is the former apical surface which has become overlaid by neurons and is devoid of basal lamina or specialized membrane junctions. This surface provides the interaction with the neuronal network which underlies plasticity (Chung and Barres, 2011; Eroglu and Barres, 2010; Goldberg et al., 2010; Henneberger and Rusakov, 2010; Raisman and Field, 1973; Rouach et al., 2008). It carries the ligand–receptor membrane interactions involved in axonal extension (Doherty and Walsh, 1996; Zhang et al., 2008) and is an essential partner in synaptic transmission (Araque, 2008; Haydon, 2001). For the growth of axons, this surface must be configured so as to provide a continuous pathway through the astrocytic network.

2 RESPONSES TO INJURY

2.1 Neurons

All damage to the nervous system affects both neurons and glia. Yet, up to now, the majority of efforts at repair still concentrate largely on neurons. However, the long-standing dogma that the adult central neurons are unable to regenerate is challenged by numerous studies showing

1. that central neurons are capable of forming new pathways and new-synaptic connections after injury (e.g., Bareyre et al., 2004; Barritt et al., 2006; Liu et al., 2010; Raineteau and Schwab, 2001; Raisman, 1969; Ylera et al., 2009);
2. that severed CNS axons are capable of long-distance regeneration after transplantation of peripheral glia (e.g., Kuhlengel et al., 1990; Li and Raisman, 1994; Villegas-Pérez et al., 1988); and
3. that regenerating axons can cross a lesion, reenter the astrocytic territory of the CNS, and form new functional connections after transplantation of glia from the olfactory system (e.g., Li et al., 1997, 1998; Lu et al., 2001; Ramón-Cueto et al., 2000).

None of this supports the view that adult central neurons are intrinsically unable to grow axons or make new synaptic connections. Equally, it does not support the view that either the astrocytic or the oligodendrocytic environments of the adult CNS are intrinsically inhibitory to axon growth. In circumstances where transplanted embryonic neurons (Li and Raisman, 1993; Wictorin et al., 1992) or microtransplants of adult neuron suspensions (Davies et al., 1999) do not induce an astrocytic scar, the axons of the donor neurons grow for considerable distances along existing myelinated fiber tracts (Fig. 2).

The balance of evidence suggests that the failure of axon regeneration is a consequence of the local reaction of the astrocytes at the injury site and it is therefore to the astrocytic behavior at this site that repair therapies should be directed.

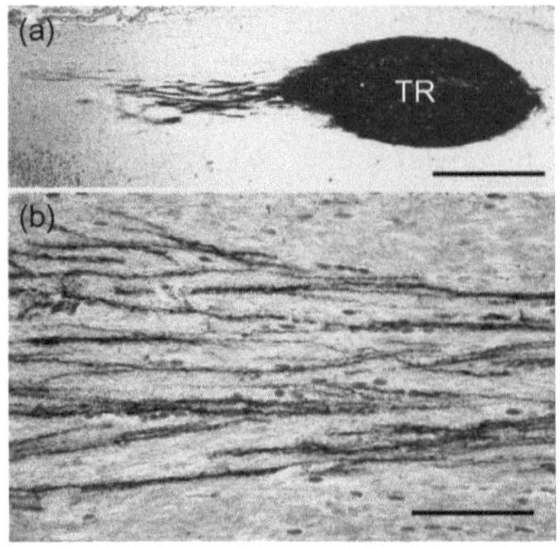

FIGURE 2

(a) A transplant of E14 hippocampal tissue (TR) generating a leash of axons projecting caudally through the host adult fimbria (Thy-1.2 allelic immunostaining, 37 day survival). (b) Enlargement of donor axons which travel freely, aligned with the host corticospinal tract, and dispersed among the adult host myelinated fibers (M6 immunostaining of E16 mouse hippocampal donor tissue, 6 day survival). Scale bars, 500 μm in (a), 200 μm in (b).

Fig. 2(a) reprinted with permission from Davies et al. (1993).

2.2 The glial scar

Injuries disrupt the glia–pial surface of the CNS (Fig. 3) and open the blood–brain barrier. The scarring response of the astrocytes takes advantage of the rapid motility of their processes (e.g., Nishida and Okabe, 2007) and alterations in the cytoskeletal intermediate filaments (Ekmark-Lewen et al., 2010; Eng et al., 2000). This probably begins within minutes of injury. The formation of the glial scar involves a structural cooperation between astrocytes and meningeal fibroblasts (Kawano et al., 2012; Stichel and Muller, 1994), in which the astrocytes reassemble so as to restore a continuous basal-lamina-lined surface sealing off the CNS.

Given the multiplicity of glial functions, it is perhaps simplistic to think of astrocytic scarring simply as a purposeless evolutionary event resulting in a perverse and self-defeating mechanism to block axon regeneration. The effect of the astrocytic scar is to reseal the breach in the coverings of the CNS, protect the neurons from mesenchymal and blood-borne cell invasion, and enable the ionic milieu for neural transmission to be maintained (Silver and Miller, 2004). Astrocytic scars are defensive; they are the CNS method of healing. However, the penalty for this life-saving response is that the configuration of the resulting scar closes off the aligned glial pathways which could otherwise provide a route for severed axons to regenerate. The progress of the sprouts formed at the tip of severed axons is blocked.

Our analysis is based on a "structural" view that axon growth requires access to continuous channels formed by the alignment of the specialized *glia to neuronal surfaces*. On this "pathway hypothesis," for a glial scar to become penetrable by

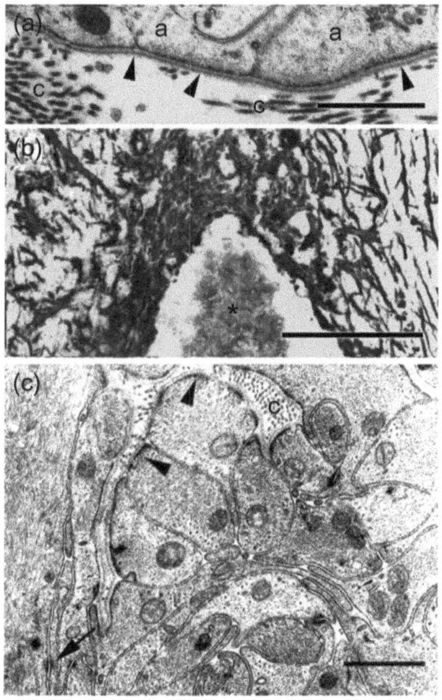

FIGURE 3

(a) High-power electron micrograph through the normal glia surface of the hypothalamus to show the intimate clothing of the astrocytic processes (a) by basal lamina (arrowheads). c, collagen fibers. (b) High power view of the dorsal tip of the astrocytic scar formed at 65 days around a lesion of the corticospinal tract (cf., low power view in Fig. 13b). GFAP immunostaining. lesion center, *. (c) Electron micrograph of an astrocytic scar, consisting of a tangle of astrocytic processes rich in cytoskeletal microfilaments and microtubules in a cross section of the optic nerve, at 7 months after a lesion. Arrow, desmosomal attachment between astrocytic processes (*glia to glial surfaces*); arrowheads, hemidesmosomes on the basal-lamina-lined surfaces facing the collagen-containing extracellular spaces (*glia to mesodermal surfaces*). Scale bars, 2 μm in (a) and (c), 50 μm in (b).

Reprinted with permission from Li et al. (2003b).

regenerating axons would necessitate the "opening" of the scar by a reconfiguration of the astrocytic membranes (Li et al., 2005, 2006; Raisman and Li, 2007).

The alternative or "molecular" view, which is largely the current view, is that injury upregulates the expression of inhibitory molecules by astrocytes (Asher et al., 2001), which coupled with myelin associated inhibitory factors (Schwab et al., 1993), prevent axon regeneration. On this view, a combination of molecular interventions to counteract both these inhibitory influences, together with molecules designed to stimulate axon growth (Pearse et al., 2004), would result in axon regeneration. This view, however, does not take into account the structure of the glial scar, or give any indication of what surfaces the axons would use to regenerate, or, indeed, how they would navigate through the tangle of astrocytic membranes (Fig. 3b and c), let alone the associated fibroblastic processes and collagen-containing extracellular matrix presented by a lesion-induced scar.

What gives hope for repair is that the astrocytic scar is not a dead structure, but an assemblage of living cells which have reconfigured into an elaborate organized structure with the vital purpose of protection. One approach to modifying the configuration of the astrocytic scar is to transplant glia, such as Schwann cells (Kuhlengel et al., 1990; Li and Raisman, 1994; Villegas-Pérez et al., 1988) or olfactory ensheathing cells (OECs; see below), from territories where the glia do not form scars, and where axon regeneration does occur after injury. Transplanting non-CNS glia into the CNS offers promise as a method for opening astrocytic scars and reconstructing the pathways needed for axon regeneration after CNS injuries.

3 THE OLFACTORY SYSTEM

The ability of the adult olfactory system to renew itself from stem cells and to grow new axons across injuries and into the astrocytic environment of the olfactory bulbs was recognized from around the 1960s (Graziadei and Montigraziadei, 1979; Holbrook et al., 1995; Murrell et al., 2005). OECs are the glia of the olfactory nerves (Doucette, 1984). They provide the pathway for olfactory axons to enter the CNS and form synaptic connections with their postsynaptic targets (Raisman, 1985).

3.1 OECs

Graziadei demonstrated that after removing their target by neonatal olfactory lobectomy, the developing olfactory nerve fibers enter the frontal cortex and form glomeruli therein (Graziadei et al., 1979). This was the first indication that the OECs of growing olfactory nerves can interact with astrocytes so as to exert a door-opening effect which allows olfactory axons to enter the astrocytic territory of a *non-olfactory* region of the CNS. With the development of culture systems for deriving OECs from adult tissue (Devon and Doucette, 1995; Ramón-Cueto and Nieto-Sampedro, 1992), it became possible to extend this heterotypic confrontation of OECs and astrocytes by transplanting cultured OECs into adult CNS lesions in distant non-olfactory areas, such as the spinal cord. This provided evidence that adult OECs can integrate into the astrocytic network in non-olfactory areas of the adult CNS. This interaction is mirrored in culture, where it was shown that OECs mingle with astrocytes, but Schwann cells do not (Lakatos et al., 2000).

Histologically, electron micrographs show that in the olfactory nerves *in situ* OECs present the same three surfaces as astrocytes—viz. a *glial to glial surface* furnished with desmosome/gap junctional complexes, a basal-lamina-lined *glia to mesodermal surface* intimately clothed by very fine encircling fibroblastic processes, and a naked *glia to neuronal surface* prolonged into fine sheet-like membranes enclosing the bundles of olfactory nerve axons (Fig. 4; Field et al., 2003).

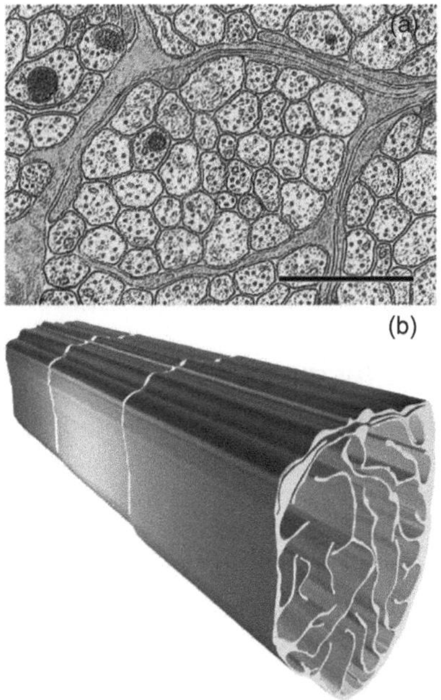

FIGURE 4

(a) High-power electron micrograph to show a cross section of a bundle of olfactory axons enclosed by the inner, smooth membranes of an OEC. (b) Schematic representation of the channels formed by a column of three aligned OECs in the normal olfactory nerves. Scale bar, 1 μm in (A).

Reprinted with permission from Ibrahim et al. (2006).

3.2 Fibroblasts

It has been customary for the CNS to be considered as a tissue apart, with the meningeal fibroblasts of the pia mater at best providing a mechanical protection, at worst an enemy to be kept out. Yet the association of fibroblastic processes with the astrocytes of the CNS, with the OECs of the olfactory system, and with the Schwann cells of peripheral nerve is intimate. All these cell types have a basal-lamina-lined *glia to mesodermal surface*. It would be strange not to think of the interaction with fibroblasts playing an important role in the formation of the astrocytic scar around CNS lesions. By the same token, the astrocytes and fibroblasts must collaborate actively in the opening of the astrocytic scar by transplantation of OECs.

Cultures of primary olfactory nerve tissue contain both OECs and olfactory nerve fibroblasts (ONFs; Ibrahim et al., 2009; Li et al., 1998, 2003a,b). Our experience with transplantation of these cultured OEC/ONF mixtures suggests that the inclusion of ONFs is required for the survival and the reparative properties of the transplants (see also Barnett and Chang, 2004) and, as will be seen below, the histology of the transplants indicates that the transplanted ONFs play a leading, probably crucial role in repair of tract injuries by OEC transplants.

4 REPAIR BY OEC TRANSPLANTS

Repair of CNS injuries by transplantation of OECs involves more than a simple signaling system from OECs to neurons. It reconfigures the astrocytic scar by formation of a complex tissue resulting from an organized interaction between axons, OECs, astrocytes, and fibroblasts. We will illustrate examples of these tissue rearrangements from our own work with transplants into corticospinal tract lesions, transected optic nerves, and reapposed avulsed spinal dorsal roots.

4.1 Corticospinal tract lesions

Mixtures of cultured OECs and ONFs transplanted into corticospinal tract lesions provide a pathway for the axons to regenerate (Li et al., 1997, 1998; Ramón-Cueto et al., 2000). At the earliest stages, we see the lesion area penetrated by an advancing wave of extremely fine fibroblastic processes all strictly aligned in the rostro–caudal axis of the tract (Figs. 5 and 6). This is closely followed by a front of regenerating axons ending in expanded growth cones and intimately ensheathed by very fine processes arising from transplanted OECs (Figs. 5c and 7).

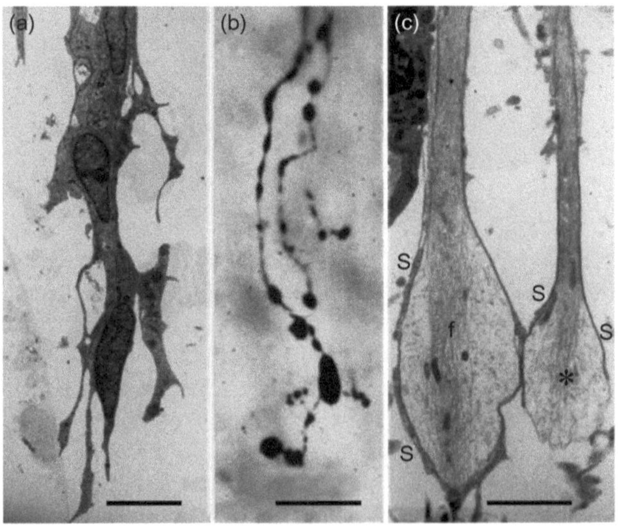

FIGURE 5

Longitudinal sections of a transplant of OECs into the corticospinal tract. (a) Electron micrograph of an advancing leash of fine processes arising from transplanted olfactory nerve fibroblasts. (b) Varicose axons sprouts in a lesion of the corticospinal tract. Anterograde labeling from a cortical injection of biotin dextran, 3 weeks survival. (c) Electron micrograph showing two axon varicosities (*, f = neurofilament core) closely ensheathed by very fine OEC processes (S), 10 days survival. Scale bars, 5 μm in (a), 20 μm in (b), 5 μm in (c).

Reprinted with permission from Li et al. (1997, 1998).

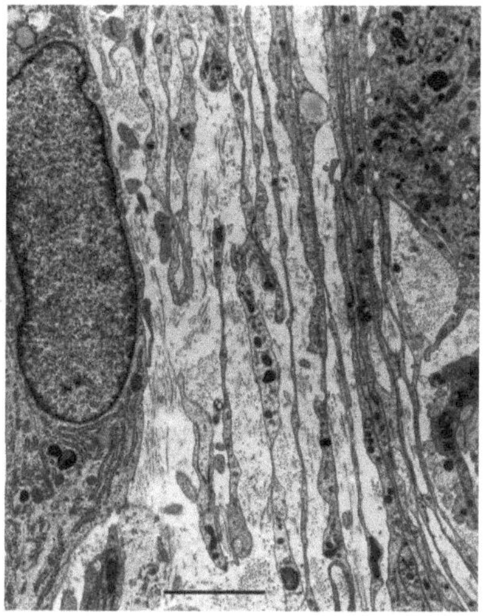

FIGURE 6

An example of fibroblasts preceding OECs in a longitudinal section of an OEC/ONF transplant in the transected corticospinal tract. Long, parallel streamers of fine fibroblastic processes travel caudally in advance of the front of regenerating axons. 10 days after lesion and OEC transplant. Scale bar, 5 μm.

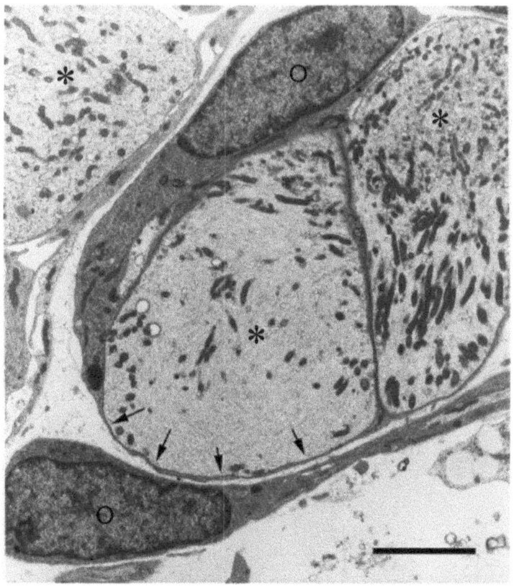

FIGURE 7

Large expanded growth cones (asterisks) containing whorls of neurofilaments and mitochondria closely invested by very fine membranes (arrows) arising from two transplanted OECs (O). Longitudinal section from the corticospinal tract 10 days after lesion and OEC transplant. Scale bar, 10 μm.

Between 1 and 2 weeks, the advancing front has crossed the lesion area and the transplanted cells display a concentrically organized structure comparable to that in peripheral nerves (Fig. 8). An outer whorl of fibroblastic processes encloses a collagen-containing extracellular space through which pass the OECs furnished on their outer surface with basal lamina. By 4 weeks, the axons passing through the graft have become myelinated with peripheral type myelin produced by the OECs (see also Franklin et al., 1996; Sasaki et al., 2004). After leaving the graft, the axons reenter the CNS and elongate through the distal tract (Figs. 9 and 10a) where they become myelinated by the host oligodendrocytes. They terminate by arborising in the medial gray matter and restore a directed forepaw reaching function (Keyvan-Fouladi et al., 2003).

4.2 Optic nerve lesions

OEC/ONF mixtures were transplanted into transections of the optic nerve 2 mm behind the disc. These transplants induced axon regeneration for some 10 mm (Fig. 10b) but were unable to maintain axon penetration throughout the whole distal

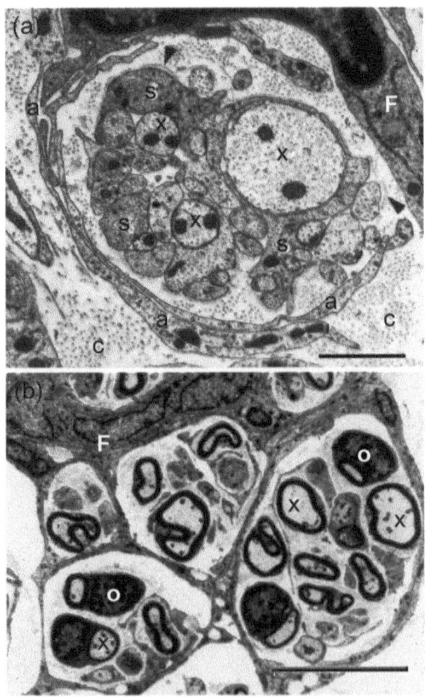

FIGURE 8

Cross section electron micrographs of the concentric, peripheral-nerve-like configuration. (a) Indicates outer sheath of processes of fibroblasts (F), enclosing a collagen-containing extracellular space (c) with OECs (s) ensheathing regenerating corticospinal axons (x) regenerating through the graft, unmyelinated at 10 days (in a), and myelinated by OECs (o) at 30 days (in b) after lesion and transplantation. Scale bar, 1 μm in (a), 10 μm in (b).

Reprinted with permission from Li et al. (1998).

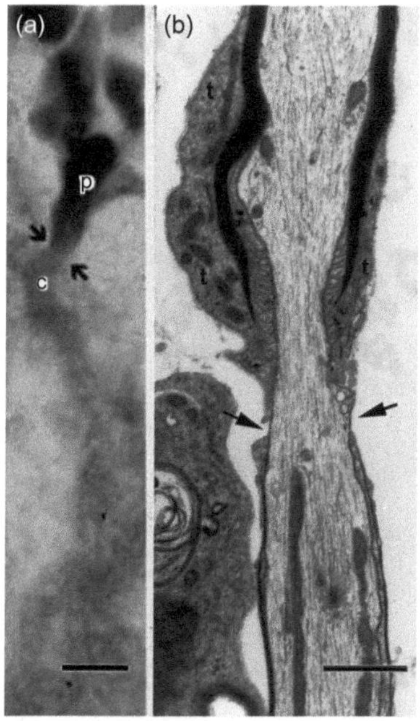

FIGURE 9

Formation of a mixed node (arrows) at the point where a regenerating corticospinal axon myelinated by peripheral type myelin leaves the OEC transplant and re-enters the distal tract where it is myelinated by central, oligodendrocytic myelin. The peripheral type myelin produced by the transplanted OECs is identified in the cryostat section (p in a) by black P0 immunostaining and in the EM (b, upper) by the thicker sheath with outer cytoplasmic tongue (t). The central type oligodendrocytic myelin of the distal tract (c in (a)) is pale purple (light gray in print version) shown by MOG immunostaining in the cryostat section. In the EM (b, lower) it has a thinner sheath, with no outer cytoplasmic tongue. Scale bars, 5 μm in (a), 2 μm in (b). (For interpretation of the references to color in this figure legend, the reader is referred to the online version of this chapter.)

Reprinted with permission from Li et al. (1998).

length of the optic tract. Serial cross-section electron micrographs (Fig. 11a) showed that multiple advancing channels formed by whorls of transplanted ONFs and containing massed collagen fibers prefigure the routes to be taken by the OECs. As in the case of the corticospinal tract, the OECs and their ensheathed axons travel into these channels (Fig. 11b and c) to form a peripheral nerve-like concentric arrangement in which the regenerating axons were ensheathed by OECs which were themselves surrounded on their outer, basal-lamina-lined *glia to mesodermal surface* by a collagen-containing extracellular space bounded by an encirclement of ONF processes (Li et al., 1998, 2003b). Thus, both in the scarring response after lesions (Kawano et al., 2012) and after transplantation of OEC/ONF mixtures, the astrocytic responses are closely coordinated with proactive fibroblast responses.

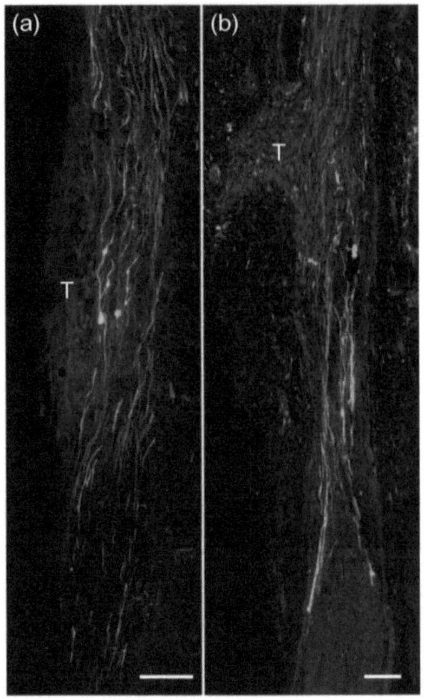

FIGURE 10

Longitudinal sections of regenerating axons (a) passing completely through an OEC transplant (T) in the corticospinal tract (green/white in print version); regenerating axons anterogradely labeled by injection of biotin dextran into the cortex; red (gray in print version) p75 immunostaining; 4 weeks survival. (b) Passing through an OEC transplant (T) and entering for 10 mm into the distal optic nerve (green/white in print version; labeled by cholera toxin B injected into the vitreous chamber, red (gray in print version), Sytox Orange nuclear counterstaining; 7 months survival). Scale bars, 200 μm in (a), 500 μm in (b). (For interpretation of the references to color in this figure legend, the reader is referred to the online version of this chapter.)

Reprinted with permission from Li et al. (1997, 2003b).

4.3 Avulsed dorsal roots

In order to retain transplanted OEC/ONF mixtures in lesion sites, we developed a method of transplanting the cells immobilized in the endogenous matrix which they generate in culture (Li et al., 2003a). In combination with a fibrin sealant, this matrix provided a surgical method for retaining the transplanted cells at the site of reapposition of avulsed spinal roots (Li et al., 2004, 2007; Raisman et al., 2011). This procedure facilitated the reexamination of dorsal root repair. The transplanted OECs formed a bridge (Fig. 12a) which allowed the regenerating dorsal root axons to grow out from the Schwann cell territory of the peripheral nerve. The OECs ensheathed the axons in a one-to-one, unmyelinated fashion (Fig. 12b,c,d) and conveyed them across the interface into the astrocytic territory of the spinal cord. Once they had entered the

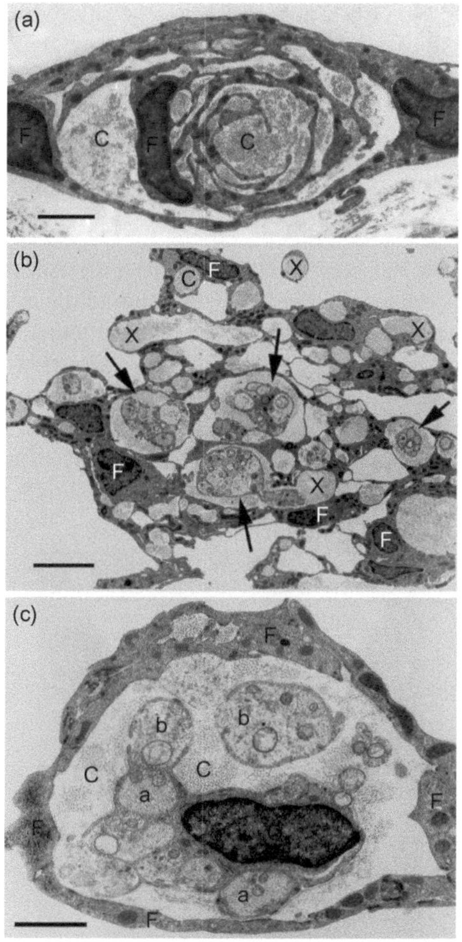

FIGURE 11

Electron micrographs of serial cross sections of a transected optic nerve 7 months after transplantation of an OEC/ONF mixture. (a) Multiple advancing channels formed by whorls of three transplanted ONFs (F) and containing massed collagen fibers (C) prefigure the routes to be taken by the OECs. (b) Some channels (e.g., arrows) contain regenerating optic nerve axons enwrapped in thin OEC processes of OECs which can be seen at high power to be lined by a basal lamina and which lie in collagen-containing fibroblast-bounded channels. Other channels (e.g., x) contain only collagen (cf., the empty channels in a). (c) Cross section of a single, collagen-containing fibroblastic channel (F) within which the nucleated perikaryon of a basal-lamina-lined OEC (O) gives rise to fine cytoplasmic sheets completely (e.g., a) or partially (e.g., b) enclosing regenerating RGC axons. Scale bars, 5 μm (a,b) 2 μm (c).

Reprinted with permission from Li et al. (2003b).

CNS, the axons traveled freely through the astrocytic environment of the spinal cord, and elongated for considerable distances, mingled separately among the myelinated fibers in the oligodendrocytic environment of the dorsal columns (as seen in Fig. 2b) to reach distant targets in the cuneate nuclei and restore electrophysiological transmission and behavioral recovery (Ibrahim et al., 2009).

4.4 Crossing the astrocytic scar

In the corticospinal tract and the dorsal root lesions, the common factor preventing axon regeneration was the local closed configuration of the astrocytes at the scarred interface. In both situations, the transplanted OECs interacted with the host astrocytes so as to realign the astrocytic and fibroblastic processes (Figs. 12a and 13c). This resulted in a door-opening effect which was able to provide a pathway for axons to regenerate across the interface, and which was still effective when OECs were transplanted into corticospinal tract lesions after a delay of 8 weeks, a time when the lesion scar would have been fully formed (Keyvan-Fouladi et al., 2005).

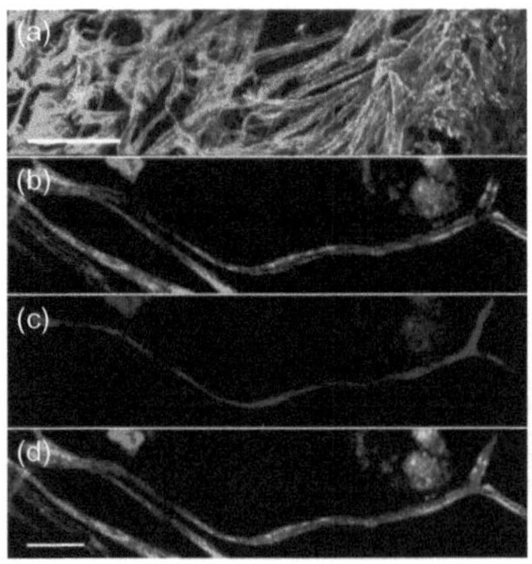

FIGURE 12

(a) The interface between the dorsal root entry zone of the spinal cord (on the right) and a matrix transplanted OEC graft placed at the point of apposition of the severed central end of a lower cervical dorsal rootlet (on the left) at 2 weeks survival. Outgrowing GFAP immunostained astrocytic processes (red fluorescence in digital version) from the spinal cord intermingle with the laminin immunostained peripheral nerve tissue (green fluorescence in digital version). This only occurs when OECs are included at the site of reapposition. (b) Matrix transplanted OECs (fluorescence, green in digital version; GFAP) ensheathing regenerating lower cervical dorsal root axons (red in digital version, neurofilament immunostaining in c) forming a bridge between the apposed central end of a severed dorsal root (on the left) and the former dorsal root entry point on the spinal cord (on the right). (d) Overlay 8 weeks after lesion and transplant. Scale bar, 100 μm in (a), 10 μm in (b,c,d). (For interpretation of the references to color in this figure legend, the reader is referred to the online version of this chapter.)

Reprinted with permission from Ibrahim et al. (2009) and Li et al. (2004).

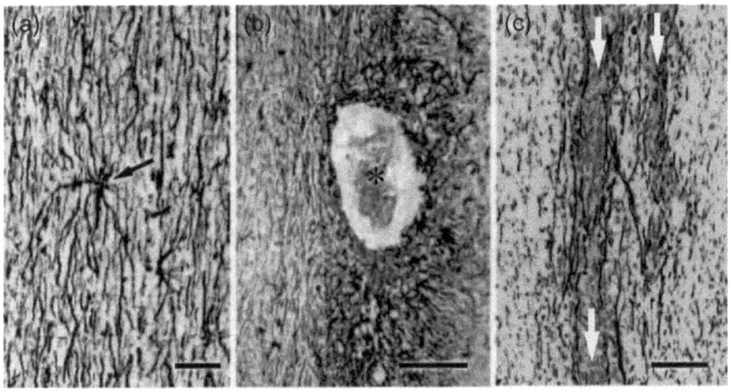

FIGURE 13

Alignment of GFAP immunostained astrocytic processes in the corticospinal tract: (a) vertical alignment of processes in normal tract (arrow indicates cell body), (b) massive accumulation of densely reduplicated scarred astrocytic processes walling off a lesion (*) at 65 days, (c) 30 days after transplantation of OECs into a lesion. The trails (white arrows) of transplanted OECs (darker gray) induce a vertical realignment of the astrocytic processes. Scale bars, 50 μm in (a), 500 μm in (b), 100 μm in (c).

Reprinted with permission (A) Li et al. (2005) and (B) Li and Raisman (1995).

Once the interface had been crossed, the undamaged astrocytic network was permissive for the regenerating axons to travel over long distances (Ibrahim et al., 2009; Keyvan-Fouladi et al., 2005). A similar ability is shown in situations in which embryonic or adult neurons are transplanted (Davies et al., 1999; Li and Raisman, 1993), or when intact axons sprout in response to an injury (Bareyre et al., 2004; Barritt et al., 2006; Li and Raisman, 1994; Raineteau and Schwab, 2001). In all these circumstances, neither normal astrocytes nor normal myelinating oligodendrocytes prevented axon growth, and the CNS retained or re-expressed sufficient of the developmental cues (Kopp et al., 2010; Lowery and Van, 2009; Maness and Schachner, 2007; Tessier-Lavigne and Goodman, 1996) for the regenerating axons to reestablish functional connections in their normal target areas.

4.5 Mechanism of OEC repair

The science of CNS repair by transplanted OECs is promising, but in its infancy. There are many unsolved questions. To understand the cellular interactions by which OEC transplants provide a route for axon regeneration, it is essential to take into account the participation of all the different cell types involved. Central to the proposed mechanism is the interaction of the transplanted OECs and ONFs with the host astrocytes. Fig. 14 assembles the known observations into a scheme, which we have called the "pathway hypothesis" (Li et al., 2005, 2006; Raisman and Li, 2007).

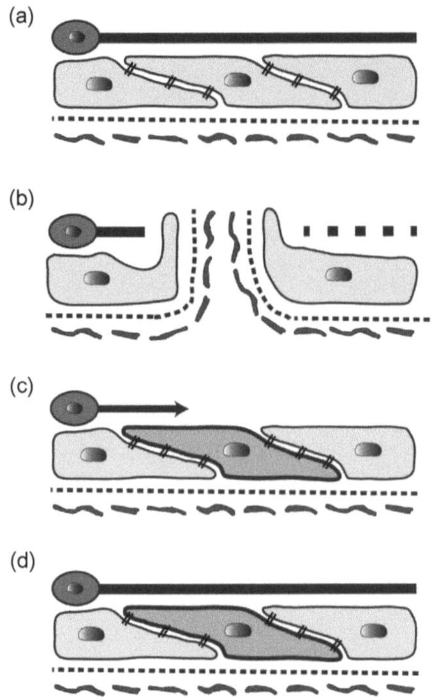

FIGURE 14

The pathway hypothesis depends on the concept of the three astrocytic surfaces as applied to (a) normal, (b) lesion, and (c) repair by transplanted OECs (darker gray filling) leading to (d) regeneration of axons. Figs. 3 and 13 illustrate various stages of reorganization of the scar membranes and incorporation of the transplanted OECs with later myelination. The presence of the desmosome/gap junctional complexes shown between the incorporated OECs (in c,d) and the astrocytes is hypothetical and would remain to be demonstrated both structurally and functionally. Symbols as in Fig. 1.

5 THE PATHWAY HYPOTHESIS

We propose that neurons retain throughout life the ability to grow and form new connections and that failure to regenerate after injury is due to the formation of an astrocytic scar. The incorporation of OECs into a lesion area reconfigures the astrocytic scar so to provide a continuous surface furnished with the ligand–receptor mechanisms needed for axon extension.

All three glial surfaces are involved in this tissue rearrangement.

- The integration of the transplanted OECs involves a *glia to glial surface* interaction with the host astrocytes leading to the door-opening effect.

- The transplanted OECs cooperate with the host astrocytes to maintain the continuity of the basal-lamina-lined *glia to mesodermal surface* facing fibroblasts and needed to preserve the glia–pial outer protection of the CNS.
- The integration of the transplanted OECs reconfigures the astrocytic scar so as to induce an alignment of the *glia to neuronal surfaces* which provides continuous channels allowing axons to enter the graft, grow through it, and re-enter the CNS (Li et al., 2005).

Once beyond the scarred interface the regenerating axons re-enter the glial network, through which they can make long distance, target directed growth, form new synapses, and restore lost functions.

References

Araque, A., 2008. Astrocytes process synaptic information. Neuron Glia Biol. 4, 3–10.

Asher, R.A., Morgenstern, D.A., Moon, L.D., Fawcett, J.W., 2001. Chondroitin sulphate proteoglycans: inhibitory components of the glial scar. Prog. Brain Res. 132, 611–619.

Bareyre, F.M., Kerschensteiner, M., Raineteau, O., Mettenleiter, T.C., Weinmann, O., Schwab, M.E., 2004. The injured spinal cord spontaneously forms a new intraspinal circuit in adult rats. Nat. Neurosci. 7, 269–277.

Barnett, S.C., Chang, L., 2004. Olfactory ensheathing cells and CNS repair: going solo or in need of a friend? Trends Neurosci. 27, 54–60.

Barritt, A.W., Davies, M., Marchand, F., Hartley, R., Grist, J., Yip, P., McMahon, S.B., Bradbury, E.J., 2006. Chondroitinase ABC promotes sprouting of intact and injured spinal systems after spinal cord injury. J. Neurosci. 26, 10856–10867.

Bennett, M.V., Contreras, J.E., Bukauskas, F.F., Saez, J.C., 2003. New roles for astrocytes: gap junction hemichannels have something to communicate. Trends Neurosci. 26, 610–617.

Chung, W.S., Barres, B.A., 2011. The role of glial cells in synapse elimination. Curr. Opin. Neurobiol. 22, 438–445.

Davies, S.J.A., Field, P.M., Raisman, G., 1993. Long fibre growth by axons of embryonic mouse hippocampal neurons micro-transplanted into the adult rat fimbria. Eur. J. Neurosci. 5, 95–106.

Davies, S.J., Goucher, D.R., Doller, C., Silver, J., 1999. Robust regeneration of adult sensory axons in degenerating white matter of the adult rat spinal cord. J. Neurosci. 19, 5810–5822.

Devon, R., Doucette, R., 1995. Olfactory ensheathing cells do not require L-ascorbic acid in vitro to assemble a basal lamina or to myelinate dorsal root ganglion neurites. Brain Res. 688, 223–229.

Doherty, P., Walsh, F.S., 1996. CAM-FGF receptor interactions: a model for axonal growth. Mol. Cell. Neurosci. 8, 99–111.

Doucette, J.R., 1984. The glial cells in the nerve fiber layer of the rat olfactory bulb. Anat. Rec. 210, 385–391.

Ekmark-Lewen, S., Lewen, A., Israelsson, C., Li, G.L., Farooque, M., Olsson, Y., Ebendal, T., Hillered, L., 2010. Vimentin and GFAP responses in astrocytes after contusion trauma to the murine brain. Restor. Neurol. Neurosci. 28, 311–321.

Eng, L.F., Ghirnikar, R.S., Lee, Y.L., 2000. Glial fibrillary acidic protein: GFAP thirty-one years (1969–2000). Neurochem. Res. 25, 1439–1451.

Eroglu, C., Barres, B.A., 2010. Regulation of synaptic connectivity by glia. Nature 468, 223–231.

Field, P.M., Li, Y., Raisman, G., 2003. Ensheathment of the olfactory nerves in the adult rat. J. Neurocytol. 32, 317–324.

Franklin, R.J.M., Gilson, J.M., Franceschini, I.A., Barnett, S.C., 1996. Schwann cell like myelination following transplantation of an olfactory bulb-ensheathing cell line into areas of demyelination in the adult CNS. Glia 17, 217–224.

Giaume, C., Koulakoff, A., Roux, L., Holcman, D., Rouach, N., 2010. Astroglial networks: a step further in neuroglial and gliovascular interactions. Nat. Rev. Neurosci. 11, 87–99.

Goldberg, M., De, P.M., Volman, V., Berry, H., Ben-Jacob, E., 2010. Nonlinear gap junctions enable long-distance propagation of pulsating calcium waves in astrocyte networks. PLoS Comput. Biol. 6, 1–14.

Graziadei, P.P.C., Montigraziadei, G.A., 1979. Neurogenesis and neuron regeneration in the olfactory system of mammals. I. Morphological aspects of differentiation and structural organization of the olfactory sensory neurons. J. Neurocytol. 8, 1–18.

Graziadei, P.P.C., Levine, R.R., Montigraziadei, G.A., 1979. Plasticity of connections of the olfactory sensory neuron: regeneration into the forebrain following bulbectomy in the neonatal mouse. Neuroscience 4, 713–727.

Haydon, P.G., 2001. GLIA: listening and talking to the synapse. Nat. Rev. Neurosci. 2, 185–193.

Henneberger, C., Rusakov, D.A., 2010. Synaptic plasticity and $Ca2^+$ signalling in astrocytes. Neuron Glia Biol. 6, 141–146.

Holbrook, E.H., Szumowski, K.E.M., Schwob, J.E., 1995. An immunochemical, ultrastructural, and developmental characterization of the horizontal basal cells of rat olfactory epithelium. J. Comp. Neurol. 363, 129–146.

Iadecola, C., Nedergaard, M., 2007. Glial regulation of the cerebral microvasculature. Nat. Neurosci. 10, 1369–1376.

Ibrahim, A., Li, Y., Li, D., Raisman, G., el Masry, W.S., 2006. Olfactory ensheathing cells: ripples of an incoming tide? Lancet Neurol. 5, 453–457.

Ibrahim, A.G., Kirkwood, P.A., Raisman, G., Li, Y., 2009. Restoration of hand function in a rat model of repair of brachial plexus injury. Brain 132, 1268–1276.

Kawano, H., Kimura-Kuroda, J., Komuta, Y., Yoshioka, N., Li, H.P., Kawamura, K., Li, Y., Raisman, G., 2012. Role of the lesion scar in the response to damage and repair of the central nervous system. Cell Tissue Res. 349, 169–180.

Keyvan-Fouladi, N., Raisman, G., Li, Y., 2003. Functional repair of the corticospinal tract by delayed transplantation of olfactory ensheathing cells in adult rats. J. Neurosci. 23, 9428–9434.

Keyvan-Fouladi, N., Raisman, G., Li, Y., 2005. Delayed repair of corticospinal tract lesions as an assay for the effectiveness of transplantation of Schwann cells. Glia 51, 306–311.

Kopp, M.A., Brommer, B., Gatzemeier, N., Schwab, J.M., Pruss, H., 2010. Spinal cord injury induces differential expression of the profibrotic semaphorin 7A in the developing and mature glial scar. Glia 58, 1748–1756.

Kuhlengel, K.R., Bunge, M.B., Bunge, R.P., Burton, H., 1990. Implantation of cultured sensory neurons and Schwann cells into lesioned neonatal rat spinal cord. II. Implant characteristics and examination of corticospinal tract growth. J. Comp. Neurol. 293, 74–91.

Lakatos, A., Franklin, R.J.M., Barnett, S.C., 2000. Olfactory ensheathing cells and Schwann cells differ in their in vitro interactions with astrocytes. Glia 32, 214–225.

Li, Y., Raisman, G., 1993. Long axon growth from embryonic neurons transplanted into myelinated tracts of the adult rat spinal cord. Brain Res. 629, 115–127.

Li, Y., Raisman, G., 1994. Schwann cells induce sprouting in motor and sensory axons in the adult rat spinal cord. J. Neurosci. 14, 4050–4063.

Li, Y., Raisman, G., 1995. Sprouts from cut corticospinal axons persist in the presence of astrocytic scarring in long-term lesions of the adult rat spinal cord. Exp. Neurol. 134, 102–111.

Li, Y., Field, P.M., Raisman, G., 1997. Repair of adult rat corticospinal tract by transplants of olfactory ensheathing cells. Science 277, 2000–2002.

Li, Y., Field, P.M., Raisman, G., 1998. Regeneration of adult rat corticospinal axons induced by transplanted olfactory ensheathing cells. J. Neurosci. 18, 10514–10524.

Li, Y., Decherchi, P., Raisman, G., 2003a. Transplantation of olfactory ensheathing cells into spinal cord lesions restores breathing and climbing. J. Neurosci. 23, 727–731.

Li, Y., Sauvé, Y., Li, D., Lund, R.D., Raisman, G., 2003b. Transplanted olfactory ensheathing cells promote regeneration of cut adult rat optic nerve axons. J. Neurosci. 23, 7922–7930.

Li, Y., Carlstedt, T., Berthold, C.-H., Raisman, G., 2004. Interaction of transplanted olfactory-ensheathing cells and host astrocytic processes provides a bridge for axons to regenerate across the dorsal root entry zone. Exp. Neurol. 188, 300–308.

Li, Y., Li, D., Raisman, G., 2005. Interaction of olfactory ensheathing cells with astrocytes may be the key to repair of tract injuries in the spinal cord: the 'pathway hypothesis'. J. Neurocytol. 34, 343–351.

Li, Y., Li, D., Raisman, G., 2006. Interaction of olfactory ensheathing cells with astrocytes may be the key to repair of tract injuries in the spinal cord: the 'pathway hypothesis'. J. Neurocytol. 34, 343–351.

Li, Y., Yamamoto, M., Raisman, G., Choi, D., Carlstedt, T., 2007. An experimental model of ventral root repair showing the beneficial effect of transplanting olfactory ensheathing cells. Neurosurgery 60, 734–740.

Liu, K., Lu, Y., Lee, J.K., Samara, R., Willenberg, R., Sears-Kraxberger, I., Tedeschi, A., Park, K.K., Jin, D., Cai, B., Xu, B., Connolly, L., Steward, O., Zheng, B., He, Z., 2010. PTEN deletion enhances the regenerative ability of adult corticospinal neurons. Nat. Neurosci. 13, 1075–1081.

Lowery, L.A., Van, V.D., 2009. The trip of the tip: understanding the growth cone machinery. Nat. Rev. Mol. Cell Biol. 10, 332–343.

Lu, J., Féron, F., Ho, S.H., Mackay-Sim, A., Waite, P.M.E., 2001. Transplantation of nasal olfactory tissue promotes partial recovery in paraplegic adult rats. Brain Res. 889, 344–357.

Maness, P.F., Schachner, M., 2007. Neural recognition molecules of the immunoglobulin superfamily: signaling transducers of axon guidance and neuronal migration. Nat. Neurosci. 10, 19–26.

Murrell, W., Féron, F., Wetzig, A., Cameron, N., Splatt, K., Bellette, B., Bianco, J., Perry, C., Lee, G., Mackay-Sim, A., 2005. Multipotent stem cells from adult olfactory mucosa. Dev. Dyn. 233, 496–515.

Nishida, H., Okabe, S., 2007. Direct astrocytic contacts regulate local maturation of dendritic spines. J. Neurosci. 27, 331–340.

Pearse, D.D., Pereira, F.C., Marcillo, A.E., Bates, M.L., Berrocal, Y.A., Filbin, M.T., Bunge, M.B., 2004. cAMP and Schwann cells promote axonal growth and functional recovery after spinal cord injury. Nat. Med. 10, 610–616.

Raineteau, O., Schwab, M.E., 2001. Plasticity of motor systems after incomplete spinal cord injury. Nat. Rev. Neurosci. 2, 263–273.

Raisman, G., 1969. Neuronal plasticity in the septal nuclei of the adult rat. Brain Res. 14, 25–48.

Raisman, G., 1985. Specialized neuroglial arrangement may explain the capacity of vomeronasal axons to reinnervate central neurons. Neuroscience 14, 237–254.

Raisman, G., Field, P.M., 1973. Time course and efficiency of collateral reinnervation of denervated synaptic sites in the septal nuclei of the rat. Anat. Rec. 175, 419.

Raisman, G., Li, Y., 2007. Repair of neural pathways by olfactory ensheathing cells. Nat. Rev. Neurosci. 8, 312–319.

Raisman, G., Carlstedt, T., Choi, D., Li, Y., 2011. Clinical prospects for transplantation of OECs in the repair of brachial and lumbosacral plexus injuries: opening a door. Exp. Neurol. 229, 168–173.

Ramón-Cueto, A., Nieto-Sampedro, M., 1992. Glial cells from adult rat olfactory bulb: immunocytochemical properties of pure cultures of ensheathing cells. Neuroscience 47, 213–220.

Ramón-Cueto, A., Cordero, M.I., Santos-Benito, F.F., Avila, J., 2000. Functional recovery of paraplegic rats and motor axon regeneration in their spinal cords by olfactory ensheathing glia. Neuron 25, 425–435.

Rouach, N., Koulakoff, A., Abudara, V., Willecke, K., Giaume, C., 2008. Astroglial metabolic networks sustain hippocampal synaptic transmission. Science 322, 1551–1555.

Sasaki, M., Lankford, K.L., Zemedkun, M., Kocsis, J.D., 2004. Identified olfactory ensheathing cells transplanted into the transected dorsal funiculus bridge the lesion and form myelin. J. Neurosci. 24, 8485–8493.

Schwab, M.E., Kapfhammer, J.P., Bandtlow, C.E., 1993. Inhibitors of neurite growth. Annu. Rev. Neurosci. 16, 565–595.

Silver, J., Miller, J.H., 2004. Regeneration beyond the glial scar. Nat. Rev. Neurosci. 5, 146–156.

Stichel, C.C., Muller, H.W., 1994. Relationship between injury-induced astrogliosis, laminin expression and axonal sprouting in the adult rat brain. J. Neurocytol. 23, 615–630.

Tessier-Lavigne, M., Goodman, C.S., 1996. The molecular biology of axon guidance. Science 274, 1123–1133.

Verkhratsky, A., Kettenmann, H., 1996. Calcium signalling in glial cells. Trends Neurosci. 19, 346–352.

Villegas-Pérez, M.P., Vidal-Sanz, M., Bray, G.M., Aguayo, A.J., 1988. Influences of peripheral nerve grafts on the survival and regrowth of axotomized retinal ganglion cells in adult rats. J. Neurosci. 8, 265–280.

Wictorin, K., Brundin, P., Sauer, H., Lindvall, O., Bjorklund, A., 1992. Long distance directed axonal growth from human dopaminergic mesencephalic neuroblasts implanted along the nigrostriatal pathway in 6-hydroxydopamine lesioned adult rats. J. Comp. Neurol. 323, 475–494.

Ylera, B., Erturk, A., Hellal, F., Nadrigny, F., Hurtado, A., Tahirovic, S., Oudega, M., Kirchhoff, F., Bradke, F., 2009. Chronically CNS-injured adult sensory neurons gain regenerative competence upon a lesion of their peripheral axon. Curr. Biol. 19, 930–936.

Zhang, Y., Yeh, J., Richardson, P.M., Bo, X., 2008. Cell adhesion molecules of the immunoglobulin superfamily in axonal regeneration and neural repair. Restor. Neurol. Neurosci. 26, 81–96.

CHAPTER

Current status of myelin replacement therapies in multiple sclerosis

12

Jeffrey K. Huang[*,†,‡,1], **Robin J.M. Franklin**[*,†,‡]

[*]*Wellcome Trust and MRC, Cambridge Stem Cell Institute, University of Cambridge, Cambridge, UK*

[†]*Department of Veterinary Medicine, University of Cambridge, Cambridge, UK*

[‡]*Department of Biology, Georgetown University, Washington, DC, USA*

[1]*Corresponding author. Tel.: +1-202-687-1741, Fax: +1-202-687-5662, e-mail address: jh1659@georgetown.edu; jeffrey.k.huang@gmail.com*

Abstract

Multiple sclerosis is an autoimmune disease of the human central nervous system characterized by immune-mediated myelin and axonal damage, and chronic axonal loss attributable to the absence of myelin sheaths. There are two aspects to the treatment of MS—first, the prevention of damage by suppressing the maladaptive immune system, and second, the long-term preservation of axons by the promotion of remyelination, a regenerative process in which new myelin are restored to demyelinated axons. Medicine has made significant progress in the first of these in recent years—there is an increasing number of ever more effective disease-modifying immunomodulatory interventions. However, there are currently no widely used regenerative therapies in MS. Conceptually, there are two approaches to remyelination therapy—transplantation of myelinogenic cells and promotion of endogenous remyelination mediated by myelinogenic cells present within the diseased tissue. In this chapter, in addition to describing why remyelination therapies are important, we review both these approaches, outlining their current status and future developments.

Keywords

stem cell, oligodendrocyte, cell transplantation, endogenous adult stem cells, CNS myelin, remyelination, regeneration, multiple sclerosis

Progress in Brain Research, Volume 201, ISSN 0079-6123, http://dx.doi.org/10.1016/B978-0-444-59544-7.00011-1

1 INFLAMMATORY DESTRUCTION OF CENTRAL NERVE CABLES

Multiple sclerosis (MS) is a common and progressively disabling neurological disorder that affects 12.5 million adults worldwide. It is typically characterized by transient episodes of inflammatory attacks on the brain, optic nerve, and spinal cord, in which lipid-rich myelin membranes and frequently the axons that myelin enwrap are damaged, and followed by sustained axonal dysfunction and chronic neurodegeneration (Compston and Coles, 2002). Until recently, treatment strategy had principally focused on reducing or preventing relapses, which were viewed as critical contributors to neurological symptoms and disability, by therapeutically targeting CNS inflammation. Although these treatments are able to limit the frequency or intensity of new episodes, they are disappointingly unable to prevent the undesirable transition to progressive MS, in which the accumulation of irreversible clinical disability from chronic neurodegeneration inevitably occurs (Compston, 2006; Trapp and Nave, 2008). The failure to prevent disease progression in MS by anti-inflammatory and immunomodulation therapies has therefore thrown chronic neurodegeneration into sharp relief as a poorly understood feature of MS pathogenesis. At present, approximately 50% of people affected by MS are at the progressive stage of the disease, and yet, there are no treatments. A strategy to protect axons and prevent neurons from progressive degeneration in contrast to the already heavily invested treatment strategy to suppress inflammation is desperately needed, but is not available.

2 MYELIN MAINTAINS AXONAL INTEGRITY

It is thought that chronic neurodegeneration in MS occurs via a noncell-autonomous mechanism, in which its pathogenesis begins from the loss of oligodendrocytes, but disease progression results from failed axonal protection, structurally and metabolically, followed by progressive neurodegeneration (Franklin and Ffrench-Constant, 2008). In the early phase of MS, clinical disabilities are generally reversible. Typically, in between episodes of attack (relapse) are periods of recovery (remission), during which tissue repair and myelin regeneration (remyelination) occur. Remyelination is spontaneous and neuroprotective, and limits the axonal dysfunction or degeneration that otherwise occurs with myelin loss. Following demyelination, oligodendrocyte precursor cells (OPCs), which are widely distributed throughout the adult CNS, are recruited to the lesion where they are able to differentiate into mature oligodendrocytes, engage with, and remyelinate the demyelinated axons. This spontaneous regenerative process following injury is crucial in CNS repair, as it reestablishes the axon–glial contact necessary for saltatory conduction, and restores neurotrophic support and axoplasmic transport, thus enabling recovery from clinical disability caused by CNS demyelination. However, over time, the ability to regenerate myelin becomes limited so that the proportion of demyelinated axons may eventually outweigh the proportion of remyelinated axons in MS. The increasing number of unhealed, denuded axons in the CNS has severe consequences, as they lose trophic support from surrounding oligodendrocyte myelin, display impaired conduction, axoplasmic transport defect, and increased oxidative stress, and thus are vulnerable to degeneration (Mahad et al., 2008; Trapp and Stys, 2009). In progressive MS, chronic axonal loss and neurodegeneration correlates with remyelination failure. Myelin is therefore needed to maintain axonal integrity, and failed remyelination is likely a cause leading to irreversible clinical decline at the later stage of the disease. A neuroprotective strategy for myelin replacement, particularly in progressive MS, would be expected to prevent neurodegeneration and limit the accumulation of clinical disability.

3 OBSTACLES FACING REMYELINATION

It remains unclear exactly why remyelination fails with chronic demyelination (Franklin, 2002b). It is possible that lingering inhibitory factors or diminished pro-regenerative factors in the chronically demyelinated environment prevent successful remyelination. For example, myelin debris, extracellular matrix deposits, and oligodendrocyte-derived inhibitory molecules have all been found to accumulate in the chronically demyelinated environment (Back et al., 2005; Charles et al., 2002; Kotter et al., 2006; Syed et al., 2011). Another possibility is that OPCs are unable to regenerate myelin because of extrinsic and intrinsic cellular changes associated with aging (Gilson and Blakemore, 1993; Rist and Franklin, 2008). Most regenerative processes slow down with age. Indeed, remyelination is very inefficient in old animals, largely due to a failure in OPC differentiation (Sim et al., 2002). Moreover, epigenetic change in OPCs, in which altered histone acetylation status favors the inhibition of OPC differentiation, is a prominent feature in the aged brain and in chronic progressive MS (Pedre et al., 2011; Shen et al., 2008). A study by Confavreux and Vukusic (2006) assessed the disease onset and course, and clinical disability score of over 1000 people with MS. They revealed that the time to disability milestones, and the age at which they are reached, appear to follow a predictable course regardless of previous relapse frequency or severity, suggesting that disease progression in MS is age dependent and not influenced by the initial course or symptoms of the disease (Confavreux and Vukusic, 2006). The result of this study suggests that the accumulation of irreversible disability is attributed to the progressive failure of myelin repair. Recently, we have found that the rate and extent of remyelination could be improved dramatically in old mice if they were experimentally conjoined by parabiosis to young animals (Ruckh et al., 2012). Upon exposure to circulatory factors and innate immune cells from young mice that have homed in to lesions of old animals, endogenous OPCs from the old mice were able to overcome age-related factors, and differentiate into mature remyelinating oligodendrocytes, suggesting that remyelination failure, despite possible changes intrinsically in OPCs with age, can be reversed under the appropriate environment.

4 REGENERATIVE MEDICINE IN MS

Regenerative medicine involves the experimental manipulation of stem or progenitor cells to replace damaged tissues and aims to restore their structure and function, and to prevent their further deterioration. Stem cells are immature, undifferentiated cell populations that have the intrinsic potential to differentiate into mature cells of multiple tissue lineages. By using stem cell therapy to promote myelin replacement around demyelinated axons in MS, it may be possible to limit, or stop neurodegeneration, particularly when intrinsic remyelination has become inadequate (Franklin and Ffrench-Constant, 2010; Martino et al., 2010). There are at least two approaches in which stem/progenitor cells could be used for myelin replacement in MS. One approach is to transplant exogenous neural stem/progenitor cells with myelinating potential directly, and the other is to mobilize endogenous neural stem/progenitor cells in the adult CNS. Studies in animal models of demyelination have demonstrated that each strategy has the potential to improve remyelination, although through different mechanisms. However, a few caveats should be considered for each strategy when developing a stem cell therapy for MS (Duncan et al., 2008; Franklin and Ffrench-Constant, 2010; Martino et al., 2010).

5 EXOGENOUS CELL THERAPY

Several cell transplantation strategies have been proposed for the treatment of MS (Table 1). To replace myelin in MS, ideally, myelinogenic cells such as OPCs should be used (although peripherally derived Schwann cells and olfactory ensheathing cells are also myelinogenic and are attractive alternatives to OPCs; see Franklin, 2002a). It is possible to derive and expand OPCs in culture from human embryonic stem cells, or induced pluripotent cells (Czepiel et al., 2011; Hu et al., 2009; Liu et al., 2011). While exogenous OPCs can readily populate lesions and replace myelin in animal models of acute demyelination, they are less likely to do so in the chronically demyelinated environment (Franklin, 2002b). A possible explanation for this is that inflammation has subsided under chronic demyelination so that the pro-regenerative stimuli that must normally accompany acute inflammation are unavailable to promote myelin debris clearance and/or OPC differentiation (Foote and Blakemore, 2005; Setzu et al., 2006). Similarly in chronic MS lesions, there is an abundance of adult OPCs that are seemingly unable to terminally differentiate despite being in contact with axons (Chang et al., 2002; Scolding et al., 1998; Wolswijk, 1998). It is therefore important that an optimal time of cell delivery is determined for efficient myelin replacement in MS. Moreover, MS lesions are multifocal, and spread out unpredictably in the CNS, thus an obstacle that needs to be considered is the route by which exogenous OPCs are administered in order to reach a specific lesion site (plaque-guided approach), or multiple lesion sites (homing and diffusion approach).

Peripherally derived stem cells such as mesenchymal stem cells have also been used for transplantation in animal models of demyelination (Uccelli et al., 2011a,b).

Table 1 Cell transplantation strategies in MS

Cell type	Route of delivery	Immunomodulation	Indirect remyelination	Direct remyelination	Clinical trials
Oligodendrocyte precursor cells	Local	No evidence	No evidence	Yes	No
Schwann cells	Local	No	No evidence	Yes	Single (unreported)
Olfactory ensheathing cells	Local	No	No evidence	Yes	No[a]
Neural stem cells	Systemic	Yes	Possible	Yes (minor)	No
Mesenchymal stem cells	Systemic	Yes	Possible	No	Several
Hematopoietic stem cells	Systemic	Yes	No evidence	No	Several

[a]*Several for spinal cord injury, none for primary demyelinating disease.*

These cells, as well as neural stem cells have been administered intravenously or intracerebroventricularly into animal models of demyelination and have been demonstrated to induce effects that may be indicative of enhanced endogenous remyelination and neuroprotection. In a recent small-scale study, in which autologous bone-marrow-derived mesenchymal stem cells were intravenously infused into patients with secondary progressive MS, improvement in CNS structure, function, and physiology, suggesting possibly enhanced neuroprotection, were observed months after the treatment (Connick et al., 2012). However, rather than through direct cell replacement, the administered stem cells likely influence the injury environment through immunomodulation and provision of trophic factors to promote tissue remodeling and neuroprotection (Cossetti et al., 2012; Freedman et al., 2010; Gosselin and Rivest, 2011; Martino et al., 2011; Uccelli et al., 2011b). Intravenous infusion of mesenchymal stem cells into patients appears to have minor adverse effects; however, although this strategy shows promise, further large-scale, longitudinal studies need to be performed before mesenchymal stem cells could be considered as a treatment option for progressive MS.

6 ENDOGENOUS CELL THERAPY

Endogenous OPCs are widely distributed in the adult CNS (Franklin and Ffrench-Constant, 2008; Gallo et al., 2008; Levine et al., 2001; Nishiyama et al., 2009; Paukert and Bergles, 2006). They interact with synapses and nodes of Ranvier and are responsible for replacing myelin after CNS injury (Zawadzka et al., 2010). OPCs migrate toward sites of demyelination, and at the site of lesion, are induced to differentiate and mature to replace myelin, but this ability fails with age and in progressive MS. A feasible strategy is to therapeutically enhance endogenous OPC differentiation by pharmacologically stimulating the intrinsic signaling pathways of OPC differentiation, or removing the nonpermissive environment in chronic demyelination (Huang et al., 2011a). Several regulators of CNS remyelination have recently been identified, including retinoid X receptor-gamma (RXRγ), LINGO1, Notch1, and Wnt. Drugs targeting these pathways are already available, or currently being developed (Fig. 1).

RXRγ is a potentially promising drug target for stimulating myelin replacement by endogenous OPCs (Huang et al., 2011b,c). In MS tissues, RXRγ is highly expressed by oligodendrocyte lineage cells, macrophages, and astrocytes in acute and remyelinating lesions, compared to chronic inactive lesions, suggesting that RXRγ is involved in CNS regeneration (Huang et al., 2011c). In RXRγ knockout mice, undifferentiated OPCs accumulate in lesions following demyelination, suggesting that RXRγ is a positive regulator of OPC differentiation. Moreover, when OPCs are exposed to retinoid X receptor (RXR) agonists in culture, oligodendrocytes differentiate and form myelin membrane-like sheets. The addition of an RXR agonist 9-*cis*-retinoic acid to *ex vivo* cerebellar slice cultures of demyelination and to focally demyelinated rats also significantly improves the extent of remyelination.

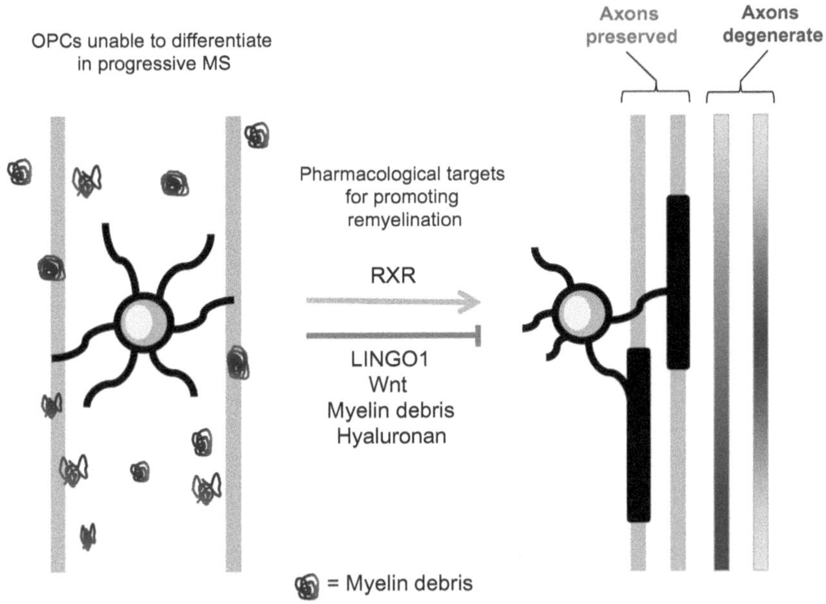

FIGURE 1

Pharmacological approach to improve CNS remyelination in progressive MS. Oligodendrocyte precursor cells (OPCs) are widely distributed in chronic lesions, however are unable to differentiate and remyelinate axons. Chronic demyelination with limited remyelination results in the dysfunction and progressive degeneration of axons. One of the major inhibitors of remyelination is myelin debris. Several gene targets have recently been identified that can positively or negatively regulate remyelination. Pharmacological agonists targeting RXR, as well as antibody against LINGO1 are available. (For color version of this figure, the reader is referred to the online version of this chapter.)

RXRγ is a member of the RXR family that also includes RXRα and RXRβ (Altucci et al., 2007; Germain et al., 2006; Lefebvre et al., 2010). RXRs exert transcriptional activity through heterodimerization with other nuclear receptors, including PPARs, LXRs, retinoic acid receptors, vitamin D receptors, and thyroid hormone receptors. Currently, RXR agonists are being tested for treatment of certain cancers and are also thought to be useful in the treatment of metabolic disorders (Huang et al., 2011a). A licensed RXR agonist Targretin (bexarotene) is already in clinical use for the treatment of cutaneous T cell lymphoma (Ballanger et al., 2010). It is known that RXR agonists can also attenuate inflammation by regulating macrophage activity (Diab et al., 2004; Xu and Drew, 2006). Moreover, bexarotene was shown to enhance the clearance of soluble Aβ, resulting in the significant reduction of Aβ plaques in a mouse model of Alzheimer's Disease (Cramer et al., 2012) and may therefore enhance the phagocytic clearance of remyelination inhibitory myelin debris (Kotter et al., 2006). Thus, bexarotene and other synthetic RXR agonists could conceivably advance MS therapy, by suppressing inflammatory damage through immunomodulation, and also preserving axons by promoting remyelination. Moreover, RXR agonists will be of value as a regenerative therapy, filling the unoccupied niche of a treatment for progressive MS.

An active area of research in MS is to identify pharmacological strategies to reduce the inhibition of OPC differentiation in the chronically demyelinated environment (Coman et al., 2005; Franklin and Ffrench-Constant, 2008; Kotter et al., 2011; Miller and Mi, 2007; Sloane et al., 2010). Several negative regulators of OPC differentiation have recently been identified, including LINGO1, Notch1, and Wnt (Fancy et al., 2009; John et al., 2002; Mi et al., 2005). One potential compound, which is currently undergoing clinical trial, is the anti-LINGO1 monoclonal antibody (Rudick et al., 2008). LINGO1 is a leucine rich repeat protein that is specifically expressed in the CNS (Mi et al., 2004). Addition of siRNAs against LINGO1, dominant negative LINGO1, or LINGO-Fc to OPCs in culture increases oligodendrocyte differentiation (Mi et al., 2005). Increased remyelination and improved functional recovery from experimental autoimmune encephalomyelitis, a model of immune-mediated demyelination, were observed in mice deficient in LINGO1, and in mice treated with the anti-LINGO1 antibody (Mi et al., 2007). Moreover, the anti-LINGO1 antibody is able to promote CNS remyelination by directly stimulating OPC differentiation in nonimmune, toxin-induced models of demyelination in rats (Mi et al., 2009). The restricted expression of LINGO1 in the CNS makes therapeutic targeting of LINGO1 potentially advantageous as it improves myelin repair without affecting nonneural tissues (Mi et al., 2008). More recently, it has been shown that small molecule inhibition of the enzymatic activity of tankyrase, a poly(ADP-ribose) polymerase, leads to a "switching off" of the OPC differentiation inhibitory Wnt pathway and hence to an enhancement of the efficiency of remyelination in experimental models (Fancy et al., 2009, 2011).

7 FUTURE OUTLOOK

The last decade has seen not only significant advances towards reliable regenerative therapies for MS but also a shift of emphasis away from cell transplantation-based approaches toward drug-based interventions that aim to boost endogenous regenerative processes. The pioneering work on cell transplantation-mediated remyelination (e.g., Archer et al. 1997; Groves et al., 1993) certainly gave the field an enormous impetus but has to date failed to make the clinical breakthrough that many predicted 20 or so years ago. The reasons for this are many but primarily involve unresolved questions about the source and nature of appropriate cells and, more especially, how cells might best be delivered to the many lesions in need of regenerative therapy within an individual patient. Intriguingly, experiments aimed at addressing the delivery of cells with remyelinating potential to multiple sites via systemic delivery led to the discovery of the immunomodulatory effects of such cells and the subsequent translational pursuit of this function of the cells rather than of their regenerative properties (see Martino et al., 2010, 2011). The significant increases in our understanding of the biology of remyelination and the key regulatory pathways controlling the differentiation of multipotent cells into remyelinating oligodendrocytes, together with the discovery that the age-associated intrinsic decline in remyelination is reversible, combine to make the prospects of regenerative medicines rather than regenerative cells a realistic proposition. This is not to say that medicine and cell approaches are mutually exclusive: once drug-based approaches have been developed, there remains the issue of optimal and targeted delivery, which may well involve the use of appropriately engineered cells. The converse may also be true. For the genetic demyelinating diseases (a discussion of which is beyond the remit of this article), where the oligodendrocyte lineage is inherently dysfunctional, cell therapies remain in the front line of myelin regenerative approaches. This is in contrast to MS. Nevertheless, the drug-based approaches being developed for MS may have value as adjuncts to cell therapies, augmenting the myelinating capacity of the transplanted cells. Myelin regenerative therapies are not in wide clinical use yet—but there are many reasons, some of which have been discussed above, why we can be optimistic about the future.

References

Altucci, L., Leibowitz, M.D., Ogilvie, K.M., de Lera, A.R., Gronemeyer, H., 2007. RAR and RXR modulation in cancer and metabolic disease. Nat. Rev. Drug Discov. 6, 793–810.

Archer, D.R., Cuddon, P.A., Lipsitz, D., Duncan, I.D., 1997. Myelination of the canine central nervous system by glial cell transplnatation: a model for repair of human myelin disease. Nat. Med. 3, 54–59.

Back, S.A., Tuohy, T.M.F., Chen, H., Wallingford, N., Craig, A., Struve, J., Luo, N.L., Banine, F., Liu, Y., Chang, A., et al., 2005. Hyaluronan accumulates in demyelinated lesions and inhibits oligodendrocyte progenitor maturation. Nat. Med. 11, 966–972.

Ballanger, F., Nguyen, J.M., Khammari, A., Dréno, B., 2010. Evolution of clinical and molecular responses to bexarotene treatment in cutaneous T-cell lymphoma. Dermatology 220, 370–375.

Chang, A., Tourtellotte, W.W., Rudick, R., Trapp, B.D., 2002. Premyelinating oligodendrocytes in chronic lesions of multiple sclerosis. N. Engl. J. Med. 346, 165–173.

Charles, P., Reynolds, R., Seilhean, D., Rougon, G., Aigrot, M.S., Niezgoda, A., Zalc, B., Lubetzki, C., 2002. Re-expression of PSA-NCAM by demyelinated axons: an inhibitor of remyelination in multiple sclerosis? Brain 125, 1972–1979.

Coman, I., Barbin, G., Charles, P., Zalc, B., Lubetzki, C., 2005. Axonal signals in central nervous system myelination, demyelination and remyelination. J. Neurol. Sci. 233, 67–71.

Compston, A., 2006. The basis for treatment in multiple sclerosis. Acta Neurol. Scand. Suppl. 183, 41–47.

Compston, A., Coles, A., 2002. Multiple sclerosis. Lancet 359, 1221–1231.

Confavreux, C., Vukusic, S., 2006. Age at disability milestones in multiple sclerosis. Brain 129, 595–605.

Connick, P., Kolappan, M., Crawley, C., Webber, D.J., Patani, R., Michell, A.W., Du, M.-Q., Luan, S.-L., Altmann, D.R., Thompson, A.J., et al., 2012. Autologous mesenchymal stem

cells for the treatment of secondary progressive multiple sclerosis: an open-label phase 2a proof-of-concept study. Lancet Neurol. 11, 150–156.
Cossetti, C., AlfaroCervello, C., Donegà, M., Tyzack, G., Pluchino, S., 2012. New perspectives of tissue remodelling with neural stem and progenitor cell-based therapies. Cell Tissue Res. 349, 321–329.
Cramer, P.E., Cirrito, J.R., Wesson, D.W., Lee, C.Y.D., Karlo, J.C., Zinn, A.E., Casali, B.T., Restivo, J.L., Goebel, W.D., James, M.J., et al., 2012. ApoE-directed therapeutics rapidly clear β-amyloid and reverse deficits in AD mouse models. Science 335, 1503–1506.
Czepiel, M., Balasubramaniyan, V., Schaafsma, W., Stancic, M., Mikkers, H., Huisman, C., Boddeke, E., Copray, S., 2011. Differentiation of induced pluripotent stem cells into functional oligodendrocytes. Glia 59, 882–892.
Diab, A., Hussain, R.Z., Lovett-Racke, A.E., Chavis, J.A., Drew, P.D., Racke, M.K., 2004. Ligands for the peroxisome proliferator-activated receptor-gamma and the retinoid X receptor exert additive anti-inflammatory effects on experimental autoimmune encephalomyelitis. J. Neuroimmunol. 148, 116–126.
Duncan, I.D., Goldman, S., Macklin, W.B., Rao, M., Weiner, L.P., Reingold, S.C., 2008. Stem cell therapy in multiple sclerosis: promise and controversy. Mult. Scler. 14, 541–546.
Fancy, S.P.J., Baranzini, S.E., Zhao, C., Yuk, D.-I., Irvine, K.-A., Kaing, S., Sanai, N., Franklin, R.J.M., Rowitch, D.H., 2009. Dysregulation of the Wnt pathway inhibits timely myelination and remyelination in the mammalian CNS. Genes Dev. 23, 1571–1585.
Fancy, S.P.J., et al., 2011. Axin2 as regulatory and therapeutic target in newborn brain injury and remyelination. Nat. Neurosci. 14, 1009–1016.
Foote, A.K., Blakemore, W.F., 2005. Inflammation stimulates remyelination in areas of chronic demyelination. Brain 128, 528–539.
Franklin, R.J.M., 2002a. Remyelination of the demyelinated CNS: the case for and against transplantation of central, peripheral and olfactory glia. Brain Res. Bull. 57, 827–832.
Franklin, R.J.M., 2002b. Why does remyelination fail in multiple sclerosis? Nat. Rev. Neurosci. 3, 705–714.
Franklin, R.J.M., Ffrench-Constant, C., 2008. Remyelination in the CNS: from biology to therapy. Nat. Rev. Neurosci. 9, 839–855.
Franklin, R.J.M., Ffrench-Constant, C., 2010. Stem cell treatments and multiple sclerosis. BMJ 340, c1387.
Freedman, M.S., Bar-Or, A., Atkins, H.L., Karussis, D., Frassoni, F., Lazarus, H., Scolding, N., Slavin, S., Le Blanc, K., Uccelli, A., 2010. The therapeutic potential of mesenchymal stem cell transplantation as a treatment for multiple sclerosis: consensus report of the International MSCT Study Group. Mult. Scler. 16, 503–510.
Gallo, V., Mangin, J.-M., Kukley, M., Dietrich, D., 2008. Synapses on NG2-expressing progenitors in the brain: multiple functions? J. Physiol. (Lond.) 586, 3767–3781.
Germain, P., Chambon, P., Eichele, G., Evans, R.M., Lazar, M.A., Leid, M., De Lera, A.R., Lotan, R., Mangelsdorf, D.J., Gronemeyer, H., 2006. International Union of Pharmacology. LXIII. Retinoid X receptors. Pharmacol. Rev. 58, 760–772.
Gilson, J., Blakemore, W.F., 1993. Failure of remyelination in areas of demyelination produced in the spinal cord of old rats. Neuropathol. Appl. Neurobiol. 19, 173–181.

Gosselin, D., Rivest, S., 2011. Immune mechanisms underlying the beneficial effects of autologous hematopoietic stem cell transplantation in multiple sclerosis. Neurotherapeutics 8, 643–649.

Groves, A.K., et al., 1993. Repair of demyelinated lesions by transplantation of purified O-2A progenitor cells. Nature 362, 453–455.

Hu, B.-Y., Du, Z.-W., Zhang, S.-C., 2009. Differentiation of human oligodendrocytes from pluripotent stem cells. Nat. Protoc. 4, 1614–1622.

Huang, J.K., Fancy, S.P.J., Zhao, C., Rowitch, D.H., Ffrench-Constant, C., Franklin, R.J.M., 2011a. Myelin regeneration in multiple sclerosis: targeting endogenous stem cells. Neurotherapeutics 8, 650–658.

Huang, J.K., Jarjour, A.A., Ffrench-Constant, C., Franklin, R.J.M., 2011b. Retinoid X receptors as a potential avenue for regenerative medicine in multiple sclerosis. Expert Rev. Neurother. 11, 467–468.

Huang, J.K., Jarjour, A.A., Nait Oumesmar, B., Kerninon, C., Williams, A., Krezel, W., Kagechika, H., Bauer, J., Zhao, C., Evercooren, A.B.-V., et al., 2011c. Retinoid X receptor gamma signaling accelerates CNS remyelination. Nat. Neurosci. 14, 45–53.

John, G.R., Shankar, S.L., Shafit-Zagardo, B., Massimi, A., Lee, S.C., Raine, C.S., Brosnan, C.F., 2002. Multiple sclerosis: re-expression of a developmental pathway that restricts oligodendrocyte maturation. Nat. Med. 8, 1115–1121.

Kotter, M.R., Li, W.-W., Zhao, C., Franklin, R.J.M., 2006. Myelin impairs CNS remyelination by inhibiting oligodendrocyte precursor cell differentiation. J. Neurosci. 26, 328–332.

Kotter, M.R., Stadelmann, C., Hartung, H.-P., 2011. Enhancing remyelination in disease—can we wrap it up? Brain 134, 1882–1900.

Lefebvre, P., Benomar, Y., Staels, B., 2010. Retinoid X receptors: common heterodimerization partners with distinct functions. Trends Endocrinol. Metab. 21, 676–683.

Levine, J.M., Reynolds, R., Fawcett, J.W., 2001. The oligodendrocyte precursor cell in health and disease. Trends Neurosci. 24, 39–47.

Liu, Y., Jiang, P., Deng, W., 2011. OLIG gene targeting in human pluripotent stem cells for motor neuron and oligodendrocyte differentiation. Nat. Protoc. 6, 640–655.

Mahad, D., Lassmann, H., Turnbull, D., 2008. Review: mitochondria and disease progression in multiple sclerosis. Neuropathol. Appl. Neurobiol. 34, 577–589.

Martino, G., Franklin, R.J.M., Baron Van Evercooren, A., Kerr, D.A., 2010. Stem cell transplantation in multiple sclerosis: current status and future prospects. Nat. Rev. Neurol. 6, 247–255.

Martino, G., Pluchino, S., Bonfanti, L., Schwartz, M., 2011. Brain regeneration in physiology and pathology: the immune signature driving therapeutic plasticity of neural stem cells. Physiol. Rev. 91, 1281–1304.

Mi, S., Lee, X., Shao, Z., Thill, G., Ji, B., Relton, J., Levesque, M., Allaire, N., Perrin, S., Sands, B., et al., 2004. LINGO-1 is a component of the Nogo-66 receptor/p75 signaling complex. Nat. Neurosci. 7, 221–228.

Mi, S., Miller, R.H., Lee, X., Scott, M.L., Shulag-Morskaya, S., Shao, Z., Chang, J., Thill, G., Levesque, M., Zhang, M., et al., 2005. LINGO-1 negatively regulates myelination by oligodendrocytes. Nat. Neurosci. 8, 745–751.

Mi, S., Hu, B., Hahm, K., Luo, Y., Kam Hui, E.S., Yuan, Q., Wong, W.M., Wang, L., Su, H., Chu, T.-H., et al., 2007. LINGO-1 antagonist promotes spinal cord remyelination and

axonal integrity in MOG-induced experimental autoimmune encephalomyelitis. Nat. Med. 13, 1228–1233.
Mi, S., Sandrock, A., Miller, R.H., 2008. LINGO-1 and its role in CNS repair. Int. J. Biochem. Cell Biol. 40, 1971–1978.
Mi, S., Miller, R.H., Tang, W., Lee, X., Hu, B., Wu, W., Zhang, Y., Shields, C.B., Zhang, Y., Miklasz, S., et al., 2009. Promotion of central nervous system remyelination by induced differentiation of oligodendrocyte precursor cells. Ann. Neurol. 65, 304–315.
Miller, R.H., Mi, S., 2007. Dissecting demyelination. Nat. Neurosci. 10, 1351–1354.
Nishiyama, A., Komitova, M., Suzuki, R., Zhu, X., 2009. Polydendrocytes (NG2 cells): multifunctional cells with lineage plasticity. Nat. Rev. Neurosci. 10, 9–22.
Paukert, M., Bergles, D.E., 2006. Synaptic communication between neurons and NG2+ cells. Curr. Opin. Neurobiol. 16, 515–521.
Pedre, X., Mastronardi, F., Bruck, W., López-Rodas, G., Kuhlmann, T., Casaccia, P., 2011. Changed histone acetylation patterns in normal-appearing white matter and early multiple sclerosis lesions. J. Neurosci. 31, 3435–3445.
Rist, J.M., Franklin, R.J.M., 2008. Taking ageing into account in remyelination-based therapies for multiple sclerosis. J. Neurol. Sci. 274, 64–67.
Ruckh, J.M., Zhao, J.-W., Shadrach, J.L., van Wijngaarden, P., Rao, T.N., Wagers, A.J., Franklin, R.J.M., 2012. Rejuvenation of regeneration in the aging central nervous system. Cell Stem Cell 10, 96–103.
Rudick, R.A., Mi, S., Sandrock Jr., A.W., 2008. LINGO-1 antagonists as therapy for multiple sclerosis: in vitro and in vivo evidence. Expert Opin. Biol. Ther. 8, 1561–1570.
Scolding, N., Franklin, R., Stevens, S., Heldin, C.H., Compston, A., Newcombe, J., 1998. Oligodendrocyte progenitors are present in the normal adult human CNS and in the lesions of multiple sclerosis. Brain 121 (Pt 12), 2221–2228.
Setzu, A., Lathia, J.D., Zhao, C., Wells, K., Rao, M.S., Ffrench-Constant, C., Franklin, R.J.M., 2006. Inflammation stimulates myelination by transplanted oligodendrocyte precursor cells. Glia 54, 297–303.
Shen, S., Sandoval, J., Swiss, V.A., Li, J., Dupree, J., Franklin, R.J.M., Casaccia-Bonnefil, P., 2008. Age-dependent epigenetic control of differentiation inhibitors is critical for remyelination efficiency. Nat. Neurosci. 11, 1024–1034.
Sim, F.J., Zhao, C., Penderis, J., Franklin, R.J.M., 2002. The age-related decrease in CNS remyelination efficiency is attributable to an impairment of both oligodendrocyte progenitor recruitment and differentiation. J. Neurosci. 22, 2451–2459.
Sloane, J.A., Batt, C., Ma, Y., Harris, Z.M., Trapp, B., Vartanian, T., 2010. Hyaluronan blocks oligodendrocyte progenitor maturation and remyelination through TLR2. Proc. Natl. Acad. Sci. U.S.A. 107, 11555–11560.
Syed, Y.A., Hand, E., Möbius, W., Zhao, C., Hofer, M., Nave, K.A., Kotter, M.R., 2011. Inhibition of CNS remyelination by the presence of semaphorin 3A. J. Neurosci. 31, 3719–3728.
Trapp, B.D., Nave, K.-A., 2008. Multiple sclerosis: an immune or neurodegenerative disorder? Annu. Rev. Neurosci. 31, 247–269.
Trapp, B.D., Stys, P.K., 2009. Virtual hypoxia and chronic necrosis of demyelinated axons in multiple sclerosis. Lancet Neurol. 8, 280–291.
Uccelli, A., Benvenuto, F., Laroni, A., Giunti, D., 2011a. Neuroprotective features of mesenchymal stem cells. Best Pract. Res. Clin. Haematol. 24, 59–64.

Uccelli, A., Laroni, A., Freedman, M.S., 2011b. Mesenchymal stem cells for the treatment of multiple sclerosis and other neurological diseases. Lancet Neurol. 10, 649–656.

Wolswijk, G., 1998. Chronic stage multiple sclerosis lesions contain a relatively quiescent population of oligodendrocyte precursor cells. J. Neurosci. 18, 601–609.

Xu, J., Drew, P.D., 2006. 9-Cis-retinoic acid suppresses inflammatory responses of microglia and astrocytes. J. Neuroimmunol. 171, 135–144.

Zawadzka, M., et al., 2010. CNS-resident glial progenitor/stem cells produce Schwann cells as well as oligodendrocytes during repair of CNS demyelination. Cell Stem Cell 6, 578–590.

CHAPTER

Stem cell-based treatments for spinal cord injury 13

Lindsey A. Wyatt[*,†,‡,§], **Hans S. Keirstead**[*,†,‡,§,1]

[*]*Reeve-Irvine Research Center, University of California at Irvine, Irvine, CA, USA*

[†]*Sue and Bill Gross Stem Cell Research Center, University of California at Irvine, Irvine, CA, USA*

[‡]*Department of Anatomy & Neurobiology, School of Medicine, University of California at Irvine, Irvine, CA, USA*

[§]*Department of Neurological Surgery, School of Medicine, University of California at Irvine, Irvine, CA, USA*

[1]*Corresponding author. Tel.: +1-949-8246213, Fax: +1-949-8249223, e-mail address: hansk@uci.edu*

Abstract

Cell replacement strategies hold great promise for the treatment of central nervous system injuries and degenerative diseases. The advancement of stem cell therapies has proven to be a viable therapeutic approach to limit secondary degeneration and restore neuronal circuitry at the site of injury. Cell replacement strategies confer phenotype-specific and neurotrophic benefits to the surrounding tissue; however, the mechanisms of transplant-mediated repair are unique to each transplant population. Here, we review stem cell-based therapies for spinal cord injury and disease, involving a number of stem cell derivates. We discuss the mechanisms by which each of these populations exert their affects and briefly discuss phenotype-specific cell replacement in these models.

Keywords

stem cells, spinal cord injury, embryonic stem cells, oligodendrocyte progenitor cells, motor neuron progenitor cells, neural stem cell, hematopoietic stem cell, mesenchymal stem cell

1 INTRODUCTION

Cell replacement strategies hold great promise for the treatment of central nervous system injuries and degenerative diseases. Actual cell replacement requires appropriate site-specific differentiation *in situ* or transplantation of lineage-committed high-purity populations of a desired cell type. Although the ultimate goal for stem

Progress in Brain Research, Volume 201, ISSN 0079-6123, http://dx.doi.org/10.1016/B978-0-444-59544-7.00012-3

cell therapies is to restore function by replacing cells and restoring neuronal circuitry, the transplantation of cells into a pathological niche critically affects their differentiation potential and function, presenting a major challenge to cell replacement strategies. However, young progenitor cells, despite their origin, derivation, and differentiation potential *in vivo*, secrete growth factors that have beneficial effects on the injury/disease environment. Thus, cell replacement strategies are inherently combination therapies, as they confer phenotype-specific benefits as well as neurotrophic benefits to surrounding tissue.

Progenitor cells secrete a combination of neurotrophins and neurotrophic cytokines, mediating recovery by providing survival factors, guidance molecules, angiogenic and immunomodulatory factors, or cues for proliferation and differentiation of endogenous stem and progenitor cells. The growth factor expression profile, and hence the mechanisms of transplant-mediated repair, is unique to each transplant population. Here, we review stem cell-based therapies for spinal cord injury (SCI) and disease, involving a number of stem cell derivates. We discuss the mechanisms by which each of these populations exert their affects and briefly discuss phenotype-specific cell replacement in these models.

2 EPIDEMIOLOGY

There are an estimated 10,000–12,000 SCIs every year in the United States. The Christopher and Dana Reeve Foundation (2009) undertook a project to gather updated, high-quality epidemiological data of individuals with SCI in 2009. This exhaustive survey indicated that nearly 2% of the population of the United States, over 5.5 million individuals, reported some sort of paralysis. SCI accounted for 0.4% or 1,275,000 of those paralysis cases.

3 PATHOGENESIS

SCI is characterized by an initial mechanical trauma to the spinal cord that results in breakdown of the blood–brain barrier, influx of peripheral inflammatory factors, activation of glial cells, excitotoxicity, and necrosis. A secondary cascade of signaling events leads to the cyclic increase in inflammatory cytokines and chemokines, leading to apoptosis, progressive loss of oligodendrocytes leading to demyelination, and axonal degeneration. The inflammatory response is characterized by fluid accumulation and the influx of immune cells—neutrophils, T-cells, macrophages, and monocytes (Anderson et al., 2004), facilitated by their expression of matrix metalloproteinases (Noble et al., 2002; Wright et al., 2011). Neutrophils and T-cells are the first to enter the CNS followed by macrophages, which together with activated microglia, remove dead cells, and debris via phagocytosis. Macrophages can aid nerve regrowth by phagocytosing myelin debris, which is known to inhibit axonal regeneration (Chen et al., 2000; David et al., 1990; McKerracher et al., 1994;

Wright et al., 2011), and may release protective cytokines, such as basic fibroblast growth factor, nerve growth factor (NGF), and neurotrophin 3, to promote neuronal regeneration and tissue repair (Rabchevsky and Striet, 1998; Wright et al., 2011). The cytokines and chemokines produced by immune cells also propagate the inflammatory response, inducing a reactive process of secondary apoptosis in the tissue that surrounds the injury site (Noble et al., 2002; Popovich et al., 2002; Taoka et al., 1997; Wright et al., 2011).

Oligodendrocytes are also victim to apoptosis and necrosis (Rhonaghi et al., 2010). Oligodendrocytes form myelin, which facilitate saltatory conduction within axons. The loss of oligodendrocytes strips myelin from intact axons in ascending and descending pathways adjacent to the injury epicenter. Although the mechanism of oligodendrocyte apoptosis is not well understood, it has been suggested that Fas receptors located on the surface of oligodendrocytes can be activated by Fas ligands expressed by activated microglia, which in turn initiates a caspase cascade triggering apoptosis (Li et al., 2002). Also, serum proteins such as thrombin have a neurotoxic effect and can facilitate cell death by activating protease-activated receptors on microglia (Xue et al., 2006).

An excessive release of neurotransmitters can cause additional damage by excitotoxicity. Glutamate is an excitatory neurotransmitter. In SCI, glutamate accumulates in the extracellular fluid, leading to an overactivation of neurotransmitter receptors. As a result, calcium enters the cell and activates enzymes (phospholipases, proteases, etc.) that subsequently damage cell structures. This damage often leads to apoptosis and free radical-mediated lipid oxidation (Liu et al., 1997).

The injured spinal cord eventually becomes gliotic. Gliosis is beneficial for the reestablishment of physical and chemical integrity of the CNS but imparts a significant obstacle for neuroregeneration (Rhonaghi et al., 2010). During gliosis, astrocytes and oligodendrocyte progenitors are activated within, and recruited to, the injury site. Astrocytes and oligodendrocytes secrete inhibitory molecules that prevent physical and functional recovery of the injured CNS. Chondroitin-sulfated (CS) proteoglycans (PGs) are important inhibitory molecules synthesized by reactive astrocytes. CSPGs consist of a protein core to which glycosaminoglycan (GAG) side chains are attached. It has been suggested that the inhibitory activity of CSPGs is derived from their CS GAG side chains, as treatments with chondroitinase ABC (which cleaves these chains) reduces CSPG inhibition to neurites *in vitro* (Wright et al., 2011; Zuo, et al., 1998) and regeneration of axons *in vivo* (Bradbury et al., 2002; Wright et al., 2011).

Other inhibitory molecules present within the glial scar include myelin-associated proteins, such as myelin-associated glycoprotein (MAG), Nogo-A, and oligodendrocyte-myelin glycoprotein (OMgp) (Chen et al., 2000; McKerracher et al., 1994; Wright et al., 2011). Oligodendrocytes and Schwann cells express MAG, which is a potent inhibitor of neurite outgrowth when used as a culture substrate (McKerracher et al., 1994; Wright et al., 2011). Oligodendrocytes also express Nogo-A and OMgp, which act as inhibitors to axonal growth (Oertle et al., 2003; Wright et al., 2011). The presence of these growth-inhibitory myelin-associated proteins is a result of myelin breakdown in the vicinity and incomplete clearance of debris.

Importantly, SCI is not without regenerative plasticity, including the formation of novel circuits through sprouting and anatomical reorganization. Axonal branching, synaptogenesis, dendritic growth, and reorganization are characteristic features of SCI, including the aberrant growth of axons that have been damaged, which can form new synaptic connections. The upregulation of endogenous neurotrophic factors and their receptors are partly responsible for regenerative plasticity, as well as axonal sparing.

Thus, SCI is characterized by a highly reactive environment that presents significant obstacles for repair, as well as for the survival and integration of transplanted cells.

4 STEM CELLS

Stem cells can be divided into two broad categories, embryonic stem cells (ESCs) and somatic stem cells. ESCs are derived from blastocysts and maintain the ability to differentiate into any cell type (pluripotency; Thomson et al., 1998). Somatic stem cells include all endogenous stem cells present in the adult, as well as fetal and neonatal tissues. Somatic stem cells are able to differentiate into a limited number of cell lineages (multipotency). A number of stem cell types including oligodendrocyte progenitor cells (OPCs), motor neuron progenitor (MNP) cells, neural stem cells (NSCs), mesenchymal stem cells (MSCs), and hematopoietic stem cells (HSCs) have been considered in stem cell-based approaches for the treatment of SCI.

5 EMBRYONIC STEM CELLS

ESCs are pluripotent cells that have the capability to differentiate into cell types of all three embryonic germ layers, the ectoderm, mesoderm, and endoderm. ESCs owe their discovery to the early work of Kleinsmith and Pierce (1964), who demonstrated that a single embryonal carcinomal cell had the capability to self-renew and to generate various mature cell types. A breakthrough experiment followed this discovery, which used ESCs to generate an entire mouse (Nagy et al., 1993). Successful isolation of human ESCs was achieved in 1998 (Thomson et al., 1998). The impressive plasticity of ESCs was of interest to both the scientific and medical community, in particular, the potential of ESCs to repair and replace damaged tissues. ESCs can be prepared from several methods including preimplantation embryos, blastocyst-stage embryos (unused embryos created during *in vitro* fertilization treatments), somatic cell nuclear transfer, or parthenogenetic activation of eggs (Cibelli et al., 2002; Vrana et al., 2003).

ESCs have traditionally been propagated using a mouse feeder layer supplemented with bovine serum. However, current approaches allow ESCs to be grown using media with few or no reagents derived from animal sources (Lee et al., 2005a; Richards et al., 2002), and, very recently, fully defined and rodent-free propagation media with physiological levels of growth factors have become available (Nistor et al., 2011). The lack of rodent reagents circumvents the potential for cellular incorporation of zoonotic constituents from the media, reducing the potential for immunological rejection following transplantation, and the physiological levels of growth factors encourages genetic stability and a balanced differentiation profile.

With appropriate culture conditions, ESCs will not undergo the genomic, mitochondrial, and epigenetic changes that lead to transformation; therefore, they will remain undifferentiated (Zeng and Rao, 2007). ESCs can be grown in essentially unlimited quantities because they retain high telomerase activity and, therefore, do not undergo senescence. Typically, telomeres shorten with aging, limiting cells to a fixed number of divisions. ESCs express a ribonucleoprotein, which adds telomere repeats to chromosome tips to preserve telomere length, corresponding with ESC longevity. The retention of high telomerase activity suggests that the self-renewal capacity of ESCs is more efficient than that of somatic cells. High levels of telomerase in germ line and embryonic tissues are remarkably different from somatic cells, which do not express telomerase at all.

Since ESCs can be expanded to a seemingly limitless supply, stably banked, and directed to differentiate into diverse cell types, they are currently a promising source for treating neurological disorders and traumas, including SCI. ESCs can maintain a normal karyotype and differentiation potential even after years of culture. For example, in one of the earliest line derivations, Thomson et al. (1998) performed karyotyping on the cells at passages 2–7 and no abnormalities were identified. Specifically, H1, H13, and H14 had a normal XY karyotype while H7 and H9 had a normal XX karyotype. Ultimately, this process produced five ESC lines from five separate embryos. Since this derivation, thousands of ESC lines have been successfully generated and banked.

Safety is paramount to the use of ESCs in therapy. Since ESCs are highly proliferative, unrestricted in their development, and sensitive to environmental cues for differentiation, tumor formation is a legitimate risk for transplanted ESC-derivates. Transplants of nondifferentiated ESCs form teratomas that consist of cells from all three embryonic germ layers (Thomson et al., 1998). Matsuda et al. (2009) studied whether tumorigenesis could be avoided by coculturing murine ESCs (mESCs) with bone marrow stromal cells (BMSCs). *In vitro* markers for mRNA expression typical of undifferentiated cells decreased in mESCs cocultured with BMSCs. Although spinal cord injured mice transplanted with mESCs-generated tumors, no tumor development was observed in mice cotransplanted with both BMSCs and mESCs. The BMSCs expressed neurotrophic factors (NGF, GDNF, and BDNF) *in vivo* following transplantation for at least 1 month, which may have contributed to mESCs differentiation, thereby suppressing tumorigenesis (Matsuda et al., 2009). An alternative approach to removing the risk of tumorigenesis is removal of undifferentiated ESCs from any transplant population. Even restriction of ESCs to multipotent neural progenitors can generate inappropriate cells, such as nestin-positive expanding cells (Roy et al., 2006). This could lead, for example, to the differentiation of transplanted cells into astrocytes (Svendsen et al., 1997). While astrocytes can act as support cells, they are also associated with exacerbated glial scarring and increased pain perception in regions of trauma (Davies et al., 2008).

Prior to transplantation, the directed differentiation of multipotent or pluripotent stem cells into committed precursor phenotypes can address cellular deficits, better predict outcomes, and limit adverse side effects (Sharp and Keirstead, 2007). Researchers have restricted differentiation of ESCs into committed neural (Bain et al., 1995; Carpenter et al., 2001; Nishimura et al., 2003; Reubinoff et al., 2001) and neuronal (Nistor et al., 2011) lineage phenotypes, which can integrate and help restore function following transplantation into regions of SCI (Deshpande et al., 2006; Rossi et al., 2010; Sharp et al., 2010). ESCs have also been directed to differentiate into high-purity oligodendrocyte progenitors (Keirstead et al., 2005; Nistor et al., 2005; Sharp et al., 2010) and MNPs (Rossi et al., 2010; Wyatt and Keirstead, 2010; Wyatt et al., 2011). Together, these reports indicate that ESCs are a reliable source of committed neural cell types for stem cell-based therapy.

6 OLIGODENDROCYTE PROGENITORS CELLS

In the forebrain, the majority of OPCs arise during late embryogenesis and early postnatal development from cells of the subventricular zones of the lateral ventricles. In the spinal cord, they arise from the ventral region of the developing spinal cord and spread laterally and dorsally. Cells migrate away from germinal zones to populate both developing white and gray matter, where they differentiate and mature into oligodendrocytes. Oligodendrocyte-lineage cells can be identified by their expression of a number of antigens, including ganglioside GD3, NG2 CSPG, galactocerebroside, CNPase, myelin basic protein (MBP), oligodendrocyte marker 4, and the platelet-derived growth factor-alpha receptor subunit. OPCs secrete multiple growth factors and oligodendrocyte-myelin permits saltatory conduction in axons.

Our laboratory generated high-purity OPCs from hESC (hOPCs) as a potential treatment for SCI, on the basis that OPCs secrete growth factors, and myelinate regions of demyelination (Cloutier et al., 2006; Faulkner and Keirstead, 2005; Keirstead et al., 2005; Nistor et al., 2005; Sharp et al., 2010). OPCs can affect neurofilament phosphorylation and improve cortical neuron survival through secretion of soluble factors such as glial-derived neurotrophic factor (GDNF) and insulin-like growth factor-1 (IGF-1; Wilkins et al., 2003). Conditioned media from hOPCs causes extensive neurite outgrowth from both cortical (Faulkner and Keirstead, 2005) and sensory (Zhang et al., 2006) neurons *in vitro*. Cortical neurons exposed to hOPC-conditioned media display increased viability and proliferation (Faulkner and Keirstead, 2005). Other studies have shown that oligodendrocyte-secreted soluble factors effect neurofilament accumulation within axons, clustering of sodium channels along axons, and maturation and stabilization of axons (Colello et al., 1994; Kaplan et al., 1997; Sanchez et al., 1996). These results suggest that OPCs generate one or more soluble, neurotrophic factors that benefit neuronal function.

Studies by Zhang et al. (2006) and Sharp et al. (2010) identified soluble factors expressed by hOPCs. Analyses of gene microarrays indicated that 49 growth factor mRNAs were highly expressed by hOPCs, and antibody arrays and ELISAs identified secreted proteins in the conditioned media. The hOPCs expressed and secreted neurotrophic factors IGF-1, brain-derived neurotrophic factor (BDNF), NT-3, NGF, hepatocyte growth factor, transforming growth factors-beta1 and -beta2, stem cell factor, vascular endothelial growth factor, activin A, and midkine. Therefore, it is possible that OPC transplants may provide trophic support and create a regenerative environment to prevent cell damage.

The *shiverer* rodent model is ideal for demonstrating myelination by transplanted hOPCs, because the model does not contain MBP or multilayered compact myelin. When hOPCs were transplanted into the *shiverer* rodent model of dysmyelination, the cells integrated into the white matter, differentiated into oligodendrocytes, formed multilayered compact myelin, and produced MBP-positive areas within the spinal cord (Nistor et al., 2005) and brain (Izrael et al., 2007). Therefore, these hOPCs demonstrated the ability to integrate and differentiate into functional oligodendrocytes.

Transplantation of hOPCs to regions of acute SCI results in pathotropism, cell survival and differentiation, sparing of endogenous tissues, enhanced remyelination, and improved locomotor outcomes (Keirstead et al., 2005; Sharp et al., 2010), without harmful effects (Cloutier et al., 2006). Thus, hOPCs may confer benefit by myelinating demyelinated axons and by providing neurotrophic support to surrounding tissue. Based on these results, a Phase I clinical trial of this stem cell-based therapy was started in mid-2010 in the United States. This was the first ESC-based clinical trial approved by a regulatory body, worldwide. hOPCs were injected into the spinal cords of five patients, 6–14 days after an acute thoracic SCI, to assess safety and efficacy in restoring neurological function. Despite a successful Phase I mid-term report issued by the sponsor (Geron Corp., 2011), the trial was recently halted for financial reasons. This set-back reflects the difficulties of developing an emerging technology in an emerging sector. The lack of precedent imposes multiple delays, such as the need for regulatory bodies to establish "white papers" articulating hESC-specific policy, safety and efficacy standards, or clinical requirements. The need to invent efficient, closed system manufacturing regimes and their associated quality assurance criteria in a cost-efficient manner adds further delays. Both the tasks and the associated delays require money, adding significantly to the "burn" of the sponsor, and thereby increasing the risk profile of product development.

7 MOTOR NEURON PROGENITORS

Spinal cord motor neurons synapse with muscle fibers to facilitate muscle contraction and with muscle spindles to modify proprioceptive sensitivity. As clustering of acetylcholine receptors in myocytes leads to muscle contraction, mature motor neurons express acetylcholine related markers, including choline acetyltransferase and vesicular acetylcholine transferase.

The motor neuron differentiation pathway is largely controlled by sonic hedgehog and retinoic acid (Miles et al., 2004). Several researchers have manipulated these pathways to differentiate MNPs at approximately 1–50% purity from mouse ESCs (Miles et al., 2004; Soundararajan et al., 2006; Wichterle et al., 2002) and hESCs (Li et al., 2005; Shin, et al., 2005; Singh Roy et al., 2005). More recently, our laboratory developed methods to differentiate hESCs to high-purity (>95%) human MNP cultures. These cultures displayed characteristic motor neuron-specific markers, a typical electrophysiological profile, functionally innervated human and rodent muscle, and secreted physiologically active growth factors leading to neurite branching and neuronal survival (Rossi et al., 2010; Wyatt and Keirstead, 2010).

Neurotrophin secretion by transplanted MNPs has been shown to affect endogenous cells via several mechanisms. MNPs promote motor neuron survival and regrowth and assist in regulating the maturation of neuromuscular synapses (Corti et al., 2010; Dreyfus et al., 1999; Rossi et al., 2010). MNPs increase the release of neurotransmitter and direct synaptic transmission via the Trk family of receptor tyrosine kinases and the p75 neurotrophin receptor (Wyatt et al., 2011) Also, MNPs enhance neuronal survival, the growth of host axons in both normal and injured spinal cords, and functional recovery (Grumbles et al., 2009; Rossi et al., 2010; Wyatt and Keirstead, 2010). Erceg et al. (2010) demonstrated that transplanted hESC-derived OPCs and MNPs promoted locomotor recovery after spinal cord transection, better than either cell type alone.

Underscoring the ability of transplant-derived neurotrophic secretions to effect functional benefit in the absence of a phenotype-specific benefit, Rossi et al. (2010) transplanted high-purity MNPs into the injured spinal cords of adult rats. As expected, no transplant-derived axons extended to peripheral targets. Nonetheless, transplantation did elicit anatomical regeneration and functional recovery, which correlated with neurotrophic secretion from transplanted cells. As transplanted cells only differentiated into motor neurons, this study clearly demonstrates that neurotrophic support from transplanted hESC-derivatives represents a viable treatment strategy for SCI.

Neurotrophin secretion has also been shown to alter glial scar formation. Reactive astrocytes, pericytes, fibroblasts, microglia, and macrophages all contribute to the formation of the glial scar. At the molecular level, growth inhibition within the glial scar is contributed to predominately by PGs. PGs are important for the function of growth factors, cell adhesion, and lipoprotein metabolism. However, high levels of CSPGs at the glial scar form a molecular barrier, preventing axon growth (Lander, 1993; Margolis and Margolis, 1993). Emerging data supports the view that neurotrophin secretion by hESC-based transplants may fundamentally alter the glial scar, thereby attenuating the growth-inhibitory environment of a SCI.

8 NEURAL STEM CELLS

Endogenous NSCs are multipotent cells with the potential to differentiate into neuronal and glial lineages within the CNS. The ineffectiveness of resident NSC neurogenesis or gliogenesis to compensate for or repair CNS disease or injury reflects a deficiency of progenitor cell number, impairment caused by the local injury environment, affected NSC genetic and epigenetic makeup, or some combination of these factors.

NSCs can be propagated *in vitro* by exposure to various growth factors, maintain the capacity for self-renewal and generate differentiated cells that can functionally integrate (Caldwell et al., 2001) and lead to repair of the damaged CNS (Cummings et al., 2005; Iwanami et al., 2005; Nunes et al., 2003; Ogawa et al., 2002). Researchers have successfully isolated and propagated human NSCs from specific regions of the developing and adult brain (Lois and Alvares-Buylla, 1993; Reynolds and Weiss, 1992; Snyder et al., 1992; Uchida et al., 2000), spinal cord (Mayer-Proschel et al., 1997), and optic nerve (Shi et al., 1998). NSCs are generally obtained from fetal or cadaveric tissue and have a decreased potential for differentiation after multiple passages, increasing the cost of large-scale manufacturing (Wright et al., 2006).

Several reports underscore the difficulty of generating neurons from NSCs in the injured spinal cord. Endogenous neural progenitor cells are inherently inefficient in differentiating toward the motor neuron lineage, in part due to the ratio of Ngn2 to Oligo2, which determines motor neuron versus oligodendrocyte differentiation; this ratio is 10-fold lower in NPCs as compared to ESCs (Lee et al., 2005b). This could potentially be addressed by overexpressing genes involved in motor neuron development including HB9, NKX6.1, and NGN2 or by coculturing NSCs with endothelial cells (Bohl et al., 2008). Transplanted NSCs have been shown to integrate in regions of tissue damage, differentiate into myelinating oligodendrocytes, and cause clinical improvement following intraventricular, intravenous, intraspinal, or intraperitoneal delivery to various demyelinating or dysmyelinating animal models (Ben-Hur et al., 2003; Bulte et al., 2003; Einstein et al., 2003; Pluchino et al., 2003). NSC transplantation into sites of SCI results in integration and functional recovery, but the transplanted cells differentiate largely into astrocytes and oligodendrocytes within minimal neuronal differentiation (Cummings et al., 2005; Enzman et al., 2006; Hooshmand et al., 2009; Karimi-Abdolrezaee et al., 2006; Yan et al., 2007). These findings may reflect the gliogenic nature of the adult injured spinal cord.

9 MESENCHYMAL AND HEMATOPOIETIC STEM CELLS

There are two categories of bone marrow stem cells, HSCs and MSCs. HSCs are multipotent stem cells that give rise to all the blood cell types from the myeloid and lymphoid lineages. MSCs are multipotent stem cells that can differentiate into various cell types, including osteoblasts, chondrocytes, and adipocytes.

HSCs and MSCs may transdifferentiate along glial and neuronal pathways. HSCs and MSCs have the ability to form glial and neuronal lineage cells in response to various types of genetic, chemical, and/or physiological induction (Brazilay et al., 2009; Chen et al., 2006; Koshizuka et al., 2004; Krabbe et al., 2005; Phinney and Isakova, 2005; Sigurjonsson et al., 2005), although in some studies, the characterization of cell phenotype was limited to the detection of lineage-specific markers with no glial or neuronal cell function apparent. Notably, glial or neuronal differentiation of HSCs and MSCs prior to their transplantation into CNS injury sites was not necessary to promote remyelination, axonal regeneration, and functional recovery (Akiyama et al., 2002a,b; Ankeny et al., 2004; Chopp et al., 2000; Cizkova et al., 2006; Himes et al., 2006; Hofstetter et al., 2002; Koshizuka et al., 2004;

Neuhuber et al., 2005; Ohta et al., 2004; Sasaki et al., 2001; Sigurjonsson et al., 2005; Sykova and Jendlova, 2005; Vaquero et al., 2006; Wu et al., 2003; Zurita and Vaquero, 2004).

While MSCs have been shown to aid in functional recovery after SCI, the mechanism of action is unclear. MSCs may create a more favorable environment by modulating the immune response thereby limiting damage (Aggarwal and Pittenger, 2005; Noel et al., 2007), expressing growth factors and cytokines (Song et al., 2004), improving vascularization, providing a permissive growth substrate, and/or suppressing cavity formation (Hofstetter et al., 2002). Both HSCs and MSCs remain attractive cell sources for cell replacement therapy, as they can be easily and reproducibly isolated from bone marrow aspirates and reintroduced into patients as autografts.

10 PURITY

Inherent to the safety profile of a transplant population is the identity of its constituents. Certainly, the transplant population should be devoid of undifferentiated hESCs or large numbers of undifferentiated cells. To more easily ascribe the mechanism of action and reduce risk profiles, high-purity cell populations are desirable.

Purity can be determined using cell counting/sorting devices with appropriate reporters. Fluorescence activated cell sorting (FACS) has the advantage of isolating live, specified cells from a mixed cell population. FACS requires that the cells be labeled with fluorescent antibodies to phenotype-specific cell surface proteins or engineered to express a fluorescent protein under the control of a promoter. However, these techniques have drawbacks. Immunodetection requires digestion of cell surface proteins to remove the antibody label while reporter gene expression requires genetic manipulation of the starting population. Regardless of the technique, the use of FACS to produce a pure population infers that the purity of a target cell population can be assessed from one or two markers that are excluded from the nontarget cell population. This can be problematic when the literature suggests more than one potential cell type for a single marker. While immunolabeled sorting methods are considered useful for research purposes, there is a tumorigenic risk posed by the use of reporter genes, making this method undesirable for clinical stem cell-based therapy as per current U.S. Food and Drug Administration (FDA) guidelines. Also, sorting methods produce a limited yield, thereby reducing their commercial or clinical viability. FACS is a powerful tool for the isolation of subpopulations of cells from a mixed population; however, clinical applications require refined methods of directed differentiation.

11 CHALLENGES FOR CLINICAL TRANSLATION

Stringent guidelines exist for the translation of stem cell-based therapies for SCI (Fawcett et al., 2007; Lammertse et al., 2007; Steeves et al., 2007; Tuszynski et al., 2007). While regulatory bodies do not require an understanding of the underlying molecular mechanism of action, they demand extensive demonstration of safety and a plausibility of functional benefit. Preclinical studies aim to address these requirements. Preclinical safety studies assess tumorigenesis, biodistribution, immunological complications, allodynia, toxicity associated with dying cells, and complications associated with an unexpected change in phenotype of the transplanted cells.

Tumorigenesis is a concern for transplantation of ESCs; however, the risk decreases as ESCs differentiate. ESCs share several phenotypes with tumor cells including rapid replication rate, genetic instability, and telomerase activity (Ben-David and Benvenisty, 2011). Upon transplantation into immunodeficient mice, ESCs inevitably form teratomas, which consist of endodermal, mesodermal, and ectodermal lineages (Thomson et al., 1998). Somatic stem cells are considered less tumorigenic, because they are less proliferative and less multipotent. Thus, hESC-derivates should be devoid of undifferentiated cells, and lineage committed, prior to transplantation.

Another concern is the rejection of transplanted cells, in which xenogeneic or allogeneic antigens are recognized as foreign by the recipient's immune system. If the immune system destroys the transplanted cells, it would clearly eliminate any benefit that may have been conferred. Furthermore, the inflammatory response associated with cell rejection could cause additional harm to the patient. HLA matching of the stem cell-based transplant to the host is one method to avoid immune rejection. However, there are a limited number of clinically relevant stem cell lines, so this approach is not presently practical. Immunosuppression prevents rejection of transplanted cells but has several side effects and risks including nausea, vomiting, diarrhea, liver and kidney toxicity, lowered counts of leukocytes and platelets, and increased susceptibility to infections and malignancies (Habwe, 2006). It is conceivable that transplant rejection may be overcome by tolerizing recipients prior to transplantation (Rosengard and Turka, 2001; Salama et al., 2001), by generating stem cells that are immunologically compatible with recipients, or by developing a "universal donor" ESC line (Lengerke et al., 2007). It remains to be seen if these later approaches will be successful or commercially viable.

Differentiation of transplanted cells to undesired phenotypes is a major safety concern for SCI. A study by Hofstetter et al. (2005) underscored the risk of pain formation following aberrant differentiation of transplanted NSCs. NSCs were transplanted into the low-thoracic spinal cords of rats 1-week post-injury. Functional recovery was observed in the affected hind limbs, but abnormal and painful sensitivity developed in the forepaws (which had been unaffected by the injury). While the transplanted cells had the potential to produce neurons and oligodendrocytes, histology indicated that they had differentiated *in situ* to a predominantly astrocytic phenotype, ultimately promoting sprouting of sensory fibers within the spinal cord that were associated with allodynia. When the NSCs were predifferentiated into oligodendrocyte progenitors, functional recovery was also attained in the hind limbs but allodynia was avoided. These results were later verified by other researchers (Hendricks et al., 2006).

While a less differentiated cell may better respond to environmental cues and show enhanced capacity for migration and growth, a more differentiated cell may mitigate the risks of inappropriate differentiation and neoplasia. Lineage restriction to the progenitor stage is essential to limit tumor formation and differentiation of inappropriate cell types from hESCs. Lineage restriction prior to transplantation partially overcomes signals of the injury environment, which may provide inappropriate differentiation cues (Shihabuddin et al., 2000; Song et al., 2002).

Preclinical studies will aid decisions on the best transplant location within the spinal cord, the number of transplant sites, the appropriate cell number to transplant, and the optimal timing of the transplant relative to the onset of SCI. While careful preclinical studies critical to the development of stem cell-based therapies, it should be appreciated that adverse events occur in human clinical trials, despite exhaustive preclinical development.

12 REGULATORY AGENCIES

In the United States, clinical trials designed to demonstrate the safety and effectiveness of stem cell therapies are regulated by the U.S. FDA. The FDA has established clear guidelines for stem cell therapies that span pharmacology and toxicology, manufacture of cell and tissue based products, xeno- and allogeneic transplantation, and surgical methods. Stem cell transplants must be defined in terms of sterility, purity, potency, identity, stability, safety, and efficacy.

To determine whether it is reasonable to grant permission for a clinical trial to proceed, the FDA evaluates potential risk based on results derived from analytical assessment of product characteristics, preclinical proof-of-concept, and safety testing (Fink, 2009). Furthermore, standard operating procedures must be compiled to standardize the methods used in preparing and delivering a stem cell-based therapy. Surgical teams will need to be assembled and well trained in every aspect of the transplantation. Equipment may be subject to additional regulatory requirements prior to conducting human clinical trials. Although several regulatory and manufacturing hurdles still exist for the commercial and clinical viability of stem cell-based therapy for SCI, their translation into the clinical setting will provide a lasting infrastructure and regulatory pathway.

Acknowledgments

The authors would like to acknowledge Margaret Coutts and Jason Sharp for their valuable input. H. S. K. is Chairman of the Scientific Advisory Board of California Stem Cell, Inc.

References

Aggarwal, S., Pittenger, M.F., 2005. Human mesenchymal stem cells modulate allogeniec immune cell responses. Blood 105, 1815–1822.

Akiyama, Y., Radtke, C., Homou, O., Kocsis, J.D., 2002a. Remyelination of the spinal cord following intravenous delivery of bone marrow stromal cells. Glia 39, 229–236.

Akiyama, Y., Radtke, C., Kocsis, J.D., 2002b. Remyelination of th rat spinal cord by transplantation of identified bone marrow stromal cells. J. Neurosci. 22, 6623–6630.

Anderson, A.J., Robert, S., Huang, W., Young, W., Cotman, C.W., 2004. Activation of complement pathways after contusion-induced spinal cord injury. J. Neurotrauma 21 (12), 1831–1846.

Ankeny, D.P., McTigue, D.M., Jakeman, L.B., 2004. Bone marrow transplants provide tissue protection and directional guidance for axons after contusive spinal cord injury in rats. Exp. Neurol. 190, 17–31.

Bain, G., Kitchens, D., Yao, M., Huettner, J.E., Gottlieb, D.I., 1995. Embryonic stem cells express neuronal properties in vitro. Dev. Biol. 168, 342–357.

Ben-David, U., Benvenisty, N., 2011. The tumorigenicity of human embryonic and induced pluripotent stem cells. Nat. Rev. Cancer 11, 268–277.

Ben-Hur, T., Einstein, O., Mizrachi-Kol, R., Ben-Menachem, O., Reinhartz, E., Karussis, D., Abramsky, O., 2003. Transplanted multipotential neural precursor cells migrate into the inflamed white matter in response to experimental autoimmune encephalomyelitis. Glia 41, 73–80.

Bohl, D., Liu, S., Blanchard, S., Hocquemiller, M., Haase, G., Heard, J.M., 2008. Directed evolution of motor neurons from genetically engineered neural precursors. Stem Cells 26, 2564–2575.

Bradbury, R.J., King, V.R., Bennet, G.S., 2002. Chondroitinase ABS promotes functional recovery after spinal cord injury. Nature 416, 636–640.

Brazilay, R., Melamed, E., Offen, D., 2009. Introducing transcription factors to multipotent mesenchymal stem cells: making transdifferentiation possible (review). Stem Cells 27, 2509–2515.

Bulte, J.W., Ben-Hurr, T., Miller, B.R., Mizrachi-Kol, R., Einstein, O., Reinhartz, E., Zywicke, H.A., Douglas, T., Frank, J.A., 2003. MR microscopy of magnetically labeled neurospheres transplanted into the Lewis EAE rat brain. Magn. Reson. Med. 50, 201–205.

Caldwell, M.A., He, X., Wilkie, N., Pollack, S., Marshall, G., Wafford, K.A., Svendsen, C.N., 2001. Growth factors regulate the survival and fate of cells derived from human neurospheres. Nat. Biotechnol. 19, 475–479.

Carpenter, M.K., Inokuma, M.S., Denham, J., Mujtaba, T., Chiu, C.P., Rao, M.S., 2001. Enrichment of neurons and neural precursors from human embryonic stem cells. Exp. Neurol. 172, 383–397.

Chen, M.S., Huber, A.B., van der Harr, M.E., Frank, M., Schnell, L., Spillman, A.A., Christ, F., Schwab, M.E., 2000. Nogo-A is a myelin associated neurite outgrowth inhibitor and an antigen for monoclonal antibody IN-1. Nature 403, 434–439.

Chen, Y., Teng, F.Y., Tang, B.L., 2006. Coaxing bone marrow stromal mesenchymal stem cells towards neuronal differentiation: progress and uncertainties (review). Cell. Mol. Life Sci. 63, 1649–1657.

Chopp, M., Zhang, X.H., Li, Y., Wang, L., Chen, J., Lu, D., Lu, M., Rosenblum, M., 2000. Spinal cord injury in rat: treatment with bone marrow stromal cell transplantation. Neuroreport 11, 3001–3005.

Christopher and Dana Reeve Foundation, 2009. One degree of separation | paralysis and spinal cord injury in the United States. Retrieved February 14, 2012 from http://www.christopherreeve.org/atf/cf/{3d83418f-b967-4c18-8ada-adc2e5355071}/8112REPTFINAL.PDF.

Cibelli, J.B., Grant, K.A., Chapman, K.B., Cunniff, K., Worst, T., Green, H.L., Walker, S.J., Gutin, P.H., Vilner, L., Tabar, V., Dominiko, T., Kane, J., Wettstein, P.J., Lanza, R.P., Studer, L., Vrana, K.E., West, M.D., 2002. Parthenogentic stem cells in nonhuman primates. Science 295, 819.

Cizkova, D., Rosocha, J., Vanicky, I., Jergova, S., Cizek, M., 2006. Transplantation of human mesenchymal stem cells improve functional recovery after spinal cord injury in the rat. Cell. Mol. Neurobiol. 26, 1167–1180.

Cloutier, F., Siegenthaler, M.M., Nistor, G., Keirstead, H.S., 2006. Transplantation of human embryonic stem cell-derived oligodendrocyte progenitors into rat spinal cord injuries does not cause harm. Regen. Med. 1, 469–479.

Colello, R.J., Pott, U., Schwab, M.E., 1994. The role of oligodendrocytes and myelin on axon maturation in the developing rat ratinofugal pathway. J. Neurosci. 14, 2594–2605.

Corti, S., Nizzardo, M., Nardini, M., Donadoni, C., Salani, S., Ronchi, D., Simone, C., Falcone, M., Papdimitriou, D., Locatelli, F., Mezzina, N., Gianni, F., Bresolin, N., Comi, G.P., 2010. Embryonic stem cell-derived neural stem cells improve spinal muscular atrophy phenotype in mice. Brain 133, 465–481.

Cummings, B.J., Uchida, N., Tamaki, S.J., Salazar, D.L., Hooshmand, M., Summers, R., Gage, F.H., Anderson, A.J., 2005. Human neural stem cells differentiate and promote locomotor recovery in spinal cord-injured mice. Proc. Natl. Acad. Sci. U.S.A. 102, 14069–14074.

David, S., Bouchard, C., Tsatas, O., Giftochristos, N., 1990. Macrophages can modify the nonpermissive nature of the adult mammalian CNS. Neuron 5, 463–469.

Davies, J.E., Proschel, C., Zhang, N., Noble, M., Mayer-Proschel, M., Davies, S.J.A., 2008. Transplanted astrocytes derived from BMP- or CNTF- treated glial-restricted precursors have opposite effects on recovery and allodynia after spinal cord injury. J. Biol. 7, 24.

Deshpande, D.M., Kim, Y.S., Martinex, T., Carmen, J., Dike, S., Shats, I., Rubin, L.L., Drummond, J., Krishnan, C., Hoke, A., Maragakis, N., Shefner, J., Rothstein, J.D., Kerr, D.A., 2006. Recovery from paralysis in adult rats using embryonic stem cells. Ann. Neurol. 60, 32–44.

Dreyfus, C.F., Dai, X., Lercher, L.D., Racey, B.R., Friedman, W.J., Black, I.B., 1999. Expression of neurotrophins in the adult spinal cord in vivo. J. Neurosci. Res. 56, 1–7.

Einstein, O., Karussis, D., Grigoriadis, N., Mizrachi-Kol, R., Reinhartz, E., Abramsky, O., Ben-Hur, T., 2003. Intraventricular transplantation of neural precursor cell spheres attenuates acute experimental allergic encephalomyelitis. Mol. Cell. Neurosci. 24, 1074–1082.

Enzman, G.U., Benton, R.L., Talbott, J.F., Cao, Q., Wittemore, S.R., 2006. Functional considerations of stem cell transplantation therapy for spinal cord repair. J. Neurotrauma 23, 47–495.

Erceg, S., Ronaghi, M., Oria, M., Rosello, M.G., Arago, M.A.P., Lopez, M.G., Radojevic, I., Moreno-Manzano, V., Rodriguez-Jimenez, F.J., Bhattacharya, S.S., Cordoba, J., Stojkovic, M., 2010. Transplanted oligodendrocytes and motoneuron progenitor cells derived from human embryonic stem cells express neurotrophic factors. Stem Cells 15, 943–952.

Faulkner, J., Keirstead, H.S., 2005. Human embryonic stem cell-derived oligodendrocyte progenitors for the treatment of spinal cord injury. Transpl. Immunol. 15, 131–142.

Fawcett, J.W., Curt, A., Steeves, J.D., Coleman, W.P., Tuszynski, M.H., Lammertse, D., Bartlett, P.F., Blight, A.R., Dietz, V., Ditunno, J., Dobkin, B.H., Havton, L.A.,

Ellaway, P.H., Fehlings, M.G., Privat, A., Grossman, R., Guest, J.D., Kleitman, N., Nakamura, M., Gaviria, M., Short, D., 2007. Guidelines for the conduct of clinical trials for spinal cord injury as developed by the ICCP panel: spontaneous recovery after spinal cord injury and statistical power needed for therapeutic clinical trials. Spinal Cord 45, 190–205.

Fink Jr., D.W., 2009. FDA regulation of stem cell-based products. Science 324, 1662–1663.

Geron Corporation: information for patients, clinical trial details. http://www.geron.com/patients/clinicaltrials/hESC.aspx.

Grumbles, R.M., Sesodia, S., Wood, P.M., Thomas, C.K., 2009. Neurotrophic factors improve motoneuron survival and function of muscle innervated by embryonic neurons. J. Neuropathol. Exp. Neurol. 68, 736–746.

Habwe, V.Q., 2006. Posttransplantation quality of life: more than graft function. Am. J. Kidney Dis. 47, S98–S110.

Hendricks, W.A., Pak, E.S., Owensby, J.P., Menta, K.J., Glazova, M., Moretto, J., Hollis, S., Brewer, K.L., Murashov, A.K., 2006. Predifferentiated embryonic stem cells prevent chronic pain behaviours and restore sensory function following spinal cord injury in mice. Mol. Med. 12, 34–46.

Himes, B.T., Neuhuber, B., Coleman, C., Kushner, R., Swanger, S.A., Kopen, G.C., Wagner, J., Shumsky, J.S., Fischer, I., 2006. Recovery of function following grafting of human bone marrow-derived stromal cells into the injured spinal cord. Neurorehabil. Neural Repair 20, 278–296.

Hofstetter, C.P., Schwarz, E.J., Hess, D., Widenfalk, J., El Manira, A., Prockop, D.J., Olson, L., 2002. Marrow stromal cells form guiding strands in the injured spinal cord and promote recovery. Proc. Natl. Acad. Sci. U.S.A. 99, 2199–2204.

Hofstetter, C.P., Holmstrom, N.A., Lilja, J.A., Schweinhardt, P., Hao, J., Spenger, C., Wiesenfeld-Hallin, Z., Kurpad, S.N., Frisen, J., Olson, L., 2005. Allodynia limits the usefulness of intrapspinal neural stem cell grafts; directed differentiation improves outcome. Nat. Neurosci. 8, 346–353.

Hooshmand, M.J., Sontag, C.J., Uchida, N., Tamaki, S., Anderson, S.J., Cummings, B.J., 2009. Analysis of host-mediated repair mechanisms after human CNS-stem cell transplantation for spinal cord injury: correlation of engraftment with recovery. PLoS One 4, e5871.

Iwanami, A., Kaneko, S., Nakamura, M., Kanemura, Y., Mori, H., Kobayashi, S., Yamasaki, M., Momoshima, S., Ishii, H., Ando, K., Tanioka, Y., Tamaoki, N., Nomura, T., Toyama, Y., Okano, H., 2005. Transplantation of human neural stem cells for spinal cord injury in primates. J. Neurosci. Res. 80 (2), 182–190.

Izrael, M., Zhang, P., Kaufman, R., Shinder, V., Ella, R., Amit, M., Itskovitz-Eldor, J., Chebath, J., Revel, M., 2007. Human oligodendrocytes derived from embryonic stem cells: effect of noggin on phenotypic differentiation in vitro and on myelination in vivo. Mol. Cell. Neurosci. 34, 310–323.

Kaplan, M.R., Meyer-Franke, A., Lambert, S., Bennett, V., Duncan, I.D., Levinson, S.R., Barres, B.A., 1997. Induction of sodium channel clustering by oligodendrocytes. Nature 386, 724–728.

Karimi-Abdolrezaee, S., Eftekharpour, E., Wang, J., Morshead, C.M., Fehlings, M.G., 2006. Delayed transplantation of adult neural precursor cells promotes remyelination and functional neurological recovery after spinal cord injury. J. Neurosci. 26, 3377–3389.

Keirstead, H.S., Nistor, G., Bernal, G., Totoiu, M., Cloutier, F., Sharp, K., Steward, O., 2005. Human embryonic stem cell-derived oligodendrocyte progenitor cell transplants remyelinate and restore locomotion after spinal cord injury. J. Neurosci. 25, 4694–4705.

Kleinsmith, L.J., Pierce Jr., G.B., 1964. Multipotentiality of single embryonal carcinoma cells. Cancer Res. 24, 1544.

Koshizuka, S., Okada, S., Okawa, A., Koda, M., Murasawa, M., Hashimoto, M., Kamada, T., Yoshinaga, K., Murakami, M., Moriya, H., Yamazaki, M., 2004. Transplanted hematopoietic stem cells from bone marrow differentiate into neural lineage cells and promote functional recovery after spinal cord injury in mice. J. Neuropathol. Exp. Neurol. 63, 64–72.

Krabbe, C., Zimmer, J., Meyer, M., 2005. Neural transdifferentiation of mesenchymal stem cells—a critical review. APMIS 113, 831–844.

Lammertse, D., Tuszynski, M.H., Steeves, J.D., Curt, A., Fawcett, J.W., Rask, C., Ditunno, J.F., Fehlings, M.G., Guest, J.D., Ellaway, P.H., Kleitman, N., Blight, A.R., Dobkin, B.H., Grossman, R., Katoh, H., Privat, A., Kalichman, M., 2007. Guidelines for the conduct of clinical trials for spinal cord injury as developed by the ICCP panel: clinical trial design. Spinal Cord 45, 232–242.

Lander, A.D., 1993. Proteoglycans in the nervous system. Curr. Opin. Neurobiol. 3, 716–723.

Lee, J.B., Lee, J.E., Park, J.H., Kim, S.J., Kim, M.K., Roh, S.I., Yoon, H.S., 2005a. Establishment and maintenance of human embryonic stem cell lines on human feeder cells derived from uterine endometrium under serum-free condition. Biol. Reprod. 72, 42–49.

Lee, S.K., Lee, B., Ruiz, E.C., Pfaff, S.L., 2005b. Oligo2 and Ngn2 function in opposition to modulate gene expression in motor neuron progenitor cells. Genes Dev. 19, 282–292.

Lengerke, C., Kim, K., Lerou, P., Daley, G.Q., 2007. Differentiation potential of histocopatible parthenogenic embryonic stem cells. Ann. N. Y. Acad. Sci. 1106, 209–218.

Li, W., Maeda, Y., Ming, X., Cook, S., Chapin, J., Husar, W., Dowling, P., 2002. Apoptotic death following fas activation in human oligodendrocyte hybrid cultures. J. Neurosci. Res. 69, 189–196.

Li, X.J., Du, Z.W., Zarnowska, E.D., Pankratz, M., Hansen, L.O., Pearce, R.A., Zhang, S.C., 2005. Specification of motorneurons from human embryonic stem cells. Nat. Biotechnol. 23, 215–221.

Liu, X.Z., Xu, X.M., Hu, R., Du, C., Zhang, S.X., McDonald, J.W., Dong, H.X., Wu, Y.J., Fan, G.S., Jacquin, M.F., Hsu, C.Y., Choi, D.W., 1997. Neuronal and glial apoptosis after traumatic spinal cord injury. J. Neurosci. 17, 5395–5406.

Lois, C., Alvares-Buylla, A., 1993. Proliferating subventricular zone cells in the adult mammalian forebrain can differentiate into neurons and glia. Proc. Natl. Acad. Sci. U.S.A. 90, 2074–2077.

Margolis, R.K., Margolis, R.U., 1993. Nervous tissue proteoglycans. Experientia 49, 429–446.

Matsuda, R., Yoshikawa, M., Kimura, H., Ouji, Y., Nakase, H., Nishimura, F., Nonaka, J., Toriumi, H., Yamada, S., Nishiofuku, M., Moriya, K., Ishizaka, S., Nakamura, M., Sakaki, T., 2009. Cotransplantation of mouse embryonic stem cells and bone marrow stromal cells following spinal cord injury suppresses tumor development. Cell Transplant. 18, 39–54.

Mayer-Proschel, M., Kalyani, A.J., Mugtaba, T., Rao, M.S., 1997. Isolation of lineage restricted neuronal precursors from multipotent neuroepithelial stem cells. Neuron 19, 773–785.

McKerracher, L., David, S., Jackson, D.L., Kottis, V., Dunn, R.J., Braun, P.E., 1994. Identification of myelin-associated glycoprotein as a major myelin-derived inhibitor or neurite outgrowth. Neuron 13, 805–811.

Miles, G.B., Yohn, D.C., Wichterle, H., Jessell, T.M., Rafuse, V.F., Brownstone, R.M., 2004. Functional properties of motoneurons derived from mouse embryonic stem cells. J. Neurosci. 24, 7848–7858.

Nagy, A., Rossant, J., Nagy, R., Abramow-Newerly, W., Roder, J.C., 1993. Derivation of completely cell culture-derived mice from early-passage embryonic stem cells. Proc. Natl. Acad. Sci. U.S.A. 90, 8424–8428.

Neuhuber, B., Himes, B.T., Shumsky, J.S., Gallo, G., Fischer, I., 2005. Axon growth and reovery of function supported by human bone marrow stromal cells in the injured spinal cord exhibit donor variations. Brain Res. 1035, 73–85.

Nishimura, F., Yoshikawa, M., Kanda, S., Nonaka, M., Yokota, H., Shiroi, A., Nakase, H., Hirabayashi, H., Ouji, Y., Birumachi, J., Ishizaka, S., Sakaki, T., 2003. Potential use of embryonic stem cells form the treatment of mouse parkinsonian models: improved behavior by transplantation of in vitro differentiated dopaminergic neurons from embryonic stem cells. Stem Cells 21, 171–180.

Nistor, G.I., Totoiu, M.O., Haque, N., Carpenter, M.K., Keirstead, H.S., 2005. Human embryonic stem cells differentiate into oligodendrocytes in high purity and myelinate after spinal cord transplantation. Glia 49, 385–396.

Nistor, G., Siegenthaler, M.M., Poirier, S.N., Rossi, S., Poole, A.J., Charlton, M.E., McNeish, J.D., Airriess, C.N., Keirstead, H.S., 2011. Derivation of high purity neuronal progenitors from human embryonic stem cells. PLoS One 6 (6), e20692.

Noble, L.J., Donovan, F., Igarashi, T., Goussev, S., Werb, Z., 2002. Matrix metalloproteinases limit functional recovery after spinal cord injury by modulation of early vasculature events. J. Neurosci. 22, 7526–7535.

Noel, D., Djouad, F., Bouffi, C., Mrugala, D., Jorgensen, C., 2007. Multipotent mesenchymal stromal cells and immune tolerance. Leuk. Lymphoma 48, 1283–1289.

Nunes, M.C., Roy, N.S., Keyoung, H.M., Goodman, R.R., McKhann II, G., Jiang, L., Kang, J., Nedergaard, M., Goldman, S.A., 2003. Identification and isolation of multipotential neural progenitor cells from the subcortical white matter of the adult human brain. Nat. Med. 9, 439–447.

Oertle, T., van der Haas, M.E., Bandtlow, C.E., Robeva, A., Burfeind, P., Buss, A., Huber, A.B., Simonen, M., Schell, L., Brosamle, C., Klemens, K., Rudiger, V., Schwab, M.E., 2003. Nogo-A inhibits neurite outgrowth and cell spreading with three distinct regions. J. Neurosci. 23, 5393–5406.

Ogawa, Y., Sawamoto, K., Miyata, T., Miyao, S., Watanabe, M., Nakamura, M., Bregman, B.S., Koiki, M., Uchiyama, Y., Toyama, Y., Okano, H., 2002. Transplantation of in vitro-expanded fetal neural progenitor cells results in neurogenesis and functional recovery after spinal cord contusion injury in adult rats. J. Neurosci. Res. 69, 925–933.

Ohta, M., Suzuki, Y., Noda, T., Ejiri, Y., Dezawa, M., Kataoka, K., Chou, H., Ishikawa, N., Matsumoto, N., Iwashita, Y., Mizuta, E., Kuno, S., Ide, C., 2004. Bone marrow stromal cells infused into the rat cerebrospinal fluid promote functional recovery of the injured rat spinal cord with reduced cavity formation. Exp. Neurol. 187, 266–278.

Phinney, D.G., Isakova, I., 2005. Plasticity and therapeutic potential of mesenchymal stem cells in the nervous system (review). Curr. Pharm. Des. 11, 1255–1265.

Pluchino, S., Quattrini, A., Brambilla, E., Gritti, A., Salani, G., Dina, G., Galli, R., Del Carro, U., Amadio, S., Bergami, A., Furlan, R., Comi, G., Vescovi, A.L., Martino, G., 2003. Injection of adult neurospheres induces recovery in a chronic model of multiple sclerosis. Nature 422, 688–694.

Popovich, P.G., Guan, Z., McGuaghy, V., Fishcer, L., Hickey, W.F., Basso, D.M., 2002. The neuropathological behavioural consequences of intraspinal microglial/macrophage activation. J. Neuropathol. Exp. Neurol. 61, 623–633.

Rabchevsky, A.G., Striet, W.J., 1998. Role of microglia in post injury repair and regeneration of the CNS. Ment. Retard. Dev. Disabil. Res. Rev. 4, 187–192.
Reubinoff, B.E., Itsykson, P., Turetsky, T., Pera, M.F., Reinhartz, E., Itzik, A., Ben-Hur, T., 2001. Neural progenitors from human embryonic stem cells. Nat. Biotechnol. 19, 1134–1140.
Reynolds, B.A., Weiss, S., 1992. Generation of neurons and astrocytes from isolated cells of the adult mammalian central nervous system. Science 255, 1707–1710.
Rhonaghi, M., Erceg, S., Moreno-Manzano, V., Stojkovic, M., 2010. Challengers of stem cell therapy for spinal cord injury: human embryonic stem cells, endogenous neural stem cells, or induced pluripotent stem cells? Stem Cells 28, 93–99.
Richards, M., Fong, C.Y., Chan, W.K., Wong, P.C., Bongso, A., 2002. Human feeders support prolonged undifferentiated growth of human inner cell masses and embryonic stem cells. Nat. Biotechnol. 20, 933–936.
Rosengard, B.R., Turka, L.A., 2001. The tolerant recipient: looking great in someone else's genes. J. Clin. Invest. 107, 33–34.
Rossi, S.L., Nistor, G., Wyatt, T.J., Yin, H.Z., Pool, A.J., Weiss, J.H., Gardener, M.J., Dijkstra, S., Fischer, D.F., Keirstead, H.S., 2010. Histological and functional benefit following transplantation of motor neuron progenitors to the injured rat spinal cord. PLoS One 5, e11852.
Roy, N.S., Cleren, C., Singh, S.K., Yang, L., Beal, M.F., Goldman, S.A., 2006. Functional engraftment of human ES cell-derived dopaminergic neurons enriched by coculture with telomerase-immortalized midbrain astrocytes. Nat. Med. 12, 1259–1268.
Salama, A.D., Remuzzi, G., Harmon, W.E., Sayegh, M.G., 2001. Challenges to achieving clinical transplantation tolerance. J. Clin. Invest. 108, 943–948.
Sanchez, I., Hassinger, L., Paskevich, P.A., Shine, H.D., Nixon, R.A., 1996. Oligodendroglia regulate the regional expansion of axon caliber and local accumulation of neurofilaments during development independently of myelin formation. J. Neurosci. 16, 5095–5105.
Sasaki, M., Honmou, O., Akiyama, Y., Uede, T., Hashi, K., Kocsis, J.D., 2001. Transplantation of an acutely osilated bone marrow fraction repairs demyelinated adult rat spinal cord axons. Glia 35, 26–34.
Sharp, J., Keirstead, H.S., 2007. Therapuetic applications of oligodendrocyte precursors derived from human embryonic stem cells. Curr. Opin. Biotechnol. 18, 434–440.
Sharp, J., Frame, J., Siegenthaler, M., Nistor, G., Keirstead, H.S., 2010. Human embryonic stem cell-derived oligodendrocyte progenitor cell transplants improve recovery after cervical spinal cord injury. Stem Cells 28, 152–163.
Shi, J., Marinovich, A., Barres, B.A., 1998. Purification and characterization of adult oligodendrocyte precursor cells from the rat optic nerve. J. Neurosci. 18, 4627–4636.
Shihabuddin, L.S., Horner, P.J., Ray, K., Gage, F.H., 2000. Adult spinal cord stem cells generate neurons after transplantation in the adult dentate gyrus. J. Neurosci. 20, 8727–8735.
Shin, S., Dalton, S., Stice, S.L., 2005. Human motor neuron differentiation from human embryonic stem cells. Stem Cells Dev. 14, 266–269.
Sigurjonsson, O.E., Perreault, M.-C., Egeland, T., Glover, J.C., 2005. Adult human hematopoietic stem cells produce meurons efficiently in the regeneration chicken embryo spinal cord. Proc. Natl. Acad. Sci. U.S.A. 102, 5227–5232.

Singh Roy, N., Nakano, T., Xuing, L., Kang, J., Nedergaard, M., Goldman, S.A., 2005. Enhacner-specified GFP-based FACS purification of human spinal motor neurons from embryonic stem cells. Exp. Neurol. 196, 224–234.

Snyder, E.Y., Deitcher, D.L., Walsh, C., Arnold-Aldea, S., Hartwieg, E.A., Cepko, C.L., 1992. Multipotent neural cell lines can engraft and participate in development of mouse cerebellum. Cell 68, 33–51.

Song, H., Stevens, C.F., Gage, F.H., 2002. Astroglia induce neurogenesis from adult neural stem cells. Nature 417, 39–44.

Song, S., Kamath, S., Mosquera, D., Zigova, T., Sanberg, P., Vesely, D.L., Sanchez-Ramos, J., 2004. Expression of brain natriuretic peptide by human bone marrow stromal cells. Exp. Neurol. 185, 191–197.

Soundararajan, P., Miles, G.B., Rubin, L.L., Brownstone, R.M., Rafuse, V.F., 2006. Motoneurons derived from embryonic stem cells express transcription factors and develop phenotypes chracteristic of meidal motor column neurons. J. Neurosci. 26, 3256–3268.

Steeves, J.D., Lammertse, D., Curt, A., Fawcett, J.W., Tuszynski, M.H., Ditunno, J.F., Ellaway, P.H., Fehlings, M.G., Guest, J.D., Kleitman, N., Barlett, P.F., Blight, A.R., Dietz, V., Dobkin, B.H., Grossman, R., Short, D., Nakamura, M., Coleman, W.P., Gaviria, M., Privat, A., 2007. Guidelines for the conduct of clinical trials for spinal cord injury (SCI) as developed by the ICCP panel: clinical trial outcome measures. Spinal Cord 45, 206–221.

Svendsen, C.N., Caldwell, M.A., Shen, J., ter Borg, M.G., Rosser, A.E., Tyers, P., Karmiol, S., Dunnett, S.B., 1997. Long-term survival of human central nervous system progenitor cells transplanted into a rat model of Parkinson's disease. Exp. Neurol. 148, 135–146.

Sykova, E., Jendlova, P., 2005. Magnetic resonance tracking of implanted adult and embryonic stem cells in injured brain and spinal cord. Ann. N. Y. Acad. Sci. 1049, 146–160.

Taoka, Y., Okajuma, K., Uchiba, M., Murakami, K., Kushimoto, S., Johno, M., Naruo, M., Okabe, H., Takatsuki, K., 1997. Role of neutrophils in spinal cord injury in the rat. Neuroscience 74, 1177–1182.

Thomson, J.A., Itskovitz-Eldor, J., Shapiro, S.S., Waknitz, M.A., Swiergiel, J.J., Marshall, V.S., Jones, J.M., 1998. Embryonic stem cell lines derived form human blastocysts. Science 282, 1145–1147.

Tuszynski, M.H., Steeves, J.D., Fawcett, J.W., Lammertse, D., Kalichman, M., Rask, C., Curt, A., Ditunno, J.F., Fehlings, M.G., Guest, J.D., Ellaway, P.H., Kleitman, N., Bartlett, P.F., Blight, A.R., Dietz, V., Dobkin, B.H., Grossman, R., Privat, A., 2007. Guidelines for the conduct of clinical trials for spinal cord injury as developed by the ICCP panel: clinical trial inclusion/exclusion criteria and ethics. Spinal Cord 45, 222–231.

Uchida, N., Buck, D.W., He, D., Reitsma, M.J., Masek, M., Phan, T.V., Tsukamoto, A.S., Gage, F.G., Weissman, I.L., 2000. Direct isolation of human central nervous system stem cells. Proc. Natl. Acad. Sci. U.S.A. 97, 14720–14725.

Vaquero, J., Zurita, M., Oya, S., Santos, M., 2006. Cell therapy using bone marrow stromal cells in chronic paraplegic rats: systemic or local administration? Neurosci. Lett. 398, 129–134.

Vrana, K.E., Hipp, J.D., Goss, A.M., McCool, B.A., Riddle, D.R., Walker, S.K., Wettstein, P.J., Studer, L.P., Tabar, V., Cunniff, K., Chapman, K., Vilner, L., West, M.D., Grant, K.D., Cibelli, J.B., 2003. Nonhuman primate parthenogenetic stem cells. Proc. Natl. Acad. Sci. U.S.A. 100, 2088–2093.

Wichterle, H., Lieberam, I., Porter, J.A., Jessell, T.M., 2002. Directed differentiation of embryonic stem cells into motor neurons. Cell 110, 385–397.

Wilkins, A., Majed, H., Layfield, R., Compston, A., Chandran, S., 2003. Oligodendrocytes promote neuronal survival and axonal length by distinct intracellular mechanisms: a novel role for oligodendrocyte-derived glial cell line-derived neurotrophic factor. J. Neurosci. 23, 4967–4974.

Wright, L.S., Prowse, K.R., Wallace, K., Linskens, M.H., Svendsen, C.N., 2006. Human progenitor cells isolated from the developing cortex undergo decreased neurogenesis and eventual senescence following expansion in vitro. Exp. Cell Res. 312, 2107–2120.

Wright, K.T., Masri, W.E., Osman, A., Chowdhury, J., Johnson, W., 2011. Consise review: bone marrow for the treatment of spinal cord injury: mechanisms and clinical applications. Stem Cells 29, 169–178.

Wu, S., Suzuki, Y., Ejiri, Y., Noda, T., Bai, H., Kitada, M., Kataoka, K., Ohta, M., Chou, H., Ide, C., 2003. Bone marrow stromal cells enhance differentiation of co-cultured neurosphere cells and promote regernation of the injured spinal cord. J. Neurosci. Res. 72, 343–351.

Wyatt, T.J., Keirstead, H.S., 2010. Stem cell-derived neurotrophic support for the neuromuscular junction in spinal muscular atrophy. Expert Opin. Biol. Ther. 10, 1587–1594.

Wyatt, T.J., Rossi, S.L., Siegenthaler, M.M., Frame, J., Robles, R., Nistor, G., Keirstead, H.S., 2011. Human motor neuron progenitor transplantation leads to endogenous neuronal sparing in 3 models of motor neuron loss. Stem Cells Int. 2011, 11.

Xue, M., Hollenberg, M.D., Yong, V.W., 2006. Combination of thrombin and matrix metalloproteinase-9 exacerbates neurotoxicity in cell culture and intracerebral hemorrhage in mice. J. Neurosci. 40, 10281–10291.

Yan, J., Xu, L., Welsh, A.M., Hatfield, G., Hazel, T., Johe, K., Koliatsos, V.E., 2007. Extensive neuronal differentiation of human neural stem cell grafts in adult rat spinal cord. PLoS Med. 4, e39.

Zeng, X., Rao, M.S., 2007. Human embryonic stem cells: long term stability, absence of senescence and a potential cell source for neural replacement. Neuroscience 145, 1348–1358.

Zhang, Y.W., Denham, J., Thies, R.S., 2006. Oligodendrocyte progenitor cells derived from human embryonic stem cells express neurotrophic factors. Stem Cells Dev. 15, 943–952.

Zuo, J., Neubauer, D., Dyess, K., Ferguson, T.A., Muir, D., 1998. Degradation of chondroitin sulphate proteoglycan enhances the neurite promoting potential of spinal cord tissue. Exp. Neurol. 154, 654–666.

Zurita, M., Vaquero, J., 2004. Functional recovery in chronic paraplegia after bone marrow stromal cells transplantation. Neuroreport 15, 1105–1108.

CHAPTER

14 The challenges of long-distance axon regeneration in the injured CNS

Daniel J. Chew, James W. Fawcett, Melissa R. Andrews[1,2]

Centre for Brain Repair, University of Cambridge, Forvie Site, Robinson Way, Cambridge, UK

[1]*Current address: School of Medicine University of St Andrews, Medical and Biological Sciences Building, North Haugh, St Andrews, UK*

[2]*Corresponding author. Tel.: +44-1334-463558, Fax: +44-1334-463393, e-mail address: mra5@st-andrews.ac.uk*

Abstract

Injury to the central nervous system (CNS) that results in long-tract axonal damage typically leads to permanent functional deficits in areas innervated at, and below, the level of the lesion. The initial ischemia, inflammation, and neurodegeneration are followed by a progressive generation of scar tissue, dieback of transected axons, and demyelination, creating an area inhibitory to regrowth and recovery. Two ways to combat this inhibition is to therapeutically target the extrinsic and intrinsic properties of the axon–scar environment. Scar tissue within and surrounding the lesion site can be broken down using an enzyme known as chondroitinase. Negative regulators of adult neuronal growth, such as Nogo, can be neutralized with antibodies. Both therapies greatly improve functional recovery in animal models. Alternatively, modifying the intrinsic growth properties of CNS neurons through gene therapy or pharmacotherapy has also shown promising axonal regeneration after injury. Despite these promising therapies, the main challenge of long-distance axon regeneration still remains; achieving a level of functional and organized connectivity below the level of the lesion that mimics the intact CNS.

Keywords

CNS regeneration, glial scar, PTEN, spinal cord injury, integrins, Nogo

Abbreviations

AAV	adeno-associated virus
BBB	blood–brain barrier
BDNF	brain-derived neurotrophic factor
Ca^{2+}	Calcium
cAMP	cyclic adenosine monophosphate

Progress in Brain Research, Volume 201, ISSN 0079-6123, http://dx.doi.org/10.1016/B978-0-444-59544-7.00013-5

ChABC	chondroitinase ABC
ChAC	chondroitinase AC
CNS	central nervous system
CSPG	chondroitin sulfate proteoglycan
CST	corticospinal tract
DC	dorsal column
DRG	dorsal root ganglia
ECM	extracellular matrix
FAK	focal adhesion kinase
GAG	glycosoaminoglycan
GAP-43	growth-associated protein 43 kDa
GEF	guanine nucleotide exchange factor
GFAP	glial fibrillary acidic protein
GSK-3β	glycogen synthase kinase 3 beta
IL	interleukin
K^+	Potassium
LPS	lipopolysaccharide
MAG	myelin associated glycoprotein
MAPKKK	mitogen-activated protein kinase kinase kinase
MMP	matrix metalloproteinase
MT	microtubule
mTOR	mammalian target of rapomycin
Nrp1	neuropilin-1
NT	neurotrophins/neurotrophic factors
OMgp	oligodendrocyte myelin glycoprotein
OPC	oligodendrocyte precursor cell
PI3K	phosphoinositide 3-kinase
PIP2	phosphatidylinositol (4,5) bisphosphate
PIP3	phosphatidylinositol (3,4,5) trisphosphate
PNN	perineuronal net
PNS	peripheral nervous system
PTEN	Phosphatase and tensin homolog
PTPσ	protein tyrosine phosphate sigma
RGC	retinal ganglion cells
Rheb	Ras homolog enriched in brain
ROCK	Rho-associated kinase
SCI	spinal cord injury
Sema	semaphorin
SOCS3	suppressor of cytokine signaling 3
TNF-α	tumor necrosis factor-alpha
TSC1/2	tuberous sclerosis complex 1/2
α	alpha (subunit)
β	Beta (subunit)

1 SPINAL CORD INJURY DEMOGRAPHICS AND GENERAL INFORMATION

According to the National Spinal Cord Injury Statistical Center, the annual incidence of spinal cord injury (SCI) in the worldwide population is estimated at 40 people per million. A conservative estimate suggests 2.5 million people are currently suffering the debilitating consequences of living with SCI. The injury typically affects young people due to the activity based nature of the etiology; including motor vehicle accidents (40%) and sporting (8%). Additionally, there are also increasing numbers of older people with cervical spinal stenosis that sustain SCI trauma from minor falls (28%), as well as victims of violence and war (15%). The number of people living with paralysis from all causes is much higher (almost 6 million in the US alone), when other central nervous system (CNS) disorders such as stroke, multiple sclerosis, and traumatic brain injury are included (Cahill et al., 2009).

Prior to the two World Wars, the management of SCI patients resulted in an 80% mortality rate within the first few weeks (reviewed by Ditunno and Formal, 1994). This was mainly due to a lack of treatment for bladder and pressure sore management, ventilator dependence, and the incidence of respiratory or systemic infection. Since then, and especially past the 1970s, acute survival rates have continually improved to above 90% (Frankel et al., 1998). However, life expectancy is still considerably lower compared to the normal population (51.4 years in 2004), particularly in ventilator-dependent patients, with the most common causes of death being pneumonia and septicaemia (Strauss et al., 2006). Methylprednisolone has been recommended for acute neuroprotection, however, it is now generally thought that the adverse effects outweigh any advantages from neuroprotection (George et al., 1995; Gerhart et al., 1995; Hurlbert, 2000). There is currently no efficacious nonsurgical clinical treatment available for SCI that can aid recovery, and people afflicted can expect to live with functional impairment indefinitely, although several promising clinical trials involving Minocycline, Riluzole, Cethrin, and ATI355 (against human Nogo) are underway (Baptiste and Fehlings, 2006; Fehlings et al., 2011). Additionally, the financial burden of looking after a person suffering the consequences of SCI is large. The annual costs have been recorded as >$9 billion in the United States, with a single high-tetraplegic patient (C1–C4) costing >$900,000 in the first year, and >$170,000 in each subsequent year.

The typical outcome of traumatic SCI is functional deficits at and below the level of the lesion. The outcome severity is dependent upon lesion completeness and level. The commonest lesions occur just above or below the rib cage producing quadriplegic or paraplegic patients, respectively. All patients with functionally complete lesions lose bladder, bowel, and sexual function, and following cervical lesions, autonomic function is greatly impaired. For severe injuries that are localized in the high cervical spinal cord, severe breathing difficulties result, and many patients become dependent on artificial ventilation. Encouragingly, up to 70% of patients spontaneously recover at least one motor level in the spinal cord (Steeves et al., 2011). Some patients show enhanced lower limb excitability (Calancie et al., 2002), and up to 80% of SCI patients suffer from debilitating neuropathic pain that develops over time, at and below the level of injury (Siddall et al., 2003). Palliative treatments range from non-steroidal anti-inflammatory drugs, GABA agonists, tricyclic antidepressants, anticonvulsants, or opioid medications of which not all patients respond (Baastrup and Finnerup, 2008), with many becoming physiologically tolerant to the most effective treatments over time (Mayer et al., 1999).

There are currently two challenges for functional repair of the chronically damaged spinal cord. The first is protection of the lesion site, with growth of blood vessels and control of the local glial reaction. The second is restoration of connections; this can be through regrowth of axons, through the lesion site to reinnervate the intact spinal cord, or through local sprouting; in both cases there must be reformation of target specific functional synapses. Encouragingly, according to observations from animal models of recovery after SCI, only 10–15% of functioning fiber tracts are required to provide basic locomotor function (Dietz and Curt, 2006). However, even this level of regeneration is challenging as the site of injury is inherently nonpermissive to axon regrowth (reviewed by Fawcett and Asher, 1999). In addition, unlike peripheral nervous system (PNS) axons which generally initiate regeneration and grow axons, injured corticospinal tract (CST) and other spinal cord axons have an intrinsic predisposition to form dystrophic retraction bulbs after axotomy (reviewed by Bradke et al., 2012; Erturk et al., 2007; Tom et al., 2004). This is made all the more challenging as the axonal regeneration required for functional recovery can be up to a meter in length, although more local regeneration to relay neurones near the lesion may return function. Fortunately, most clinical SCI lesions are anatomically incomplete with some sparing of lateral white matter tracts, which provides an alternative method of enhancing functional recovery. Circuitry that is stimulated appropriately can bypass the lesion site through plasticity of local interneuronal circuits and/or intact descending systems (Ghosh et al., 2010; Schnell et al., 2011). Therapeutic intervention that takes advantage of the spinal cord's ability to reorganize rather than regenerate has led to promising results in preclinical models, for example, recovery of respiratory function (Alilain et al., 2011; Kajana and Goshgarian, 2008) and task specific recovery of delicate forelimb function (Garcia-Alias et al., 2009; Wang et al., 2011; Yamamoto et al., 2009). Humans too have an innate ability for plasticity after SCI, with the most notable being the development of reflex bladder voiding in response to interruption of the ascending and descending systems that control normal micturition (de Groat and Yoshimura, 2012). There are, however, negative effects of atypical rewiring of the spinal cord, including spasticity, spasms, locomotor exhaustion, and electromyographic (EMG) reduction in lower limb musculature, in the chronic phase of SCI (reviewed by Dietz, 2010, 2012). Pharmacological treatments, such as Baclofen and Diazepam, along with rehabilitative therapy are currently being trialed in SCI patients to manage these symptoms (reviewed by Elbasiouny et al., 2010). With the problems associated with atypical neuronal sprouting, it is ideally regeneration of long-tract CST axons, and restoration of normal functional synaptic connections that is the greatest challenge for spinal cord repair.

2 PATHOLOGY OF SCI: ACUTE

Traumatic injuries to the spinal cord are usually a mixture of compressive and contusive in etiology and rarely anatomically complete. A persistent compressive component can be treated surgically through decompressive laminectomy (Fehlings and Tator, 1999; Furlan et al., 2011). The acute contusive trauma of SCI leads to pathological changes that can be separated into three phases: the acute phase, the phase of secondary tissue loss, and the chronic phase of SCI (Bareyre and Schwab, 2003; Tator, 1995).

After the initial and direct shearing injury to neuronal tracts and blood vessels, a number of events occur that lead to the exacerbation of neurodegeneration and the impediment of regeneration. In the acute phase, the contusive impact causes immediate mechanical damage to soft tissue and instantaneous necrosis predominately in the gray matter. Over the next few minutes, ectopic action potential barrages are initiated in and around the lesion in surviving neurones and spared axons, due to electrolytic shifts of cations lasting for up to 24 h (Young and Koreh, 1986); these may lead to energetic exhaustion and death of neurones. This is clinically classified as spinal shock (Hiersemenzel et al., 2000) and only the H-reflex of spinal activity can be elicited (Dietz, 2010). Gradually as spinal shock subsides and ionic gradients are restored, the subacute stage of SCI is reached, and early spinal reflexes of large amplitude are elicited along with a late spinal reflex of small amplitude (Dietz, 2010).

Severance of posterior and anterior spinal veins and arteries that lie on the surface of the cord supplying the interior gray matter causes an ischaemic insult to the targeted segment as well as below the level of the lesion (Sandler and Tator, 1976). There is loss of the blood–brain barrier (BBB), internal hemorrhage (Jaeger and Blight, 1997; Maikos and Shreiber, 2007), altered blood flow (Senter and Venes, 1978), and endothelial cell death (Benton et al., 2008) at the site of injury. These events further prolong and exacerbate ischaemia and render the BBB permeable. The leaky BBB will prolong the ionic fluctuation disturbances within the injury site with increased Calcium (Ca^{2+}) and decreased Potassium (K^{+}) ions leading to excitotoxicity (Chesler et al., 1994; Stokes et al., 1983; Young and Koreh, 1986). Paradoxically, spared tissue surrounding the primary infarct site will undergo elevated K^{+} and decreased Ca^{2+} ions, prompting speculation of an intrinsic protective mechanism of the spinal cord to reduce collateral damage (Young and Koreh, 1986). At around 4 days postinjury in the rat, there is an attempt to reform the BBB with a period of neovascularization and repair at the lesion epicenter (Imperato-Kalmar et al., 1997). This response is short-lived and by 1 week postinjury the number of blood vessels in the lesion core again declines (Imperato-Kalmar et al., 1997). Recently, studies focused on exploiting the neuroprotective potential of angiogenesis within spinal cord lesions have shown that treatment with $\alpha v\beta 3$ integrin peptide (C16) or an angiopoietin-1 mimetic can reduce inflammation, rescue lesion site blood vessels and white matter, and promote locomotor recovery after contusion injury in mice (Han and Whelan, 2010).

After SCI, there are reactive changes and proliferation of microglia and astrocytes (Hains and Waxman, 2006) and microglia rapidly extend, retract, and reorient their processes toward the injury site (Davalos et al., 2005; Fig. 1). The immediate

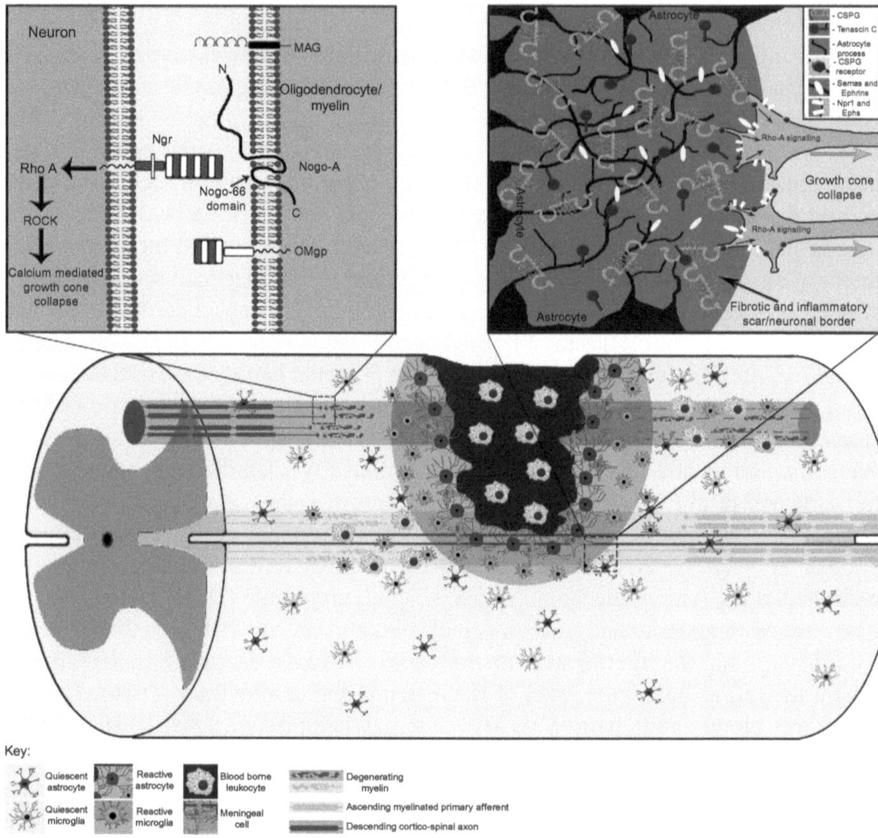

FIGURE 1

Extrinsic factors that impede successful axonal regeneration. After trauma to the spinal cord, a cystic cavity forms containing inflammatory cells and mediators. Surrounding this, a fibrotic and glial scar develops, consisting of reactive astrocytes, microglia, and meningeal cells, to primarily wall off the cavity. This scar forms both a physical barrier through interlocking astrocytic processes, and a chemical barrier through CSPG, tenascin-C, semaphorin, and ephrin expression, to regenerating axons. As a result of the injury, ascending and descending neuronal tracts become severed. The result of retarded Wallerian degeneration of transected axons is excessive myelin debris, which expresses MAG, OMgp, and Nogo-A complexes that interact with the Nogo receptor (Ngr). The RhoA intracellular signaling pathway is one of the main downstream effectors of growth cone collapse and thwarted regeneration. (For color version of this figure, the reader is referred to the online version of this chapter.)

role of these cells is phagocytosis of neuronal debris, proinflammatory responses, and beginning the structural maintenance of the injury site (David and Kroner, 2011). This upregulation is not only localized to the site of injury, but also along the entire axis of the dorsal spinal cord, termed “gliopathy,” which may contribute to the development of chronic pain by erratic neuromodulation of spared neurones (Coull et al., 2005; Hains and Waxman, 2006; Milligan and Watkins, 2009). Disruption of the BBB during and after SCI also aids the extravasation of leukocytes from the blood stream including, neutrophils (24 h peak), macrophages (14 day peak), and T- and B-lymphocytes (14 days and beyond; Carlson et al., 1998; Popovich et al., 1997) making up the innate immune response that includes the release of cytokines (Tumor Necrosis Factor-alpha, TNF-α, and interleukin, IL-6), chemokines, IL-1β, myeloperoxidase (MPO), and matrix metalloproteinase (MMPs; Stammers et al., 2012). These cells can also release damaging mediators such as free radicals, eicosanoids, glutamate, and NOS, that can lead to neuropathic pain (Thacker et al., 2007), and further exacerbate neuronal cell death (Beattie, 2004; Han and Whelan, 2010). Paradoxically, the innate immune response of macrophages has the ability to promote regeneration as well as impede and exacerbate neurodegeneration, depending on the form of activation (David and Kroner, 2011; Gensel et al., 2009; Kigerl et al., 2009). However, removal of glia after injury is not helpful; selective ablation of astrocytes leads to poor functional recovery after SCI (Faulkner et al., 2004), and microglia play an essential role in the reorganization of neuronal circuits in CNS injury (reviewed by David and Kroner, 2011; Neumann et al., 2009), leaving the field still unsure about the best therapeutic strategy to influence the innate immune response after SCI.

Another damaging result of SCI is delayed secondary cell loss through apoptotic mechanisms of oligodendrocytes (Buss et al., 2005; Casha et al., 2001; Shuman et al., 1997), oligodendrocyte precursor cells (OPCs; Horky et al., 2006; McTigue et al., 2001), astrocytes (Grossman et al., 2001), microglia (Shuman et al., 1997), and neurones (Crowe et al., 1997; Grossman et al., 2001; Huang et al., 2007; Liu et al., 1997). Apoptosis is controlled by ionic disturbances (increased $exCa2+$), free radial expression (NOS and COX2), and inflammatory cytokines (IL-1α/β, IL-6, IL-8, and TNF-α). There are distinct acute and delayed waves of apoptosis in the spinal cord after SCI (Liu et al., 1997) which corresponds with progressive loss of neurones in the dorsal, intermediate, and ventral horns within the lesion site (Huang et al., 2007). The neuronal loss at the level of the lesion, as long as it is away from areas of lower motor neurone pools, will not be the major cause of loss of function to an area of the body; however, local loss of motor neurones can produce a local lower motor neurone flaccid paralysis. Paralysis comes from the severance of the descending axons supplying the body of lower motor neurones below the lesion and also from a loss of sensation and proprioception due to severance of ascending fibers in the white matter tracts. The disruption to spinal cord axons leads to an initial growth response which varies depending on the axon. Many CNS axons lack the intrinsic ability to form a new growth cone and are also inhibited by molecules in the environment (reviewed by Bradke et al., 2012; Erturk et al., 2007; Tom et al., 2004). Axonal dieback through Wallerian degeneration (George and Griffin, 1994; Guth et al., 1999) can last over several days, and lead to axonal tips retracting several segments away from the initial lesion. This coupled to abortive regrowth, results in a characteristic morphology of retracted axons including reactive end bulbs, dystrophic endings, and axons which turn away from the lesion (Andrews and Stelzner, 2004; McPhail et al., 2004). It has also been suggested that both the level of the injury as well as the type of axon influence the amount of axon retraction as rubrospinal, vestibulospinal, and reticulospinal axons have little dieback after a cervical hemisection lesion (Houle and Jin, 2001), whereas CST fibers have been observed to retract up to 2.5 mm after a thoracic injury (Pallini et al., 1988).

In contrast to the general inability of CNS axons to regenerate, damaged peripheral nerve axons have been shown to regenerate successfully. Wallerian degeneration within the PNS is characterized by the efficient clearing of axonal and myelin debris released from damaged axons in the injury site (Allt, 1975) by Schwann cells and invading macrophages (Perry et al., 1987). Regenerating neurones express high levels of growth-promoting proteins, such as growth-associated protein 43 (GAP-43; Bisby and Tetzlaff, 1992), which helps to extend growth cones into the injury site, while Schwann cell columns ("bands of Bungner") guide regenerating axons from the proximal stump into the distal stump, with eventual remyelination by the Schwann cells. This is not to say that Wallerian degeneration does not occur in the CNS, but it does not, however, facilitate axon growth to the same extent as in the PNS. For example, after a dorsal rhizotomy lesion, the clearing of myelin and axonal debris which occurs within 30 days in the PNS is still ongoing at 90 days postlesion in the dorsal columns (DCs; George and Griffin, 1994). Additionally, other known degenerative processes such as the breakdown of the axoplasm and macrophage infiltration in the CNS are substantially delayed relative to the PNS. Interestingly, in transgenic mice in which Wallerian degeneration is significantly delayed (WldS, a naturally occurring mutant strain), Steward and colleagues demonstrated that wound healing after a primary DC lesion is delayed along with a prolonged inflammatory response and denser fibrotic scar formation (Zhang et al., 1996). This demonstrates that however inefficient, Wallerian degeneration is necessary for healing and stabilizing CNS lesions.

3 PATHOLOGY OF SCI: SUBACUTE/CHRONIC

3.1 Cyst formation

As a result of persistent inflammation and axonal dieback within the lesion epicenter, there is a progressively expanding cystic cavity. Studies in rat spinal cord after clip compression injury demonstrate a direct correlation between the size of the cystic cavity and the severity of the injury (Poon et al., 2007). Attempts to reduce the size of the cystic cavity postlesion have focussed on repairing the dura directly or transplanting cadaveric dura mater allografts onto rat spinal cord fixed with fibrin glue following a dorsal hemisection (Iannotti et al., 2006; Zhang et al., 2004). In these studies, attenuation in the size of the cystic cavity was found, along with a decrease in amount of connective tissue scarring, improved flow of cerebrospinal fluid, and a reduction of ED-1-positive macrophages/microglia at the lesion site (Iannotti et al., 2006; Zhang et al., 2004). Of interest in the field of experimental SCI are the differences in response to CNS damage between species. For example, in response to SCI in the rat, a cystic cavity forms around the lesion epicenter that is filled with degenerative debris stabilizing by 3 weeks postinjury, whereas a similar lesion in a mouse leads instead to an area full of dense connective tissue but void of necrosis or debris-filled cavitation (Inman and Steward, 2003). Studies examining the relation of spinal cord size to cavitational response have revealed that this response is unique to mouse spinal cord; comparative studies indicate that the spinal cord in an animal with a similar size, the hamster, responds to SCI with a fluid-filled cyst with minimal debris (Inman and Steward, 2003).

3.2 Glial scar and chondroitin sulfate proteoglycans

At the site of injury, a reactive mechanism occurs to form a protective barrier, known as the glial scar, to inhibit further spread of damage. Although it serves a protective function, providing trophic and metabolic support (reviewed by Rolls et al., 2009), the glial scar is a hallmark feature which contributes to CNS regenerative failure (Fawcett and Asher, 1999; Silver and Miller, 2004; Windle et al., 1952). Astrocytes, microglia, invading macrophages, meningeal cells, and OPCs that have entered the lesion persist on or near the borders of the injury contributing to the scar that by the end of the second week post injury acts as a formidable barrier surrounding the damaged tissue (Fawcett and Asher, 1999; Silver and Miller, 2004; Fig. 1). The glial scar also serves to repair the BBB thereby further minimizing the inflammatory response and continued cell death at the lesion site (Faulkner et al., 2004). The initial formation of the glial scar involves a cascade of events during which astrocytes become hypertrophic and reactive, upregulating expression of intermediate filaments such as glial fibrillary acidic protein (GFAP) and vimentin in addition to secreting chondroitin sulfate proteoglycans (CSPGs, discussed below). Astrocytes have been shown to proliferate in response to injury, upregulating immature markers such as nestin and vimentin (Fawcett and Asher, 1999). Along with the increased density of reactive astrocytes at the site of injury, astrocytes also extend processes to join adjacent astrocytes creating the barrier to surrounding tissue (Reier and Houle, 1988). Additionally, glial scar astrocytes have been documented to act as scaffolds for lesion site neovasculature as well as for removing potential destructive excitatory amino acids such as glutamate (reviewed by Rolls et al., 2009).

The glial scar is very inhibitory to axon regeneration, due to the expression of several inhibitory molecules including proteoglycans, ephrins (discussed in a later section), and the chemorepellent molecules slits and netrins. Slits and netrins are secreted proteins important for axon guidance during development as secreted proteins providing repellent cues to extending axons. Following brain or SCI, slit-2 upregulation by reactive astrocytes (Hagino et al., 2003), slit-1, slit-3, and netrin-1 upregulation by lesion-invading macrophages and meningeal fibroblasts (Wehrle et al., 2005) suggests they may contribute additional inhibitory or repellent signals for sprouting fibers. Proteoglycans, the main inhibitory molecules expressed and/or secreted after injury are extracellular matrix (ECM) molecules consisting of a protein core with covalently linked glycosaminoglycan (GAG) side chains (Fawcett and Asher, 1999). The GAG side chains are extensive, linear chains of repeated disaccharide units which serve as the major hindrance of axonal growth (Bovolenta et al., 1993). Within the family of CSPGs, there are a variety of differences in size and length of the GAG chains ranging from one GAG chain as in decorin to over 100 GAG chains as in aggrecan (Fawcett and Asher, 1999). The principal proteoglycans of the glial scar are the CSPGs (Silver and Miller, 2004) of which there are four major groups; the lecticans which include neurocan, versican, brevican, and aggregan; small leucine-rich proteoglycans which include bicorin and decorin; phosphacan/receptor-type-protein-tyrosine phosphatase β; and other CSPGs including NG2 (Fawcett and Asher, 1999). Not all CSPGs have been shown to be inhibitory for regrowth; in fact, CSPGs are known to play an important role during development to help direct axonal growth/elongation (Hsu et al., 2005). Most CSPGs are secreted into the ECM while some are inserted into the plasma membrane. In addition, heparan, dermatan, and keratan sulfate proteoglycans (Johnson-Green et al., 1991), as well as tenascin-C (Zhang et al., 1997), are also produced by astrocytes and contribute to the inhibition within CNS injuries. In addition to contributing to the glial scar, CSPGs exert further inhibitory influence into the CNS ECM by decreasing plasticity with the formation of perineuronal nets (PNNs). PNNs are specialized ECM structures that form at the end of the critical period, essentially turning off plasticity (Pizzorusso et al., 2002). PNNs surround the soma and dendrites of several classes of neurons and consist of hyaluronan (including hyaluronan synthase and receptor), tenascin-R, CSPGs (aggrecan, phophacan, neurocan, and slightly less brevican and versican), and link proteins (CRTL1, BRAL1, and BRAL2; reviewed by Galtrey and Fawcett, 2007). Functionally, in nearby intact tissue, the presence of PNNs restricts regenerative sprouting into the lesion epicenter serving as an additional barrier to CNS repair.

3.3 Fibrotic scar

In addition to the glial reaction, the core of the lesion contains a "fibrotic" or collagenous scar, derived from proliferating pericytes and meningeal cells, and also containing macrophages and OPCs (Goritz et al., 2011; Hermanns et al., 2001). This presents an effective barrier to regenerating axons but also serves a wound healing function (Hermanns et al., 2001). For example, following the initial insult, there is an influx of fibroblasts, leptomeningeal cells, and macrophages/microglia that populate the borders of the region. They secrete collagen (mainly type IV), laminin, fibronectin (Hermanns et al., 2001), and semaphorins (Sema; De Winter et al., 2002), molecules which are normally only found at low levels in intact spinal cord. Histological evidence from both rat and human SCI demonstrate that these ECM molecules can be found as early as 2 days postinjury but reside at the injury site chronically (Hausmann and Betz, 2000; Tate et al., 2007).

3.4 Chronic demyelination and myelin debris

In human SCI, pathologists have documented myelin debris persisting years after the insult (Buss et al., 2004), demonstrating the ineffective and incomplete myelin clearance within the CNS. Wallerian degeneration of damaged CNS axons results in axonal and myelin debris in and caudal to the injury, contributing to the persistence of inhibitory myelin proteins including myelin associated glycoprotein (MAG; Mukhopadhyay et al., 1994), oligodendrocyte myelin glycoprotein (OMgp; Mikol et al., 1990), and Nogo-A (Schweigreiter and Bandtlow, 2006), all of which have been shown to inhibit axonal growth. Nogo-A (Schnell and Schwab, 1990), first described by Schwab and colleagues, is found specifically in the CNS (Chen et al., 2000), whereas other myelin proteins, MAG and OMgp, are found both in the PNS and CNS (reviewed by Trapp, 1990).

At chronic time points, there is evidence to suggest the occurrence of a secondary phase of demyelination which may affect nearby intact axons (Totoiu and Keirstead, 2005) leading to more cumulative axonal loss (Bjartmar et al., 2001). Additionally, oligodendrocytes lack an associated basal lamina (reviewed by Colognato et al., 2005), a feature found within PNS Schwann cells which aids successful regeneration and remyelination. Although many oligodendrocytes die acutely, others survive but remain in a quiescent state which may later undergo apoptosis (Crowe et al., 1997). NG2-positive OPCs which, during development, can give rise to oligodendrocytes, have been found to repopulate the lesion area during the first 14 days postlesion with some proof of remyelination of sprouting fibers (McTigue et al., 2001). There are documented cases of spontaneous remyelination, however, in most cases it has been described as incomplete, abortive, and unstable (Blakemore and Crang, 1985; Totoiu and Keirstead, 2005). Additionally, infiltrating Schwann cells have been shown to remyelinate CNS axons after injury, in both rat (Black et al., 1986) and human (Guest et al., 2005).

Experiments performed to enhance myelin clearance have tested whether increasing phagocytosis with systemic treatment using lipopolysaccharide (LPS) can clear away myelin debris (Vallières et al., 2006). In this study, it was indeed shown that myelin could be cleared more efficiently from a mouse dorsal hemisection lesion with LPS injection; however, there was no associated functional recovery (Vallières et al., 2006).

4 CASES OF SUCCESSFUL CNS REGENERATION

4.1 Invertebrates and lower vertebrate models of CNS regeneration

One of the many characteristics separating lower vertebrates and invertebrates from mammals is in their extensive CNS regenerative ability after injury. Some lower vertebrates and invertebrates possess a capacity to regrow fibers and recover lost function following damage, potentially providing clues about which features facilitate and which features hinder regenerative growth.

Caenorhabditis elegans (*C. elegans*) and others within the phylum annelida, have a fairly simple nervous system, with 300 compared to the 100 billion neurones within human brain. The advantage of *C. elegans* is the ease of genetic modification allowing screens for genes that influence regeneration. Laser axotomy can be performed in *C. elegans*, following which efficient regrowth occurs on the order of a few days. The proximal stump responds after injury by retracting and extending filopodia within 3–6 h, with a growth cone detectable by 6 h postaxotomy, giving rise to relatively complete fiber extension over the course of 2–3 days (Ghosh-Roy et al., 2010; Wu et al., 2007). Interestingly, almost all neurones in adult *C. elegans* survive axotomy; however, regrowth postinjury is slower than during development and the accuracy of targeting decreases with age (Wu et al., 2007). Another well-studied invertebrate, the *aplysia californica*, has shown comparable mechanisms of regeneration to *C. elegans*. Similar to mammals, axonal injury and regrowth in *C. elegans* and *A. californica* is dependent on an increase in Ca^{2+} levels. After axonal injury, there is a rapid and transient rise in intracellular Ca^{2+} helping to seal the membrane of the damaged stumps and subsequently leading to new growth cone formation and extension involving microtubule (MT) depolymerization (reviewed by Bradke et al., 2012; Ghosh-Roy et al., 2010). Ca^{2+} is also needed for mammalian axon regeneration, whereby removing Ca^{2+} from the culture medium, prevents axotomized dorsal root ganglia (DRG) neurones from regenerating (reviewed by Bradke et al., 2012).

Interestingly, two characteristic contributors of regenerative failure in mammalian CNS are largely absent in *C. elegans*. *C. elegans* lack myelin and have no detectable signs of glial scarring around the lesion site even though CSPGs are present within their nervous system. Regardless of these differences, *C. elegans* and mammalian CNS share the major pathways signaling axon guidance, including those involving Robo, Slit, netrins, ephrins, and Sema (reviewed by Chen and Chisholm, 2011). Additionally, the mitogen-activated protein kinase kinase kinase (MAPKKK) signaling pathway (DLK-1) has been shown to regulate synapse formation (Nakata et al., 2005), be essential for axon regrowth (although not for development), and may have effects on the MT cytoskeleton via local or transcription control (Ghosh-Roy et al., 2010; Hammarlund et al., 2009; Yan et al., 2009). Further, p38 MAPK3, CCAAT/enhancer-binding protein 1 (CEBP-1), and kinase glh-binding protein 1 (a c-Jun amino-terminal kinase) have also been shown to be crucial for axon regeneration (Ghosh-Roy et al., 2010; Nix et al., 2011; Yan et al., 2009). Additionally, in a close relative of the *C. elegans*, the medicinal leech (Hirudo medicinalis), in response to axotomy, there is a large influx of microglia that surround the lesion and secrete the ECM molecule laminin into the area surrounding the lesion (Masuda-Nakagawa et al., 1993). It is likely that this response further enhances the repair potential by providing a growth-permissive substrate for regenerating fibers.

Fish also afford a great deal of insight into successful CNS regeneration. For example, in zebrafish (danio rerio) most retinal ganglion cells (RGCs) do not undergo axotomy-induced cell death, as observed in mammals. Additionally, zebrafish are capable of regenerating any cell type in the retina from a population of resident stem cells within either the ciliary marginal zone or from Müller glia which have the ability to reenter the cell cycle (Fimbel et al., 2007), however, functional regeneration as observed in goldfish retina is delayed (Matsukawa et al., 2004). Interestingly, after optic nerve crush in adult zebrafish, there is an upregulation of molecules from development including GAP-43, L1, and NCAM. Myelin within the fish nervous system does not possess the same inhibitory effects as mammalian myelin, including lacking Nogo-A (Diekmann et al., 2005). For example, cultured rat cortical neurones grow well on myelin derived from goldfish, but are strongly inhibited on rodent CNS myelin (Caroni and Schwab, 1988b). The reverse is also true, goldfish retinal explants successfully grow on goldfish myelin, but not on rodent CNS myelin (Bastmeyer et al., 1991). Fish axons successfully regenerate and remyelinate after injury, as their oligodendrocytes are able to dedifferentiate and redifferentiate after injury (Ankerhold and Stuermer, 1999). Additionally, there is wound closure devoid of scar tissue within zebrafish telencephalon after stab lesion and robust neurogenesis after injury (Baumgart et al., 2012). Interestingly, in another species, the xenopus laevis, despite the successful optic nerve regeneration they undergo postinjury, the spinal cord is not permissive for regrowth after the larval stage (Ferretti et al., 2003). In addition, it has been shown that myelin within xenopus spinal cord, but not optic nerve myelin, is responsive to IN-1 antibody (which neutralizes CNS myelin inhibitors) suggesting the presence of these inhibitors specifically in their spinal cord myelin (Lang et al., 1995).

5 EXTRINSIC FACTORS THAT IMPEDE AXON REGENERATION IN THE INJURY SPINAL CORD

5.1 Myelin inhibition

Lingering myelin debris persisting around the lesion site after SCI express several growth inhibitory molecules as discussed previously (including MAG, OMgp, and Nogo-A; see Fig. 1). The other splice variants of Nogo-A, Nogo-B and -C, have currently unknown functions, with the latter expressed only in muscle. However, Nogo-B, which is also expressed in neurones, may provide a compensatory mechanism in Nogo-A single knock-out mice (Kim et al., 2003). Nogo-A activity inhibits growth cone function through its C-terminal domain and Nogo-66, via the receptor subunit NgR present on axons (Fournier et al., 2001). NgR also interacts with MAG and OMgp (reviewed by Filbin, 2003) enhancing the inhibitory effect on neurite growth through the RhoA/Rho-associated kinase (ROCK) intracellular pathway (Duffy et al., 2009; Fournier et al., 2003; Niederost et al., 2002). Several ways of manipulating the myelin-axon inhibitory pathway to improve anatomical and functional outcomes have been assessed in rodent models of SCI. These include antibody sequestration to Nogo-66, knock-out models (nogo-ab$^{trap/trap}$, nogo-a$^{elll/elll}$, nogo-ab$^{atg/atg}$), and enzyme inhibition of intra-axonal signaling pathways. Inhibition of

Nogo-66 with either intraventricular or intrathecal application of neutralizing antibodies (IN-1; Brosamle et al., 2000; Caroni and Schwab, 1988a; Fouad et al., 2004; Freund et al., 2006), or receptor antagonism with peptide administration intrathecally or systemically (NEP1–40; GrandPre et al., 2002), has led to promising regenerative outcomes in animal models. The proven efficacy of anti-Nogo-A in nonhuman primate models of SCI (Fouad et al., 2004; Freund et al., 2006, 2007) led to a Phase 1 clinical trial by Novartis in Europe, assessing primarily safety and feasibility of the application of the antibody to human Nogo, called ATI355, as an acute treatment for SCI patients. Both long-distance regeneration of CST (Freund et al., 2006; Liebscher et al., 2005; Maier et al., 2009; Schnell and Schwab, 1990) as well as plasticity of interneuronal circuits (Freund et al., 2007; Maier et al., 2009) have been noted with Nogo inhibition. However, the efficacy of anti-Nogo-A is dependent upon the time of therapeutic intervention postinjury, as delayed anti-Nogo treatment post-SCI is progressively less effective (Gonzenbach et al., 2012). Also, anti-Nogo-A treatment is detrimental when combined inappropriately with rehabilitation, producing abnormal and dysfunctional movements (Maier et al., 2009). Nogo knock-out mice (Nogo-A$^{-/-}$, Nogo-A/B$^{-/-}$, Nogo-A/B/C$^{-/-}$) produce puzzlingly different regenerative phenotypes, with some positive (Kim et al., 2003; Simonen et al., 2003), others mostly negative (Lee et al., 2009; Steward et al., 2008; Zheng et al., 2003) even though the myelin produced from all of these mutants was significantly less inhibitory than wild type. The debate into reasons behind this alarming discrepancy, including axon labeling techniques, mouse back-strain dependence, mutant allele derivation, and age, continues (Steward et al., 2007; Woolf, 2003). A triple knock-out of all myelin inhibitory molecules (MAG, OMgp, and Nogo-A) also produces conflicting evidence in different laboratories. Some authors suggest the effect of global myelin inhibitory protein deletion acts synergistically to restrict axon growth (Cafferty et al., 2010), while others report failure (Lee et al., 2010).

5.2 Scar inhibition

Uncleared inhibitory myelin is not the only obstacle impeding successful regeneration of the damaged nervous system. Additional inhibition of axon growth also occurs through extracellular inhibitory molecules, namely CSPGs expressed and secreted mainly by astrocytes and meningeal cells after injury (reviewed by Kwok et al., 2008) and also ephrins and Sema3A released from the damaged cord (Fig. 1). Semas are a family of proteins that are both membrane-associated and secreted, and play an important role in axon guidance and synapse formation during development (reviewed by Verma et al., 2008). Sema3A, the best characterized, is constitutively expressed in motor neurones, and guides descending supraspinal pathways and reflex pathways from sensory afferents that express the Sema receptor neuropilin-1 (Nrp1) through repellent signaling (Gavazzi et al., 2000; Giger et al., 1998). Sema3A is greatly upregulated and secreted by meningeal fibroblasts after injury to the spinal cord, and may contribute to poor axonal regeneration (De Winter et al., 2002). A proteoglycan, decorin, has been recently shown to suppress Sema3A (as well as several growth inhibitory CSPGs) expression in fibroblasts in a rat cortical lesion, as well as significantly increasing growth support of DRGs on fibroblasts in culture (Minor et al., 2011). An amino acid peptide, called ASNKL, has shown similar suppression of Sema3A inhibition *in vitro*, converting repulsion into attraction and preventing growth cone collapse (Castellani et al., 2004). However, pharmacological blockade of Sema3A–Nrp1 signaling with the peptide ASNKL *in vivo*, lead to no significant improvement in motor function after SCI in mice (Mire et al., 2008), suggesting the milieu of other inhibitory signals within the glial scar compensate and maintain a growth inhibitory state. Certainly, another Sema7A has been shown to be upregulated in reactive astrocytes after SCI, contributing to the development and maintenance of the glial scar (Kopp et al., 2010).

A bacterial enzyme first isolated from *Proteus vulgaris* in 1968 (Yamagata et al., 1968), chondroitinase ABC (ChABC), has provided an alternative approach for therapeutic intervention after SCI (reviewed by Bradbury and Carter, 2011). ChABC degrades CSPGs by cleaving the long repeating disaccharide units that make up the highly growth inhibitory GAG chains (Crespo et al., 2007; Yamagata et al., 1968). An acute injection of ChABC abolishes the upregulation of CSPGs, produced after injury to the CNS (Lin et al., 2008). Over the past 10 years, ChABC has been shown in numerous experiments to improve the regeneration of cut axons or sprouting of intact axons *in vitro* or after CNS damage through digestion of CSPGs (Alilain et al., 2011; Barritt et al., 2006; Bradbury et al., 2002; Garcia-Alias et al., 2009; Hunanyan et al., 2010; Massey et al., 2006; Moon et al., 2001; Tom et al., 2009; Wang et al., 2011). The digestion of CS–GAG chains by ChABC has been hypothesized to result in neuroregeneration and repair in a number of ways (reviewed by Bradbury and Carter, 2011). First, ChABC reduces the interaction of CS–GAG on neuronal terminals expressing CSPG receptors (protein tyrosine phosphate sigma, PTPσ), that inhibit axonal growth through GTPase Rho downstream signaling pathways (Fry et al., 2010; Monnier et al., 2003; Shen et al., 2009). PTPσ receptor deficient mice show greatly improved CST regeneration compared to wild-type after SCI (Fry et al., 2010). Second, ChABC overcomes the CSPG-mediated masking of growth-promoting molecules such as laminin that have become locked within the glial scar (McKeon et al., 1991; Zuo et al., 1998). Third, CSPG degradation products such as disaccharides, may have anti-inflammatory and neuroprotective effects (Rolls et al., 2009), and M2 (phenotypically anti-inflammatory) macrophages may be given greater access to axon tips after CSPG degradation, aiding neuroprotection and regeneration (Kigerl et al., 2009). Other hypotheses yet to be clarified *in vivo* include, ChABC-induced liberation of sequestered neurotrophic (NT) factors from within the glial scar, making them available to growth cones (Nandini et al., 2004), and the removal of cellular stressors that may impede regeneration (Crespo et al., 2007).

The major limitation of ChABC is that it is a bacterial enzyme specific to Prokaryotes and therefore, potentially immunogenic. It also has the added limitation of instability at body temperature, and the enzyme has shown *in vivo* activity limited to 10 days postinjury (Lin et al., 2008), followed by the time taken (possibly a week or more) for the return of inhibitory CSPGs. Most experimental treatments have used repeated intrathecal injections and/or injections into the cord parenchyma. Therefore, there is interest in alternative methods for applying ChABC therapy. Two forms of chondroitinase have been delivered by viral vector, (in the case of ChABC, modification was required of several *N*-glycosylation sites through directed mutagenesis; Muir et al., 2010), or Tet-On adenoviral vector transfection of chondroitinase AC (ChAC; Curinga et al., 2007). Following, lentiviral vector introduction of this ChABC-enzyme into mammalian cells, injured CST axons of the rat have been shown to express ChABC for up to 4 weeks (Zhao et al., 2011) and 8 weeks in the case of ChAC (Jin et al., 2011) postlesion, respectively, alongside the promotion of axonal regeneration (Zhao et al., 2011). However, long-term expression is likely to produce a damaging immune response, although this has not been observed in the cited studies. However, Bellamkonda and colleagues have demonstrated a method for stabilizing ChABC with trehalose, and delivering it in slow-release liposomes, giving up to 6 weeks activity postadministration, providing greater penetration into spinal tissue, and greater CSPG digestion (Lee et al., 2010).

The effects of ChABC on long-distance axon regeneration are fairly modest and its main effect is on local sprouting and plasticity (reviewed by Kwok et al., 2008). For these effects, it is important to combine the ChABC with rehabilitation tasks to drive plasticity (Alilain et al., 2011; Garcia-Alias et al., 2009; Wang et al., 2011). The effectiveness of these combined regimes has made it possible to explore a critical issue for translation to the clinic, the therapeutic time window of ChABC treatment. There is evidence that the effect of ChABC on axon regeneration is time-dependent (Garcia-Alias et al., 2008; Mingorance et al., 2006). However, the effects on behavioral recovery decrease little if at all over time. A recent study (Wang et al., 2011) shows effective recovery of function following chronic ChABC administration 1 month post-SCI. In this study, there was no detectable long-distance regeneration of CST axons, only local sprouting. Another recent study has examined the recovery of respiratory drive to the diaphragm one year after injury, using a combination of ChABC and behavioral reinforcement with low oxygen (Alilain et al., 2011). If effective therapy can be administered to patients whose injury and function has stabilized after injury, the practical problems of conducting clinical trials will be greatly reduced. Time-dependence and quantity of administration of enzymes into the spinal cord lesion that can break up the scar needs to be carefully assessed, as alongside detrimental effects to neuroregeneration there are important benefits to the glial scar. The early glial scar provides an environment for the reestablishment of the BBB, a barrier to inflammatory leukocytes (Voskuhl et al., 2009), and the generation of a structurally sound scaffold that is required for the preservation of function (Faulkner et al., 2004), and even neuroprotection (Spence et al., 2011).

5.3 Inflammation

Macrophages and microglia exhibit a neuroprotective/neuroregenerative phenotype as well as a proinflammatory/neurotoxic phenotype in response to environmental signals produced by CNS injury (reviewed by David and Kroner, 2011). Reactive microglia release a variety of chemicals and cytokines, including peroxynitrite and nitric oxide, that kills neurones (Kaushal et al., 2007). Preclinical models of SCI have shown that blocking or reducing macrophage/microglial activity and accumulation at the injury site using silica dust (Blight, 1994) or clodronate (Popovich et al., 1999) to deplete macrophages, NT factors such as ProBDNF (brain-derived neurotrophic factor; Wong et al., 2010) to inhibit their infiltration and migration, or anti-inflammatory drugs such as minocycline (Stirling et al., 2004) and FK506 (Lopez-Vales et al., 2005) to reduce macrophage reactivity, is beneficial to functional recovery, preventing neuronal or oligodendrocyte cell death and axonal die-back. In addition, inhibition of integrins responsible for leukocyte recruitment with antibodies has shown promising efficacy after SCI in reducing inflammation. Neutrophils and macrophages express integrins, such as the β2 subunit integrins (CD11b and CD11d/CD18) and the integrin heterodimer α4β1, on their cell surface, and use them to bind to vascular cell adhesion molecule-1 (VCAM-1) and intercellular adhesion molecule 3 (ICAM-3) on blood vessels, to initiate extravasation from the blood into the spinal cord parenchyma after SCI (Grayson et al., 1998; Van der Vieren et al., 1995). Weaver and colleagues have shown that treatment with a monoclonal antibody against these integrins, reduces cell death and free radical expression in the lesion site (Bao et al., 2005), provides neuroprotection (Saville et al., 2004), and restores lost serotonergic projections (Oatway et al., 2005), improving sensory, autonomic and motor function in mice and rats with SCI (Fleming et al., 2008; Gris et al., 2004). Neutralization of integrins is most effective in the acute stage of SCI at 6–48 h and ineffective at subacute/chronic time points (7 days; Fleming et al., 2009). Although this looks like a promising new anti-inflammatory therapy for SCI, when the initial study (Gris et al., 2004) was replicated (Hurtado et al., 2012) treatment efficacy could not be found, advocating caution before translation to clinical trials.

Conversely to the proinflammatory neurodegenerative mechanisms, macrophages have been shown to have an important role in neuroregeneration. The Schwartz lab have shown that macrophages previously exposed to peripheral nerve segments and then implanted into transected spinal cord or optic nerve actually aid anatomical and functional repair in rats, as they become preprimed for phagocytosis. The converse is true for macrophages exposed to optic nerve segments, that showed minimal phagocytic potential (Schwartz et al., 1999), providing an indication as to why PNS axons recover better after injury than CNS axons. Importantly, these studies indicate a double-edged opposing action on the processes of inflammation after SCI from the same subset of immune cells. It is only recently that two distinct macrophage lineages have been identified, that confer either neurotoxic (defined as M1 macrophage) or neuroprotective/neuroregenerative (defined as M2 macrophages) responses to injury (Kigerl et al., 2009). Most macrophages acquire predominantly the former polarization in response to SCI, explaining the poor intrinsic regeneration observed. M1 activated macrophages have been shown to induce retraction of damaged axons through direct cell–cell interaction (Horn et al., 2008), and secretion of MMP-9 that degrade ECM molecules such as laminin and fibronectin that are essential for growth cone extension (Busch et al., 2009), as well as produce CSPGs in a concentration 17-fold higher than M2 cells (Martinez et al., 2006). These macrophage processes, including microglial–neuronal interactions, occur in part through CX3CR1 (fractalkine receptor), with CX3CR1-deficient mice showing greater anatomical and functional recovery than wild type after SCI (Donnelly et al., 2011). Undifferentiated macrophages do not confer these same properties. The most likely growth-promoting action of M2 cells on regeneration is indirect, including immunity, anti-inflammatory processes, phagocytosis, pro-healing, and iron sequestration from red blood cells; however, growth stimulation is also a possibility, although this has yet to be shown in SCI models (reviewed by David and Kroner, 2011).

Although very little is currently known about the role of the polarized macrophage in the CNS after SCI, it is an attractive area for translational SCI research. Through modulating the microenvironment of the injured spinal cord, the polarization of macrophages from the phagocytic inflammatory M1 phenotype in the acute stage of injury, can be shifted toward the regenerative M2 phenotype in the chronic stage.

6 INTRINSIC FACTORS THAT LIMIT THE NEURONAL GROWTH RESPONSE AFTER INJURY

Intrinsically, the potential for regenerative growth of an adult CNS neurone is low while embryonic axons have a much greater growth potential. Several lines of research have been examining ways to restore the regenerative vigor of adult neurones by applying knowledge from lower vertebrate and invertebrate CNS, recapitulating developmental states, or turning off inhibitory processes within the cell.

6.1 Conditioning lesions

Conditioning is a means of priming a neurone for CNS regrowth by modifying its intrinsic growth potential through exogenous means, typically by a peripheral injury. Conditioning lesions have been studied extensively because of their ability to enhance regeneration of the central branches of sensory axons within the CNS. Although not a viable clinical therapy, they do provide anatomical and cellular insight into potential intrinsic mechanisms to promote regeneration. These peripheral lesions have the most successful effect on regrowth in the CNS when performed at the same time or 1 week prior to the central injury (Neumann and Woolf, 1999; Richardson and Issa, 1984). Recent studies however have shown that a conditioning lesion can enhance regeneration of primary afferents through a chronic DC injury, only if an additional central lesion is applied caudally (Ylera et al., 2009). It must also be kept in mind that peripheral conditioning lesions only apply to two types of neurones, DRGs and ascending DC axons, and RGCs and optic nerve axons (lens injury). Additionally, conditioning lesions of the DRG affect mainly the large diameter fibers and not the small unmyelinated ones (Kalous and Keast, 2010). Besides axotomy, inflammation induced by macrophage implantation or injection of the inflammatory agent, Corynebacterium parvum (Lu and Richardson, 1991), or zymosan (prior to injury; Steinmetz et al., 2005) into the DRG also enhances the regrowth of dorsal root axons. Conversely, forced macrophage depletion within the DRG prevents regenerative growth induced by the conditioning lesion indicating the macrophage requirement for this response (Salegio et al., 2011). These results correlate with studies evaluating optic nerve regrowth showing that a macrophage-derived growth factor, oncomodulin, stimulates RGC regrowth in a cyclic adenosine monophosphate (cAMP)-dependent manner (Yin et al., 2006). Likewise, studies that have included the addition of dibutyryl cAMP at the DRG prior to the central injury have demonstrated the regrowth of DC axons into the lesion site (Neumann et al., 2002) and slightly beyond (Qiu et al., 2002). cAMP levels decline with the onset of development corresponding to a loss of regenerative ability. Treatment with cAMP in the DRG not only increases DC axon regrowth but has been shown to directly overcome inhibition by MAG (Neumann et al., 2002) a response which initially involves PKA (protein kinase A) phosphorylation (Qiu et al., 2002). Blocking the degradation of cAMP by a phosphodiesterase inhibitor, rolipram, shows promising regeneration following SCI (Nikulina et al., 2004). Conditioning lesions, however, induce a greater regenerative response than cAMP treatment likely due to the recruitment of a larger variety of signaling mechanisms involved in the growth response (Blesch et al., 2012). Still other studies that have included intrathecal infusion of the cytokine IL-6 near the injury site have seen enhanced DC regeneration similar to that observed with a preconditioning lesion (Cao et al., 2006). In conjunction with that study, DC regrowth following a preconditioning lesion, failed in IL-6 knockout mice (Cafferty et al., 2004). All of these studies indicate that a variety of factors may be involved in the conditioning effect on DC axons, all of which upregulate the same proteins, including GAP-43, known to be involved in axonal regeneration after peripheral nerve injury (Bisby and Tetzlaff, 1992) and found in DC axons after a conditioning lesion (Schreyer and Skene, 1991).

6.2 Phosphatase and tensin homolog

Studies examining the role of PTEN (phosphatase and tensin homolog) inhibition in the nervous system have recently demonstrated significant increases in CNS regenerative ability in both the visual system and spinal cord. PTEN is a tumor suppressor gene that is mutated in many cancers. It functions as both a lipid and protein phosphatase, and PTEN null animals die during embryogenesis (Di Cristofano et al., 1998). Targeted PTEN knockout animals (using an NSE promoter driven CRE transgenic mouse line) in areas of cortex and hippocampus have been characterized by abnormal dendritic and axon growth, impaired social and sensory behavior, as well as neuronal hypertrophy and resultant macrocephaly (Kwon et al., 2006). PTEN has multiple downstream effectors including mTOR (mammalian target of rapomycin), GSK-3β (glycogen synthase kinase 3 beta), and 3′ PIs (3′ phosphorylated phosphoinositides; reviewed by Park et al., 2010). PTEN opposes PI3K (phosphoinositide 3-kinase) action via receptor tyrosine kinases which regulate the phosphorylation

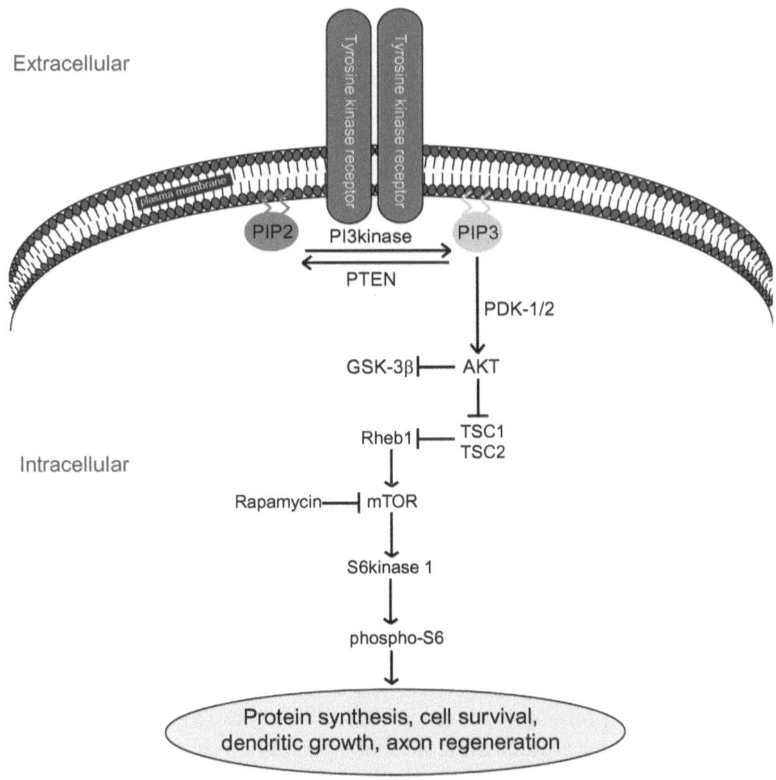

FIGURE 2

Schematic of PTEN/PI3K signaling pathway. PTEN dephosphorylates PIP3 to produce PIP2, acting in opposition to PI3kinase, the kinase that phosphorylates and converts PIP2 into PIP3. In a PTEN knockout, PIP3 accumulates thereby activating AKT. Upon activation of AKT, several downstream effectors become phosphorylated including GSK-3β and TSC1 and 2. mTOR is downstream of AKT and upon its activation, it phosphorylates and activates S6kinase1 which in turn phosphorylates ribosomal protein S6. Phospho-S6 has been shown to be involved in activating protein translation and stimulating cell growth; it has also been correlated with increased axonal regeneration in PTEN knockout mice (reviewed by Park et al., 2010). (For color version of this figure, the reader is referred to the online version of this chapter.)

of PIP2 (phosphatidylinositol (4,5) bisphosphate) to PIP3 (phosphatidylinositol (3,4,5) trisphosphate; Fig. 2), therefore a knockout of PTEN will lead to accumulation of PIP3. PIP3 downstream signaling activates several pathways, notably the AKT pathway via activation through PI-dependent kinase-1/2 (PDK-1/2) and inhibition of GSK-3β which have both been linked to axon outgrowth (Eickholt et al., 2002) as well as inhibition of the tuberous sclerosis complex 1/2 (TSC1/2) and downstream Rheb (Ras homolog enriched in brain). PI3K signaling in primary cultured neurons has been shown to also regulate cell survival, dendritic growth, and arborisation (Klesse and Parada, 1998). In studies examining the regenerative ability of PTEN deficient animals, an adeno-associated virus (AAV)-expressing Cre injected into PTEN floxed mice either intravitreally for optic nerve studies (Park et al., 2008) or into sensorimotor cortex for CST studies after SCI (Liu et al., 2010) have shown promising results. Knockout of PTEN in RGCs led to significant regrowth of axotomized optic nerve fibers, through protection from axotomy-induced phospho-S6 decline and sustaining mTOR activity (Park et al., 2008). In studies post-SCI, equally impressive regeneration was observed in the CST into and beyond the lesion following PTEN deletion in sensorimotor cortex from birth or from 4 weeks postnatally, with similarly increased levels of phospho-S6 (Liu et al., 2010). Further studies in the optic nerve with PTEN deletion have included a combinatorial approach of the JAK/STAT pathway via deletion of its negative regulator, SOCS3 (suppressor of cytokine signaling 3). SOCS3 deletion alone has been shown to enhance optic nerve fiber regeneration (Smith et al., 2009). The dual deletion of PTEN and SOCS3 led to significant regeneration in the optic nerve in a delayed treatment paradigm when compared with each deletion separately (Sun et al., 2011).

6.3 Taxol/Paclitaxel

Recent evidence has demonstrated that taxol (Paclitaxel, a clinically approved anticancer drug) increases neurite outgrowth and axon regeneration in the visual system and spinal cord (Erturk et al., 2007; Hellal et al., 2011; Sengottuvel et al., 2011). Taxol stabilizes MTs by interfering with the normal breakdown of MTs during cell division. Taxol treatment within the nervous system has been shown to have dose-dependent effects. At low levels, taxol promotes polymerization at the plus end of MTs which can prevent formation of axon retraction bulbs and axon tip swelling after SCI (Erturk et al., 2007), allow axons of cerebellar granule neurons to overcome myelin inhibition (Erturk et al., 2007), and increase neurite length of cultured RGCs (Sengottuvel et al., 2011). At higher concentrations, however, these growth-promoting properties are reduced or reversed as it is thought that high concentrations of taxol hyperstabilises MTs, blocking MT dynamics completely (Derry et al., 1995). Following optic nerve crush and administration of taxol at the injury site, there was delayed scar formation and reduction of macrophages, suppressed CSPG expression, as well as regeneration beyond the injury site, a response that was further enhanced by a conditioning lens injury (Sengottuvel et al., 2011). After SCI and intrathecal administration of taxol, there was a significant reduction in fibrosis (levels of collagen IV, fibronectin, and laminin) at the lesion site, descending serotonergic axons were significantly increased caudally to a dorsal hemisection lesion with indications of growth cones at the growing tips and, in addition, DC axons were found at the lesion cavity edge and beyond the lesion (when combined with a conditioning lesion; Hellal et al., 2011). Functional recovery was noted at 6–8 weeks postinjury indicated by significantly better paw placement and grid walk results (Hellal et al., 2011). Further analysis biochemically suggests that through interference with TGF-β1 signaling, taxol reduces the CSPG GAG fraction at the lesion site and additionally leads to an increase in the number of detyrosinated MTs as well as kinesin-1, suggesting interference in TGF-β1-mediated fibrosis (Hellal et al., 2011).

Taxol-induced recovery of function and enhanced regeneration represents a promising and novel approach to treating SCI, as the drug already has FDA-approval. However, it must be recognized that in patients undergoing cancer therapy with Paclitaxel at a higher dose than was used for SCI experiments, there are substantial side effects which include peripheral sensory neuropathy (Lee and Swain, 2006), presenting typically as tingling, burning, joint, and muscle pain, as well as numbness, loss of vibratory sensation and decreases in deep tendon reflexes and proprioception (Postma et al., 1995). Fortunately, many sufferers found that the pain is abolished upon ending the treatment (Pignata et al., 2006). Peters and colleagues have shown that after intravenous taxol treatment, ATF3 (activating transcription factor 3) expression is upregulated in large diameter neurons and Schwann cells in the sciatic nerve, and GFAP immunoreactivity is increased in satellite cells around the DRG neurons as well as in the dorsal horn of the spinal cord (Peters et al., 2007), both key markers of neuropathic pain (Hu and McLachlan, 2000; Tsujino et al., 2000). In addition, there was also an increase in CD11-positive microglia in lamina III–VI in the spinal cord and an increase in CD68-positive macrophages in the DRG (Peters et al., 2007). These issues may not be a problem at the low dose used for modification of axons after SCI but will clearly have to be monitored.

6.4 Rho/ROCK

The Rho family of GTPases (including Rho, Rac, and Cdc42), have been shown to transduce extracellular signals to the actin cytoskeleton in order to modify axon outgrowth and growth cone motility. RhoA has partly opposing actions to Rac and Cdc42, in that RhoA activation results in growth cone collapse whereas activation of Rac and Cdc42 results in growth cone attraction and possible extension due to F-actin polymerization in filopodia and lamellipodia (reviewed by Zurn and Bandtlow, 2006). RhoA has been of particular interest, because several CNS inhibitors activate it and cause effects on growth cones through it, including Nogo-A, Nogo-66, MAG, and the CSPG versican V2 (Niederost et al., 2002; Schweigreiter et al., 2004). Experiments aimed at inactivating RhoA using the enzyme clostridal toxin C3 transferase, or inactivating ROCK, a downstream effector of Rho, with recombinant Y27632 protein (an ATP competitive antagonist) have been successful at significantly regenerating optic nerve axons after lesion (Lehmann et al., 1999) and CST fibers along with concurrent increases in GAP-43 mRNA in motor neurones post-SCI (Dergham et al., 2002). RhoA inhibition has the added benefit of neuroprotection after SCI through the inhibition of apoptosis (Dubreuil et al., 2003). These promising results have led to the initiation of a clinical trial using a RhoA antagonist, a drug known as Cethrin® (Fehlings et al., 2011). Initial results from this trial indicate good safety and tolerability of the drug with promising levels of motor recovery in many of the patients although there has only been a small cohort to date, but further trials are planned (Fehlings et al., 2011).

6.5 Ephs/ephrins

Intimately related to the previously described Rho family of GTPases are the Ephs. Ephs are the largest subfamily of transmembrane receptor tyrosine kinases and they bind membrane bound ligands called Ephrins. The Eph receptors can be subdivided into two groups EphAs, those that bind GPI-anchored ephrin-A ligands, and EphBs, those that bind transmembrane ephrin-B ligands. They regulate axon guidance through growth cone collapse and are highly expressed during development, but are present at lower levels in the adult (reviewed by Goldshmit et al., 2006). Upon ligand binding, the receptor undergoes a conformational change. However, in order to induce downstream signaling events (such as cell adhesion, migration, proliferation, synaptic plasticity, and neurite outgrowth) several Eph/Ephrin bound complexes must be clustered. Eph/Ephrin complexes can stimulate intracellular signaling in the cell containing the Eph receptor or the cell containing the Ephrin ligand, either through forward or reverse signaling, respectively. Forward signaling affecting neurite outgrowth has been shown to involve Rho, Rac, or Cdc42 (reviewed by Dickson, 2001). More specifically, RhoA is activated upon interaction with the EphA4 GEF (guanine nucleotide exchange factor) known as ephexin, highly enriched in the CNS. Following EphA4 binding of EphrinA1, ephexin is phosphorylated and the growth cone collapses (Shamah et al., 2001). Additionally, it has been shown that Vav2 (a GEF) also modulates growth cone collapse via EphrinA binding, distinct from the ephexin pathway. Vav2 is recruited and activated by autophosphorylated Ephs following EphrinA binding leading to Rac1-mediated endocytosis of the Ephrin/Eph complexes (Cowan et al., 2005). Growth cone collapse is hindered if Vav2 is blocked or if ephexin is mutated (Cowan et al., 2005; Shamah et al., 2001).

There is an extensive upregulation of Ephs following SCI in astrocytes, oligodendrocytes, and in motor neurones (not EphA5) including EphA3–A8, EphB3, and ephrinB2 (Willson et al., 2002, 2003), with an upregulation of EphB2 at the lesion site by meningeal fibroblasts (Bundesen et al., 2003). EphA4 has also been shown to accumulate in severed CST axons (Fabes et al., 2006). Further, examination of EphA4 deficient mice has revealed a striking decrease in astrogliosis postinjury with a concomitant decrease in glial scar formation (Goldshmit et al., 2004). Likewise, blocking EphA4 with a specific blocking peptide enhanced sprouting of CST axons rostral to the injury site (Fabes et al., 2006). Additionally, ephrinB3 expression postnatally has been shown to be expressed in myelinating oligodendrocytes and importantly, EphA4-positive cortical neurones inhibit neurite outgrowth in the presence of ephrinB3 implicating this ligand–receptor complex as an additional myelin inhibitory factor (Benson et al., 2005).

6.6 Integrins

Promising results have recently been demonstrated in the field of integrin biology in relation to CNS repair. Integrins are heterodimeric transmembrane receptors that exist in every cell of the body except red blood cells. They consist of α and β subunits that dimerize upon ligand binding with an ECM ligand such as collagen, fibronectin, laminin, or tenascin-C. Integrin function is controlled by activation, either by extracellular cations or through inside-out signaling mediated by talin and kindlin. Upon dimerization, integrin receptors may become activated and undergo conformational changes which induce intracellular signaling to modulate cell motility, survival, proliferation, and neurite outgrowth (Fig. 3). An *in vitro* study directly comparing embryonic DRGs with adult DRGs, demonstrated that both grew well on high levels of laminin (a normally permissive substrate), however, on low levels of laminin or low levels of fibronectin, only the embryonic DRGs extended neurites (Condic, 2001). Significant neurite outgrowth could be induced in the adult DRGs, however, following forced expression of α5 or α1 integrin subunits, the fibronectin and laminin receptors, respectively (Condic, 2001). It was also shown that the embryonic DRGs have the innate capacity to upregulate their integrin expression to grow on these otherwise inhibitory substrates, whereas the adult DRGs did not (Condic et al., 1999). The main ECM glycoprotein in CNS is tenascin-C, and the integrin receptor which binds to it is α9β1, an integrin absent in adult CNS. Experiments by our lab overexpressing this receptor in PC12 cells and adult DRGs plated on tenascin-C demonstrated a significant amount of neurite outgrowth, similar to outgrowth observed on laminin (Andrews et al., 2009). Further, *in vivo*, forced expression of α9 in DRG neurones modestly enhanced dorsal root axon growth through the dorsal root entry zone (DREZ) after a dorsal root crush injury, as well as modestly enhanced growth of DC axons into a DC crush spinal lesion (Andrews et al., 2009). The modest regrowth observed *in vivo* compared with our *in vitro* studies led us to examine the role that integrin activation had on enhancing neurite outgrowth and axon regeneration as others have shown that inhibitors, including those from myelin, can deactivate integrins (Hu and Strittmatter, 2008). Antibodies which activate β1 integrin have been efficacious in overcoming the inhibitory effects of the proteoglycan, aggrecan (Tan et al., 2011). In addition, intracellularly, an adaptor protein known as kindlin, which binds the β integrin subunit, has been shown to enhance the activation state of integrins to induce downstream signaling events (Moser et al., 2008). Kindlins are FERM-domain containing proteins with three family members, kindlin-1, -2, and -3, of which kindlin-1 is absent in CNS (Ussar et al., 2006). Adult DRGs do not extend processes on aggrecan. However, forced expression of kindlin-1 rescues neurite outgrowth to control levels and is integrin-dependent, as demonstrated by increased phospho-FAK (focal adhesion kinase) expression (Tan et al., 2012). This effect carries over *in vivo* as forced expression of kindlin-1 in DRGs increases dorsal root axon growth through the DREZ and into the spinal cord after dorsal root crush injury (Tan et al., 2012).

Integrins play a crucial role in the inflammatory response post-SCI, whereas blocking certain subunits has neuroprotective and growth-promoting effects as discussed in an earlier section. Additionally, integrins have also been implicated in myelination during development, mainly the β1 integrins (Barros et al., 2009), as well as remyelination within the lesion site, as αv integrin is expressed on OPCs that are recruited to the injury (Zhao et al., 2009). Despite its role during myelination, remyelination can occur in the absence of β1 integrin (Benninger et al., 2006),

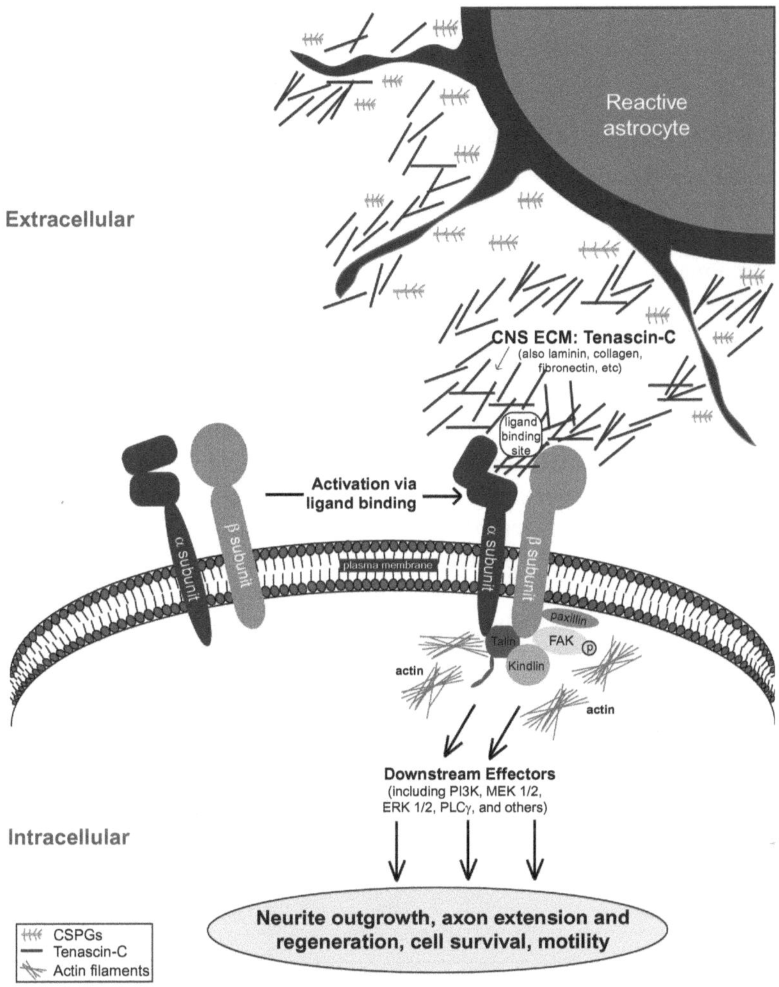

FIGURE 3

Schematic of integrin activation and signaling. Single transmembrane α and β subunits exist in an inactive state. Upon ligand binding, α and β subunits dimerize and undergo a conformational change, thus becoming activated. Intracellularly, cytoplasmic adaptor proteins (for example: talin, kindlin, paxillin, and focal adhesion kinase (FAK)) are recruited to sites of activation forming a complex linked to the actin cytoskeleton. Binding and regulation of these molecules through various dephosphorylation/phosphorylation events, including the phosphorylation of FAK, leads to downstream signaling involving ERK 1/2, MEK1/2, PLCγ, PI3kinase, and others, which modulates cell survival, motility, and in the case of neurons, neurite outgrowth and axon regeneration. With respect to the CNS, reactive astrocytes surrounding a spinal cord lesion secrete the ECM component tenascin-C, however, neurons/axons do not express the paired integrin receptor α9β1, thus leading to regenerative failure. (For color version of this figure, the reader is referred to the online version of this chapter.)

although studies using dominant negative β1 integrin in oligodendrocytes indicate a significant delay in myelination (Camara et al., 2009).

6.7 Growth and NT factors

NT factors have various roles in development; in the adult CNS many neurones continue to express NT receptors, and the NTs can interact with these to promote axon regrowth (reviewed by Bo et al., 2011; Hendriks et al., 2004; Lu and Tuszynski, 2008). Treatment with NT factors post-SCI has led to promising results with respect to anatomical regeneration/sprouting and functional recovery. For example, treatment with BDNF has led to enhancement of rubrospinal, reticulospinal, vestibulospinal, and raphé-spinal regrowth along with recovery of forelimb function after lateral hemisection lesion (Jin et al., 2002; Liu et al., 1999). NGF (nerve growth factor) treatment in SCI has promoted regrowth of damaged TH-positive (tyrosine hydroxylase, coerulo-spinal) fibers (Grill et al., 1997), while treatment with NT-3 (neurotrophin-3) has promoted regrowth of CST fibers (Grill et al., 1997). Additionally, GDNF (glial-derived neurotrophic factor) administration post-SCI induces remyelination along with the increased growth of both motor and sensory axons (Blesch and Tuszynski, 2003). Due to extensive research centering on NT factors, the field has expanded to include several different means of NT administration. For example these include: *ex vivo* therapy, where cells are transplanted that are transfected/transduced with the NT factor, either by plasmid or with viral vectors; *in vivo* gene therapy, where a viral vector is injected into the targeted cells to drive their own expression of the NT factor; or injection or infusion of the NT factor directly to the lesion site. Many researchers are using NT factors in combination with other promising approaches for repair including cellular therapies (Ruitenberg et al., 2005; Taylor et al., 2006), chondroitinase (Massey et al., 2008), and anti-Nogo antibody (Schnell et al., 2011).

7 CONCLUSIONS

Long-distance regeneration is not presently achievable in mammalian CNS due to a myriad of extrinsic and intrinsic factors limiting regrowth after injury. Several attempts have been made to enhance regeneration with the most promising therapies stemming from NogoA-blocking treatments, chondroitinase, Cethrin, NTs, and taxol studies.

It is clear that both intrinsic and extrinsic factors play essential roles in the inhibition of axonal regrowth in the lesioned mammalian CNS. A multitude of ways to manipulate these factors has produced promising effects on sprouting of intact spared fibers, on regeneration of transected axons, and on functional recovery. Even with this success, functional restoration is partial, and seldom approaches the normal behavior mediated by the undamaged spinal cord. This is because regeneration of long-tract axons, even when intrinsically manipulated into a regenerative growth state and forced through a glial scar, is still incomplete in both the number and distance, and reconnection with the appropriate motor units is imprecise. Nevertheless, the various treatments that have been described will form the basis of the first generation of spinal repair therapies that will become available to patients. The current challenge is to work out how we can manage to achieve a level of functional connectivity that mimics the undamaged CNS. It seems probable from current work that this will involve combinatorial therapies.

Acknowledgments

This work was supported by the Christopher and Dana Reeve Foundation, the EPSRC, the Plasticise European Research Consortium (Programme Framework 7), and the International Foundation for Research in Paraplegia (M. R. A.).

References

Alilain, W.J., Horn, K.P., Hu, H., Dick, T.E., Silver, J., 2011. Functional regeneration of respiratory pathways after spinal cord injury. Nature 475, 196–200.

Allt, G., 1975. The node of Ranvier in experimental allergic neuritis: an electron microscope study. J. Neurocytol. 4, 63–76.

Andrews, M.R., Stelzner, D.J., 2004. Modification of the regenerative response of dorsal column axons by olfactory ensheathing cells or peripheral axotomy in adult rat. Exp. Neurol. 190, 311–327.

Andrews, M.R., Czvitkovich, S., Dassie, E., Vogelaar, C.F., Faissner, A., Blits, B., Gage, F.H., ffrench-Constant, C., Fawcett, J.W., 2009. Alpha9 integrin promotes neurite outgrowth on tenascin-C and enhances sensory axon regeneration. J. Neurosci. 29, 5546–5557.

Ankerhold, R., Stuermer, C.A., 1999. Fate of oligodendrocytes during retinal axon degeneration and regeneration in the goldfish visual pathway. J. Neurobiol. 41, 572–584.

Baastrup, C., Finnerup, N.B., 2008. Pharmacological management of neuropathic pain following spinal cord injury. CNS Drugs 22, 455–475.

Bao, F., Dekaban, G.A., Weaver, L.C., 2005. Anti-CD11d antibody treatment reduces free radical formation and cell death in the injured spinal cord of rats. J. Neurochem. 94, 1361–1373.

Baptiste, D.C., Fehlings, M.G., 2006. Pharmacological approaches to repair the injured spinal cord. J. Neurotrauma 23, 318–334.

Bareyre, F.M., Schwab, M.E., 2003. Inflammation, degeneration and regeneration in the injured spinal cord: insights from DNA microarrays. Trends Neurosci. 26, 555–563.

Barritt, A.W., Davies, M., Marchand, F., Hartley, R., Grist, J., Yip, P., McMahon, S.B., Bradbury, E.J., 2006. Chondroitinase ABC promotes sprouting of intact and injured spinal systems after spinal cord injury. J. Neurosci. 26, 10856–10867.

Barros, C.S., Nguyen, T., Spencer, K.S., Nishiyama, A., Colognato, H., Muller, U., 2009. Beta1 integrins are required for normal CNS myelination and promote AKT-dependent myelin outgrowth. Development 136, 2717–2724.

Bastmeyer, M., Beckmann, M., Schwab, M.E., Stuermer, C.A., 1991. Growth of regenerating goldfish axons is inhibited by rat oligodendrocytes and CNS myelin but not by goldfish optic nerve tract oligodendrocytelike cells and fish CNS myelin. J. Neurosci. 11, 626–640.

Baumgart, E.V., Barbosa, J.S., Bally-Cuif, L., Gotz, M., Ninkovic, J., 2012. Stab wound injury of the zebrafish telencephalon: a model for comparative analysis of reactive gliosis. Glia 60, 343–357.

Beattie, M.S., 2004. Inflammation and apoptosis: linked therapeutic targets in spinal cord injury. Trends Mol. Med. 10, 580–583.

Benninger, Y., Colognato, H., Thurnherr, T., Franklin, R.J., Leone, D.P., Atanasoski, S., Nave, K.A., ffrench-Constant, C., Suter, U., Relvas, J.B., 2006. Beta1-integrin signaling mediates premyelinating oligodendrocyte survival but is not required for CNS myelination and remyelination. J. Neurosci. 26, 7665–7673.

Benson, M.D., Romero, M.I., Lush, M.E., Lu, Q.R., Henkemeyer, M., Parada, L.F., 2005. Ephrin-B3 is a myelin-based inhibitor of neurite outgrowth. Proc. Natl. Acad. Sci. U.S.A. 102, 10694–10699.

Benton, R.L., Maddie, M.A., Worth, C.A., Mahoney, E.T., Hagg, T., Whittemore, S.R., 2008. Transcriptomic screening of microvascular endothelial cells implicates novel molecular regulators of vascular dysfunction after spinal cord injury. J. Cereb. Blood Flow Metab. 28, 1771–1785.

Bisby, M.A., Tetzlaff, W., 1992. Changes in cytoskeletal protein synthesis following axon injury and during axon regeneration. Mol. Neurobiol. 6, 107–123.

Bjartmar, C., Kinkel, R.P., Kidd, G., Rudick, R.A., Trapp, B.D., 2001. Axonal loss in normal-appearing white matter in a patient with acute MS. Neurology 57, 1248–1252.

Black, J.A., Waxman, S.G., Sims, T.J., Gilmore, S.A., 1986. Effects of delayed myelination by oligodendrocytes and Schwann cells on the macromolecular structure of axonal membrane in rat spinal cord. J. Neurocytol. 15, 745–761.

Blakemore, W.F., Crang, A.J., 1985. The use of cultured autologous Schwann cells to remyelinate areas of persistent demyelination in the central nervous system. J. Neurol. Sci. 70, 207–223.

Blesch, A., Tuszynski, M.H., 2003. Cellular GDNF delivery promotes growth of motor and dorsal column sensory axons after partial and complete spinal cord transections and induces remyelination. J. Comp. Neurol. 467, 403–417.

Blesch, A., Lu, P., Tsukada, S., Alto, L.T., Roet, K., Coppola, G., Geschwind, D., Tuszynski, M.H., 2012. Conditioning lesions before or after spinal cord injury recruit broad genetic mechanisms that sustain axonal regeneration: superiority to camp-mediated effects. Exp. Neurol. 235, 162–173.

Blight, A.R., 1994. Effects of silica on the outcome from experimental spinal cord injury: implication of macrophages in secondary tissue damage. Neuroscience 60, 263–273.

Bo, X., Wu, D., Yeh, J., Zhang, Y., 2011. Gene therapy approaches for neuroprotection and axonal regeneration after spinal cord and spinal root injury. Curr. Gene Ther. 11, 101–115.

Bovolenta, P., Wandosell, F., Nieto-Sampedro, M., 1993. Characterization of a neurite outgrowth inhibitor expressed after CNS injury. Eur. J. Neurosci. 5, 454–465.

Bradbury, E.J., Carter, L.M., 2011. Manipulating the glial scar: chondroitinase ABC as a therapy for spinal cord injury. Brain Res. Bull. 84, 306–316.

Bradbury, E.J., Moon, L.D., Popat, R.J., King, V.R., Bennett, G.S., Patel, P.N., Fawcett, J.W., McMahon, S.B., 2002. Chondroitinase ABC promotes functional recovery after spinal cord injury. Nature 416, 636–640.

Bradke, F., Fawcett, J.W., Spira, M.E., 2012. Assembly of a new growth cone after axotomy: the precursor to axon regeneration. Nat. Rev. Neurosci. 13, 183–193.

Brosamle, C., Huber, A.B., Fiedler, M., Skerra, A., Schwab, M.E., 2000. Regeneration of lesioned corticospinal tract fibers in the adult rat induced by a recombinant, humanized IN-1 antibody fragment. J. Neurosci. 20, 8061–8068.

Bundesen, L.Q., Scheel, T.A., Bregman, B.S., Kromer, L.F., 2003. Ephrin-B2 and EphB2 regulation of astrocyte-meningeal fibroblast interactions in response to spinal cord lesions in adult rats. J. Neurosci. 23, 7789–7800.

Busch, S.A., Horn, K.P., Silver, D.J., Silver, J., 2009. Overcoming macrophage-mediated axonal dieback following CNS injury. J. Neurosci. 29, 9967–9976.

Buss, A., Brook, G.A., Kakulas, B., Martin, D., Franzen, R., Schoenen, J., Noth, J., Schmitt, A.B., 2004. Gradual loss of myelin and formation of an astrocytic scar during Wallerian degeneration in the human spinal cord. Brain 127, 34–44.

Buss, A., Pech, K., Merkler, D., Kakulas, B.A., Martin, D., Schoenen, J., Noth, J., Schwab, M.E., Brook, G.A., 2005. Sequential loss of myelin proteins during Wallerian degeneration in the human spinal cord. Brain 128, 356–364.

Cafferty, W.B., Gardiner, N.J., Das, P., Qiu, J., McMahon, S.B., Thompson, S.W., 2004. Conditioning injury-induced spinal axon regeneration fails in interleukin-6 knock-out mice. J. Neurosci. 24, 4432–4443.

Cafferty, W.B., Duffy, P., Huebner, E., Strittmatter, S.M., 2010. MAG and OMgp synergize with Nogo-A to restrict axonal growth and neurological recovery after spinal cord trauma. J. Neurosci. 30, 6825–6837.

Cahill, A., Fredine, H., Zilberman, L., Ibanez, B., Fox, M., Winneberger, D., 2009. One degree of separation: paralysis and spinal cord injury in the United States. Christopher and Dana Reeve Paralysis Resource Center, pp. 1–28.

Calancie, B., Molano, M.R., Broton, J.G., 2002. Interlimb reflexes and synaptic plasticity become evident months after human spinal cord injury. Brain 125, 1150–1161.

Camara, J., Wang, Z., Nunes-Fonseca, C., Friedman, H.C., Grove, M., Sherman, D.L., Komiyama, N.H., Grant, S.G., Brophy, P.J., Peterson, A., ffrench-Constant, C., 2009. Integrin-mediated axoglial interactions initiate myelination in the central nervous system. J. Cell Biol. 185, 699–712.

Cao, Z., Gao, Y., Bryson, J.B., Hou, J., Chaudhry, N., Siddiq, M., Martinez, J., Spencer, T., Carmel, J., Hart, R.B., Filbin, M.T., 2006. The cytokine interleukin-6 is sufficient but not necessary to mimic the peripheral conditioning lesion effect on axonal growth. J. Neurosci. 26, 5565–5573.

Carlson, S.L., Parrish, M.E., Springer, J.E., Doty, K., Dossett, L., 1998. Acute inflammatory response in spinal cord following impact injury. Exp. Neurol. 151, 77–88.

Caroni, P., Schwab, M.E., 1988a. Antibody against myelin-associated inhibitor of neurite growth neutralizes nonpermissive substrate properties of CNS white matter. Neuron 1, 85–96.

Caroni, P., Schwab, M.E., 1988b. Two membrane protein fractions from rat central myelin with inhibitory properties for neurite growth and fibroblast spreading. J. Cell Biol. 106, 1281–1288.

Casha, S., Yu, W.R., Fehlings, M.G., 2001. Oligodendroglial apoptosis occurs along degenerating axons and is associated with FAS and p75 expression following spinal cord injury in the rat. Neuroscience 103, 203–218.

Castellani, V., Falk, J., Rougon, G., 2004. Semaphorin3A-induced receptor endocytosis during axon guidance responses is mediated by L1 CAM. Mol. Cell. Neurosci. 26, 89–100.

Chen, L., Chisholm, A.D., 2011. Axon regeneration mechanisms: insights from C. elegans. Trends Cell Biol. 21, 577–584.

Chen, M.S., Huber, A.B., van der Haar, M.E., Frank, M., Schnell, L., Spillmann, A.A., Christ, F., Schwab, M.E., 2000. Nogo-A is a myelin-associated neurite outgrowth inhibitor and an antigen for monoclonal antibody IN-1. Nature 403, 434–439.

Chesler, M., Young, W., Hassan, A.Z., Sakatani, K., Moriya, T., 1994. Elevation and clearance of extracellular K^+ following graded contusion of the rat spinal cord. Exp. Neurol. 125, 93–98.

Colognato, H., ffrench-Constant, C., Feltri, M.L., 2005. Human diseases reveal novel roles for neural laminins. Trends Neurosci. 28, 480–486.

Condic, M.L., 2001. Adult neuronal regeneration induced by transgenic integrin expression. J. Neurosci. 21, 4782–4788.

Condic, M.L., Snow, D.M., Letourneau, P.C., 1999. Embryonic neurons adapt to the inhibitory proteoglycan aggrecan by increasing integrin expression. J. Neurosci. 19, 10036–10043.

Coull, J.A., Beggs, S., Boudreau, D., Boivin, D., Tsuda, M., Inoue, K., Gravel, C., Salter, M.W., De Koninck, Y., 2005. BDNF from microglia causes the shift in neuronal anion gradient underlying neuropathic pain. Nature 438, 1017–1021.

Cowan, C.W., Shao, Y.R., Sahin, M., Shamah, S.M., Lin, M.Z., Greer, P.L., Gao, S., Griffith, E.C., Brugge, J.S., Greenberg, M.E., 2005. Vav family GEFs link activated Ephs to endocytosis and axon guidance. Neuron 46, 205–217.

Crespo, D., Asher, R.A., Lin, R., Rhodes, K.E., Fawcett, J.W., 2007. How does chondroitinase promote functional recovery in the damaged CNS? Exp. Neurol. 206, 159–171.

Crowe, M.J., Bresnahan, J.C., Shuman, S.L., Masters, J.N., Beattie, M.S., 1997. Apoptosis and delayed degeneration after spinal cord injury in rats and monkeys. Nat. Med. 3, 73–76.

Curinga, G.M., Snow, D.M., Mashburn, C., Kohler, K., Thobaben, R., Caggiano, A.O., Smith, G.M., 2007. Mammalian-produced chondroitinase AC mitigates axon inhibition by chondroitin sulfate proteoglycans. J. Neurochem. 102, 275–288.

Davalos, D., Grutzendler, J., Yang, G., Kim, J.V., Zuo, Y., Jung, S., Littman, D.R., Dustin, M.L., Gan, W.B., 2005. ATP mediates rapid microglial response to local brain injury *in vivo*. Nat. Neurosci. 8, 752–758.

David, S., Kroner, A., 2011. Repertoire of microglial and macrophage responses after spinal cord injury. Nat. Rev. Neurosci. 12, 388–399.

de Groat, W.C., Yoshimura, N., 2012. Plasticity in reflex pathways to the lower urinary tract following spinal cord injury. Exp. Neurol. 235, 123–132.

Dergham, P., Ellezam, B., Essagian, C., Avedissian, H., Lubell, W.D., McKerracher, L., 2002. Rho signaling pathway targeted to promote spinal cord repair. J. Neurosci. 22, 6570–6577.

Derry, W.B., Wilson, L., Jordan, M.A., 1995. Substoichiometric binding of taxol suppresses microtubule dynamics. Biochemistry (Mosc.) 34, 2203–2211.

De Winter, F., Oudega, M., Lankhorst, A.J., Hamers, F.P., Blits, B., Ruitenberg, M.J., Pasterkamp, R.J., Gispen, W.H., Verhaagen, J., 2002. Injury-induced class 3 semaphorin expression in the rat spinal cord. Exp. Neurol. 175, 61–75.

Dickson, B.J., 2001. Rho GTPases in growth cone guidance. Curr. Opin. Neurobiol. 11, 103–110.

Di Cristofano, A., Pesce, B., Cordon-Cardo, C., Pandolfi, P.P., 1998. Pten is essential for embryonic development and tumour suppression. Nat. Genet. 19, 348–355.

Diekmann, H., Klinger, M., Oertle, T., Heinz, D., Pogoda, H.M., Schwab, M.E., Stuermer, C.A., 2005. Analysis of the reticulon gene family demonstrates the absence of the neurite growth inhibitor Nogo-A in fish. Mol. Biol. Evol. 22, 1635–1648.

Dietz, V., 2010. Behavior of spinal neurons deprived of supraspinal input. Nat. Rev. Neurol. 6, 167–174.

Dietz, V., 2012. Neuronal plasticity after a human spinal cord injury: positive and negative effects. Exp. Neurol. 235, 1406–1412.

Dietz, V., Curt, A., 2006. Neurological aspects of spinal-cord repair: promises and challenges. Lancet Neurol. 5, 688–694.

Ditunno Jr., J.F., Formal, C.S., 1994. Chronic spinal cord injury. N. Engl. J. Med. 330, 550–556.

Donnelly, D.J., Longbrake, E.E., Shawler, T.M., Kigerl, K.A., Lai, W., Tovar, C.A., Ransohoff, R.M., Popovich, P.G., 2011. Deficient CX3CR1 signaling promotes recovery after mouse spinal cord injury by limiting the recruitment and activation of Ly6Clo/iNOS+ macrophages. J. Neurosci. 31, 9910–9922.

Dubreuil, C.I., Winton, M.J., McKerracher, L., 2003. Rho activation patterns after spinal cord injury and the role of activated Rho in apoptosis in the central nervous system. J. Cell Biol. 162, 233–243.

Duffy, P., Schmandke, A., Sigworth, J., Narumiya, S., Cafferty, W.B., Strittmatter, S.M., 2009. Rho-associated kinase II (ROCKII) limits axonal growth after trauma within the adult mouse spinal cord. J. Neurosci. 29, 15266–15276.

Eickholt, B.J., Walsh, F.S., Doherty, P., 2002. An inactive pool of GSK-3 at the leading edge of growth cones is implicated in Semaphorin 3A signaling. J. Cell Biol. 157, 211–217.

Elbasiouny, S.M., Moroz, D., Bakr, M.M., Mushahwar, V.K., 2010. Management of spasticity after spinal cord injury: current techniques and future directions. Neurorehabil. Neural Repair 24, 23–33.

Erturk, A., Hellal, F., Enes, J., Bradke, F., 2007. Disorganized microtubules underlie the formation of retraction bulbs and the failure of axonal regeneration. J. Neurosci. 27, 9169–9180.

Fabes, J., Anderson, P., Yanez-Munoz, R.J., Thrasher, A., Brennan, C., Bolsover, S., 2006. Accumulation of the inhibitory receptor EphA4 may prevent regeneration of corticospinal tract axons following lesion. Eur. J. Neurosci. 23, 1721–1730.

Faulkner, J.R., Herrmann, J.E., Woo, M.J., Tansey, K.E., Doan, N.B., Sofroniew, M.V., 2004. Reactive astrocytes protect tissue and preserve function after spinal cord injury. J. Neurosci. 24, 2143–2155.

Fawcett, J.W., Asher, R.A., 1999. The glial scar and central nervous system repair. Brain Res. Bull. 49, 377–391.

Fehlings, M.G., Tator, C.H., 1999. An evidence-based review of decompressive surgery in acute spinal cord injury: rationale, indications, and timing based on experimental and clinical studies. J. Neurosurg. 91, 1–11.

Fehlings, M.G., Theodore, N., Harrop, J., Maurais, G., Kuntz, C., Shaffrey, C.I., Kwon, B.K., Chapman, J., Yee, A., Tighe, A., McKerracher, L., 2011. A phase I/IIa clinical trial of a recombinant Rho protein antagonist in acute spinal cord injury. J. Neurotrauma 28, 787–796.

Ferretti, P., Zhang, F., O'Neill, P., 2003. Changes in spinal cord regenerative ability through phylogenesis and development: lessons to be learnt. Dev. Dyn. 226, 245–256.

Filbin, M.T., 2003. Myelin-associated inhibitors of axonal regeneration in the adult mammalian CNS. Nat. Rev. Neurosci. 4, 703–713.

Fimbel, S.M., Montgomery, J.E., Burket, C.T., Hyde, D.R., 2007. Regeneration of inner retinal neurons after intravitreal injection of ouabain in zebrafish. J. Neurosci. 27, 1712–1724.

Fleming, J.C., Bao, F., Chen, Y., Hamilton, E.F., Relton, J.K., Weaver, L.C., 2008. Alpha4-beta1 integrin blockade after spinal cord injury decreases damage and improves neurological function. Exp. Neurol. 214, 147–159.

Fleming, J.C., Bao, F., Chen, Y., Hamilton, E.F., Gonzalez-Lara, L.E., Foster, P.J., Weaver, L.C., 2009. Timing and duration of anti-alpha4beta1 integrin treatment after spinal cord injury: effect on therapeutic efficacy. J. Neurosurg. Spine 11, 575–587.

Fouad, K., Klusman, I., Schwab, M.E., 2004. Regenerating corticospinal fibers in the Marmoset (Callitrix jacchus) after spinal cord lesion and treatment with the anti-Nogo-A antibody IN-1. Eur. J. Neurosci. 20, 2479–2482.

Fournier, A.E., GrandPre, T., Strittmatter, S.M., 2001. Identification of a receptor mediating Nogo-66 inhibition of axonal regeneration. Nature 409, 341–346.

Fournier, A.E., Takizawa, B.T., Strittmatter, S.M., 2003. Rho kinase inhibition enhances axonal regeneration in the injured CNS. J. Neurosci. 23, 1416–1423.

Frankel, H.L., Coll, J.R., Charlifue, S.W., Whiteneck, G.G., Gardner, B.P., Jamous, M.A., Krishnan, K.R., Nuseibeh, I., Savic, G., Sett, P., 1998. Long-term survival in spinal cord injury: a fifty year investigation. Spinal Cord 36, 266–274.

Freund, P., Schmidlin, E., Wannier, T., Bloch, J., Mir, A., Schwab, M.E., Rouiller, E.M., 2006. Nogo-A-specific antibody treatment enhances sprouting and functional recovery after cervical lesion in adult primates. Nat. Med. 12, 790–792.

Freund, P., Wannier, T., Schmidlin, E., Bloch, J., Mir, A., Schwab, M.E., Rouiller, E.M., 2007. Anti-Nogo-A antibody treatment enhances sprouting of corticospinal axons rostral to a unilateral cervical spinal cord lesion in adult macaque monkey. J. Comp. Neurol. 502, 644–659.

Fry, E.J., Chagnon, M.J., Lopez-Vales, R., Tremblay, M.L., David, S., 2010. Corticospinal tract regeneration after spinal cord injury in receptor protein tyrosine phosphatase sigma deficient mice. Glia 58, 423–433.

Furlan, J.C., Noonan, V., Cadotte, D.W., Fehlings, M.G., 2011. Timing of decompressive surgery of spinal cord after traumatic spinal cord injury: an evidence-based examination of pre-clinical and clinical studies. J. Neurotrauma 28, 1371–1399.

Galtrey, C.M., Fawcett, J.W., 2007. The role of chondroitin sulfate proteoglycans in regeneration and plasticity in the central nervous system. Brain Res. Rev. 54, 1–18.

Garcia-Alias, G., Lin, R., Akrimi, S.F., Story, D., Bradbury, E.J., Fawcett, J.W., 2008. Therapeutic time window for the application of chondroitinase ABC after spinal cord injury. Exp. Neurol. 210, 331–338.

Garcia-Alias, G., Barkhuysen, S., Buckle, M., Fawcett, J.W., 2009. Chondroitinase ABC treatment opens a window of opportunity for task-specific rehabilitation. Nat. Neurosci. 12, 1145–1151.

Gavazzi, I., Stonehouse, J., Sandvig, A., Reza, J.N., Appiah-Kubi, L.S., Keynes, R., Cohen, J., 2000. Peripheral, but not central, axotomy induces neuropilin-1 mRNA expression in adult large diameter primary sensory neurons. J. Comp. Neurol. 423, 492–499.

Gensel, J.C., Nakamura, S., Guan, Z., van Rooijen, N., Ankeny, D.P., Popovich, P.G., 2009. Macrophages promote axon regeneration with concurrent neurotoxicity. J. Neurosci. 29, 3956–3968.

George, R., Griffin, J.W., 1994. Delayed macrophage responses and myelin clearance during Wallerian degeneration in the central nervous system: the dorsal radiculotomy model. Exp. Neurol. 129, 225–236.

George, E.R., Scholten, D.J., Buechler, C.M., Jordan-Tibbs, J., Mattice, C., Albrecht, R.M., 1995. Failure of methylprednisolone to improve the outcome of spinal cord injuries. Am. Surg. 61, 659–663 (discussion 663–654).

Gerhart, K.A., Johnson, R.L., Menconi, J., Hoffman, R.E., Lammertse, D.P., 1995. Utilization and effectiveness of methylprednisolone in a population-based sample of spinal cord injured persons. Paraplegia 33, 316–321.

Ghosh, A., Haiss, F., Sydekum, E., Schneider, R., Gullo, M., Wyss, M.T., Mueggler, T., Baltes, C., Rudin, M., Weber, B., Schwab, M.E., 2010. Rewiring of hindlimb corticospinal neurons after spinal cord injury. Nat. Neurosci. 13, 97–104.

Ghosh-Roy, A., Wu, Z., Goncharov, A., Jin, Y., Chisholm, A.D., 2010. Calcium and cyclic AMP promote axonal regeneration in Caenorhabditis elegans and require DLK-1 kinase. J. Neurosci. 30, 3175–3183.

Giger, R.J., Urquhart, E.R., Gillespie, S.K., Levengood, D.V., Ginty, D.D., Kolodkin, A.L., 1998. Neuropilin-2 is a receptor for semaphorin IV: insight into the structural basis of receptor function and specificity. Neuron 21, 1079–1092.

Goldshmit, Y., Galea, M.P., Wise, G., Bartlett, P.F., Turnley, A.M., 2004. Axonal regeneration and lack of astrocytic gliosis in EphA4-deficient mice. J. Neurosci. 24, 10064–10073.

Goldshmit, Y., McLenachan, S., Turnley, A., 2006. Roles of Eph receptors and ephrins in the normal and damaged adult CNS. Brain Res. Rev. 52, 327–345.

Gonzenbach, R.R., Zoerner, B., Schnell, L., Weinmann, O., Mir, A.K., Schwab, M.E., 2012. Delayed anti-nogo-a antibody application after spinal cord injury shows progressive loss of responsiveness. J. Neurotrauma 29, 567–578.

Goritz, C., Dias, D.O., Tomilin, N., Barbacid, M., Shupliakov, O., Frisen, J., 2011. A pericyte origin of spinal cord scar tissue. Science 333, 238–242.

GrandPre, T., Li, S., Strittmatter, S.M., 2002. Nogo-66 receptor antagonist peptide promotes axonal regeneration. Nature 417, 547–551.

Grayson, M.H., Van der Vieren, M., Sterbinsky, S.A., Michael Gallatin, W., Hoffman, P.A., Staunton, D.E., Bochner, B.S., 1998. alphadbeta2 integrin is expressed on human eosinophils and functions as an alternative ligand for vascular cell adhesion molecule 1 (VCAM-1). J. Exp. Med. 188, 2187–2191.

Grill, R., Murai, K., Blesch, A., Gage, F.H., Tuszynski, M.H., 1997. Cellular delivery of neurotrophin-3 promotes corticospinal axonal growth and partial functional recovery after spinal cord injury. J. Neurosci. 17, 5560–5572.

Gris, D., Marsh, D.R., Oatway, M.A., Chen, Y., Hamilton, E.F., Dekaban, G.A., Weaver, L.C., 2004. Transient blockade of the CD11d/CD18 integrin reduces secondary damage after spinal cord injury, improving sensory, autonomic, and motor function. J. Neurosci. 24, 4043–4051.

Grossman, S.D., Rosenberg, L.J., Wrathall, J.R., 2001. Temporal-spatial pattern of acute neuronal and glial loss after spinal cord contusion. Exp. Neurol. 168, 273–282.

Guest, J.D., Hiester, E.D., Bunge, R.P., 2005. Demyelination and Schwann cell responses adjacent to injury epicenter cavities following chronic human spinal cord injury. Exp. Neurol. 192, 384–393.

Guth, L., Zhang, Z., Steward, O., 1999. The unique histopathological responses of the injured spinal cord. Implications for neuroprotective therapy. Ann. N. Y. Acad. Sci. 890, 366–384.

Hagino, S., Iseki, K., Mori, T., Zhang, Y., Hikake, T., Yokoya, S., Takeuchi, M., Hasimoto, H., Kikuchi, S., Wanaka, A., 2003. Slit and glypican-1 mRNAs are coexpressed in the reactive astrocytes of the injured adult brain. Glia 42, 130–138.

Hains, B.C., Waxman, S.G., 2006. Activated microglia contribute to the maintenance of chronic pain after spinal cord injury. J. Neurosci. 26, 4308–4317.

Hammarlund, M., Nix, P., Hauth, L., Jorgensen, E.M., Bastiani, M., 2009. Axon regeneration requires a conserved MAP kinase pathway. Science 323, 802–806.

Han, P., Whelan, P.J., 2010. Tumor necrosis factor alpha enhances glutamatergic transmission onto spinal motor neurons. J. Neurotrauma 27, 287–292.

Hausmann, R., Betz, P., 2000. The time course of the vascular response to human brain injury—an immunohistochemical study. Int. J. Leg. Med. 113, 288–292.

Hellal, F., Hurtado, A., Ruschel, J., Flynn, K.C., Laskowski, C.J., Umlauf, M., Kapitein, L.C., Strikis, D., Lemmon, V., Bixby, J., Hoogenraad, C.C., Bradke, F., 2011. Microtubule stabilization reduces scarring and causes axon regeneration after spinal cord injury. Science 331, 928–931.

Hendriks, W.T., Ruitenberg, M.J., Blits, B., Boer, G.J., Verhaagen, J., 2004. Viral vector-mediated gene transfer of neurotrophins to promote regeneration of the injured spinal cord. Prog. Brain Res. 146, 451–476.

Hermanns, S., Klapka, N., Muller, H.W., 2001. The collagenous lesion scar—an obstacle for axonal regeneration in brain and spinal cord injury. Restor. Neurol. Neurosci. 19, 139–148.

Hiersemenzel, L.P., Curt, A., Dietz, V., 2000. From spinal shock to spasticity: neuronal adaptations to a spinal cord injury. Neurology 54, 1574–1582.

Horky, L.L., Galimi, F., Gage, F.H., Horner, P.J., 2006. Fate of endogenous stem/progenitor cells following spinal cord injury. J. Comp. Neurol. 498, 525–538.

Horn, K.P., Busch, S.A., Hawthorne, A.L., van Rooijen, N., Silver, J., 2008. Another barrier to regeneration in the CNS: activated macrophages induce extensive retraction of dystrophic axons through direct physical interactions. J. Neurosci. 28, 9330–9341.

Houle, J.D., Jin, Y., 2001. Chronically injured supraspinal neurons exhibit only modest axonal dieback in response to a cervical hemisection lesion. Exp. Neurol. 169, 208–217.

Hsu, J.Y., Stein, S.A., Xu, X.M., 2005. Temporal and spatial distribution of growth-associated molecules and astroglial cells in the rat corticospinal tract during development. J. Neurosci. Res. 80, 330–340.

Hu, P., McLachlan, E.M., 2000. Distinct sprouting responses of sympathetic and peptidergic sensory axons proximal to a sciatic nerve transection in guinea pigs and rats. Neurosci. Lett. 295, 59–63.

Hu, F., Strittmatter, S.M., 2008. The N-terminal domain of Nogo-A inhibits cell adhesion and axonal outgrowth by an integrin-specific mechanism. J. Neurosci. 28, 1262–1269.

Huang, W.L., George, K.J., Ibba, V., Liu, M.C., Averill, S., Quartu, M., Hamlyn, P.J., Priestley, J.V., 2007. The characteristics of neuronal injury in a static compression model of spinal cord injury in adult rats. Eur. J. Neurosci. 25, 362–372.

Hunanyan, A.S., Garcia-Alias, G., Alessi, V., Levine, J.M., Fawcett, J.W., Mendell, L.M., Arvanian, V.L., 2010. Role of chondroitin sulfate proteoglycans in axonal conduction in Mammalian spinal cord. J. Neurosci. 30, 7761–7769.

Hurlbert, R.J., 2000. Methylprednisolone for acute spinal cord injury: an inappropriate standard of care. J. Neurosurg. 93, 1–7.

Hurtado, A., Marcillo, A., Frydel, B., Bunge, M.B., Bramlett, H.M., Dietrich, W.D., 2012. Anti-CD11d monoclonal antibody treatment for rat spinal cord compression injury. Exp. Neurol. 233, 606–611.

Iannotti, C., Zhang, Y.P., Shields, L.B., Han, Y., Burke, D.A., Xu, X.M., Shields, C.B., 2006. Dural repair reduces connective tissue scar invasion and cystic cavity formation after acute spinal cord laceration injury in adult rats. J. Neurotrauma 23, 853–865.

Imperato-Kalmar, E.L., McKinney, R.A., Schnell, L., Rubin, B.P., Schwab, M.E., 1997. Local changes in vascular architecture following partial spinal cord lesion in the rat. Exp. Neurol. 145, 322–328.

Inman, D.M., Steward, O., 2003. Physical size does not determine the unique histopathological response seen in the injured mouse spinal cord. J. Neurotrauma 20, 33–42.

Jaeger, C.B., Blight, A.R., 1997. Spinal cord compression injury in guinea pigs: structural changes of endothelium and its perivascular cell associations after blood-brain barrier breakdown and repair. Exp. Neurol. 144, 381–399.

Jin, Y., Fischer, I., Tessler, A., Houle, J.D., 2002. Transplants of fibroblasts genetically modified to express BDNF promote axonal regeneration from supraspinal neurons following chronic spinal cord injury. Exp. Neurol. 177, 265–275.

Jin, Y., Ketschek, A., Jiang, Z., Smith, G., Fischer, I., 2011. Chondroitinase activity can be transduced by a lentiviral vector *in vitro* and *in vivo*. J. Neurosci. Methods 199, 208–213.

Johnson-Green, P.C., Dow, K.E., Riopelle, R.J., 1991. Characterization of glycosaminoglycans produced by primary astrocytes *in vitro*. Glia 4, 314–321.

Kajana, S., Goshgarian, H.G., 2008. Spinal activation of the cAMP-PKA pathway induces respiratory motor recovery following high cervical spinal cord injury. Brain Res. 1232, 206–213.

Kalous, A., Keast, J.R., 2010. Conditioning lesions enhance growth state only in sensory neurons lacking calcitonin gene-related peptide and isolectin B4-binding. Neuroscience 166, 107–121.

Kaushal, V., Koeberle, P.D., Wang, Y., Schlichter, L.C., 2007. The Ca^{2+}-activated K^+ channel KCNN4/KCa3.1 contributes to microglia activation and nitric oxide-dependent neurodegeneration. J. Neurosci. 27, 234–244.

Kigerl, K.A., Gensel, J.C., Ankeny, D.P., Alexander, J.K., Donnelly, D.J., Popovich, P.G., 2009. Identification of two distinct macrophage subsets with divergent effects causing either neurotoxicity or regeneration in the injured mouse spinal cord. J. Neurosci. 29, 13435–13444.

Kim, J.E., Li, S., GrandPre, T., Qiu, D., Strittmatter, S.M., 2003. Axon regeneration in young adult mice lacking Nogo-A/B. Neuron 38, 187–199.

Klesse, L.J., Parada, L.F., 1998. p21 ras and phosphatidylinositol-3 kinase are required for survival of wild-type and NF1 mutant sensory neurons. J. Neurosci. 18, 10420–10428.

Kopp, M.A., Brommer, B., Gatzemeier, N., Schwab, J.M., Pruss, H., 2010. Spinal cord injury induces differential expression of the profibrotic semaphorin 7A in the developing and mature glial scar. Glia 58, 1748–1756.

Kwok, J.C., Afshari, F., Garcia-Alias, G., Fawcett, J.W., 2008. Proteoglycans in the central nervous system: plasticity, regeneration and their stimulation with chondroitinase ABC. Restor. Neurol. Neurosci. 26, 131–145.

Kwon, C.H., Luikart, B.W., Powell, C.M., Zhou, J., Matheny, S.A., Zhang, W., Li, Y., Baker, S.J., Parada, L.F., 2006. Pten regulates neuronal arborization and social interaction in mice. Neuron 50, 377–388.

Lang, D.M., Rubin, B.P., Schwab, M.E., Stuermer, C.A., 1995. CNS myelin and oligodendrocytes of the Xenopus spinal cord—but not optic nerve—are nonpermissive for axon growth. J. Neurosci. 15, 99–109.

Lee, J.J., Swain, S.M., 2006. Peripheral neuropathy induced by microtubule-stabilizing agents. J. Clin. Oncol. 24, 1633–1642.

Lee, J.K., Chan, A.F., Luu, S.M., Zhu, Y., Ho, C., Tessier-Lavigne, M., Zheng, B., 2009. Reassessment of corticospinal tract regeneration in Nogo-deficient mice. J. Neurosci. 29, 8649–8654.

Lee, J.K., Geoffroy, C.G., Chan, A.F., Tolentino, K.E., Crawford, M.J., Leal, M.A., Kang, B., Zheng, B., 2010. Assessing spinal axon regeneration and sprouting in Nogo-, MAG-, and OMgp-deficient mice. Neuron 66, 663–670.

Lehmann, M., Fournier, A., Selles-Navarro, I., Dergham, P., Sebok, A., Leclerc, N., Tigyi, G., McKerracher, L., 1999. Inactivation of Rho signaling pathway promotes CNS axon regeneration. J. Neurosci. 19, 7537–7547.

Liebscher, T., Schnell, L., Schnell, D., Scholl, J., Schneider, R., Gullo, M., Fouad, K., Mir, A., Rausch, M., Kindler, D., Hamers, F.P., Schwab, M.E., 2005. Nogo-A antibody improves regeneration and locomotion of spinal cord-injured rats. Ann. Neurol. 58, 706–719.

Lin, R., Kwok, J.C., Crespo, D., Fawcett, J.W., 2008. Chondroitinase ABC has a long-lasting effect on chondroitin sulphate glycosaminoglycan content in the injured rat brain. J. Neurochem. 104, 400–408.

Liu, X.Z., Xu, X.M., Hu, R., Du, C., Zhang, S.X., McDonald, J.W., Dong, H.X., Wu, Y.J., Fan, G.S., Jacquin, M.F., Hsu, C.Y., Choi, D.W., 1997. Neuronal and glial apoptosis after traumatic spinal cord injury. J. Neurosci. 17, 5395–5406.

Liu, Y., Kim, D., Himes, B.T., Chow, S.Y., Schallert, T., Murray, M., Tessler, A., Fischer, I., 1999. Transplants of fibroblasts genetically modified to express BDNF promote regeneration of adult rat rubrospinal axons and recovery of forelimb function. J. Neurosci. 19, 4370–4387.

Liu, K., Lu, Y., Lee, J.K., Samara, R., Willenberg, R., Sears-Kraxberger, I., Tedeschi, A., Park, K.K., Jin, D., Cai, B., Xu, B., Connolly, L., Steward, O., Zheng, B., He, Z., 2010. PTEN deletion enhances the regenerative ability of adult corticospinal neurons. Nat. Neurosci. 13, 1075–1081.

Lopez-Vales, R., Garcia-Alias, G., Fores, J., Udina, E., Gold, B.G., Navarro, X., Verdu, E., 2005. FK 506 reduces tissue damage and prevents functional deficit after spinal cord injury in the rat. J. Neurosci. Res. 81, 827–836.

Lu, X., Richardson, P.M., 1991. Inflammation near the nerve cell body enhances axonal regeneration. J. Neurosci. 11, 972–978.

Lu, P., Tuszynski, M.H., 2008. Growth factors and combinatorial therapies for CNS regeneration. Exp. Neurol. 209, 313–320.

Maier, I.C., Ichiyama, R.M., Courtine, G., Schnell, L., Lavrov, I., Edgerton, V.R., Schwab, M.E., 2009. Differential effects of anti-Nogo-A antibody treatment and treadmill training in rats with incomplete spinal cord injury. Brain 132, 1426–1440.

Maikos, J.T., Shreiber, D.I., 2007. Immediate damage to the blood-spinal cord barrier due to mechanical trauma. J. Neurotrauma 24, 492–507.

Martinez, F.O., Gordon, S., Locati, M., Mantovani, A., 2006. Transcriptional profiling of the human monocyte-to-macrophage differentiation and polarization: new molecules and patterns of gene expression. J. Immunol. 177, 7303–7311.

Massey, J.M., Hubscher, C.H., Wagoner, M.R., Decker, J.A., Amps, J., Silver, J., Onifer, S.M., 2006. Chondroitinase ABC digestion of the perineuronal net promotes functional collateral sprouting in the cuneate nucleus after cervical spinal cord injury. J. Neurosci. 26, 4406–4414.

Massey, J.M., Amps, J., Viapiano, M.S., Matthews, R.T., Wagoner, M.R., Whitaker, C.M., Alilain, W., Yonkof, A.L., Khalyfa, A., Cooper, N.G., Silver, J., Onifer, S.M., 2008. Increased chondroitin sulfate proteoglycan expression in denervated brainstem targets following spinal cord injury creates a barrier to axonal regeneration overcome by chondroitinase ABC and neurotrophin-3. Exp. Neurol. 209, 426–445.

Masuda-Nakagawa, L.M., Muller, K.J., Nicholls, J.G., 1993. Axonal sprouting and laminin appearance after destruction of glial sheaths. Proc. Natl. Acad. Sci. U.S.A. 90, 4966–4970.

Matsukawa, T., Arai, K., Koriyama, Y., Liu, Z., Kato, S., 2004. Axonal regeneration of fish optic nerve after injury. Biol. Pharm. Bull. 27, 445–451.

Mayer, D.J., Mao, J., Holt, J., Price, D.D., 1999. Cellular mechanisms of neuropathic pain, morphine tolerance, and their interactions. Proc. Natl. Acad. Sci. U.S.A. 96, 7731–7736.

McKeon, R.J., Schreiber, R.C., Rudge, J.S., Silver, J., 1991. Reduction of neurite outgrowth in a model of glial scarring following CNS injury is correlated with the expression of inhibitory molecules on reactive astrocytes. J. Neurosci. 11, 3398–3411.

McPhail, L.T., Stirling, D.P., Tetzlaff, W., Kwiecien, J.M., Ramer, M.S., 2004. The contribution of activated phagocytes and myelin degeneration to axonal retraction/dieback following spinal cord injury. Eur. J. Neurosci. 20, 1984–1994.

McTigue, D.M., Wei, P., Stokes, B.T., 2001. Proliferation of NG2-positive cells and altered oligodendrocyte numbers in the contused rat spinal cord. J. Neurosci. 21, 3392–3400.

Mikol, D.D., Gulcher, J.R., Stefansson, K., 1990. The oligodendrocyte-myelin glycoprotein belongs to a distinct family of proteins and contains the HNK-1 carbohydrate. J. Cell Biol. 110, 471–479.

Milligan, E.D., Watkins, L.R., 2009. Pathological and protective roles of glia in chronic pain. Nat. Rev. Neurosci. 10, 23–36.

Mingorance, A., Sole, M., Muneton, V., Martinez, A., Nieto-Sampedro, M., Soriano, E., del Rio, J.A., 2006. Regeneration of lesioned entorhino-hippocampal axons *in vitro* by combined degradation of inhibitory proteoglycans and blockade of Nogo-66/NgR signaling. FASEB J. 20, 491–493.

Minor, K.H., Bournat, J.C., Toscano, N., Giger, R.J., Davies, S.J., 2011. Decorin, erythroblastic leukaemia viral oncogene homologue B4 and signal transducer and activator of transcription 3 regulation of semaphorin 3A in central nervous system scar tissue. Brain 134, 1140–1155.

Mire, E., Thomasset, N., Jakeman, L.B., Rougon, G., 2008. Modulating Sema3A signal with a L1 mimetic peptide is not sufficient to promote motor recovery and axon regeneration after spinal cord injury. Mol. Cell. Neurosci. 37, 222–235.

Monnier, P.P., Sierra, A., Schwab, J.M., Henke-Fahle, S., Mueller, B.K., 2003. The Rho/ROCK pathway mediates neurite growth-inhibitory activity associated with the chondroitin sulfate proteoglycans of the CNS glial scar. Mol. Cell. Neurosci. 22, 319–330.

Moon, L.D., Asher, R.A., Rhodes, K.E., Fawcett, J.W., 2001. Regeneration of CNS axons back to their target following treatment of adult rat brain with chondroitinase ABC. Nat. Neurosci. 4, 465–466.

Moser, M., Nieswandt, B., Ussar, S., Pozgajova, M., Fassler, R., 2008. Kindlin-3 is essential for integrin activation and platelet aggregation. Nat. Med. 14, 325–330.

Muir, E.M., Fyfe, I., Gardiner, S., Li, L., Warren, P., Fawcett, J.W., Keynes, R.J., Rogers, J.H., 2010. Modification of N-glycosylation sites allows secretion of bacterial chondroitinase ABC from mammalian cells. J. Biotechnol. 145, 103–110.

Mukhopadhyay, G., Doherty, P., Walsh, F.S., Crocker, P.R., Filbin, M.T., 1994. A novel role for myelin-associated glycoprotein as an inhibitor of axonal regeneration. Neuron 13, 757–767.

Nakata, K., Abrams, B., Grill, B., Goncharov, A., Huang, X., Chisholm, A.D., Jin, Y., 2005. Regulation of a DLK-1 and p38 MAP kinase pathway by the ubiquitin ligase RPM-1 is required for presynaptic development. Cell 120, 407–420.

Nandini, C.D., Mikami, T., Ohta, M., Itoh, N., Akiyama-Nambu, F., Sugahara, K., 2004. Structural and functional characterization of oversulfated chondroitin sulfate/dermatan sulfate hybrid chains from the notochord of hagfish. Neuritogenic and binding activities for growth factors and neurotrophic factors. J. Biol. Chem. 279, 50799–50809.

Neumann, S., Woolf, C.J., 1999. Regeneration of dorsal column fibers into and beyond the lesion site following adult spinal cord injury. Neuron 23, 83–91.

Neumann, S., Bradke, F., Tessier-Lavigne, M., Basbaum, A.I., 2002. Regeneration of sensory axons within the injured spinal cord induced by intraganglionic cAMP elevation. Neuron 34, 885–893.

Neumann, H., Kotter, M.R., Franklin, R.J., 2009. Debris clearance by microglia: an essential link between degeneration and regeneration. Brain 132, 288–295.

Niederost, B., Oertle, T., Fritsche, J., McKinney, R.A., Bandtlow, C.E., 2002. Nogo-A and myelin-associated glycoprotein mediate neurite growth inhibition by antagonistic regulation of RhoA and Rac1. J. Neurosci. 22, 10368–10376.

Nikulina, E., Tidwell, J.L., Dai, H.N., Bregman, B.S., Filbin, M.T., 2004. The phosphodiesterase inhibitor rolipram delivered after a spinal cord lesion promotes axonal regeneration and functional recovery. Proc. Natl. Acad. Sci. U.S.A. 101, 8786–8790.

Nix, P., Hisamoto, N., Matsumoto, K., Bastiani, M., 2011. Axon regeneration requires coordinate activation of p38 and JNK MAPK pathways. Proc. Natl. Acad. Sci. U.S.A. 108, 10738–10743.

Oatway, M.A., Chen, Y., Bruce, J.C., Dekaban, G.A., Weaver, L.C., 2005. Anti-CD11d integrin antibody treatment restores normal serotonergic projections to the dorsal, intermediate, and ventral horns of the injured spinal cord. J. Neurosci. 25, 637–647.

Pallini, R., Fernandez, E., Sbriccoli, A., 1988. Retrograde degeneration of corticospinal axons following transection of the spinal cord in rats. A quantitative study with anterogradely transported horseradish peroxidase. J. Neurosurg. 68, 124–128.

Park, K.K., Liu, K., Hu, Y., Smith, P.D., Wang, C., Cai, B., Xu, B., Connolly, L., Kramvis, I., Sahin, M., He, Z., 2008. Promoting axon regeneration in the adult CNS by modulation of the PTEN/mTOR pathway. Science 322, 963–966.

Park, K.K., Liu, K., Hu, Y., Kanter, J.L., He, Z., 2010. PTEN/mTOR and axon regeneration. Exp. Neurol. 223, 45–50.

Perry, V.H., Brown, M.C., Gordon, S., 1987. The macrophage response to central and peripheral nerve injury. A possible role for macrophages in regeneration. J. Exp. Med. 165, 1218–1223.

Peters, C.M., Jimenez-Andrade, J.M., Kuskowski, M.A., Ghilardi, J.R., Mantyh, P.W., 2007. An evolving cellular pathology occurs in dorsal root ganglia, peripheral nerve and spinal cord following intravenous administration of paclitaxel in the rat. Brain Res. 1168, 46–59.

Pignata, S., De Placido, S., Biamonte, R., Scambia, G., Di Vagno, G., Colucci, G., Febbraro, A., Marinaccio, M., Lombardi, A.V., Manzione, L., Carteni, G., Nardi, M., Danese, S., Valerio, M.R., de Matteis, A., Massidda, B., Gasparini, G., Di Maio, M., Pisano, C., Perrone, F., 2006. Residual neurotoxicity in ovarian cancer patients in clinical remission after first-line chemotherapy with carboplatin and paclitaxel: the Multicenter Italian Trial in Ovarian cancer (MITO-4) retrospective study. BMC Cancer 6, 5.

Pizzorusso, T., Medini, P., Berardi, N., Chierzi, S., Fawcett, J.W., Maffei, L., 2002. Reactivation of ocular dominance plasticity in the adult visual cortex. Science 298, 1248–1251.

Poon, P.C., Gupta, D., Shoichet, M.S., Tator, C.H., 2007. Clip compression model is useful for thoracic spinal cord injuries: histologic and functional correlates. Spine 32, 2853–2859 (Phila Pa 1976).

Popovich, P.G., Wei, P., Stokes, B.T., 1997. Cellular inflammatory response after spinal cord injury in Sprague-Dawley and Lewis rats. J. Comp. Neurol. 377, 443–464.

Popovich, P.G., Guan, Z., Wei, P., Huitinga, I., van Rooijen, N., Stokes, B.T., 1999. Depletion of hematogenous macrophages promotes partial hindlimb recovery and neuroanatomical repair after experimental spinal cord injury. Exp. Neurol. 158, 351–365.

Postma, T.J., Vermorken, J.B., Liefting, A.J., Pinedo, H.M., Heimans, J.J., 1995. Paclitaxel-induced neuropathy. Ann. Oncol. 6, 489–494.

Qiu, J., Cai, D., Dai, H., McAtee, M., Hoffman, P.N., Bregman, B.S., Filbin, M.T., 2002. Spinal axon regeneration induced by elevation of cyclic AMP. Neuron 34, 895–903.

Reier, P.J., Houle, J.D., 1988. The glial scar: its bearing on axonal elongation and transplantation approaches to CNS repair. Adv. Neurol. 47, 87–138.

Richardson, P.M., Issa, V.M., 1984. Peripheral injury enhances central regeneration of primary sensory neurones. Nature 309, 791–793.

Rolls, A., Shechter, R., Schwartz, M., 2009. The bright side of the glial scar in CNS repair. Nat. Rev. Neurosci. 10, 235–241.

Ruitenberg, M.J., Levison, D.B., Lee, S.V., Verhaagen, J., Harvey, A.R., Plant, G.W., 2005. NT-3 expression from engineered olfactory ensheathing glia promotes spinal sparing and regeneration. Brain 128, 839–853.

Salegio, E.A., Pollard, A.N., Smith, M., Zhou, X.F., 2011. Macrophage presence is essential for the regeneration of ascending afferent fibres following a conditioning sciatic nerve lesion in adult rats. BMC Neurosci. 12, 11.

Sandler, A.N., Tator, C.H., 1976. Review of the effect of spinal cord trama on the vessels and blood flow in the spinal cord. J. Neurosurg. 45, 638–646.

Saville, L.R., Pospisil, C.H., Mawhinney, L.A., Bao, F., Simedrea, F.C., Peters, A.A., O'Connell, P.J., Weaver, L.C., Dekaban, G.A., 2004. A monoclonal antibody to CD11d reduces the inflammatory infiltrate into the injured spinal cord: a potential neuroprotective treatment. J. Neuroimmunol. 156, 42–57.

Schnell, L., Schwab, M.E., 1990. Axonal regeneration in the rat spinal cord produced by an antibody against myelin-associated neurite growth inhibitors. Nature 343, 269–272.

Schnell, L., Hunanyan, A.S., Bowers, W.J., Horner, P.J., Federoff, H.J., Gullo, M., Schwab, M.E., Mendell, L.M., Arvanian, V.L., 2011. Combined delivery of Nogo-A antibody, neurotrophin-3 and the NMDA-NR2d subunit establishes a functional 'detour' in the hemisected spinal cord. Eur. J. Neurosci. 34, 1256–1267.

Schreyer, D.J., Skene, J.H., 1991. Fate of GAP-43 in ascending spinal axons of DRG neurons after peripheral nerve injury: delayed accumulation and correlation with regenerative potential. J. Neurosci. 11, 3738–3751.

Schwartz, M., Moalem, G., Leibowitz-Amit, R., Cohen, I.R., 1999. Innate and adaptive immune responses can be beneficial for CNS repair. Trends Neurosci. 22, 295–299.

Schweigreiter, R., Bandtlow, C.E., 2006. Nogo in the injured spinal cord. J. Neurotrauma 23, 384–396.

Schweigreiter, R., Walmsley, A.R., Niederost, B., Zimmermann, D.R., Oertle, T., Casademunt, E., Frentzel, S., Dechant, G., Mir, A., Bandtlow, C.E., 2004. Versican V2 and the central inhibitory domain of Nogo-A inhibit neurite growth via p75NTR/NgR-independent pathways that converge at RhoA. Mol. Cell. Neurosci. 27, 163–174.

Sengottuvel, V., Leibinger, M., Pfreimer, M., Andreadaki, A., Fischer, D., 2011. Taxol facilitates axon regeneration in the mature CNS. J. Neurosci. 31, 2688–2699.

Senter, H.J., Venes, J.L., 1978. Altered blood flow and secondary injury in experimental spinal cord trauma. J. Neurosurg. 49, 569–578.

Shamah, S.M., Lin, M.Z., Goldberg, J.L., Estrach, S., Sahin, M., Hu, L., Bazalakova, M., Neve, R.L., Corfas, G., Debant, A., Greenberg, M.E., 2001. EphA receptors regulate growth cone dynamics through the novel guanine nucleotide exchange factor ephexin. Cell 105, 233–244.

Shen, Y., Tenney, A.P., Busch, S.A., Horn, K.P., Cuascut, F.X., Liu, K., He, Z., Silver, J., Flanagan, J.G., 2009. PTPsigma is a receptor for chondroitin sulfate proteoglycan, an inhibitor of neural regeneration. Science 326, 592–596.

Shuman, S.L., Bresnahan, J.C., Beattie, M.S., 1997. Apoptosis of microglia and oligodendrocytes after spinal cord contusion in rats. J. Neurosci. Res. 50, 798–808.

Siddall, P.J., McClelland, J.M., Rutkowski, S.B., Cousins, M.J., 2003. A longitudinal study of the prevalence and characteristics of pain in the first 5 years following spinal cord injury. Pain 103, 249–257.

Silver, J., Miller, J.H., 2004. Regeneration beyond the glial scar. Nat. Rev. Neurosci. 5, 146–156.

Simonen, M., Pedersen, V., Weinmann, O., Schnell, L., Buss, A., Ledermann, B., Christ, F., Sansig, G., van der Putten, H., Schwab, M.E., 2003. Systemic deletion of the myelin-associated outgrowth inhibitor Nogo-A improves regenerative and plastic responses after spinal cord injury. Neuron 38, 201–211.

Smith, P.D., Sun, F., Park, K.K., Cai, B., Wang, C., Kuwako, K., Martinez-Carrasco, I., Connolly, L., He, Z., 2009. SOCS3 deletion promotes optic nerve regeneration *in vivo*. Neuron 64, 617–623.

Spence, R.D., Hamby, M.E., Umeda, E., Itoh, N., Du, S., Wisdom, A.J., Cao, Y., Bondar, G., Lam, J., Ao, Y., Sandoval, F., Suriany, S., Sofroniew, M.V., Voskuhl, R.R., 2011. Neuroprotection mediated through estrogen receptor-alpha in astrocytes. Proc. Natl. Acad. Sci. U.S.A. 108, 8867–8872.

Stammers, A.T., Liu, J., Kwon, B.K., 2012. Expression of inflammatory cytokines following acute spinal cord injury in a rodent model. J. Neurosci. Res. 90, 782–790.

Steeves, J.D., Kramer, J.K., Fawcett, J.W., Cragg, J., Lammertse, D.P., Blight, A.R., Marino, R.J., Ditunno Jr., J.F., Coleman, W.P., Geisler, F.H., Guest, J., Jones, L., Burns, S., Schubert, M., van Hedel, H.J., Curt, A., 2011. Extent of spontaneous motor recovery after traumatic cervical sensorimotor complete spinal cord injury. Spinal Cord 49, 257–265.

Steinmetz, M.P., Horn, K.P., Tom, V.J., Miller, J.H., Busch, S.A., Nair, D., Silver, D.J., Silver, J., 2005. Chronic enhancement of the intrinsic growth capacity of sensory neurons combined with the degradation of inhibitory proteoglycans allows functional regeneration of sensory axons through the dorsal root entry zone in the mammalian spinal cord. J. Neurosci. 25, 8066–8076.

Steward, O., Zheng, B., Banos, K., Yee, K.M., 2007. Response to: Kim et al., "axon regeneration in young adult mice lacking Nogo-A/B." Neuron 38, 187–199. Neuron 54, 191–195.

Steward, O., Sharp, K., Yee, K.M., Hofstadter, M., 2008. A re-assessment of the effects of a Nogo-66 receptor antagonist on regenerative growth of axons and locomotor recovery after spinal cord injury in mice. Exp. Neurol. 209, 446–468.

Stirling, D.P., Khodarahmi, K., Liu, J., McPhail, L.T., McBride, C.B., Steeves, J.D., Ramer, M.S., Tetzlaff, W., 2004. Minocycline treatment reduces delayed oligodendrocyte death, attenuates axonal dieback, and improves functional outcome after spinal cord injury. J. Neurosci. 24, 2182–2190.

Stokes, B.T., Fox, P., Hollinden, G., 1983. Extracellular calcium activity in the injured spinal cord. Exp. Neurol. 80, 561–572.

Strauss, D.J., Devivo, M.J., Paculdo, D.R., Shavelle, R.M., 2006. Trends in life expectancy after spinal cord injury. Arch. Phys. Med. Rehabil. 87, 1079–1085.

Sun, F., Park, K.K., Belin, S., Wang, D., Lu, T., Chen, G., Zhang, K., Yeung, C., Feng, G., Yankner, B.A., He, Z., 2011. Sustained axon regeneration induced by co-deletion of PTEN and SOCS3. Nature 480, 372–375.

Tan, C.L., Kwok, J.C., Patani, R., Ffrench-Constant, C., Chandran, S., Fawcett, J.W., 2011. Integrin activation promotes axon growth on inhibitory chondroitin sulfate proteoglycans by enhancing integrin signaling. J. Neurosci. 31, 6289–6295.

Tan, C.L., Andrews, M.R., Kwok, J.C., Heintz, T.G.P., Gumy, L.F., Fassler, R., Fawcett, J.W., 2012. Kindlin-1 enhances axon growth on inhibitory CSPGs and promotes sensory axon regeneration. J. Neurosci. 32, 7325–7335.

Tate, C.C., Tate, M.C., LaPlaca, M.C., 2007. Fibronectin and laminin increase in the mouse brain after controlled cortical impact injury. J. Neurotrauma 24, 226–230.

Tator, C.H., 1995. Update on the pathophysiology and pathology of acute spinal cord injury. Brain Pathol. 5, 407–413.

Taylor, L., Jones, L., Tuszynski, M.H., Blesch, A., 2006. Neurotrophin-3 gradients established by lentiviral gene delivery promote short-distance axonal bridging beyond cellular grafts in the injured spinal cord. J. Neurosci. 26, 9713–9721.

Thacker, M.A., Clark, A.K., Marchand, F., McMahon, S.B., 2007. Pathophysiology of peripheral neuropathic pain: immune cells and molecules. Anesth. Analg. 105, 838–847.

Tom, V.J., Steinmetz, M.P., Miller, J.H., Doller, C.M., Silver, J., 2004. Studies on the development and behavior of the dystrophic growth cone, the hallmark of regeneration failure, in an *in vitro* model of the glial scar and after spinal cord injury. J. Neurosci. 24, 6531–6539.

Tom, V.J., Kadakia, R., Santi, L., Houle, J.D., 2009. Administration of chondroitinase ABC rostral or caudal to a spinal cord injury site promotes anatomical but not functional plasticity. J. Neurotrauma 26, 2323–2333.

Totoiu, M.O., Keirstead, H.S., 2005. Spinal cord injury is accompanied by chronic progressive demyelination. J. Comp. Neurol. 486, 373–383.

Trapp, B.D., 1990. Myelin-associated glycoprotein. Location and potential functions. Ann. N. Y. Acad. Sci. 605, 29–43.

Tsujino, H., Kondo, E., Fukuoka, T., Dai, Y., Tokunaga, A., Miki, K., Yonenobu, K., Ochi, T., Noguchi, K., 2000. Activating transcription factor 3 (ATF3) induction by axotomy in sensory and motor neurons: a novel neuronal marker of nerve injury. Mol. Cell. Neurosci. 15, 170–182.

Ussar, S., Wang, H.V., Linder, S., Fassler, R., Moser, M., 2006. The Kindlins: subcellular localization and expression during murine development. Exp. Cell Res. 312, 3142–3151.

Vallières, N., Berard, J.L., David, S., Lacroix, S., 2006. Systemic injections of lipopolysaccharide accelerates myelin phagocytosis during Wallerian degeneration in the injured mouse spinal cord. Glia 53, 103–113.

Van der Vieren, M., Le Trong, H., Wood, C.L., Moore, P.F., St John, T., Staunton, D.E., Gallatin, W.M., 1995. A novel leukointegrin, alpha d beta 2, binds preferentially to ICAM-3. Immunity 3, 683–690.

Verma, P., Garcia-Alias, G., Fawcett, J.W., 2008. Spinal cord repair: bridging the divide. Neurorehabil. Neural Repair 22, 429–437.

Voskuhl, R.R., Peterson, R.S., Song, B., Ao, Y., Morales, L.B., Tiwari-Woodruff, S., Sofroniew, M.V., 2009. Reactive astrocytes form scar-like perivascular barriers to leukocytes during adaptive immune inflammation of the CNS. J. Neurosci. 29, 11511–11522.

Wang, D., Ichiyama, R.M., Zhao, R., Andrews, M.R., Fawcett, J.W., 2011. Chondroitinase combined with rehabilitation promotes recovery of forelimb function in rats with chronic spinal cord injury. J. Neurosci. 31, 9332–9344.

Wehrle, R., Camand, E., Chedotal, A., Sotelo, C., Dusart, I., 2005. Expression of netrin-1, slit-1 and slit-3 but not of slit-2 after cerebellar and spinal cord lesions. Eur. J. Neurosci. 22, 2134–2144.

Willson, C.A., Irizarry-Ramirez, M., Gaskins, H.E., Cruz-Orengo, L., Figueroa, J.D., Whittemore, S.R., Miranda, J.D., 2002. Up-regulation of EphA receptor expression in the injured adult rat spinal cord. Cell Transplant. 11, 229–239.

Willson, C.A., Miranda, J.D., Foster, R.D., Onifer, S.M., Whittemore, S.R., 2003. Transection of the adult rat spinal cord up-regulates EphB3 receptor and ligand expression. Cell Transplant. 12, 279–290.

Windle, W.F., Clemente, C.D., Chambers, W.W., 1952. Inhibition of formation of a glial barrier as a means of permitting a peripheral nerve to grow into the brain. J. Comp. Neurol. 96, 359–369.

Wong, I., Liao, H., Bai, X., Zaknic, A., Zhong, J., Guan, Y., Li, H.Y., Wang, Y.J., Zhou, X.F., 2010. ProBDNF inhibits infiltration of ED1 + macrophages after spinal cord injury. Brain Behav. Immun. 24, 585–597.

Woolf, C.J., 2003. No Nogo: now where to go? Neuron 38, 153–156.
Wu, Z., Ghosh-Roy, A., Yanik, M.F., Zhang, J.Z., Jin, Y., Chisholm, A.D., 2007. Caenorhabditis elegans neuronal regeneration is influenced by life stage, ephrin signaling, and synaptic branching. Proc. Natl. Acad. Sci. U.S.A. 104, 15132–15137.
Yamagata, T., Saito, H., Habuchi, O., Suzuki, S., 1968. Purification and properties of bacterial chondroitinases and chondrosulfatases. J. Biol. Chem. 243, 1523–1535.
Yamamoto, M., Raisman, G., Li, D., Li, Y., 2009. Transplanted olfactory mucosal cells restore paw reaching function without regeneration of severed corticospinal tract fibres across the lesion. Brain Res. 1303, 26–31.
Yan, D., Wu, Z., Chisholm, A.D., Jin, Y., 2009. The DLK-1 kinase promotes mRNA stability and local translation in C. elegans synapses and axon regeneration. Cell 138, 1005–1018.
Yin, Y., Henzl, M.T., Lorber, B., Nakazawa, T., Thomas, T.T., Jiang, F., Langer, R., Benowitz, L.I., 2006. Oncomodulin is a macrophage-derived signal for axon regeneration in retinal ganglion cells. Nat. Neurosci. 9, 843–852.
Ylera, B., Erturk, A., Hellal, F., Nadrigny, F., Hurtado, A., Tahirovic, S., Oudega, M., Kirchhoff, F., Bradke, F., 2009. Chronically CNS-injured adult sensory neurons gain regenerative competence upon a lesion of their peripheral axon. Curr. Biol. 19, 930–936.
Young, W., Koreh, I., 1986. Potassium and calcium changes in injured spinal cords. Brain Res. 365, 42–53.
Zhang, Z., Fujiki, M., Guth, L., Steward, O., 1996. Genetic influences on cellular reactions to spinal cord injury: a wound-healing response present in normal mice is impaired in mice carrying a mutation (WldS) that causes delayed Wallerian degeneration. J. Comp. Neurol. 371, 485–495.
Zhang, Y., Winterbottom, J.K., Schachner, M., Lieberman, A.R., Anderson, P.N., 1997. Tenascin-C expression and axonal sprouting following injury to the spinal dorsal columns in the adult rat. J. Neurosci. Res. 49, 433–450.
Zhang, Y.P., Iannotti, C., Shields, L.B., Han, Y., Burke, D.A., Xu, X.M., Shields, C.B., 2004. Dural closure, cord approximation, and clot removal: enhancement of tissue sparing in a novel laceration spinal cord injury model. J. Neurosurg. 100, 343–352.
Zhao, C., Fancy, S.P., Franklin, R.J., ffrench-Constant, C., 2009. Up-regulation of oligodendrocyte precursor cell alphaV integrin and its extracellular ligands during central nervous system remyelination. J. Neurosci. Res. 87, 3447–3455.
Zhao, R.R., Muir, E.M., Alves, J.N., Rickman, H., Allan, A.Y., Kwok, J.C., Roet, K.C., Verhaagen, J., Schneider, B.L., Bensadoun, J.C., Ahmed, S.G., Yanez-Munoz, R.J., Keynes, R.J., Fawcett, J.W., Rogers, J.H., 2011. Lentiviral vectors express chondroitinase ABC in cortical projections and promote sprouting of injured corticospinal axons. J. Neurosci. Methods 201, 228–238.
Zheng, B., Ho, C., Li, S., Keirstead, H., Steward, O., Tessier-Lavigne, M., 2003. Lack of enhanced spinal regeneration in Nogo-deficient mice. Neuron 38, 213–224.
Zuo, J., Neubauer, D., Dyess, K., Ferguson, T.A., Muir, D., 1998. Degradation of chondroitin sulfate proteoglycan enhances the neurite-promoting potential of spinal cord tissue. Exp. Neurol. 154, 654–662.
Zurn, A.D., Bandtlow, C.E., 2006. Regeneration failure in the CNs: cellular and molecular mechanisms. Adv. Exp. Med. Biol. 557, 54–76.

CHAPTER

Schwann cell transplantation: A repair strategy for spinal cord injury?

15

Ryan R. Wiliams*, Mary Bartlett Bunge*,†,‡,1

*The Miami Project to Cure Paralysis, University of Miami Miller School of Medicine, The Lois Pope LIFE Center, Miami, FL, USA

†Department of Cell Biology, University of Miami Miller School of Medicine, Rosenthiel Medical Sciences Building, Miami, FL, USA

‡Department of Neurological Surgery, University of Miami Miller School of Medicine, The Lois Pope LIFE Center, Miami, FL, USA

1Corresponding author. Tel.: +1-305-243-4596, Fax: +1-305-243-3923, e-mail address: mbunge@miami.edu

Abstract

Schwann cells (SCs), when implanted in the injured spinal cord, support regeneration of axons, myelinate or ensheathe regenerated axons in a normal way, reduce cyst formation in the injured tissue, reduce secondary damage of tissue around the initial injury site, and modestly improve limb movements. If SC transplantation is combined with additional treatments such as methylprednisolone, neurotrophins, GDNF, olfactory ensheathing cells, chondroitinase, or elevation of cAMP levels, more axons (including those from neurons in the brainstem) regenerate into and out of the SC implant and further improve locomotion. Recent work to improve SC migration from the implant into the spinal cord by polysialylating NCAM on the SC surface has led to the novel finding that corticospinal axon growth is promoted by SCs. Recent studies are cited showing that when astrocytes extend slender processes into an implant instead of forming a sharp boundary they are permissive rather than inhibitory to axonal regrowth. The interfaces that comprise the "on-ramps" and the "off-ramps" are key to the success of a SC implant to span the injury site and to foster axon regeneration across the injury.

Keywords

neurotrophins, PSA–NCAM, Schwann cell combination therapy, reactive astrocytes, Schwann cell/astrocyte boundary assay, Schwann cell/spinal cord interface

Progress in Brain Research, Volume 201, ISSN 0079-6123, http://dx.doi.org/10.1016/B978-0-444-59544-7.00014-7

1 INTRODUCTION

The loss of tissue and consequent formation of a cyst following many types of injury to the spinal cord necessitate a repair strategy that introduces a new substratum for regenerative growth. Axons do not extend in fluid but require a structured path for growth. After injury and axonal degeneration, Schwann cells (SCs), harbored within their tunnels of basal lamina (bands of Bungner), provide a supportive framework for new axonal growth in the peripheral nervous system (PNS). This is clearly a suitable environment for axon regrowth. Because SCs are not present in the central nervous system (CNS) parenchyma, except in relation to certain innervated blood vessels, why not, then, introduce SCs into the CNS cyst to replace the fluid and to provide an environment to promote new nerve fiber growth following injury? Testing the efficacy of SC transplantation in the spinal cord has been underway for many years.

This chapter will discuss earlier findings briefly as numerous reviews have appeared (e.g., Bunge, 2008; Bunge and Wood, 2012; Fortun et al., 2009; Oudega and Xu, 2006; Pearse and Bunge, 2006; Tetzlaff et al., 2011). More recent efforts will be emphasized, with a focus on host spinal cord/implanted SC interfaces as they are key to the entrance of regenerated axons into and exit from the grafts that span the injury site. The effectiveness of transplanted SCs that are initially less differentiated or more immature than those obtained from adult peripheral nerve and that have been generated from sources other than adult peripheral nerve are discussed in the review by Bunge and Wood (2012). The preparation of human SCs and their use in a clinical trial also are described in the 2012 review.

2 EARLIER STUDIES OF SC TRANSPLANTATION

In the early 1970s, methods were being developed to extirpate and purify populations of the cells that comprise the PNS, dorsal root ganglion neurons, SCs, and fibroblasts. The intent was to be able to study their molecular interactions *in vitro*, far easier to do than to dissect them *in vivo*. It was found, for example, that a molecule on the surface of axons regulates SC proliferation (Wood and Bunge, 1975), that SCs generate extracellular matrix such as a number of types of collagen (Bunge et al., 1980), and that basal lamina formation is required for SCs to ensheathe and myelinate axons (Bunge et al., 1982; Eldridge et al., 1989).

The ability to acquire purified populations of SCs in culture prompted Bunge (1975) to propose that these cells as well as central glia could be introduced into the CNS for repair. The spinal cord-injured person's own SCs could be used after their procurement from a piece of peripheral nerve in the leg and expansion in number in culture over a period of a few weeks, thus avoiding the need for immune suppression. This vision has led to investigation of SCs and their roles in spinal cord repair over decades, culminating in an application to the FDA for a human autologous SC clinical trial. A team at the Miami Project to Cure Paralysis has just obtained FDA approval (http://miamiproject.miami.edu). As many as 10 million cells are available from a single sural nerve biopsy. A standard operating procedure for manufacturing a human SC product suitable for transplantation into humans has been developed; this product can supply up to 50–100 million cells within 3–4 weeks. A SC clinical trial is already underway in Iran (Saberi et al., 2008, 2011).

In our laboratory, two injury models have been studied in adult Fischer rats, the complete transection/SC bridge model to be able to detect with certainty truly regenerated axons (Xu et al., 1995a, 1997) and a contusion model utilizing the MASCIS/NYU impactor (Takami et al., 2002). The contusion model is more clinically relevant due to cyst formation, although some axon sparing primarily around the spinal cord periphery complicates the distinction between regenerated, spared, and sprouted (from uninjured) axons. Conclusions from many experiments testing SC transplantation are that SCs (i) support regeneration of axons from both stumps of the transected spinal cord into the SC bridge, (ii) myelinate or ensheathe in a normal way axons that regenerate into the bridge, (iii) reduce cyst formation, (iv) reduce secondary damage of spinal cord tissue, and (v) improve limb movements to some degree (Schaal et al., 2007; Takami et al., 2002).

Techniques have been worked out to obtain and purify large numbers of SCs from adult rats (Meijs et al., 2004; Morrissey et al., 1991), pigs, nonhuman primates, and humans relatively easily (Casella et al., 1996; Levi, 1996; Levi et al., 1995; Rutkowski et al., 1992). They can be transduced to generate neurotrophins (e.g., Tuszynski et al., 1998) or be modified genetically in other ways, to be discussed below. The ability to genetically change SCs for transplantation is highly advantageous.

Whereas the transplantation of SCs with no additional treatments leads to the improvements listed earlier, up to now, brainstem axons have not been observed to regenerate into the SC implant, and regenerated axons do not exit the transplant. This is undoubtedly due to many factors. First, SCs manifest very limited migratory ability into astrocyte-containing CNS tissue (Blakemore and Franklin, 2000; Golden et al., 2007; Hill et al., 2007; Iwashita et al., 2000; Pearse et al., 2007). This contributes to the formation of a boundary between the grafted SCs and the apposed astrocytes that, in response to SCs, undergo hypertrophy and upregulate inhibitory molecules (Lakatos et al., 2003; Plant et al., 2001). Second, there is considerable distance between thoracic injuries and the parent neuronal somata in the brainstem. Richardson et al. (1984) demonstrated that neuronal growth responses are increased when peripheral nerve is introduced into cervical compared to lower regions of the spinal cord. That distance is an important factor also was suggested by the study of Schaal et al. (2007) who determined that, after a severe cervical rather than thoracic contusion injury, SC transplantation alone led to the presence of numerous reticulospinal axons in the lesion/SC implant site. The distance between the injury to the axon and its neuronal soma influences the intrinsic ability to initiate axonal growth. Third, the tissue reactions to spinal cord injury are multifarious and deleterious. To repair the injured spinal cord, we need to (i) halt the spread of secondary tissue damage, (ii) save as many nerve cells as possible, (iii) curb inflammation, (iv) modify scar formation, (v) neutralize inhibitory factors, (vi) awaken nerve cells to regrow axons, (vii) provide sustenance to nerve cells, (viii) promote axon regeneration across the injury, (ix) guide axon growth to appropriate areas, and (x) enable formation of functional connections.

Thus, it was soon realized that SC transplantation needed to be combined with additional interventions to be more efficacious. Using the complete transection model, we have combined SC implantation with the administration of methylprednisolone, IGF-1 + PDGF, BDNF and/or NT-3 in various ways including introducing their genes into the SCs to be transplanted and olfactory ensheathing cells injected into the stumps next to the SC bridge with or without chondroitinase. In the contusion model, we have combined SCs with olfactory ensheathing cells and IL-10/methylprednisolone, a bifunctional neurotrophin D15A (with both BDNF and NT-3 activity; Urfer et al., 1994), and elevation of cAMP. All of these combinatorial strategies were more effective than SCs alone (reviewed in Bunge 2008; Bunge and Wood, 2012; Fortun et al., 2009; Tetzlaff et al., 2011).

Numerous outcome measures are employed to assess efficacy. Although not all of them are used in every experiment, the following are usually assessed: (i) lesion and spared tissue volumes, (ii) number of SCs, (iii) species of axons as determined by immunostaining, (iv) neuroanatomical tracing to identify and detect the course of regenerated axons and to determine the location of the neuronal somata from which they regenerated, (v) number of myelinated and total axons in the SC implant, and (vi) visualization of astrocytes (and the scar), macrophages/microglia, and chondroitin sulfate proteoglycans (CSPGs) by immunostaining. With the availability of more reliable cell markers, such as the viral vector-delivered gene for enhanced green-fluorescent protein (EGFP; Golden et al., 2007; Hill et al., 2007), it is now realized that host SCs migrate into the implants of labeled SCs (Hill et al., 2007). It is not yet known to what extent the endogenous SCs contribute to the improvements observed.

Progress noted with the combinatorial interventions listed above are increases in the number of regenerated axons in the implant, including those from neurons in the brainstem, more descending axons caudal to the SC implant, and improvement in locomotion. Growth factors were effective in increasing the number of axons in the implant (Xu et al., 1995b). When SCs were combined with D15A, the implant and the number of SCs were fivefold greater and a total of about 75,000 axons were found in the implant (Golden et al., 2007). Myelinated axons numbered more than 18,000 or 26,000, depending upon the viral vector used, compared to 4718 or 4140 in control grafts. There was also a significant increase in serotonergic, noradrenergic, and sensory fiber growth in the presence of D15A. The lack of improvement in BBB scores (Basso et al., 1995, 1996), however, suggested that there was little exit of axons from the graft. Clearly, an additional intervention is needed. cAMP levels fall in the injured spinal cord. When levels were increased, myelinated axons numbered up to 10,000; more spared and/or regenerated axons from the reticular formation and raphe nuclei were observed and some from the raphe were caudal to the implants as well (Pearse et al., 2004). BBB scores were significantly improved and footfall errors on a ladder were reduced. Substantial locomotor improvement also was achieved with SCs/OECs + chondroitinase (Fouad et al., 2005). A later study of the sequelae of this combination strategy found axons in the caudal cord, not only from the raphe nuclei but also from the vestibular and reticulospinal nuclei (Vavrek et al., 2007). Methylprednisolone helped to preserve tissue as well as foster brainstem axon growth into the SC bridge (Chen et al., 1996). Olfactory ensheathing cells placed in the stumps beside the SC bridge were effective in promoting lengthy 5HT axon growth in the caudal cord (Ramón-Cueto et al., 1998).

3 MORE RECENT SC TRANSPLANTATION STUDIES

A variety of approaches have been well tested in a model in which a portion of the lateral half of the spinal cord has been removed and replaced by a hemipolymer channel containing a SC bridge (Xu et al., 1999). Results with the SC hemibridge were similar to the complete transection/SC bridge model in that propriospinal and sensory axons regenerated into the bridge. But, in contrast, axons from as many as 19 brainstem regions also regrew into the hemibridge. Some regenerating axons were observed to directly cross the caudal interface and to grow up to 3.5 mm into the caudal cord. These improvements might have resulted from improved stability and less severe damage to the cord by the hemisection rather than the complete transection. Also, the dura mater could be closed around the hemisection/implant; this enabled CSF flow to be reestablished across the lesion.

Combinatorial strategies have been investigated in this lateral hemisection model. The infusion of BDNF and NT-3 into the spinal cord beyond the graft/host cord interface was tested and found to increase the number of axons crossing the caudal interface to grow farther into the caudal spinal cord (6 mm; Bamber et al., 2001). Chau et al. (2004) found that, when chondroitinase was delivered from an infusion tip-implanted caudal to the SC bridge, CSPG expression was decreased in the caudal spinal cord (but not in the SC bridge). Without the enzyme, however, CSPG deposits were observed at the caudal host/graft interfaces. Propriospinal axons entered all bridges and exited some of them only when chondroitinase was delivered.

In another combination, the hemi-SC bridge contained not only SCs and Matrigel but also glial cell line-derived neurotrophic factor (GDNF; Iannotti et al., 2003). In this case, compared to bridges without GDNF, propriospinal axon regeneration was promoted eightfold and the SC graft/host interface was improved as indicated by reduced reactive gliosis (GFAP expression), less macrophage infiltration, and lessened cavitation. In a recent extension of this work, concentrating on a study of the host/graft interfaces, Deng et al. (2011) found both robust axonal regeneration and migration of many host astrocytes into the SC bridge. Elongated processes of astrocytes extended parallel to the graft axis and in close alignment with regenerated axons in the bridge. GDNF had influenced the interface astrocytes to be permissive for regenerative axon growth. Simultaneous work done *in vitro* utilizing a SC/astrocyte boundary assay (see below) suggested that the GDNF effect was mediated through SC–astrocyte interactions (Deng et al., 2011).

Chondroitinase ABC has been studied in many laboratories (reviewed in Bradbury and Carter, 2011) to good effect as it prevents the accumulation of CSPGs that inhibit axon growth. The enzyme, which cleaves the inhibitory sulfated glycosaminoglycan chains from CSPGs, has been shown to increase SC intermingling with astrocytes in an *in vitro* assay (Grimpe et al., 2005). In one of our recent studies (Kanno et al., 2012), we tested the novel combination of D15A and chondroitinase-secreting SCs in the rat contusion model. The combination, more effective than with either the D15A or the SCs/chondroitinase alone, led to more propriospinal axons in the SC graft and caudal to it and more corticospinal tract (CST) fibers around the graft, a greater brainstem neuron response to grow axons, and improvement in locomotion and mechanical and thermal allodynia.

Results following a compression injury were improved by designing a strategy to promote SC migration (Papastefanaki et al., 2007). The polysialylated (polysialic acid, PSA) form of the neural cell adhesion molecule, NCAM, is present on SCs (as well as on oligodendrocyte precursor cells and reactive astrocytes) for the first 2 weeks after spinal cord demyelination. Thinking that PSA, expression on NCAM present on SCs would be beneficial, early postnatal GFP–SCs were engineered to express on their surfaces the PSA form of NCAM. The enzyme, sialyl-transferase X (STX), responsible for transferring PSA to NCAM, was introduced by a retroviral vector. In an *in vitro* confrontation assay, 11× more STX–GFP–SCs entered astrocyte domains than did control SCs (Papastefanaki et al., 2007; see also Luo et al., 2011). *In vivo*, STX–GFP–SCs implanted just rostral to the mouse thoracic compression injury promoted faster and significantly greater functional recovery (BMS scale, Basso et al., 2006) compared to control engineered SCs or no cells. The earlier improvement in locomotor recovery was correlated with earlier enhanced myelination by the implanted STX–GFP–SCs, increased myelination by host SCs, and enhanced recruitment/myelination by oligodendrocyte precursor cells. More serotonergic axons were observed (in very localized areas) caudal to the lesion site in the STX–GFP–SC group than in the control-GFP–SC group. Thus, STX–GFP–SC grafting led to more axonal growth and myelination, and locomotor improvement (Papastefanaki et al., 2007).

The regeneration of sensory axons also has been promoted by PSA expression in two separate *in vivo* experiments but without SC transplantation (Zhang et al., 2007a,b). In the first (Zhang et al., 2007a), PSA expression was induced in the lesion site (dorsal hemisection) by means of a lentiviral-polysialyl transferase (PST) vector. Astrocytes and oligodendrocytes were transduced; GFAP+/PSA+ astrocyte processes extended into the lesion cavity. PSA expression increased astrocyte infiltration and penetration of regenerating sensory axons across the caudal border of the lesion and into the lesion cavity. In PSA-expressing animals with an additional peripheral nerve conditioning lesion, 20× more axons grew into the lesion cavity than in LV/GFP-injected animals with a conditioning lesion; some axons grew across the cavity and entered the rostral spinal cord. Expression of PSA in the scar reduced its inhibitory property for axonal growth.

The second study (Zhang et al., 2007b) differed in that vectors carrying PST or GFP were injected into the uninjured lumbar dorsal horn and into transected and reapposed dorsal roots to determine if axons in the central processes of DRG neurons could regenerate across the normally inhibitory barrier of the dorsal root entry zone (DREZ). The presence of PSA led to numerous axons growing past the transection site toward the spinal cord; with the addition of a sciatic nerve conditioning lesion, many regenerating axons could be followed through the DREZ and into the spinal cord. Axons growing into the spinal cord were associated with, in part, GFAP+ astrocytes and blood vessels. Thus it was PSA expression in the dorsal horn and DREZ, in combination with a conditioning lesion, that markedly promoted the regrowth of axons into the spinal cord. Large numbers of SCs migrated into the spinal cord with LV/PST-injected animals with dorsal root injury, with or without a conditioning lesion.

A valuable new study by Ghosh et al. (2012) further confirms the efficacy of PSA expression in combination with SC transplantation and extends the earlier studies by using a clinically relevant contusion model and showing that PSA-overexpressing SCs support regeneration of corticospinal tract axons. Growth of this tract does not usually occur in the peripheral nerve or SC milieu. Following the transplantation of PST-overexpressing GFP–SCs or GFP–SCs into a thoracic contusion injury, only PST–SCs migrated across the implant/host cord interfaces and from the rostral and caudal borders up to 4.4 mm within adjacent host tissue. Intermingling of PST–GFP–SCs with host astrocytes was seen many millimeters from the lesion. In fact, the extensive migration of PST–GFP–SCs into the surrounding host cord left the injury epicenter with few implanted SCs although endogenous SCs filled the injury site. There was also extensive serotonergic and CST axon growth across the implant and beyond the caudal edge of the lesion in contrast to GFP–SCs. Implantation of PST–SCs improved locomotor outcomes in the open field (BBB) and on a horizontal ladder. A linear correlation analysis revealed that SC caudal migration and CST axon caudal growth were significantly correlated with the number of ladder footfall errors.

4 THE SC GRAFT/HOST SPINAL CORD INTERFACE

> *The simplistic and uniformly negative viewpoint is no longer tenable and it is clear that there is a normal process of reactive astrogliosis that exerts many beneficial functions and does not do harm.*
>
> **Sofroniew, 2009, p. 644**

The interfaces that form between SC implants and the host spinal cord will largely determine their success, as in all implantation strategies utilized for repair. The cellular bridge needs to support growth of axons across the injury and enable these axons to reenter the rostral and caudal spinal cord beyond to stimulate the central pattern generator and/or make new connections. SCs do not usually enter areas where astrocytes reside and, where they appose astrocytes, there is upregulation of CSPGs on the astrocytes, resulting in the formation of a boundary that halts axon growth (Afshari and Fawcett, 2012). If the SCs were able to exit the bridge and accompany growing axons, would departure of axons from the bridge, the "off-ramp," be more successful? Astrocytes that react to injury by, in part, overexpressing GFAP, generally, are considered to inhibit axon growth (Fawcett and Asher, 1999; Silver and Miller, 2004). Under what circumstances can they be modified to be more supportive of axon growth (Noble et al., 2011) as in development (Smith et al., 1986)? A variety of molecules that inhibit growth, among them CSPGs, accumulate in relation to implants (Jones et al., 2003; Lakatos et al., 2003; Plant et al., 2001; see also Tang et al., 2003). How can these be reduced or eliminated? As mentioned above, several interventions have been successful in improving axonal growth into the SC bridge, across the "on-ramp." Because achieving axon growth across the off-ramp has been more challenging, new protocols are being developed to modify the caudal interface. Fundamental to this challenge is the state of the astrocytes at this interface.

There are by now a number of reports regarding treatments that lead to morphological changes in astrocytes that appear to support axon growth. At the edges of the spinal cord injury or at the interfaces of the spinal cord with a SC implant, some astrocytes form not a distinct sharp border but a highly irregular one due to their extension of long finger-like processes into the lesion or the SC implant. These contrasting images are clearly portrayed in a cartoon included in a review by White and Jakeman (2008, Fig. 4). These slender GFAP+ processes may interdigitate with SCs, thereby enabling axon growth.

Irregular host–graft borders were observed in a study of human SC bridges in completely transected nude rat spinal cord (Guest et al., 1997). In cases where SCs and astrocyte processes were interdigitated at the interfaces, a variety of axon species, including those from the brainstem, were observed to have regenerated into the bridge. The authors speculated that in these cases there had been better dural repair and less damage to the interfaces during animal movement after surgery. When hemichannels containing SCs were implanted into a gap in the hemitransected spinal cord, astrocyte processes as well as regenerated brainstem axons were observed in the bridge (Xu et al., 1999); moreover, the close association of GFAP+ processes with regenerating axons arising from the rostral interface was illustrated. This may possibly have been due to a reduced inflammatory response, increased graft stability, less damage to the cord stump, and/or duraplasty to reestablish CSF circulation. Propriospinal axon regeneration from both stumps was significantly enhanced when the hemichannels containing both SCs and GDNF were transplanted (Iannotti et al., 2003), suggesting that both interfaces had become more permissive. In a more recent study using the same model with GDNF (Deng et al., 2011), as described in the section above, scrutiny of the SC graft/spinal cord tissue interfaces revealed striking arrays of slender parallel GFAP+ processes aligned with axons arising from the interfaces to extend into the SC bridge.

A similar image of ascending regenerating axons extending beyond an irregularly bordered transplant associated with BMP-4-treated astrocytes was shown in a report by Davies et al. (2006). There was a correlation between the extent of axonal regeneration and how well the astrocytes filled the lesion site. In areas bereft of astrocytes, the axons exhibited dystrophic endings at the lesion border. The transplanted astrocytes were not only well aligned but also led to better alignment of endogenous astrocytes near lesion borders. Close alignment of astrocytes and regenerating axons also was observed in acellular, fibrin-filled, microfiber-based poly-L-lactic acid conduits grafted into a complete transection gap (Hurtado et al., 2011). In these conduits, long distance rostrocaudal axonal regeneration occurred from propriospinal neurons and supraspinal neurons in the reticular formation, and red, raphe, and vestibular nuclei.

A study just completed by Williams et al. (2012) found similar GFAP+ extensions. Transplantation of a polymer channel containing SCs and Matrigel resulted in the formation of host spinal cord/graft interfaces with variations between sharp and irregular GFAP-positive boundaries. Regions of the interfaces that were irregular exhibited long astrocyte processes that entered the bridge, sometimes up to 1 mm (Fig. 1A), whereas, at sharp interfaces, astrocyte processes in a tight meshwork

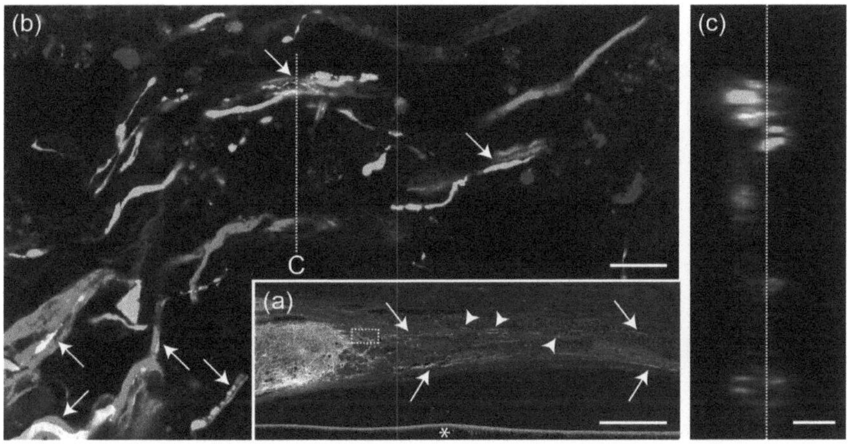

FIGURE 1

GFAP-positive processes cross the rostral host spinal cord/SC bridge interface together with brainstem axons. (a) Six weeks after the complete transection of the thoracic spinal cord, numerous EGFP-labeled brainstem axons (green, arrows), and GFAP-positive astrocyte processes (pink, arrowheads) have entered a SC bridge. SCs expressing S100 (blue) were mixed with Matrigel and injected as a fluid into a polymer channel (asterisk) placed between the severed spinal cord stumps. Scale bar = 0.5 mm. (b) A single *z*-plane confocal image near the rostral interface outlined by a dashed box in a. Arrows depict the association of SCs (blue), GFAP-positive processes (pink), and EGFP-labeled brainstem axons (green), which become white when in close proximity. Note how they elongate together, as a trio, from the rostral spinal cord/SC bridge interface. Scale bar = 5 μm. (c) A single *y*-plane of the region indicated by the dashed line in b demonstrates how the SCs and astrocyte processes bundle together and appear to form a tube that enables the brainstem axons to regenerate. The dashed line indicates the *z*-plane in which the image in (b) was taken. Scale bar = 2 μm (R.R. Williams, D.D. Pearse, and M.B. Bunge, unpublished observations). (For interpretation of the references to color in this figure legend, the reader is referred to the online version of this chapter.)

did not extend into the bridge (or injury, as illustrated in White and Jakeman, 2008). The insertion of initially fluid rather than pregelled Matrigel/SCs (initiated by Fouad et al., 2005) led to the formation of the more permissive irregular interfaces, possibly because the flow to the stumps enabled a more secure attachment between the graft and the spinal cord, with less infiltration of fibroblasts and inflammatory cells in the region of attachment. The amount of brainstem axon regeneration was directly associated with the number of astrocyte processes that entered the bridge from the rostral spinal cord/SC bridge interface. Immunostaining revealed that the GFAP-positive processes in the SC bridge were longitudinally oriented and in close proximity to S100-positive SCs and brainstem axons. The linear GFAP-positive processes appeared to form permissive tubes for the axons to emerge from the scar tissue at the interface and cross the SC bridge (Fig. 1b and c); the astrocytes, SCs, and regenerating brainstem axons were seen by electron microscopy to be contained together within a continuous basal lamina (Fig. 2). Images of astrocyte-like processes associated with axons and SCs bounded by basal lamina have been observed after CNS injury in electron micrographs by numerous investigators over many years,

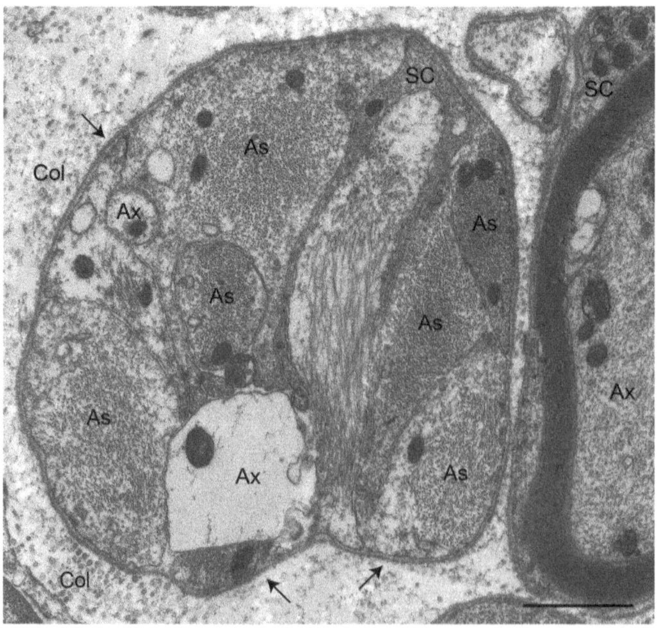

FIGURE 2

At the rostral host spinal cord/SC bridge interface, axons, SCs, and astrocyte processes are enclosed together within a continuous basal lamina. Six weeks after complete transection and implantation of a SC bridge, an electron micrograph shows axons (Ax), astrocyte processes (As), and SCs enclosed together within a continuous basal lamina (arrows). The basal lamina appears to form a tube surrounded by longitudinally oriented collagen fibers (Col). Scale bar=0.2 μm (Williams et al., 2012).

including Matthews et al. (1979) and Bunge et al. (1994). Thus, when irregular interfaces were present, SCs fostered brainstem axon regeneration into the bridge without additional interventions.

Although astrocyte processes may extend into a SC bridge from both the rostral and caudal interfaces after complete transection of the thoracic spinal cord, descending axons have not been directly observed to cross the SC bridge/spinal cord interface and regenerate significant distances into the caudal cord (Williams et al., 2012). The growth of descending axons across the caudal interface after complete transection/SC bridge implantation is challenging for many reasons. (i) These implants are relatively unstable compared to incomplete injury paradigms (Xu et al., 1999), which may cause increased scar formation at the interfaces. (ii) As discussed in Oudega et al. (1999), the caudal interface may sustain greater tissue loss than the rostral interface due to relative differences in blood supply after T8 transections. (iii) Given that CNS scar formation is progressive, descending axons may not exit the SC bridge due to temporal changes at the caudal interface, such as increased deposition of CSPGs (Plant et al., 2001). After axonal transection, growth may not be initiated until the injured neurons change their intrinsic state; this potential for delay is especially true for supraspinal axons whose somata are located more distant from the lesion. Because axons may regenerate up to 1 mm/day (Steward et al., 2003), by the time they cross relatively large lesion sites, such as complete transections and, more importantly, human lesions, the caudal interface is likely to be in a more chronic or inhibitory state. Interestingly, the paradigm utilized by Houle et al. (2006) allows axons to regenerate through the peripheral nerve graft before the distal end is implanted into the spinal cord. In this way, the regenerating axons encounter an acutely injured interface and are able to enter the spinal cord (in combination with chondroitinase). (iv) In addition to the expression of numerous growth inhibitory factors, chronic changes at the caudal interface may include an inability of the regenerating axons and their associated SCs to align with astrocyte processes extending from the caudal spinal cord. This may result in the deflection of axons away from the interface, as was observed for some brainstem axons that reached the caudal interface (R.R. Williams et al., unpublished results). (v) Alternatively, when descending axons and SCs reach the distal interface, they may cease growing and form synapses. In sum, the state of the scar found at the on-ramp, where axons enter the SC bridge, may be much different from the more chronically injured scar tissue that axons encounter when they reach the off-ramp. Understanding and manipulating factors that contribute to these differences will be paramount to enable descending or ascending axons to exit the SC bridge. New approaches as exemplified by the hemichannel/GDNF (Deng et al., 2011), peripheral nerve (Houle et al., 2006), chondroitinase (Bradbury and Carter, 2011), bone marrow stromal cells/NT-3 (Alto et al., 2009), and SCs/PSA (Ghosh et al., 2012) show promise in achieving this goal.

An area similar to the caudal interface of SC bridge/astrocyte-containing spinal cord tissue is the DREZ where the astrocytes/glia limitans preclude the mixing of SCs and astrocytes; SCs are prevented from entering the CNS. Axons that regenerate in the dorsal root also do not cross this zone (reviewed by Fraher, 1999; Golding et al., 1997; Ramer et al., 2001a) unless there are interventions such as neurotrophin application (McPhail et al., 2005; Ramer et al., 2000, 2001b, 2002), chondroitinase and/or macrophage activation with zymosan (Steinmetz et al., 2005), a conditioning lesion (Richardson and Issa, 1984), or local expression of PSA–NCAM (Zhang et al., 2007b; see above). Interestingly, when astrocytes are depleted from the spinal cord during development, SCs are able to migrate into the spinal cord subsequently (Gilmore and Sims, 1993).

Valuable *in vitro* models exist to detect changes in conditions that enable SCs and astrocytes to mix. The assumption is that if they can be treated to mix rather than to maintain their separate domains *in vitro*, then SCs will be able to cross the astrocyte boundaries after an injury *in vivo*. Usually, astrocytes and SCs remain in separate territories *in vitro*, maintaining a clear boundary between them unless treatments are introduced (Afshari et al., 2010a,b; Lakatos et al., 2000; Wilby et al., 1999). One is the boundary or confrontation assay in which a coculture is prepared using astrocytes and SCs occupying different territories with only a small gap in between. As the cells divide and migrate, the two populations become apposed. This allows assessment of the mixing behavior of these two cell types when they collide (Afshari and Fawcett, 2012; Afshari et al., 2011). Another technique is the inverted coverslip assay to study the migration of SCs from the edges of small glass fragments that have been positioned face down on astrocyte monolayers. Some of the modulating molecules identified using these techniques include N-cadherin (Wilby et al., 1999), CSPGs (Afshari et al., 2010b), FGF/heparin, Eph/Ephrins (Afshari et al., 2010a, 2011), GDNF (Deng et al., 2011), and polysialylated NCAM (Luo et al., 2011; Papastefanaki et al., 2007). These *in vitro* assays are important in suggesting new directions for creating more permissive SC/astrocyte interfaces *in vivo* and to better understand the mechanisms underlying their behavior.

In sum, recent work has revealed several interventions that improve axon regrowth into, across, and out of SC implants to better repair the spinal cord after injury. These interventions are to (i) transplant SCs that will migrate into astrocyte-containing host tissue territory, (ii) modify astrocytes to make them permissive for axon growth, and (iii) reduce or remove molecules inhibitory to axon growth that lurk in the scar. Many additions to protocols led to enhancing the permissivity of the off-ramp as well as the on-ramp for axon growth. The modification of NCAM to be polysialylated improved SC migration that was accompanied by increased axon growth including CST axons. Improved stability/more secure junctions between host tissue and implant; increased cAMP levels; and the addition of GDNF, neurotrophins, or chondroitinase were found to be efficacious. It appears that neurotrophins need to be introduced not only into implants but also into sites beyond to promote growth through the off-ramp and into the spinal cord (Alto et al., 2009; Taylor et al., 2006). Neurotrophin gradients may be required. Reduction in levels of inhibitory molecules such as CSPGs, by introducing chondroitinase ABC, also has been demonstrated to be critical for axon growth across interfaces. It will be both an exciting and important challenge to develop new findings such as these for clinical application.

Acknowledgments

Our experimental rat work and the preparation of SCs for clinical trials have been inspired by the vision of Dr. Richard P. Bunge. Many persons have contributed to this work; their contributions are acknowledged in the initial publications. We greatly appreciated their hard work and dedication. The Core Facilities in The Miami Project to Cure Paralysis have greatly facilitated our experiments. This work has been funded by The Miami Project to Cure Paralysis, continuous NIH funding over many years, The Christopher and Dana Reeve Foundation International Research Consortium, the Buoniconti Fund, and the Hollfelder and Heumann Foundations. M. B. B. is the Christine E. Lynn Distinguished Professor of Neuroscience. Expert word processing by Jenissia Jeanty is gratefully acknowledged.

References

Afshari, F.T., Fawcett, J.W., 2012. Astrocyte-Schwann-cell coculture systems. Methods Mol. Biol. 814, 381–391.

Afshari, F.T., Kwok, J.C., Fawcett, J.W., 2010a. Astrocyte-produced ephrins inhibit Schwann cell migration via VAV2 signaling. J. Neurosci. 30, 4246–4255.

Afshari, F.T., Kwok, J.C., White, L., Fawcett, J.W., 2010b. Schwann cell migration is integrin-dependent and inhibited by astrocyte-produced aggrecan. Glia 58, 857–869.

Afshari, F.T., Kwok, J.C., Fawcett, J.W., 2011. Analysis of Schwann-astrocyte interactions using in vitro assays. J. Vis. Exp. 47, 2214.

Alto, L.T., Havton, L.A., Conner, J.M., Hollis 2nd, E.R., Blesch, A., Tuszynski, M.H., 2009. Chemotropic guidance facilitates axonal regeneration and synapse formation after spinal cord injury. Nat. Neurosci. 12, 1106–1113.

Bamber, N.I., Li, H., Lu, X., Oudega, M., Aebischer, P., Xu, X.M., 2001. Neurotrophins, BDNF and NT-3 promote axonal re-entry into the distal host spinal cord through Schwann cell-seeded mini-channels. Eur. J. Neurosci. 13, 257–268.

Basso, D.M., Beattie, M.S., Bresnahan, J.C., 1995. A new, sensitive and reliable locomotor scale for open field testing in rats. J. Neurotrauma 12, 1–21.

Basso, D.M., Beattie, M.S., Bresnahan, J.C., Anderson, D.K., Faden, A.I., Gruner, J.A., Holford, T.R., Hsu, C.Y., Noble, L.J., Nockels, R., Perot, P.L., Salzman, S.K., Young, W., 1996. MASCIS evaluation of open field locomotor scores: effects of experience and teamwork on reliability. Multicenter Animal Spinal Cord Injury Study. J. Neurotrauma 13, 343–359.

Basso, D.M., Fisher, L.C., Anderson, A.J., Jakeman, L.B., McTigue, D.M., Popovich, P.G., 2006. Basso Mouse Scale for locomotion detects differences in recovery after spinal cord injury in five common mouse strains. J. Neurotrauma 23, 635–659.

Blakemore, W.F., Franklin, R.J., 2000. Transplantation options for therapeutic central nervous system remyelination. Cell Transplant. 9, 289–294.

Bradbury, E.J., Carter, L.M., 2011. Manipulating the glial scar: chondroitinase ABC as a therapy for spinal cord injury. Brain Res. Bull. 84, 306–316.

Bunge, R.P., 1975. Changing uses of nerve tissue culture 1950–1975. In: Tower, D.B. (Ed.), The Nervous System, Volume 1, The Basic Neuroscience. Raven Press, New York, pp. 31–42.

Bunge, M.B., 2008. Novel combination strategies to repair the injured mammalian spinal cord. J. Spinal Cord Med. 31, 262–269.

Bunge, M.B., Wood, P.M., 2012. Realizing the maximum potential of Schwann cells to promote recovery from spinal cord injury. In: Verhaagan, J., McDonald, J.W. (Eds.), third ed. Handbook of Clinical Neurology, Spinal Cord Trauma, vol. 109. Elsevier, New York (Chapter 34), p. 300.

Bunge, M.B., Williams, A.K., Wood, P.M., Uitto, J., Jeffrey, J.J., 1980. Comparison of nerve cell and nerve cell plus Schwann cell cultures, with particular emphasis on basal lamina and collagen formation. J. Cell Biol. 84, 184–202.

Bunge, M.B., Williams, A.K., Wood, P.M., 1982. Neuron-Schwann cell interaction in basal lamina formation. Dev. Biol. 92, 449–460.

Bunge, M.B., Holets, V.R., Bates, M.L., Clarke, T.S., Watson, B.D., 1994. Characterization of photochemically induced spinal cord injury in the rat by light and electron microscopy. Exp. Neurol. 127, 76–93.

Casella, G.T.B., Bunge, R.P., Wood, O.M., 1996. Improved method for harvesting human Schwann cells from mature peripheral nerve and expansion in vitro. Glia 17, 327–338.

Chau, C.H., Shum, D.K., Li, H., Pei, J., Lui, Y.Y., Wirthlin, L., Chan, Y.S., Xu, X.M., 2004. Chondroitinase ABC enhances axonal regrowth through Schwann cell-seeded guidance channels after spinal cord injury. FASEB J. 18, 194–196.

Chen, A., Xu, X.M., Kleitman, N., Bunge, M.B., 1996. Methylprednisolone administration improves axonal regeneration into Schwann cell grafts in transected adult rat thoracic spinal cord. Exp. Neurol. 138, 261–276.

Davies, J.E., Huang, C., Proschel, C., Noble, M., Mayer-Proschel, M., Davies, S.J., 2006. Astrocytes derived from glial-restricted precursors promote spinal cord repair. J. Biol. 5, 7.

Deng, L.X., Hu, J., Liu, N., Wang, X., Smith, G.M., Wen, X., Xu, X.M., 2011. GDNF modifies reactive astrogliosis allowing robust axonal regeneration through Schwann cell-seeded guidance channels after spinal cord injury. Exp. Neurol. 229, 238–250.

Eldridge, C.F., Bunge, M.B., Bunge, R.P., 1989. Differentiation of axon-related Schwann cells in vitro: II. Control of myelin formation by basal lamina. J. Neurosci. 9, 625–638.

Fawcett, J.W., Asher, R.A., 1999. The glial scar and central nervous system repair. Brain Res. Bull. 49, 377–391.

Fortun, J., Hill, C.E., Bunge, M.B., 2009. Combinatorial strategies with Schwann cell transplantation to improve repair of the injured spinal cord. Neurosci. Lett. 456 (3), 124–132.

Fouad, K., Schnell, L., Bunge, M.B., Schwab, M.E., Liebscher, T., Pearse, D.D., 2005. Combining Schwann cell bridges and olfactory ensheathing glia grafts with chondroitinase promotes locomotor recovery after complete transection of the spinal cord. J. Neurosci. 25, 1169–1178.

Fraher, J.P., 1999. The transitional zone and CNS regeneration. J. Anat. 194, 161–182.

Ghosh, M., Tuesta, L.M., Puentes, R., Patel, S., Melendez, K., El Maarouf, A., Rutishauser, U., Pearse, D.D., 2012. Extensive cell migration, axon regeneration, and improved function with polysialic acid-modified Schwann cells after spinal cord injury. Glia 60, 979–992.

Gilmore, S.A., Sims, T.J., 1993. Patterns of Schwann cell myelination of axons within the spinal cord. J. Chem. Neuroanat. 6, 191–199.

Golden, K.L., Pearse, D.D., Blits, B., Garg, M., Oudega, M., Wood, P.M., Bunge, M.B., 2007. Transduced Schwann cells promote axon growth and myelination after spinal cord injury. Exp. Neurol. 207, 203–217.

Golding, J., Shewan, D., Cohen, J., 1997. Maturation of the mammalian dorsal root entry zone-from entry to no entry. Trends Neurosci. 20, 303–308.

Grimpe, B., Pressman, Y., Lupa, M.D., Horn, K.P., Bunge, M.B., Silver, J., 2005. The role of proteoglycans in Schwann cell/astrocyte interactions and in regeneration failure at PNS/CNS interfaces. Mol. Cell. Neurosci. 28, 18–29.

Guest, J.D., Rao, A., Olson, L., Bunge, M.B., Bunge, R.P., 1997. The ability of human Schwann cell grafts to promote regeneration in the transected nude rat spinal cord. Exp. Neurol. 148, 502–522.

Hill, C.E., Hurtado, A., Blits, B., Bahr, B.A., Wood, P.M., Bunge, M.B., 2007. Early necrosis and apoptosis of Schwann cells transplanted into the injured rat spinal cord. Eur. J. Neurosci. 26, 1433–1445.

Houle, J.D., Tom, V.J., Mayes, D., Wagoner, G., Phillips, N., Silver, J., 2006. Combining an autologous peripheral nervous system "bridge" and matrix modification by chondroitinase allows robust, functional regeneration beyond a hemisection lesion of the adult rat spinal cord. J. Neurosci. 26, 7405–7415.

Hurtado, A., Cregg, J.M., Wang, H.B., Wendell, D.F., Oudega, M., Gilbert, R.J., McDonald, J.W., 2011. Robust CNS regeneration after complete spinal cord transection using aligned poly-L-lactic acid microfibers. Biomaterials 32, 6068–6079.

Iannotti, C., Li, H.Y., Yan, P., Lu, X.B., Wirthlin, L., Xu, X.M., 2003. Glial cell line-derived neurotrophic factor-enriched bridging transplants promote propriospinal axonal regeneration and enhance myelination after spinal cord injury. Exp. Neurol. 183, 379–393.

Iwashita, Y., Fawcett, J.W., Crang, A.J., Franklin, R.J., Blakemore, W.F., 2000. Schwann cells transplanted into normal and X-irradiated adult white matter do not migrate extensively and show poor long-term survival. Exp. Neurol. 164, 292–302.

Jones, L.L., Margolis, R.U., Tuszynski, M.H., 2003. The chondroitin sulfate proteoglycans neurocan, brevican, phosphacan, and versican are differentially regulated following spinal cord injury. Exp. Neurol. 182, 399–411.

Kanno, H., Pressman, Y., Moody, A., Berg, R., Pearse, D.D., Bunge, M.B., 2012. Combined engineering of Schwann cell implants to secrete neurotrophin and chondroitinase promotes axonal regeneration and locomotion after spinal cord injury. In: The 36th Annual Meeting of the Society for Neuroscience.

Lakatos, A., Franklin, R.J., Barnett, S.C., 2000. Olfactory ensheathing cells and Schwann cells differ in their in vitro interactions with astrocytes. Glia 32, 214–225.

Lakatos, A., Barnett, S.C., Franklin, R.J., 2003. Olfactory ensheathing cells induce less host astrocyte response and chondroitin sulphate proteoglycan expression than Schwann cells following transplantation into adult CNS white matter. Exp. Neurol. 184, 237–246.

Levi, A.D.O., 1996. Characterization of the technique involved in isolating Schwann cells from adult human peripheral nerve. J. Neurosci. Methods 68, 21–26.

Levi, A.D.O., Bunge, R.P., Lofgren, J.A., Meima, L., Hefti, F., Nikolics, K., Sliwkowski, M.X., 1995. The influence of heregulins on human Schwann cells proliferation. J. Neurosci. 15, 1329–1340.

Luo, J., Bo, X., Wu, D., Yeh, J., Richardson, P.M., Zhang, Y., 2011. Promoting survival, migration, and integration of transplanted Schwann cells by over-expressing polysialic acid. Glia 59, 424–434.

Matthews, M.A., St Onge, M.F., Faciane, C.L., Gelderd, J.B., 1979. Axon sprouting into segments of rat spinal cord adjacent to the site of a previous transection. Neuropathol. Appl. Neurobiol. 5, 181–196.

McPhail, L.T., Plunet, W.T., Das, P., Ramer, M.S., 2005. The astrocytic barrier to axonal regeneration at the dorsal root entry zone is induced by rhizotomy. Eur. J. Neurosci. 21, 267–270.
Meijs, M.F.L., Timmers, L., Pearse, D.D., Tresco, P.A., Bates, M.L., Joosten, E.A.J., Bunge, M.B., Oudega, M., 2004. Basic fibroblast growth factor promotes neuronal survival but not behavioral recovery in the transected and Schwann cell implanted rat thoracic spinal cord. J. Neurotrauma 21, 1415–1430.
Morrissey, T.K., Kleitman, N., Bunge, R.P., 1991. Isolation and functional characterization of Schwann cells derived from adult peripheral nerve. J. Neurosci. 11, 2433–2442.
Noble, M., Davies, J.E., Mayer-Pröschel, M., Pröschel, C., Davies, S.J., 2011. Precursor cell biology and the development of astrocyte transplantation therapies: lessons from spinal cord injury. Neurotherapeutics 8, 677–693.
Oudega, M., Xu, X.M., 2006. Schwann cell transplantation for repair of the adult spinal cord. J. Neurotrauma 23, 453–467.
Oudega, M., Vargas, C.G., Weber, A.B., Kleitman, N., Bunge, M.B., 1999. Long-term effects of methylprednisolone following transection of adult rat spinal cord. Eur. J. Neurosci. 11, 2453–2464.
Papastefanaki, F., Chen, P., Lavdas, A.A., Thornaidou, D., Schachner, M., Matsas, R., 2007. Grafts of Schwann cells engineered to express PSA-NCAM promote functional recovery after spinal cord injury. Brain 130, 2159–2174.
Pearse, D.D., Bunge, M.B., 2006. Designing cell- and gene-based regeneration strategies to repair the injured spinal cord. J. Neurotrauma 23, 438–452.
Pearse, D.D., Pereira, F.C., Marcillo, A.E., Bates, M.L., Berrocal, Y.A., Filbin, M.T., Bunge, M.B., 2004. cAMP and Schwann cells promote axonal growth and functional recovery after spinal cord injury. Nat. Med. 10, 610–616.
Pearse, D.D., Sanchez, A.R., Pereira, F.C., Andrade, C.M., Puzis, R., Pressman, Y., Golden, K.L., Kitay, B.M., Blits, B., Wood, P.M., Bunge, M.B., 2007. Transplantation of Schwann cells and/or olfactory ensheathing glia into the contused spinal cord: survival, migration, axon association and functional recovery. Glia 55, 976–1000.
Plant, G.W., Bates, M.L., Bunge, M.B., 2001. Inhibitory proteoglycan immunoreactivity is higher at the caudal than the rostral Schwann cell graft-transected spinal cord interface. Mol. Cell. Neurosci. 17, 471–487.
Ramer, M.S., Priestley, J.V., McMahon, S.B., 2000. Functional regeneration of sensory axons into the adult spinal cord. Nature 403, 312–316.
Ramer, M.S., McMahon, S.B., Priestley, J.V., 2001a. Axon regeneration across the dorsal root entry zone. Prog. Brain Res. 132, 621–639.
Ramer, M.S., Duraisingam, I., Priestley, J.V., McMahon, S.B., 2001b. Two-tiered inhibition of axon regeneration at the dorsal root entry zone. J. Neurosci. 21, 2651–2660.
Ramer, M.S., Bishop, T., Dockery, P., Mobarak, M.S., O'Leary, D., Fraher, J.P., Priestley, J.V., McMahon, S.B., 2002. Neurotrophin-3-mediated regeneration and recovery of proprioception following dorsal rhizotomy. Mol. Cell. Neurosci. 19, 239–249.
Ramón-Cueto, A., Plant, G.W., Avila, J., Bunge, M.B., 1998. Long-distance axonal regeneration in the transected adult rat spinal cord is promoted by olfactory ensheathing glia transplants. J. Neurosci. 18, 3803–3815.
Richardson, P.M., Issa, V.M., 1984. Peripheral injury enhances central regeneration of primary sensory neurones. Nature 309, 791–793.
Richardson, P.M., Issa, V.M., Aguayo, A.J., 1984. Regeneration of long spinal axons in the rat. J. Neurocytol. 13, 165–182.

Rutkowski, J.L., Tenekoon, G.I., McGillicuddy, J.E., 1992. Selective culture of mitotically active human Schwann cells from adult sural nerves. Ann. Neurol. 31, 580–586.

Saberi, H., Moshayedi, P., Aghayan, H.R., Arjmand, B., Hosseini, S.K., Emami-Razavi, S.H., Rahimi-Movaghar, V., Raza, M., Firouzi, M., 2008. Treatment of chronic thoracic spinal cord injury patients with autologous Schwann cell transplantation: an interim report on safety considerations and possible outcomes. Neurosci. Lett. 443, 46–50.

Saberi, H., Firouzi, M., Habibi, Z., Moshayedi, P., Aghayan, H.R., Arjmand, B., Hosseini, K., Razavi, H.E., Yekaninejad, M.S., 2011. Safety of intramedullary Schwann cell transplantation for postrehabilitation spinal cord injuries: 2-year follow-up of 33 cases. J. Neurosurg. Spine 15, 515–525.

Schaal, S.M., Kitay, B.M., Cho, K.S., Lo, T.P., Barakat, D.J., Marcillo, A.E., Sanchez, A.R., Andrade, C.M., Pearse, D.D., 2007. Schwann cell transplantation improves reticulospinal axon growth and forelimb strength after severe cervical spinal cord contusion. Cell Transplant. 16, 207–228.

Silver, J., Miller, J.H., 2004. Regeneration beyond the glial scar. Nat. Rev. Neurosci. 5, 146–156.

Smith, G.M., Miller, R.H., Silver, J., 1986. Changing role of forebrain astrocytes during development, regenerative failure, and induced regeneration upon translation. J. Comp. Neurol. 251, 23–43.

Sofroniew, M.V., 2009. Molecular dissection of reactive astrogliosis and glial scar formation. Trends Neurosci. 32, 638–647.

Steinmetz, M.P., Horn, K.P., Tom, V.J., Miller, J.H., Busch, S.A., Nair, D., Silver, D.J., Silver, J., 2005. Chronic enhancement of the intrinsic growth capacity of sensory neurons combined with the degradation of inhibitory proteoglycans allows functional regeneration of sensory axons through the dorsal root entry zone in the mammalian spinal cord. J. Neurosci. 25, 8066–8076.

Steward, O., Zheng, B., Tessier-Lavigne, M., 2003. False resurrections: distinguishing regenerated from spared axons in the injured central nervous system. J. Comp. Neurol. 459, 1–8.

Takami, T., Oudega, M., Bates, M.L., Wood, P.M., Kleitman, N., Bunge, M.B., 2002. Schwann cell but not olfactory ensheathing glia transplants improve hindlimb locomotor performance in the moderately contused adult rat thoracic spinal cord. J. Neurosci. 22, 6670–6681.

Tang, X., Davies, J.E., Davies, S.J., 2003. Changes in distribution, cell associations, and protein expression levels of NG2, neurocan, phosphacan, brevican, versican V2, and tenascin-C during acute to chronic maturation of spinal cord scar tissue. J. Neurosci. Res. 71, 427–444.

Taylor, L., Jones, L., Tuszynski, M.H., Blesch, A., 2006. Neurotrophin-3 gradients established by lentiviral gene delivery promote short-distance axonal bridging beyond cellular grafts in the injured spinal cord. J. Neurosci. 26, 9713–9721.

Tetzlaff, W., Okon, E.B., Karimi-Abdolrezaee, S., Hill, C.E., Sparling, J.S., Plemel, J.R., Plunet, W.T., Tsai, E.C., Baptiste, D., Smithson, L.J., Kawaja, M.D., Fehlings, M.G., Kwon, B.K., 2011. A systematic review of cellular transplantation therapies for spinal cord injury. J. Neurotrauma 28, 1611–1682.

Tuszynski, M.H., Weidner, N., McCormack, M., Miller, I., Powell, H., Conner, J., 1998. Grafts of genetically modified Schwann cells to the spinal cord: survival, axon growth, and myelination. Cell Transplant. 7, 187–196.

Urfer, R., Tsoulfas, P., Soppet, D., Escandón, E., Parada, L.F., Presta, L.G., 1994. The binding epitopes of neurotrophin-3 to its receptors trkC and gp75 and the design of a multifunctional human neurotrophin. EMBO J. 13, 5896–5909.

Vavrek, R., Pearse, D.D., Fouad, K., 2007. Neuronal populations capable of regeneration following a combined treatment in rats with spinal cord transection. J. Neurotrauma 24, 1667–1673.

White, R.E., Jakeman, L.B., 2008. Don't fence me in: harnessing the beneficial roles of astrocytes for spinal cord repair. Restor. Neurol. Neurosci. 26, 197–214.

Wilby, M.J., Muir, E.M., Fok-Seang, J., Gour, B.J., Blaschuk, O.W., Fawcett, J.W., 1999. N-Cadherin inhibits Schwann cell migration on astrocytes. Mol. Cell. Neurosci. 14, 66–84.

Williams, R.R., Pearse, D.D., Bunge, M.B., 2012. The importance of implanted Schwann cells and host astrocyte interactions in providing a permissive hindlimb movements after SCI. In. The 36th Annual Meeting of the Society for Neuroscience.

Wood, P.M., Bunge, R.P., 1975. Evidence that sensory axons are mitogenic for Schwann cells. Nature 256, 662–664.

Xu, X.M., Guénard, V., Kleitman, N., Bunge, M.B., 1995a. Axonal regeneration into Schwann cell–seeded guidance channels grafted into transected adult rat spinal cord. J. Comp. Neurol. 351, 145–160.

Xu, X.M., Guénard, V., Kleitman, N., Aebischer, P., Bunge, M.B., 1995b. Combination of BDNF and NT-3 promotes supraspinal axonal regeneration into Schwann cell grafts in adult rat thoracic spinal cord. Exp. Neurol. 134, 261–272.

Xu, X.M., Chen, A., Guénard, V., Kleitman, N., Bunge, M.B., 1997. Bridging Schwann cell transplants promote axonal regeneration from both the rostral and caudal stumps of transected adult rat spinal cord. J. Neurocytol. 26, 1–16.

Xu, X.M., Zhang, S.X., Li, H.Y., Aebischer, P., Bunge, M.B., 1999. Regrowth of axons into the distal spinal cord through a Schwann cell seeded mini-channel implanted into hemisected adult rat spinal cord. Eur. J. Neurosci. 11, 1723–1740.

Zhang, Y., Ghadiri-Sani, M., Zhang, X., Richardson, P.M., Yeh, J., Bo, X., 2007a. Induced expression of polysialic acid in the spinal cord promotes regeneration of sensory axons. Mol. Cell. Neurosci. 35, 109–119.

Zhang, Y., Zhang, X., Wu, D., Verhaagen, J., Richardson, P.M., Yeh, J., Bo, X., 2007b. Lentiviral-mediated expression of polysialic acid in spinal cord and conditioning lesion promote regeneration of sensory axons into spinal cord. Mol. Ther. 15, 1796–1804.

CHAPTER

Generation of motor neurons from pluripotent stem cells

16

Peter H. Chipman, Jeremy S. Toma, Victor F. Rafuse[1]

Department of Anatomy and Neurobiology, Dalhousie University, Halifax, NS, Canada

[1]Corresponding author. Tel.: +1-902 494-3609, Fax: +1-902-4941212, e-mail address: vrafuse@dal.ca

Abstract

Alpha motor neurons (also known as lower or skeletal motor neurons) have been studied extensively for over 100 years. Motor neurons control the contraction of skeletal muscles and thus are the final common pathway in the nervous system responsible for motor behavior. Muscles become paralyzed when their innervating motor neurons die because of injury or disease. Motor neuron diseases (MNDs), such as Amyotrophic Lateral Sclerosis, progressively destroy motor neurons until those inflicted succumb to the illness due to respiratory failure. One strategy being explored to study and treat muscle paralysis due to motor neuron loss involves deriving surrogate motor neurons from pluripotent stem cells. Guided by decades of research on the development of the spinal cord, recent advances in neurobiology have shown that functional motor neurons can be derived from mouse and human embryonic stem (ES) cells. Furthermore, ES cell-derived motor neurons restore motor behavior when transplanted into animal models of motor dysfunction. The recent discovery that mouse and human motor neurons can be derived from induced pluripotent stem (iPS) cells (i.e., somatic cells converted to pluripotency) has set the stage for the development of patient-specific therapies designed to treat movement disorders. Indeed, there is now hope within the scientific community that motor neurons derived from pluripotent stem cells will be used to treat MNDs through cell transplantation and/or to screen molecules that will prevent motor neuron death. In this chapter, we review the journey that led to the generation of motor neurons from ES and iPS cells, how stem cell-derived motor neurons have been used to treat/study motor dysfunction, and where the technology will likely lead to in the future.

Keywords

embryonic stem cells, induced pluripotent stem cells, motor neuron disease, motor neurons, transplantation

Progress in Brain Research, Volume 201, ISSN 0079-6123, http://dx.doi.org/10.1016/B978-0-444-59544-7.00015-9

1 INTRODUCTION

It seems fitting that alpha motor neurons (also known as lower or skeletal motor neurons) were among the first neuronal cell type to be intentionally derived from pluripotent stem cells (Wichterle et al., 2002). Alpha motor neurons have been studied for over 100 years, making them the most extensively researched cells in the mammalian central nervous system (CNS; Kernell, 2006). Our long-standing obsession with this small group of neurons is founded not only on their unique properties and attributes but also on our ability to analyze them. Motor neurons are the final common pathway in the induction of movement (Sherrington, 1906). This basic and easily quantifiable function is essential for the existence of all organisms possessing motor neurons (Arendt et al., 2008). In addition, motor neurons are relatively simple structures that are readily accessible for experimental analysis. In vertebrates, motor neurons are topographically mapped to the muscle targets they innervate (Hollyday et al., 1977; Landmesser, 1978a,b) making them ideal for the study of the assembly of neuronal circuits (Sharma and Izpisúa Belmonte, 2001). Motor neurons are also the target of diseases such as amyotrophic lateral sclerosis (ALS) and spinal muscular atrophy (SMA). These motor neuron diseases (MNDs) are devastating disorders that progressively destroy motor neurons, leading to increased muscle weakness and paralysis. There is no cure for MNDs, and even with advancements in supportive treatments most people with ALS die 3–5 years after diagnosis (Czaplinski et al., 2006). This grim reality has catalyzed stem cell biologists, MND specialists, and motor neuron scientists to work together in a collaborative effort to devise methodologies to derive motor neurons from pluripotent stem cells. There is great anticipation within the scientific community that motor neurons derived from pluripotent stem cells will be used to treat MNDs through cell transplantation therapies (Gowing and Svendsen, 2011) and to screen small molecules that will increase the survival/function of the diseased neurons (Vitale et al., 2011).

In the following sections, we review key developmental studies leading to the discovery of morphogens and transcription factors required for motor neuron differentiation and show how these discoveries were exploited to generate functional motor neurons from mouse and human embryonic stem (ES) cells. Second, we introduce induced pluripotent stem (iPS) cells and discuss how motor neurons derived from them can be used as a source of cells for transplantation therapy and as a model system to study MNDs. Finally, we consider current technologies that may circumvent some of the problems and concerns associated with the use of stem cell-derived neurons as a tool to treat CNS injuries and neurodegenerative diseases.

2 TOWARD THE GENESIS OF MOTOR NEURONS FROM STEM CELLS

One would be remiss to discuss the derivation of motor neurons from pluripotent cells without first reviewing some of the key developmental studies that provided the necessary insights, and tools, required for the task. Motor neurons are among the first cell type to differentiate from neuronal precursor cells in the ventral spinal cord (known as the neural tube) during early embryogenesis (Hollyday and Hamburger, 1977). Prior to motoneurogenesis, the ventral region contains a specialized group of epithelial cells known as the floor plate that transiently occupies the ventral midline of the neural tube (Baker, 1927; Kingsbury, 1930; reviewed by Wilson and Maden, 2005). During its brief existence, the floor plate directs a multitude of functions, one of which is to induce motor neuron differentiation. Dodd and colleagues (Placzek et al., 1991) elegantly discovered this instructional role in chick embryos by inducing floor plates in ectopic regions in the dorsal half of the neural tube. The misplaced floor plates induced surrounding neuronal precursor cells to become motor neurons indicating that the floor plate was both necessary and sufficient to induce motor neuron differentiation. Although the regulatory molecules were not identified at the time, the authors postulated that the floor plate cells must release a diffusible factor that induces motoneurogenesis (Placzek et al., 1991; Yamada et al., 1991). A clue to the factor's identity came a few years later when McMahon and colleagues (Echelard et al., 1993) discovered three genes in mice related to the *Drosophila* segment polarity gene, hedgehog (Nüsslein-Volhard and Wieschaus, 1980). One of these genes, which they termed sonic hedgehog (Shh), was expressed in the floor plate, as well as the notochord (a transient mesenchymal structure underlying the floor plate) and zone of polarizing activity (ZPA; a medial posterior region of the developing limb bud). Because ectopic expression of Shh in the mouse CNS led to activation of floor plate genes in neighboring cells, they concluded that Shh is a morphogen that patterns the ventral neural tube (Echelard et al., 1993). Several investigators, including those working with Jessell, McMahon, and Beachy, went on to show that Shh not only patterns the ventral neural tube but also induces neuronal precursor cells to differentiate into motor neurons (Chiang et al., 1996; Martí et al., 1995; Roelink et al., 1994, 1995; Tanabe et al., 1995). The Shh saga culminated when Ericson et al. (1996) elegantly showed that Shh acts as a graded signal to ventralize the neural tube by inducing progenitor cells to acquire specific neuronal fates (motor neurons or discrete subclasses of interneurons) based on their concentration threshold for Shh (reviewed by Briscoe and Ericson, 2001). That is, progenitor cells near the floor plate acquire a motor neuron identity because they are exposed to a high concentration of Shh and have a high threshold for Shh-mediated induction. Conversely, progenitor cells near the sulcus limitans differentiate into discrete interneuron subtypes because they have a low threshold for Shh-mediated signaling and are exposed to a low level of Shh (Briscoe and Ericson, 2001).

Ectodermal cells must first differentiate into neuroectoderm and form the neural tube prior to the ventralizing activity of Shh. Our knowledge of this process is limited compared to our understanding of the Shh signaling pathway. However, numerous studies indicate that ectodermal cells in the early embryo, as well as cultured ES cells, differentiate into neural progenitors cells by inhibiting BMP and regulating Wnt/FGF signaling (Chambers et al., 2009; Pera et al., 2004; Smith et al., 2008; Wilson et al., 2001). The first neural progenitor cells born during this initial phase of neurogenesis express genetic markers typical of forebrain cells (Wilson et al., 2000). Caudal neurons, such as those found in the spinal cord, therefore appear to differentiate from the rostral cells through the action of additional morphogens (Wilson & Edlund, 2001). One morphogen believed to be involved in this caudalization is *all-trans*-retinoic acid (RA). This hypothesis is based on the fact that active retinoids are localized in the posterior region of early neurula stage embryos (Chen et al., 1994; Wagner et al., 1992), and their active retinoid receptors are found throughout the embryo during gastrulation. In addition, *Xenopus* embryos treated with RA during neural tube formation form enlarged spinal cords at the expense of forebrain tissue (Durston et al., 1989). Finally, while mutant mice lacking RA receptors have abnormal skeletal and CNS structures anterior to the cervical vertebrae, the structures posterior to this region are relatively normal (Lohnes et al., 1994). Taken together, it appears that a sequence of events occurs in the generation of motor neurons during development. First, ectodermal cells differentiate into rostral neuroectoderm through the actions of BMP, Wnt, and FGF. The posterior region of the neuroectoderm is then caudalized by RA. Finally, neural progenitors cells near the floor plate acquire a motor neuron identity through the ventralizing actions of Shh (Briscoe & Ericson, 2001; Wichterle et al., 2002).

3 DERIVATION OF MOTOR NEURONS FROM ES CELLS

In 2002, Hynek Wichterle in Tom Jessell's laboratory showed for the first time that mouse ES cells can be directed to differentiate into motor neurons by exposing them to RA and an agonist of the Shh signaling pathway (Wichterle et al., 2002). In brief, Wichterle et al. (2002) cultured ES cells as free-floating clusters, which he termed embryoid bodies, for 2 days before adding RA and the agonist. Five days later, the embryoid bodies expressed several transcription factors found in motor neurons including NeuN, Lhx3, Isl1, and Hb9 (the latter being unique to postmitotic motor neurons; Arber et al., 1999). To determine whether the neurons exhibited behavior typical of embryonic motor neurons, Wichterle et al. (2002) transplanted ES cell-derived motor neurons into the spinal cord of chick embryos at the time of motoneurogenesis. Remarkably, they found that the grafted motor neurons extended numerous axons out of the spinal cord through the ventral root where they continued to project to peripheral muscles. Intramuscular axons from the ES cell-derived motor neurons contained synaptic vesicles appropriate for neurotransmission and induced clustering of postsynaptic acetylcholine receptors on the developing chick muscle fibers. Taken together, this landmark study showed that mouse ES cells rapidly develop into neurons with the molecular identity and behavioral traits typical of fully differentiated motor neurons when exposed to only two morphogens. Surprisingly, exogenous manipulation of BMP, Wnt, and FGF signaling was not required to differentiate ES cells into neuroectodermal cells prior to motor neuron differentiation. Although the reasons for this are not known, it likely occurs because BMP, Wnt, and FGF signaling takes place spontaneously within the microenvironment of the free-floating embryoid bodies (Chambers et al., 2009).

Most studies deriving motor neurons from pluripotent cells are based on the original method developed by Wichterle et al. (2002). Two elements in his protocol have proven to be invaluable for studying motor neurons and thus are worthy of further discussion. First, Wichterle et al. (2002) used ES cells generated from the inner cell mass of blastocysts harvested from transgenic mice expressing enhanced green fluorescent protein (eGFP) under the control of the Hb9 promoter (Wichterle et al., 2002). The decision to develop this ES cell line was particularly insightful in light of the fact that only 20–40% of ES cells treated with RA and an activator of the Shh signaling pathway differentiate into motor neurons. The other 60–80% of the cells differentiate into glutamatergic neurons and glia (Miles et al., 2004). eGFP expression is therefore an ideal noninvasive biomarker to identify ES cell-derived motor neurons for anatomical and electrophysiological analysis. In addition, a pure population of $eGFP^{+}$ motor neurons can be obtained, if needed, using fluorescent activated cell-sorting techniques (e.g., Soundararajan et al., 2007; Wichterle et al., 2002).

Although recombinant Shh protein activates Shh signaling in neuronal precursor cells *ex vivo* (Ericson et al., 1996), Wichterle et al. (2002) chose to use a synthetic small molecule, known as Hh-Ag, to activate the signaling pathway. This second important element in his protocol was critical because recombinant Shh protein is a very poor activator of Shh signaling in ES cells. This is likely due to the fact that recombinant Shh protein is not posttranslationally modified by cholesterol and fatty acids, both of which increase the potency of native Shh by 15- to 160-fold (Pepinsky et al., 1998; Taipale et al., 2000). Thus, while RA and native Shh differentiate ES cells into motor neurons, RA and recombinant Shh do not (Soundararajan et al., 2007). Hh-Ag is a potent activator of Shh signaling because it binds directly to the Shh pathway activator Smoothened (Smo; Frank-Kamenetsky et al., 2002), which is downstream of the hedgehog receptor Patched. This activation is sufficient to induce motor neuron differentiation when combined with RA. Smo agonists are now commercially available and are comparable to Hh-Ag in their capacity to promote motor neuron differentiation from ES cells.

3.1 Functional properties of mouse ES cell-derived motor neurons

In 2004, we sought to determine whether mouse ES cell-derived motor neurons acquire the same electrophysiological properties as young endogenous motor neurons (Miles et al., 2004). Using the same differentiation protocol developed by Wichterle et al. (2002), we found that cultured mouse ES cell-derived motor neurons expressed functionally appropriate GABA, glycine, and glutamate receptors as well as voltage-activated Na^{+}, K^{+}, and Ca^{2+} ion channels (Miles et al., 2004). Furthermore, when injected with depolarizing current, ES cell-derived motor neurons fired repetitive action potentials with a spike frequency adaption profile typical of endogenous motor neurons. Finally, we showed that ES cell-derived motor neurons make functional connections with muscle fibers when cocultured *in vitro* (Miles et al., 2004; see also Harper et al., 2004). Together with the chick embryo transplantation studies performed by Wichterle et al. (2002), these results indicate that mouse ES cell-derived motor neurons acquire most, if not all, of the morphological, behavioral, and electrophysiological properties of their endogenous counterparts.

RA and Smo agonist-treated mouse ES cells appear to differentiate into a specific subclass of motor neurons whose molecular identity, behavior, and physiology resemble those of cells that reside in the medial aspect of the medial motor column (MMC_m). These motor neurons selectively project axons to epaxial muscles lining the vertebral column (Soundararajan et al., 2006, 2007, 2010; Wichterle et al., 2002). This conclusion is based on our observations that RA and Hh-Ag treated mouse ES cell-derived motor neurons migrate to the MMC_m, where they develop the same passive membrane properties as their endogenous MMC_m counterparts when transplanted into the neural tube of chick embryos at the time of motoneurogenesis (Soundararajan et al., 2006). Furthermore, we found that the grafted neurons selectively projected axons to epaxial muscles lining the vertebral column because they express guidance cues typical of MMC_m motor neurons (Soundararajan et al., 2006, 2007). Taken together, these results indicate that specific subclasses of neurons can be generated from ES cells and that additional morphogens are likely required to produce limb innervating motor neurons.

3.2 Human ES cell-derived motor neurons

Three years after motor neurons were generated from mouse ES cells, Zhang and colleagues (Li et al., 2005) and later Singh Roy et al. (2005) demonstrated that motor neurons could be derived from human ES cells. As one might expect, deriving motor neurons from human ES cells is more laborious, takes longer, and requires additional factors compared to the derivation of motor neurons from mouse ES cells (Li et al., 2005; Shin et al., 2005; Wada et al., 2009). Typically, human ES cells are cultured for 1 week on an adherent substrate to form rosettes of neuroectodermal cells that are later harvested/isolated and treated with RA and a Smo agonist to induce motor neuron differentiation. One to two weeks later, the cultures of differentiated motor neuron progenitor cells are treated with a cocktail of trophic factors (BDNF, GDNF, and CNTF, or IGF-1) known to potentiate motor neuron differentiation and survival (Zurn et al., 1996). Like motor neurons derived from mouse ES cells, human ES cell-derived motor neurons express motor neuron-specific transcription factors and receptors, acquire synaptic machinery appropriate for neurotransmission, fire spontaneous action potentials (Lee et al., 2007), and induce postsynaptic acetylcholine receptor clustering when cocultured with myotubes (Li et al., 2005; Singh Roy et al., 2005; Wada et al., 2009). Thus, while more expertise, patience, time, money, and perseverance are required to generate motor neurons from human ES cells, the final product is remarkably similar to endogenous motor neurons and those derived from mouse ES cells.

3.3 ES cell-derived motor neurons: Preclinical studies

Cell replacement therapies, using ES cell-derived motor neurons, may restore motor function in individuals who have lost motor neurons due to injury or disease provided the transplanted cells integrate with the host's own circuitry. To determine whether this can occur, Kerr and colleagues transplanted mouse ES cell-derived motor neurons into the spinal cord of adult rats paralyzed with a neuroadapted Sindbis virus (Deshpande et al., 2006; Harper et al., 2004). Neuroadapted Sindbis virus is a neuronotropic virus that specifically targets motor neurons in the spinal cord (Griffin et al., 1994; Jackson et al., 1987), causing hindlimb paralysis due to the loss of ~80% of the axons in the ventral root (Kerr et al., 2002, 2003). Although many transplantation strategies were used, the best outcomes occurred when GDNF secreting neural stem cells were injected into the ventral roots near the injection site and the rats were administered a phosphodiesterase 4 inhibitor and dibutyryl cyclic adenosine monophosphate (dbcAMP; Deshpande et al., 2006). The phosphodiesterase 4 inhibitor and dbcAMP were administered to improve axonal growth in the inhibitory, myelin-enriched environment of the spinal cord (Pearse et al., 2004), while GDNF-secreting cells were injected into the target nerve to entice axonal outgrowth into the periphery. Using this multitargeted approach, Deshpande et al. (2006) showed that ~15% of the ES cell-derived motor neurons were still alive 6 months after transplantation and many extended axons out through the ventral roots where they eventually formed neuromuscular junctions with denervated muscle fibers. Finally, rats receiving ES cell-derived motor neurons, GDNF secreting cells, and growth enhancers showed significantly better motor function based on hind limb grip strength and blinded locomotor assessments. Taken together, these preclinical studies suggest that cell replacement therapies using ES cell-derived motor neurons may be a viable treatment strategy for restoring meaningful movement in patients with motor neuron loss provided they extend axons out of the cord and innervate distal muscle fibers.

In 2007, Lee et al. (2007) showed that human ES cell-derived motor neurons survive transplantation and extend axons into the periphery via the ventral root when grafted into the developing spinal cord of chick embryos. These results indicate that human ES cell-derived motor neurons, like their mouse ES counterparts, respond to endogenous cues that guide their axons to appropriate muscle targets. To ascertain whether human ES cell-derived motor neurons survive and integrate in a postnatal spinal cord, Studer and colleagues transplanted 80,000 RA/Shh-treated human ES cells into the cervical spinal cord of adult rats (Lee et al., 2007). Several ChAT-expressing cells remained 6 weeks after transplantation, and extensive fiber outgrowth was observed outside the graft site. Outgrowth into the spinal nerves was not examined, but seems unlikely considering that GDNF-secreting cells were not transplanted into the spinal nerve, as described by Deshpande et al. (2006). Disease onset was delayed, and motor neuron survival improved, in rodent models of ALS (López-González et al., 2009) and SMA (Corti et al., 2010) when ES cell-derived motor neurons were transplanted into the spinal cords. However, these improvements were attributed to the release of neuroprotective factors from the transplanted cells rather than to functional integration with host tissue (Corti et al., 2010).

Transplanting motor neurons into the spinal cord is flushed with technical challenges including poor survival and growth. Even if the transplanted motor neurons innervate numerous muscles, it seems unlikely that meaningful motor function will be restored unless they receive synaptic input that is appropriate for the muscle groups innervated (e.g., flexor vs. extensor muscle groups). These challenges can be avoided if ES cell-derived motor neurons are transplanted close to the target muscles. For example, inspired by transplantation studies using fetal spinal cord tissue (Thomas et al., 2000), we grafted ES cell-derived motor neurons into the distal end of transected peripheral mouse nerves and quantified muscle function 3–18 months later (Yohn et al., 2008; see also Craff et al., 2007; Kubo et al., 2009). We found that the ES cell-derived motor neurons grew axons into the denervated muscles where they formed functional synapses. Furthermore, when we electrically stimulated the grafted ES cell-derived motor neurons, the reinnervated muscles produced $\sim$40% of their original prelesion force (Yohn et al., 2008). Surprisingly, we found that the number of slow muscle fibers increased, and denervation-associated muscle atrophy was reduced in the reinnervated muscles. These latter results indicated that transplanted ES cell-derived motor neurons integrated with the host tissue to the extent that they changed the phenotype of the muscle fibers they innervated (Craff et al., 2007; Yohn et al., 2008).

Can motor neurons transplanted into a peripheral nerve be used clinically? In mice, ES cell-derived motor neurons survive for at least 18 months if they make functional connections with previously denervated muscle fibers (Yohn et al., 2008). If ES cell-derived motor neurons survive and function equally well when transplanted into human peripheral nerves, they could be used to restore meaningful function to paralyzed muscles through exogenous electrical stimulation. This concept is based, in part, on studies restoring hand function in persons with tetraplegia using neuroprosthesis (Gan et al., 2012; Peckham et al., 2002). This technology uses externally controlled electrical stimulation of endogenous nerves through implanted electrodes to activate paralyzed flexor and extensor muscles to open and close the hand. This rehabilitation strategy is only feasible, however, if meaningful force is generated upon electrical nerve stimulation. Unfortunately, many spinal cord injuries induce motor neuron cell death around the trauma site. As a result, many muscles in individuals with a spinal cord injury become permanently denervated and incapable of producing force even when the peripheral nerve is electrically stimulated (Thomas et al., 1997). Under such conditions, ES cell-derived motor neurons transplanted into the peripheral nerve could become surrogate motor neurons that would produce meaningful contractions upon electrical stimulation. A similar cell replacement strategy could be used in instances where motor nerves were lost due to a peripheral nerve injury or were surgically excised due to a peripheral neuroma.

4 iPS CELL-DERIVED MOTOR NEURONS

Developing clinical therapies involving neurons derived from human ES cells is burdened with social and logistical concerns because it requires embryonic tissue that has, or at the very least had, the potential to generate life. Furthermore, allografts usually require ongoing immunosuppression to prevent tissue rejection, although this may be less of an issue when cells are transplanted into the CNS (e.g., Mendez et al.,

2008). Many of these societal concerns and technical challenges would likely be resolved if transplanted neurons were derived from the recipient's own somatic cells rather than embryonic tissue.

In a landmark study, designed to define the smallest number of factors needed to bestow the pluripotency of ES cells onto terminally differentiated somatic cells, Takahashi and Yamanaka (2006) showed that mouse embryonic and adult fibroblasts become pluripotent when forced to express four ES cell-specific genes. These four genes, transcription factors octamer 3/4 (*Oct4*), SRY box–containing gene 2 (*Sox2*), Kruppel-like factor 4 (*Klf4*), and *c-Myc* are now commonly known as "Yamanaka" factors (Liu et al., 2008), and the pluripotent cells derived from their forced expression are termed iPS cells (Takahashi and Yamanaka, 2006; Fig. 1). In their initial study, Takahashi and Yamanaka (2006) showed that mouse iPS cells were truly pluripotent because they formed teratomas (differentiated into cell types found in all three embryonic germ layers) when transplanted into immunodeficient mice. However, they differed from ES cells in their DNA methylation pattern. By refining their reprogramming protocol further, Yamanaka and colleagues overcame these issues and showed that adult mouse fibroblasts can be converted into iPS cells that contribute to adult chimeras (Okita et al., 2007). Shortly afterward, Yamanaka and two other groups generated iPS cells from human fibroblasts (Park et al., 2008; Takahashi et al., 2007; Yu et al., 2007), setting the stage for the generation of patient-specific pluripotent stem cells that can be used for autologous tissue transplantation or for modeling neurodegenerative diseases (Kiskinis and Eggan, 2010).

In 2008, Kevin Eggan and colleagues showed that iPS cells could be generated from a skin biopsy collected from an individual diagnosed with a familial form of ALS (Dimos et al., 2008). Using the same differentiation protocol developed for human ES cells, Dimos et al. (2008) went on to show that human iPS cells could be directed to differentiate into neurons expressing motor neuron restricted transcription factors. A year later, Clive Svendsen and colleagues generated cells with the morphological and genetic signature of motor neurons from iPS cells derived from a patient with SMA (Ebert et al., 2008). Furthermore, they found that SMA protein was increased in iPS cell-derived motor neurons when treated with the SMA-inducing compounds, valproic acid and tobramycin. These latter results suggest that iPS cell-derived motor neurons retain biochemical traits similar to those found in the host's own nervous system. More recently, Bilican et al. (2012) derived motor neurons from iPS cells generated from fibroblasts collected from an individual with a TDP-43 M337V mutation (another familial form of ALS; Sreedharan et al., 2008). The iPS cell-derived motor neurons fired TTX-sensitive repetitive action potentials when depolarized (see also Karumbayaram et al., 2009) and, like their endogenous counterparts, displayed biochemical features of TDP-43 misaccumulation (Bilican et al., 2012). Thus, iPS cell-derived motor neurons, derived from biopsies taken from individuals with SMA or a familial form of ALS, acquire key pathological characteristics associated with the disease presumably because they were genetically imprinted by the somatic cells from which they were derived.

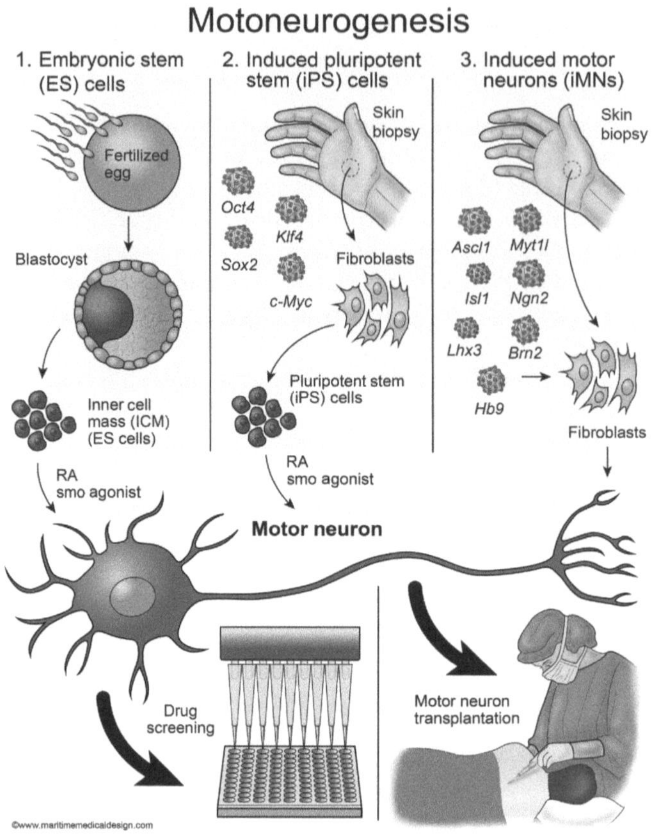

FIGURE 1

Illustration showing the derivation of motor neurons using three distinct techniques. (1) ES cells isolated from the inner cell mass of a blastocyst differentiate into motor neurons when treated with RA and an Smo agonist. (2) Skin cells are converted into iPS cells by forced expression of specific genes (typically Oct4, Klf4, Sox2, and c-Myc). iPS cells later differentiate into motor neurons when exposed to RA and an Smo agonist. (3) Skin cells are directly converted into motor neurons by forced expression of seven genes. (For color version of this figure, the reader is referred to the online version of this chapter.)

Nonhereditary neurodegenerative disorders such as sporadic ALS may be attributed to a combination of oligogenic inheritance and environmental signals (Andersen and Al-Chalabi, 2011). If true, motor neurons derived from iPS cells generated from individuals with sporadic ALS should be genetically predisposed to manifest the disease. This would make iPS cell-derived neurons an ideal model system to screen compounds to treat ALS and other sporadic neurodegenerative disorders such as Parkinson's disease. This view is substantiated by recent reports demonstrating that iPS cell-derived neurons recapitulate several disease phenotypes (Carvajal-Vergara et al., 2010; Lee et al., 2009; Urbach et al., 2010; reviewed by Grskovic et al., 2011).

5 DIRECT CONVERSION OF FIBROBLASTS INTO INDUCED MOTOR NEURONS

Inspired by Yamanaka and colleagues pioneering studies on iPS cells (Takahashi and Yamanaka, 2006), several groups have shown that it is possible to reprogram fibroblasts into postmitotic neurons using distinct sets of transcriptions factors (Caiazzo et al., 2011; Pfisterer et al., 2011; Vierbuchen et al., 2010). More recently, Kevin Eggan and colleagues used a similar approach to directly convert mouse and human fibroblasts into a type of motor neuron they termed induced motor neurons (iMNs; Son et al., 2011). They found that seven factors (Ascl1, Brn2, Myt1l, Lhx3, Hb9, Isl1, and Ngn2) were sufficient to induce mouse fibroblasts to become motor neurons while an additional eighth factor (NEUROD1) was required for successful induction of human fibroblasts (Son et al., 2011). iMNs displayed electrophysiological characteristics and acquired gene expression patterns typical of endogenous motor neurons. In addition, they formed functional synapses with muscle fibers in culture and extended axons into the periphery when transplanted into the developing chick spinal cord. Like endogenous (Haidet-Phillips et al., 2011) and ES cell-derived motor neurons, iMNs degenerated when cocultured with astrocytes expressing a mutated form of superoxide dismutase 1 (Son et al., 2011). Finally, Son et al. (2011) demonstrated that fibroblasts converted into iMNs do not transit through a proliferative state. This latter observation is particularly important with respect to cell replacement therapies because grafts containing iMNs will not contain proliferative-competent cells that could become tumorigenic. The same is not true for transplants containing ES and iPS cell-derived motor neurons (Victor Rafuse, unpublished results). While this approach for generating motor neurons is exciting, it remains to be determined whether iMNs have the capacity to form functional connections with denervated muscle fibers when transplanted into an adult host (e.g., Yohn et al., 2008).

6 SUMMARY AND FUTURE DIRECTIONS

There has been tremendous progress since Wichterle et al. (2002) showed that motor neurons could be derived from mouse ES cells using RA and an Smo agonist (Fig. 1, left column #1). Since then, ES cell-derived motor neurons have been shown to resemble their endogenous counterparts anatomically, physiologically, biochemically, and behaviorally. Furthermore, ES cell-derived motor neurons partially restore motor behavior when transplanted into animal models of motor dysfunction. More recently, iPS cells have been shown to readily differentiate into motor neurons when treated with RA and an Smo agonist (Fig. 1, middle column #2). Although not as well studied, iPS cell-derived motor neurons have many of the same anatomical, biochemical, and physiological properties as their ES cell-derived equivalents. Motor neurons can also be derived directly from somatic cells (Fig. 1, right column #3). Unlike those derived from ES and iPS cells, motor neurons derived by this latter technique do not require morphogens or the generation of pluripotent progenitor cells. Despite their origin and methods of derivation, iMNs have anatomical and electrophysiological properties that are typical of young endogenous motor neurons.

There are now three novel ways to generate motor neurons using techniques that were virtually unheard of just over a decade ago (Fig. 1). Future research must now focus on their application for studying and treating motor neuron disorders due to injury and/or disease. For example, transplantation studies will need to pioneer better ways to prolong the survival of the transplants and promote better outgrowth from the graft. In addition, meaningful movement will only be restored if the transplanted motor neurons are activated in a manner that is appropriate for the muscles they reinnervate. Consequently, future *in vivo* studies will have to evaluate the type of synaptic inputs the transplants receive and determine whether they are suitable for regulating voluntary and involuntary movements. Because denervated muscle fibers progressively lose their ability to be reinnervated over time (Fu and Gordon, 1995), future transplantation studies will have to be performed to determine when stem cell-derived motor neurons should be transplanted for best functional outcomes.

In addition to cell replacement therapies, stem cell-derived motor neurons and iMNs are also well suited for studying the pathophysiology causing MNDs. Both iMNs and iPS cell-derived motor neurons can be derived from somatic cells collected from mice models of MNDs and from individuals with a MND. These motor neurons exhibit much of the same pathophysiology that their endogenous counterparts do (e.g., Bilican et al., 2012; Son et al., 2011) making them ideal for rapidly screening thousands of compounds that may promote better survival. However, motor neurons become dysfunctional long before they perish in MNDs (Balice-Gordon et al., 2000). Consequently, novel *in vitro* model systems will have to be developed in order to rapidly screen motor neuron dysfunction and not simply cell death. Finally, it is well established that the largest motor neurons die first in MNDs (Fischer et al., 2004; Theys et al., 1999) while other motor neurons such as those of the Onuf nucleus are relatively resistant to the disease (Carvalho et al., 1995). ES and iPS cells typically differentiate into smaller postural motor neurons within the MMC_m when treated with RA and an Smo agonist (Soundararajan et al., 2006; Wichterle et al., 2002). Consequently, future research should examine whether pluripotent cells, or iMNs, can be directed to differentiate into larger motor neurons, and if so, whether they exhibit different pathophysiology compared to ES cell-derived MMC_m motor neurons. It is likely that important insights into MNDs will be gained by systematically comparing motor neurons that are vulnerable to the disease with those that are not. In summary, the last 25 years have been witness to a remarkable journey of discoveries in the field of motoneurogenesis. In less than a quarter of a century, we have gone from determining that the floor plate regulates dorsoventral patterning of the neural tube to identifying a small set of genes that can convert adult skin cells into functional motor neurons. With thisamazing progress in science, it is hard to imagine what the next 25 years will have in store. Hopefully, they will include the discovery of effective strategies to treat, or even cure, MNDs as well as motor dysfunction due to injury.

References

Andersen, P.M., Al-Chalabi, A., 2011. Clinical genetics of amyotrophic lateral sclerosis: what do we really know? Nat. Rev. Neurol. 7, 603–615.

Arber, S., Han, B., Mendelsohn, M., Smith, M., Jessell, T.M., Sockanathan, S., 1999. Requirement for the homeobox gene Hb9 in the consolidation of motor neuron identity. Neuron 23, 659–674.

Arendt, D., Denes, A.S., Jékely, G., Tessmar-Raible, K., 2008. The evolution of nervous system centralization. Philos. Trans. R. Soc. Lond. B Biol. Sci. 363, 1523–1528.

Baker, R.C., 1927. The early development of the ventral part of the neural plate of amblystoma. J. Comp. Neurol. 44, 1–27.

Balice-Gordon, R.J., Smith, D.B., Goldman, J., Cork, L.C., Shirley, A., Cope, T.C., Pinter, M.J., 2000. Functional motor unit failure precedes neuromuscular degeneration in canine motor neuron disease. Ann. Neurol. 47, 596–605.

Bilican, B., Serio, A., Barmada, S.J., Nishimura, A.L., Sullivan, G.J., Carrasco, M., Phatnani, H.P., Puddifoot, C.A., Story, D., Fletcher, J., Park, I.H., Friedman, B.A., Daley, G.Q., Wyllie, D.J., Hardingham, G.E., Wilmut, I., Finkbeiner, S., Maniatis, T., Shaw, C.E., Chandran, S., 2012. Mutant induced pluripotent stem cell lines recapitulate aspects of TDP-43 proteinopathies and reveal cell-specific vulnerability. Proc. Natl. Acad. Sci. U.S.A. 109, 5803–5808.

Briscoe, J., Ericson, J., 2001. Specification of neuronal fates in the ventral neural tube. Curr. Opin. Neurobiol. 11, 43–49.

Caiazzo, M., Dell'Anno, M.T., Dvoretskova, E., Lazarevic, D., Taverna, S., Leo, D., Sotnikova, T.D., Menegon, A., Roncaglia, P., Colciago, G., Russo, G., Carninci, P., Pezzoli, G., Gainetdinov, R.R., Gustincich, S., Dityatev, A., Broccoli, V., 2011. Direct generation of functional dopaminergic neurons from mouse and human fibroblasts. Nature 476, 224–227.

Carvajal-Vergara, X., Sevilla, A., D'Souza, S.L., Ang, Y.S., Schaniel, C., Lee, D.F., Yang, L., Kaplan, A.D., Adler, E.D., Rozov, R., Ge, Y., Cohen, N., Edelmann, L.J., Chang, B., Waghray, A., Su, J., Pardo, S., Lichtenbelt, K.D., Tartaglia, M., Gelb, B.D., Lemischka, I.R., 2010. Patient-specific induced pluripotent stem-cell-derived models of LEOPARD syndrome. Nature 465, 808–812.

Carvalho, M., Schwartz, M.S., Swash, M., 1995. Involvement of the external anal sphincter in amyotrophic lateral sclerosis. Muscle Nerve 18, 848–853.

Chambers, S.M., Fasano, C.A., Papapetrou, E.P., Tomishima, M., Sadelain, M., Studer, L., 2009. Highly efficient neural conversion of human ES and iPS cells by dual inhibition of SMAD signaling. Nat. Biotechnol. 27, 275–280.

Chen, Y., Huang, L., Solursh, M., 1994. A concentration gradient of retinoids in the early *Xenopus laevis* embryo. Dev. Biol. 161, 70–76.

Chiang, C., Litingtung, Y., Lee, E., Young, K.E., Corden, J.L., Westphal, H., Beachy, P.A., 1996. Cyclopia and defective axial patterning in mice lacking Sonic hedgehog gene function. Nature 383, 407–413.

Corti, S., Nizzardo, M., Nardini, M., Donadoni, C., Salani, S., Ronchi, D., Simone, C., Falcone, M., Papadimitriou, D., Locatelli, F., Mezzina, N., Gianni, F., Bresolin, N., Comi, G.P., 2010. Embryonic stem cell-derived neural stem cells improve spinal muscular atrophy phenotype in mice. Brain 133, 465–481.

Craff, M.N., Zeballos, J.L., Johnson, T.S., Ranka, M.P., Howard, R., Motarjem, P., Randolph, M.A., Winograd, J.M., 2007. Embryonic stem cell-derived motor neurons preserve muscle after peripheral nerve injury. Plast. Reconstr. Surg. 119, 235–245.

Czaplinski, A., Yen, A.A., Simpson, E.P., Appel, S.H., 2006. Slower disease progression and prolonged survival in contemporary patients with amyotrophic lateral sclerosis: is the natural history of amyotrophic lateral sclerosis changing? Arch. Neurol. 63, 1139–1143.

Deshpande, D.M., Kim, Y.S., Martinez, T., Carmen, J., Dike, S., Shats, I., Rubin, L.L., Drummond, J., Krishnan, C., Hoke, A., Maragakis, N., Shefner, J., Rothstein, J.D., Kerr, D.A., 2006. Recovery from paralysis in adult rats using embryonic stem cells. Ann. Neurol. 60, 32–44.

Dimos, J.T., Rodolfa, K.T., Niakan, K.K., Weisenthal, L.M., Mitsumoto, H., Chung, W., Croft, G.F., Saphier, G., Leibel, R., Goland, R., Wichterle, H., Henderson, C.E., Eggan, K., 2008. Induced pluripotent stem cells generated from patients with ALS can be differentiated into motor neurons. Science 321, 1218–1221.

Durston, A.J., Timmermans, J.P., Hage, W.J., Hendriks, H.F., de Vries, N.J., Heideveld, M., Nieuwkoop, P.D., 1989. Retinoic acid causes an anteroposterior transformation in the developing central nervous system. Nature 340, 140–144.

Ebert, A.D., Yu, J., Rose Jr., F.F., Mattis, V.B., Lorson, C.L., Thomson, J.A., Svendsen, C.N., 2009. Induced pluripotent stem cells from a spinal muscular atrophy patient. Nature 457, 277–280.

Echelard, Y., Epstein, D.J., St-Jacques, B., Shen, L., Mohler, J., McMahon, J.A., McMahon, A.P., 1993. Sonic hedgehog, a member of a family of putative signaling molecules, is implicated in the regulation of CNS polarity. Cell 75, 1417–1430.

Ericson, J., Morton, S., Kawakami, A., Roelink, H., Jessell, T.M., 1996. Two critical periods of Sonic Hedgehog signaling required for the specification of motor neuron identity. Cell 87, 661–673.

Fischer, L.R., Culver, D.G., Tennant, P., Davis, A.A., Wang, M., Castellano-Sanchez, A., Khan, J., Polak, M.A., Glass, J.D., 2004. Amyotrophic lateral sclerosis is a distal axonopathy: evidence in mice and man. Exp. Neurol. 185, 232–240.

Frank-Kamenetsky, M., Zhang, X.M., Bottega, S., Guicherit, O., Wichterle, H., Dudek, H., Bumcrot, D., Wang, F.Y., Jones, S., Shulok, J., Rubin, L.L., Porter, J.A., 2002. Small-molecule modulators of Hedgehog signaling: identification and characterization of Smoothened agonists and antagonists. J. Biol. 1, 10.

Fu, S.Y., Gordon, T., 1995. Contributing factors to poor functional recovery after delayed nerve repair: prolonged denervation. J. Neurosci. 15, 3886–3895.

Gan, L.S., Ravid, E., Kowalczewski, J.A., Olson, J.L., Morhart, M., Prochazka, A., 2012. First permanent implant of nerve stimulation leads activated by surface electrodes, enabling hand grasp and release: the stimulus router neuroprosthesis. Neurorehabil. Neural Repair 26, 335–343.

Gowing, G., Svendsen, C.N., 2011. Stem cell transplantation for motor neuron disease: current approaches and future perspectives. Neurotherapeutics 8, 591–606.

Griffin, D.E., Levine, B., Ubol, S., Hardwick, J.M., 1994. The effects of alphavirus infection on neurons. Ann. Neurol. 35 (Suppl), S23–S27.

Grskovic, M., Javaherian, A., Strulovici, B., Daley, G.Q., 2011. Induced pluripotent stem cells—opportunities for disease modelling and drug discovery. Nat. Rev. Drug Discov. 10, 915–929.

Haidet-Phillips, A.M., Hester, M.E., Miranda, C.J., Meyer, K., Braun, L., Frakes, A., Song, S., Likhite, S., Murtha, M.J., Foust, K.D., Rao, M., Eagle, A., Kammesheidt, A., Christensen, A., Mendell, J.R., BurghesKaspar, A.H., Kaspar, B.K., 2011. Astrocytes from familial and sporadic ALS patients are toxic to motor neurons. Nat. Biotechnol. 29, 824–828.

Harper, J.M., Krishnan, C., Darman, J.S., Deshpande, D.M., Peck, S., Shats, I., Backovic, S., Rothstein, J.D., Kerr, D.A., 2004. Axonal growth of embryonic stem cell-derived motoneurons in vitro and in motoneuron-injured adult rats. Proc. Natl. Acad. Sci. U.S.A. 101, 7123–7128.

Hollyday, M., Hamburger, V., 1977. An autoradiographic study of the formation of the lateral motor column in the chick embryo. Brain Res. 132, 197–208.

Hollyday, M., Hamburger, V., Farris, J.M., 1977. Localization of motor neuron pools supplying identified muscles in normal and supernumerary legs of chick embryo. Proc. Natl. Acad. Sci. U.S.A. 74, 3582–3586.

Jackson, A.C., Moench, T.R., Griffin, D.E., Johnson, R.T., 1987. The pathogenesis of spinal cord involvement in the encephalomyelitis of mice caused by neuroadapted Sindbis virus infection. Lab. Invest. 56, 418–423.

Karumbayaram, S., Kelly, T.K., Paucar, A.A., Roe, A.J., Umbach, J.A., Charles, A., Goldman, S.A., Kornblum, H.I., Wiedau-Pazos, M., 2009. Human embryonic stem cell-derived motor neurons expressing SOD1 mutants exhibit typical signs of motor neuron degeneration linked to ALS. Dis. Model. Mech. 2, 189–195.

Kernell, D., 2006. Why bother about motoneurons? In: Kernell, D. (Ed.), The Motoneurone and Its Muscle Fibres. Oxford University Press, Oxford, pp. 1–4.

Kerr, D.A., Larsen, T., Cook, S.H., Fannjiang, Y.R., Choi, E., Griffin, D.E., Hardwick, J.M., Irani, D.N., 2002. BCL-2 and BAX protect adult mice from lethal Sindbis virus infection but do not protect spinal cord motor neurons or prevent paralysis. J. Virol. 76, 10393–10400.

Kerr, D.A., Lladó, J., Shamblott, M.J., Maragakis, N.J., Irani, D.N., Crawford, T.O., Krishnan, C., Dike, S., Gearhart, J.D., Rothstein, J.D., 2003. Human embryonic germ cell derivatives facilitate motor recovery of rats with diffuse motor neuron injury. J. Neurosci. 23(12), 5131–5140.

Kingsbury, B.F., 1930. The developmental significance of the floor plate of the brain and spinal cord. J. Comp. Neurol. 50, 177–207.

Kiskinis, E., Eggan, K., 2010. Progress toward the clinical application of patient-specific pluripotent stem cells. J. Clin. Invest. 120, 51–59.

Kubo, T., Randolph, M.A., Gröger, A., Winograd, J.M., 2009. Embryonic stem cell-derived motor neurons form neuromuscular junctions in vitro and enhance motor functional recovery in vivo. Plast. Reconstr. Surg. 123, 139S–148S.

Landmesser, L., 1978a. The distribution of motoneurones supplying chick hind limb muscles. J. Physiol. 284, 371–389.

Landmesser, L., 1978b. The development of motor projection patterns in the chick hind limb. J. Physiol. 284, 391–414.

Lee, H., Shamy, G.A., Elkabetz, Y., Schofield, C.M., Harrsion, N.L., Panagiotakos, G., Socci, N.D., Tabar, V., Studer, L., 2007. Directed differentiation and transplantation of human embryonic stem cell-derived motoneurons. Stem Cells 25, 1931–1939.

Lee, G., Papapetrou, E.P., Kim, H., Chambers, S.M., Tomishima, M.J., Fasano, C.A., Ganat, Y.M., Menon, J., Shimizu, F., Viale, A., Tabar, V., Sadelain, M., Studer, L.,

2009. Modelling pathogenesis and treatment of familial dysautonomia using patient-specific iPSCs. Nature 461, 402–406.

Li, X.J., Du, Z.W., Zarnowska, E.D., Pankratz, M., Hansen, L.O., Pearce, R.A., Zhang, S.C., 2005. Specification of motoneurons from human embryonic stem cells. Nat. Biotechnol. 23, 215–221.

Liu, X., Huang, J., Chen, T., Wang, Y., Xin, S., Li, J., Pei, G., Kang, J., 2008. Yamanaka factors critically regulate the developmental signaling network in mouse embryonic stem cells. Cell Res. 18, 1177–1189.

Lohnes, D., Mark, M., Mendelsohn, C., Dollé, P., Dierich, A., Gorry, P., Gansmuller, A., Chambon, P., 1994. Function of the retinoic acid receptors (RARs) during development (I). Craniofacial and skeletal abnormalities in RAR double mutants. Development 120, 2723–2748.

López-González, R., Kunckles, P., Velasco, I., 2009. Transient recovery in a rat model of familial amyotrophic lateral sclerosis after transplantation of motor neurons derived from mouse embryonic stem cells. Cell Transplant. 18, 1171–1181.

Martí, E., Bumcrot, D.A., Takada, R., McMahon, A.P., 1995. Requirement of 19K form of Sonic hedgehog for induction of distinct ventral cell types in CNS explants. Nature 375, 322–325.

Mendez, I., Viñuela, A., Astradsson, A., Mukhida, K., Hallett, P., Robertson, H., Tierney, T., Holness, R., Dagher, A., Trojanowski, J.Q., Isacson, O., 2008. Dopamine neurons implanted into people with Parkinson's disease survive without pathology for 14 years. Nat. Med. 14, 507–509.

Miles, G.B., Yohn, D.C., Wichterle, H., Jessell, T.M., Rafuse, V.F., Brownstone, R.M., 2004. Functional properties of motoneurons derived from mouse embryonic stem cells. J. Neurosci. 24, 7848–7858.

Nüsslein-Volhard, C., Wieschaus, E., 1980. Mutations affecting segment number and polarity in Drosophila. Nature 287, 795–801.

Okita, K., Ichisaka, T., Yamanaka, S., 2007. Generation of germline-competent induced pluripotent stem cells. Nature 448, 313–317.

Park, I.H., Zhao, R., West, J.A., Yabuuchi, A., Huo, H., Ince, T.A., Lerou, P.H., Lensch, M.W., Daley, G.Q., 2008. Reprogramming of human somatic cells to pluripotency with defined factors. Nature 451, 141–146.

Pearse, D.D., Pereira, F.C., Marcillo, A.E., Bates, M.L., Berrocal, Y.A., Filbin, M.T., Bunge, M.B., 2004. cAMP and Schwann cells promote axonal growth and functional recovery after spinal cord injury. Nat. Med. 10, 61061–61066.

Peckham, P.H., Kilgore, K.L., Keith, M.W., Bryden, A.M., Bhadra, N., Montague, F.W., 2002. An advanced neuroprosthesis for restoration of hand and upper arm control using an implantable controller. J. Hand. Surg. Am. 27, 265–276.

Pepinsky, R.B., Zeng, C., Wen, D., Rayhorn, P., Baker, D.P., Williams, K.P., Bixler, S.A., Ambrose, C.M., Garber, E.A., Miatkowski, K., Taylor, F.R., Wang, E.A., Galdes, A., 1998. Identification of a palmitic acid-modified form of human Sonic hedgehog. J. Biol. Chem. 273, 14037–14045.

Pera, M.F., Andrade, J., Houssami, S., Reubinoff, B., Trounson, A., Stanley, E.G., Ward-van Oostwaard, D., Mummery, C., 2004. Regulation of human embryonic stem cell differentiation by BMP-2 and its antagonist noggin. J. Cell Sci. 117, 1269–1280.

Pfisterer, U., Kirkeby, A., Torper, O., Wood, J., Nelander, J., Dufour, A., Björklund, A., Lindvall, O., Jakobsson, J., Parmar, M., 2011. Direct conversion of human fibroblasts to dopaminergic neurons. Proc. Natl. Acad. Sci. U.S.A. 108, 10343–11038.

Placzek, M., Yamada, T., Tessier-Lavigne, M., Jessell, T., Dodd, J., 1991. Control of dorsoventral pattern in vertebrate neural development: induction and polarizing properties of the floor plate. Development (Suppl. 2), 105–122.

Roelink, H., Augsburger, A., Heemskerk, J., Korzh, V., Norlin, S., Ruiz i Altaba, A., Tanabe, Y., Placzek, M., Edlund, T., Jessell, T.M., Dodd, J., 1994. Floor plate and motor neuron induction by vhh-1, a vertebrate homolog of hedgehog expressed by the notochord. Cell 76, 761–775.

Roelink, H., Porter, J.A., Chiang, C., Tanabe, Y., Chang, D.T., Beachy, P.A., Jessell, T.M., 1995. Floor plate and motor neuron induction by different concentrations of the amino-terminal cleavage product of sonic hedgehog autoproteolysis. Cell 81, 445–455.

Sharma, K., Izpisúa Belmonte, J.C., 2001. Development of the limb neuromuscular system. Curr. Opin. Cell Biol. 13, 204–210.

Sherrington, C.S., 1906. The Integrative Action of the Nervous System. Yale University Press, New Haven.

Shin, S., Dalton, S., Stice, S.L., 2005. Human motor neuron differentiation from human embryonic stem cells. Stem Cells Dev. 14, 266–269.

Singh Roy, N., Nakano, T., Xuing, L., Kang, J., Nedergaard, M., Goldman, S.A., 2005. Enhancer-specified GFP-based FACS purification of human spinal motor neurons from embryonic stem cells. Exp. Neurol. 196, 224–234.

Smith, J.R., Vallier, L., Lupo, G., Alexander, M., Harris, W.A., Pedersen, R.A., 2008. Inhibition of Activin/Nodal signaling promotes specification of human embryonic stem cells into neuroectoderm. Dev. Biol. 313, 107–117.

Son, E.Y., Ichida, J.K., Wainger, B.J., Toma, J.S., Rafuse, V.F., Woolf, C.J., Eggan, K., 2011. Conversion of mouse and human fibroblasts into functional spinal motor neurons. Cell Stem Cell 9, 205–218.

Soundararajan, P., Miles, G.B., Rubin, L.L., Brownstone, R.M., Rafuse, V.F., 2006. Motoneurons derived from embryonic stem cells express transcription factors and develop phenotypes characteristic of medial motor column neurons. J. Neurosci. 26, 3256–3268.

Soundararajan, P., Lindsey, B.W., Leopold, C., Rafuse, V.F., 2007. Easy and rapid differentiation of embryonic stem cells into functional motoneurons using sonic hedgehog-producing cells. Stem Cells 25, 1697–1706.

Soundararajan, P., Fawcett, J.P., Rafuse, V.F., 2010. Guidance of postural motoneurons requires MAPK/ERK signaling downstream of fibroblast growth factor receptor 1. J. Neurosci. 30, 6595–6606.

Sreedharan, J., Blair, I.P., Tripathi, V.B., Hu, X., Vance, C., Rogelj, B., Ackerley, S., Durnall, J.C., Williams, K.L., Buratti, E., Baralle, F., de Belleroche, J., Mitchell, J.D., Leigh, P.N., Al-Chalabi, A., Miller, C.C., Nicholson, G., Shaw, C.E., 2008. TDP-43 mutations in familial and sporadic amyotrophic lateral sclerosis. Science 319, 1668–1672.

Taipale, J., Chen, J.K., Cooper, M.K., Wang, B., Mann, R.K., Milenkovic, L., Scott, M.P., Beachy, P.A., 2000. Effects of oncogenic mutations in Smoothened and Patched can be reversed by cyclopamine. Nature 406, 1005–1009.

Takahashi, K., Yamanaka, S., 2006. Induction of pluripotent stem cells from mouse embryonic and adult fibroblast cultures by defined factors. Cell 126, 663–676.
Takahashi, K., Tanabe, K., Ohnuki, M., Narita, M., Ichisaka, T., Tomoda, K., Yamanaka, S., 2007. Induction of pluripotent stem cells from adult human fibroblasts by defined factors. Cell 131, 861–872.
Tanabe, Y., Roelink, H., Jessell, T.M., 1995. Induction of motor neurons by Sonic hedgehog is independent of floor plate differentiation. Curr. Biol. 5, 651–658.
Theys, P.A., Peeters, E., Robberecht, W., 1999. Evolution of motor and sensory deficits in amyotrophic lateral sclerosis estimated by neurophysiological techniques. J. Neurol. 246, 438–442.
Thomas, C.K., Zaidner, E.Y., Calancie, B., Broton, J.G., Bigland-Ritchie, B.R., 1997. Muscle weakness, paralysis, and atrophy after human cervical spinal cord injury. Exp. Neurol. 148, 414–423.
Thomas, C.K., Erb, D.E., Grumbles, R.M., Bunge, R.P., 2000. Embryonic cord transplants in peripheral nerve restore skeletal muscle function. J. Neurophysiol. 84, 591–595.
Urbach, A., Bar-Nur, O., Daley, G.Q., Benvenisty, N., 2010. Differential modeling of fragile X syndrome by human embryonic stem cells and induced pluripotent stem cells. Cell Stem Cell 6, 407–411.
Vierbuchen, T., Ostermeier, A., Pang, Z.P., Kokubu, Y., Südhof, T.C., Wernig, M., 2010. Direct conversion of fibroblasts to functional neurons by defined factors. Nature 463, 1035–1041.
Vitale, A.M., Wolvetang, E., Mackay-Sim, A., 2011. Induced pluripotent stem cells: a new technology to study human diseases. Int. J. Biochem. Cell Biol. 43, 843–846.
Wada, T., Honda, M., Minami, I., Tooi, N., Amagai, Y., Nakatsuji, N., Aiba, K., 2009. Highly efficient differentiation and enrichment of spinal motor neurons derived from human and monkey embryonic stem cells. PLoS One 4, e6722.
Wagner, M., Han, B., Jessell, T.M., 1992. Regional differences in retinoid release from embryonic neural tissue detected by an in vitro reporter assay. Development 116, 55–66.
Wichterle, H., Lieberam, I., Porter, J.A., Jessell, T.M., 2002. Directed differentiation of embryonic stem cells into motor neurons. Cell 110, 385–397.
Wilson, S.I., Edlund, T., 2001. Neural induction: toward a unifying mechanism. Nat. Neurosci. 4 (Suppl), 1161–1168.
Wilson, L., Maden, M., 2005. The mechanisms of dorsoventral patterning in the vertebrate neural tube. Dev. Biol. 282, 1–13.
Wilson, S.I., Graziano, E., Harland, R., Jessell, T.M., Edlund, T., 2000. An early requirement for FGF signalling in the acquisition of neural cell fate in the chick embryo. Curr. Biol. 10, 421–429.
Wilson, S.I., Rydström, A., Trimborn, T., Willert, K., Nusse, R., Jessell, T.M., Edlund, T., 2001. The status of Wnt signalling regulates neural and epidermal fates in the chick embryo. Nature 411, 325–330.
Yamada, T., Placzek, M., Tanaka, H., Dodd, J., Jessell, T.M., 1991. Control of cell pattern in the developing nervous system: polarizing activity of the floor plate and notochord. Cell 64, 635–647.

Yohn, D.C., Miles, G.B., Rafuse, V.F., Brownstone, R.M., 2008. Transplanted mouse embryonic stem-cell-derived motoneurons form functional motor units and reduce muscle atrophy. J. Neurosci. 28, 12409–12418.
Yu, J., Vodyanik, M.A., Smuga-Otto, K., Antosiewicz-Bourget, J., Frane, J.L., Tian, S., Nie, J., Jonsdottir, G.A., Ruotti, V., Stewart, R., Slukvin, I.I., Thomson, J.A., 2007. Induced pluripotent stem cell lines derived from human somatic cells. Science 318, 1917–1920.
Zurn, A.D., Winkel, L., Menoud, A., Djabali, K., Aebischer, P., 1996. Combined effects of GDNF, BDNF, and CNTF on motoneuron differentiation in vitro. J. Neurosci. Res. 44, 133–141.

CHAPTER

Transplantation of mesenchymal stem cells in ALS

17

Letizia Mazzini*[,1], Alessandro Vercelli[†], Ivana Ferrero[‡], Marina Boido[†], Roberto Cantello*, Franca Fagioli[‡]

[*]ALS Centre, Department of Neurology, Eastern Piedmont University, "Maggiore della Carità" Hospital, Novara, Italy

[†]Neuroscience Institute of the Cavalieri Ottolenghi Foundation, University of Torino, Torino, Italy

[‡]Stem Cell Transplantation and Cellular Therapy Unit, Pediatric Onco-Hematology Division, "Regina Margherita" Children's Hospital, University of Torino, Torino, Italy

[1]Corresponding author. Tel.: +39-0321-3733834, Fax: +39-0321-3733298, e-mail address: mazzini.l@libero.it

Abstract

Amyotrophic lateral sclerosis (ALS) is a devastating incurable, neurodegenerative disease that targets motor neurons (MNs) in the primary motor cortex, brainstem, and spinal cord, leading to muscle atrophy, paralysis, and death due to respiratory failure within 2–5 years. Currently, there is no cure for ALS. The development of a therapy that can support or restore MN function and attenuate toxicity in the spinal cord provides the most comprehensive approach for treating ALS. Mesenchymal stem cells might be suitable for cell therapy in ALS because of their immunomodulatory and protective properties. In this review, the authors discuss the major challenges to the translation of *in vitro* and animal studies of MSCs therapy in the clinical setting.

Keywords

mesenchymal stem cells, neurodegeneration, neuroprotection, amyotrophic lateral sclerosis, transplantation, bone marrow

1 INTRODUCTION

Amyotrophic lateral sclerosis (ALS) is a devastating incurable neurodegenerative disease that targets motor neurons (MNs) in the primary motor cortex, brainstem, and spinal cord leading to muscle atrophy, paralysis, and death due to respiratory failure within 2–5 years. The incidence is 2–3 cases per 100,000 general population, and the prevalence is around four to six per 100,000 (Chiò et al., 2009).

Progress in Brain Research, Volume 201, ISSN 0079-6123, http://dx.doi.org/10.1016/B978-0-444-59544-7.00016-0

Approximately 90% of all cases are classified as sporadic (SALS), defined as having no family history of the disease. The remaining cases are dominantly inherited, that is, familial (FALS). Until very recently, only 30% of FALS cases had been accounted for known genetic mutations. The recent identification of the C9 or f72 hexanucleotide repeat on chromosome 9p21 appears responsible for 30–60% FALS cases, and about 8% of apparently sporadic cases (Dejesus-Hernandez et al., 2011; Renton et al., 2011). Currently, there is no cure for ALS. Riluzole is the only drug approved by the Food and Drug Administration for ALS treatment, but its outcome is very limited, since it increases survival by only 2–3 months compared with placebo (Bensimon et al., 1994).

ALS was traditionally considered a pure motor disorder. However, recent findings suggest that ALS is a multisystem disorder in which the MNs tend to be affected the earliest and the most severely (Ince et al., 2008). The pathogenic processes underlying ALS are multifactorial and, at present, not fully determined (Ferraiuolo et al., 2011). Recent data have implicated the microenvironment of the MNs, rather than the MN itself, as a primary target. ALS therefore has emerged as a non-cell-autonomous neurodegenerative disease, that is, not independent of mutant damage accumulated within other cell types that interact with the affected neurons. Putative mechanisms of toxicity targeting MNs include oxidative damage, accumulation of intracellular aggregates, mitochondrial dysfunction, defects in axonal transport, defective growth factor trophic support, altered glial function, aberrant RNA metabolism, and glutamate excitotoxicity. Each of these mechanisms represents a potential therapeutic target and many clinical trials have been developed, even though currently none of the candidate compounds has been demonstrated effective. A cell-based therapy may have the advantage of exerting multiple therapeutic effects (Lepore et al., 2008; Suzuki et al., 2008) at various sites and times within the lesion, as the cells respond to a particular pathological microenvironment (Liu and Martin, 2006) by protecting existing MNs from ongoing degeneration.

The development of a therapy that can support or restore MN function and attenuate toxicity in the spinal cord provides the most comprehensive approach for treating ALS.

2 STEM CELL TRANSPLANTATION AS A THERAPEUTIC STRATEGY

Cell therapy uses cell or tissue grafts to treat diseases or injury. Treatment focuses on cell replacement or providing environmental enrichment.

Lower MNs can be generated *in vitro* from stem cells of various sources. The newly transplanted neurons may then integrate, receive and make synapses, and recapitulate a neural network after transplantation into adult rats (Bohl et al., 2008; Kerr et al., 2003; Xu et al., 2009; Yan et al., 2007). However, practical issues might limit the clinical translation of direct MN replacement to humans. Grafted neurons must receive functional synapses, send axons through inhibitory white matter, and direct axons over long distances to the target muscles in order to retain neuromuscular function. Given these limitations neuronal replacement in ALS patients seems a distant goal.

Several studies have recently clarified the role of glia in ALS pathogenesis (Boillée et al., 2006; Ilieva et al., 2009). Neuron–astrocyte interactions are essential for the regulation of glutamate transmission. Without astrocytes, neurons become increasingly vulnerable to excitotoxicity. The disruption in astrocytic function can markedly promote neurodegeneration. The contribution of nonneuronal cells to the pathogenesis of MN degeneration has been studied in mutant SOD1 mice, in which the transgene was excised in specific cell types. Primary astrocyte and microglial cultures derived from the superoxide dismutase-mutant mouse model of ALS produce neurotoxic mediators in conditioned media that kill wild-type MNs (Nagai et al., 2007). Haidet-Phillips et al. (2011) showed that astrocytes derived from postmortem tissue from both FALS and SALS patients are similarly toxic to MNs. We might hypothesize that replacement or enrichment with healthy astrocytes could be a therapeutic approach for slowing or blocking the disease course. Studies with chimeric mice showed that delivering wild-type glial cells in the ALS model can improve the disease phenotype (Beers et al., 2006; Clement et al., 2003). Alternatively, astrocyte replacement has also been proposed as a potential therapy for slowing disease progression in ALS. Rodent glial-restricted precursors (GRP) transplanted into the spinal cords of mutant SOD1 rats differentiate into astrocytes, restore the levels of astrocyte glutamate transporter, and extend survival of these rats (Lepore et al., 2008). Human GRP transplants robustly survive and migrate in both gray and white matter and differentiate into astrocytes in SOD1G93A mouse spinal cord, despite ongoing disease progression. However, cervical spinal cord transplants did not result in MN protection or any therapeutic benefits on tests of functional outcome (Lepore, 2011).

3 GROWTH FACTORS AND ALS

Growth factors are naturally occurring proteins essential for neuronal survival: their deficiency could induce MN death in ALS patients (Henriques et al., 2010). Several growth factors were therapeutic in animal models but not in humans: clinical trials using brain-derived neurotrophic factor (BDNF), ciliary neurotrophic factor, and insulin-like growth factor demonstrated no significant survival benefits (Bongioanni et al., 2004; Sorenson et al., 2008; The BDNF Study Group, 1999). The reasons for these failures might be related to an inadequate route of administration. Penetration of large peptides, such as growth factors, into the CNS, in fact, is limited by the blood–brain barrier. On the other hand, stem cells transplanted into the nervous system produce and deliver neurotrophic and growth factors and their efficacy could be improved by genetic modification to secrete molecules that promote MN survival (Ericson et al., 2005; Klein et al., 2005). As such, stem cells might be able to detoxify the local environment around dying MNs by generating glial cells and delivering trophic factors after transplantation (Gowing and Svendsen, 2011; Suzuki and Svendsen, 2008).

The beneficial effects of glial cell replacement and enhancement of neurotrophic support in ALS provide the basis for the use of stem cells to treat this disease (Lunn et al., 2011).

4 MESENCHYMAL STEM CELLS

Stem cell-based therapies represent a new possible scenario for neurodegenerative diseases.

Multipotent mesenchymal stem cells (MSCs) are bone marrow (BM) cells that can be expanded *ex vivo* and will readily differentiate into mesodermal cell derivatives (Pittenger et al., 1999).

MSCs represent a small fraction (0.001–0.01%) of the BM cell population; therefore, to obtain a sufficient number of cells, they must be extensively expanded *ex vivo*. Although BM is the best characterized source of MSCs, umbilical cord blood (UCB), Wharton's jelly, placenta, adipose tissue (AT), and many others represent promising alternatives (Otto and Wright, 2011). Human amniotic fluid (hAF) is also considered an attractive source of MSCs for therapeutic transplantation (In'tAnker et al., 2003), and recently new reports confirm the importance of stem cells isolated from hAF for banking and multiple clinical applications (De Coppi et al., 2007; Mareschi et al., 2009; Sessarego et al., 2008).

MSCs show their ability to differentiate into mature neural cell types. They have been suggested to adopt "astrocytic" and "neuronal-like" cell fates (Wislet-Gendebien et al., 2005). When cultured in neural progenitor maintenance medium, hMSCs acquire new morphological characteristics, neural markers, and electrophysiological properties, which are suggestive of neural differentiation (Mareschi et al., 2006).

MSCs promote "bystander" immunomodulation, as they can release soluble molecules such as cytokines and chemokines and express immunorelevant receptors such as chemokine receptors and cell-adhesion molecules (Uccelli et al., 2008), and they can control drug-resistant graft-versus-host disease in humans. In animals, they also seem to limit damage to, or mediate repair of, CNS tissue via mechanisms other than cell replacement or transdifferentiation, probably via their paracrine function (Uccelli et al., 2011).

MSCs can be successfully isolated and expanded for clinical application without significant adverse effects and could represent a new possible scenario for neurodegenerative diseases. We have demonstrated that BM-MSCs from ALS patients maintain their peculiar characteristics and when expanded *in vitro* do not display chromosomal alterations or cellular senescence. Moreover, they acquire, under specific conditions, new morphological characteristics and neural markers which are suggestive of neural differentiation as well as those obtained in the healthy donors (Ferrero et al., 2008).

Recent studies have revealed that MSC therapeutic action is related to the release of protective factors, even far from the site of injection, rather than to replacement of degenerating neurons. Such a therapeutic effect may be provided by different classes of molecules, including trophic factors, anti-inflammatory cytokines, and immunomodulatory chemokines released from transplanted cells.

MSC therapy might indeed represent a promising strategy for the development of a new therapeutic strategy for neurodegenerative diseases, such as cell therapy in animal models.

Due to the multifarious and elusive pathogenesis of ALS, several different *in vivo* models of ALS have been identified, each of which displays some genetic and/or pathological hallmarks of the disease. The detailed description of the different animal models is not the aim of this review, but the reader can refer to Berthòd and Gros-Louis (2012) and Boido et al. (2012). In brief, the most widely used model consists in the insertion of the human copper/zinc superoxide dismutase 1 (SOD1) gene bearing a mutation in G93A (a mutation at position 93 from glycine to alanine), which has been found in FALS (Gurney et al., 1994). This characteristic dominant "gain-of-function" mutation leads to a neurologic syndrome in mice which is similar to the human disease. The same mutation together with H46R has been introduced in rats to provide a more severe animal model for experimental manipulation (Nagai et al., 2001).

For the characteristics of their neurological syndrome, and for being the first model introduced in ALS research, mostly SOD1 rodents have been used to test the effects of stem cell therapy. MSCs have been transplanted in mice either by intraparenchymal, intracerebroventricular (i.c.v.), intraperitoneal (i.p.), or intravenous (i.v.) injections. The issue of the administration route will be discussed in details below, as well as that of immunosuppression, which in general is not needed (Vercelli et al., 2008).

A relevant issue relative to MSC transplantation is the donor species: on the one hand, when cell-to-cell interactions must be studied, the donor and the host should be of the same species. On the other hand, in order to use hMSCs as medicinal products (see below) and exclude harmful side effects, xenogeneic transplants are performed.

Following treatment, the effects of cell therapy are evaluated in terms of cell fate/integration, functional outcome (motor behavioral tests), survival of the animal and histopathological outcome (i.e., survival of MNs, astrogliosis, and microglial activation). In the first studies, a great interest has been raised on the potential for transdifferentiation of MSCs into neurons and glial cells. In particular, neuronal differentiation has been hypothesized as a mean for cell replacement. Even though MSCs, isolated from BM, can transdifferentiate into neurons (Phinney and Isakova, 2005), it is unrealistic to expect that the transplanted stem cells or stem cell-derived MNs in ALS patients either in a clinical setting or in the experimental animal replace lost neurons, integrate into existing neural circuitry, and restore motor function. Rather, preventing cell death in host MNs via provision of neurotrophic, anti-inflammatory, and immunomodulatory factors by transplanted stem cells or stem cell-derived MNs is a more realistic and achievable approach.

MSCs, in fact, can deliver neurotrophic, anti-inflammatory, and immunomodulatory molecules (Caplan, 2009) to delay disease progression. Their interaction with the nervous system results in a reciprocal influence, which can modify the environment of MNs in ALS, delaying their degeneration and supporting their survival. MSCs act as minipumps which can be directed to express and produce molecules extending the survival of MNs (Sadan et al., 2009). Therefore, more recently, attention has been drawn to the effects of the diseased brain on MSCs in terms of production and delivery of immunomodulatory molecules and neurotrophic factors. Cultures of rat MSCs display an increased production of NGF and BDNF after stimulation with SOD1 rat brain extracts (Nicaise et al., 2011). Also, MSCs themselves can modify the local environment and influence the delivery of trophic factors and immunomodulatory molecules by the host cells.

We have shown that intraparenchymal injection of donor hMSCs can improve motor behavior, delay motor neuron death, and reduce neuroinflammation (Vercelli et al., 2008) (Fig. 1).

Other authors reported similar results with i.v. injection of donor hMSCs (Zhao et al., 2007). I.v. administrated MSCs interact specifically with several mechanisms involved in MN death (Uccelli et al., 2012). Intrathecal administration of murine MSCs into SOD1 mice lead to increased lifespan and decreased neuroinflammation, and grafted cells were found to transdifferentiate into astrocytes (Boucherie et al., 2009). In contrast to this, when hMSCs from patients affected by ALS were transplanted in the cisterna magna into SOD1 mice, a dose-dependent efficacy was reported (Kim et al., 2010). A study on the effects of i.c.v. administration of hMSCs gave contrasting results in SOD1 mice, such as prolonged lifespan, but only in females, not significant increase in surviving MNs and no effects on neuroinflammation (Morita et al., 2008). Combined intraparenchymal and i.v. engraftment of MSCs increased the lifespan and improved motor function in SOD1 rats (Forostyak et al., 2011).

Sadan and colleagues recently developed a two-step medium-based differentiation protocol for inducing MSCs into neurotrophic factor secreting cells (NTF+). The induced cells produce and release high amounts of NTFs, such as BDNF and GDNF (Dadon-Nachum et al., 2011; Sadan et al., 2008) which protect cultured MNs. Also MSC-NTF derived xeno-free growth media, a technology licensed to Brainstorm, markedly increase the expansion potential of MSCs (Chaddah et al., 2011). Transplantation studies in mice and rats thus far have also proved successful. The mouse MSC-NTFs transplanted into SOD1 (G93A) mice improved motor functions and survival. The i.m. transplantation of rat MSC-NTFs into a rat model for sciatic motor nerve injury preserves motor function, protects neuromuscular junctions, and accelerates regeneration (Dadon-Nachum et al., 2011).

Wild-type, and not SOD1, murine MSCs can be expanded in culture and exposed to growth factors which increase transcription and expression of the high-affinity glutamate transporter GLT-1 and make the cells responsive to riluzole triggering an upregulation of the GDNF production (Boucherie et al., 2008).

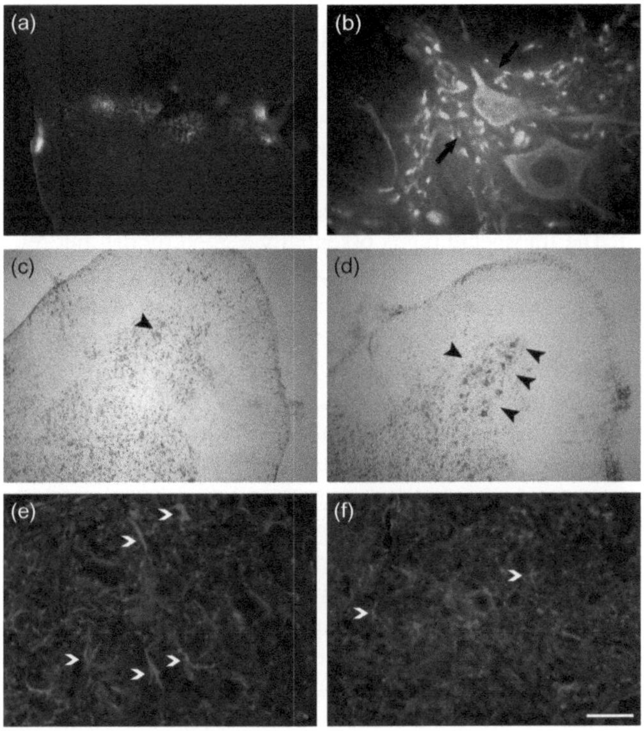

FIGURE 1

Effects of intraparenchymal transplantation of hMSCs in SOD1G93A mice (Vercelli et al., 2008). (A) Injection site in the lumbar spinal cord showing Bisbenzimide-stained hMSCs, after intraparenchymal transplantation in an ALS murine model. (B) At higher magnification, blue hMSCs (arrows) localized close to motor neurons (labeled in green by the neuronal antibody, anti-MAP2). (C and D) The number of motor neurons (arrowheads) in the lumbar spinal cord is significantly reduced in control mice (C) compared with transplanted ones (D). (E and F) Similarly, astrogliosis (anti-GFAP-immunostaining; arrowheads) results modulated in the grafted mice (F) in comparison with untreated ones (E). Scale bars: 200 μm in (A), (C), and (D); 20 μm in (B); 50 μm in (E) and (F). (For interpretation of the references to color in this figure legend, the reader is referred to the online version of this chapter.)

As reported, MSCs can be used as Trojan horses to deliver trophic factor into the diseased spinal cord, as shown by injecting i.m. MSCs transduced with a lentivirus encoding GDNF to express GDNF at high levels (see below) (Suzuki et al., 2008). This protocol is effective in asymptomatic animals, and greater effect is observed in animals with slow progression ALS.

Their biological properties independent of differentiation suggest that MSCs could have a therapeutic role in ALS through mechanisms other than tissue replacement.

5 TRANSLATION INTO THE CLINIC

Many promising results obtained in animal models have been lost in translation to the clinic. This might happen for stem cell therapy as well. In fact, transposition of data obtained in animal models to patients has some limitations. First of all, ALS lacks animal models that closely mimic human disease. Recently, a large animal model with human-like physiology, size, and genetic lesions that gives an ALS type phenotype has been proposed. The transgenic pig is potentially a good model, but rejection of stem cells and the need for immunosuppression are major drawbacks, and for efficacy studies, animals would need to be monitored for a long duration. A canine model displaying degenerative myelopathy, an adult onset neurodegenerative disease caused by a mutation in SOD1, shares many features with human ALS (Awano et al., 2009). This model could be extremely valuable for testing cell therapy strategies.

Animal models may be predictive, but often are not sufficiently reliable (Scott et al., 2008). First of all, animal models usually are conducted in a very homogeneous population, where most, if not all, cases become symptomatic and die at the same time. This is not the case for human patients. Also, humans quite often are affected by other diseases, which add to ALS. Moreover, many studies have been conducted in presymptomatic animals, which enhances the chances to change the disease course; nevertheless, in the human setting no presymptomatic diagnostic tests are available and experimental therapies can be employed only in already symptomatic patients. Finally, the most common animal model of ALS, the SOD1G93A mouse, is a quite unstable model, since the onset and the progression of the disease strongly depend on the number of the copies of the human gene, and also on the levels of the gene expression; therefore, the researcher must continuously check his colony, and also use a battery of behavioral tests to detect the exact time of onset of the disease in the individual mouse, in order to start treatment (Vercelli et al., 2010).

6 ALLOGENIC VERSUS AUTOLOGOUS MSCs FOR TRANSPLANTATION

BM-MSCs are, to date, the most common source of stem cell in the treatment of hematopoietic diseases and thus the protocols for their isolation and application are well established. MSCs may be collected from the patients themselves, thus allowing autologous transplantation, which may obviate the need for immunosuppression and also may facilitate the authorization of clinical studies. However, autologous cells might be more vulnerable to the disease and the current extensive requirements for cell manufacture and testing may render such approaches very expensive.

Ferrero et al. (2008) analyzed expanded MSCs from sporadic ALS patients and healthy donors, which did not show any evident difference in immunophenotypic and functional characteristics. However, the expansion potential varied in correlation with the age of patients. It is unclear whether these data may significantly affect the efficacy and the outcome of the treatment. Some recent studies (Koh et al., 2012), however, show that a reduction in the pluripotency and capacity to secrete trophic factors of the BM-MSCs of ALS patients correlates with poorer prognosis.

Healthy allogeneic BM-MSCs might be also considered for cell therapy in ALS patients. Mareschi et al. (2009) demonstrated that MSCs collected from amniotic fluid (AF-MSCs) have more advantageous immunophenotypic and functional characteristics compared to the BM-derived MSCs; hence, they could be proposed in clinical use.

7 CHARACTERIZATION AND MANUFACTURE OF CELL PRODUCT FOR TRANSPLANTATION

The very low density of MSCs in BM requires isolation and expansion steps before their clinical use. MSCs can easily be isolated from BM thanks to their capacity to adhere, proliferate, and expand in culture while maintaining their immunophenotypic characteristics and functions as multipotent cells. The International Society for Cellular Therapy proposed three minimal criteria to identify MSCs: the adherence to plastic, the specific surface antigen expression (positivity for CD105, CD73, CD90, and the lack of expression of CD45, CD34, CD14 or CD11b, CD79a or CD19, and HLA class II), and the multipotent capacity to differentiate into osteoblasts, adipocytes, and chondroblasts under standard conditions for *in vitro* differentiation (Dominici et al., 2006). MSCs also express variable levels of CD44, stromal antigen-1, and a group of other adhesion molecules and receptors including CD166 (vascular cell-adhesion molecule), CD54/CD102 (intracellular adhesion molecule), and CD49 (very late antigen). The variability in the expression of adult MSC surface markers could be related to the different stages of culture (Mafi et al., 2011).

Many other variables such as culture media and additives, plating density, and passaging may influence MSC culture expansion. The use of MSCs as cell therapy products requires alternative expansion methods, avoiding the use of reagent of animal origin such as fetal bovine serum (FBS). The human platelet lysate (HPL) represents a valuable and promising alternative to FBS (Bernardo et al., 2007). HPL-cultured MSCs have comparable immunomodulatory capacities to their FCS-cultured counterparts making these cells an attractive cell therapeutic tool (Flemming et al., 2011). In addition to culture media and supplements, some authors suggested that low seeding densities result in faster proliferation and a larger fraction of multipotent adult stem cells (MAPCs) (Jiang et al., 2002), while others indicated that higher densities might even be more effective for expansion of MSCs (Neuhuber et al., 2008). In accordance with other authors' data, Mareschi et al. (2012) confirmed that a low plating density results in higher yields and a faster expansion of MSCs, and may represent a good procedure for clinical use.

MSCs are considered to be advanced therapy medicinal products (ATMPs), as defined by the European Regulation N. 1394/2007 of the European Parliament on advanced therapy medicinal products, amending the 2001/83/EC Directive, that completes the regulatory setting on advanced therapies to be used in Member States. The quality and safety of ATMPs such as MSCs must be maintained throughout their production, ensuring their final use in the patient. MSCs are considered somatic-cell therapy products or tissue-engineered products depending on the source, manufacturing process, and proposed indications. The regulation N. 1394/2007 refers to the European Commission Good manufacturing practices (GMP) rules and is also in compliance with the 2004/23/EC Directive on donation, obtaining and testing of human cells and tissues. GMP ensures that products are consistently produced and controlled to the quality standards required to their intended use, from collection and manipulation of raw materials to the processing of intermediate products, the quality controls, the storage, labeling and packaging, and release.

The translation of research-based protocols into GMP-compliant procedures for large-scale production of clinical-grade MSCs requires careful analysis of all the risks and benefits to identify and control all critical aspects: source of MSCs and raw materials, culture media, supplements and disposable devices, quality control tests (Sensebé et al., 2011). Facilities, isolation methods, seeding density, growth factors, and chemicals can all influence the expansion potential and functional properties of MSCs and should be considered throughout the production process.

The quality and safety of the cell preparations should be ensured by the implementation of a quality system that guarantees the certification and traceability of every batch of material and supply utilized for the procedures, the correct utilization and cleaning of instruments, and the locations necessary for stem cell manipulation. Furthermore, the organization structure, qualification of staff with high levels of expertise, and the appropriate equipment must be implemented in dedicated clean-rooms, currently named "Cell Factories," in compliance with current GMP standards. The application of GMPs for aseptic production ensures the safety of the final cell therapy product.

8 HOW TO GET CELLS WHERE THEY ARE NEEDED?

A crucial point in transplantation is the route of delivery of the cells. A number of approaches have evolved for clinical translation of cell-based neurological therapy. Transplantation has been practiced into the CNS via different routes, i.m., i.v., or i.c.v. Also mobilization of endogenous progenitor cells has been proposed. Since ALS is a disseminated disease, the most rationale approach is to deliver cells intravenously. However, an apparent obstacle to the therapeutic plasticity of MSCs is the observation that following i.v. injection MSCs remain mostly trapped in lungs (Prockop, 2009), where they are rapidly degraded so that only small numbers enter the systemic circulation. After i.v. infusion of human MSCs into injured mice, only a small number of the infused cells (0.01%) were detected 96 h later in peripheral organs including the brain (Lee et al., 2009). However, Uccelli et al. (2012) demonstrate that i.v. injection of MSCs after disease onset in mutant

SOD1 mice results in a significant delay of symptoms and death and is associated with an improvement of several histological and biochemical parameters.

Intrathecal and intravenous transfer of autologous MSCs in patients with ALS seems safe (Karussis et al., 2010), but their therapeutic use requires an easy access to the target tissue to exert their therapeutic effect, since they respond to a particular pathological microenvironment (Liu and Martin, 2006). *In situ* administration can directly achieve this goal. The proximity of grafted cells favors the diffusion of trophic and immunomodulatory factors to MNs and surrounding glia. Most of the successful clinical outcomes with stem cells, whether neuronal or not, in ALS animal models have been achieved by direct intraspinal implantation (Corti et al., 2007; Deshpande et al., 2006; Garbuzova-Davis et al., 2001, 2002; Hemendinger et al., 2005; Lepore et al., 2008; Vercelli et al., 2008).

Although there is some evidence from preclinical studies that MSCs can firmly engraft in the CNS, robust techniques will need to be developed to track the homing of infused MSCs into the CNS (Karp et al., 2009). Many of the unsolved questions about the therapeutic effects of MSCs revolve around the question of how effectively the cells home into injured tissues. However, there have been several frustrating problems in tracking the cells *in vivo* (Karp et al., 2009; Prockop et al., 2010), such as the lack of reliable labels to identify the cells after engraftment and of definitive markers.

9 INTRAPARENCHYMAL DELIVERY

Local injections of stem cells, close to the anterior horn of the spinal cord, have the obvious advantage of placing the cells close to their therapeutic target. Transplantation in critical regions of the spinal cord involved in crucial functions such as the respiratory capacity or the control of limb movements might offer the most significant clinical benefit. Respiratory failure due to phrenic motor neuron loss is the ultimate cause of death in ALS patients (Kaplan and Hollander, 1994); hence, an efficacious strategy on respiratory function could significantly modify their prognosis. Furthermore, increasing the number of transplantation sites for achieving delivery of cells to additional regions of the spinal cord may result in improved efficacy (Xu et al., 2011). A detailed protocol for multi-segmental, intraspinal transplantation of NPCs into the cervical spinal cord ventral gray matter of neurodegenerative models such as SOD1G93A mouse had been presented by Lepore (2011). Mazzini et al. (2003, 2006, 2008, 2012) have published the results of two open-label pilot studies in which MSCs were injected with a surgical procedure into different levels of the thoracic spinal cord (T4–T5; T5–T6), of 19 ALS patients. The cells were injected using a Hamilton syringe mounted on a table fixed arm with a micrometric system which permitted a constant flow of cellular suspension. In these studies the thoracic level was chosen (i) because the risk of inducing a iatrogenic neuronal spinal cord injury is lower than at the more rostral level and (ii) because of the shorter distance between spinal cord and muscles. One milliliter of cell suspension was injected into the spinal cord in a pattern of three rows 3 mm apart, with no severe adverse events. Seventy percent of the patients manifested nonsevere events which resolved in a few weeks. No serious adverse events were seen also in the long-term (9 years follow-up) (Mazzini et al., 2012). Postsurgical MR scans revealed no pathologic intradural fluid collection and confirmed the existence of an area of hyperintensity on T2w images at the graft site in all patients, which was probably due to the MSC suspension. In the long term, T2 relaxivity relative values measured in regions of interest showed a progressive reduction of spinal cord hyperintensity. In the long term (9 years), no evidence of new masses at the injection site or anywhere else in the neuraxis was visible in any of the MRI images of the whole follow-up. There was neither syringomyelia nor pseudomeningocoele in any of the patients. DTI tractography did not detect any structural changes in the corticospinal tracts (Mazzini et al., 2012).

In another clinical study, intraspinal injection of autologous bone marrow mononuclear cells (BMNC) at thoracic level (T3–T4) was performed in 11 patients (Blanquer et al., 2012; Blanquer et al., 2010). The authors chose this injection site due to its low iatrogenic risk of surgery, to maintain the spinal column stability. Moreover, the intermediate intercostal nerves (~T3–T6) innervate the intercostal muscles with the most balanced inspiratory–expiratory function (De Troyer et al., 2005). The injection was made using a syringe attached to a conventional LP 22-gauge needle which was installed on a specially designed microtargeting micromanipulator. Two millimeters of mononucleated cells were injected in two injection sites 1 cm apart to each other. The MRI studies performed 7 days after surgery showed a transient extradural hematoma–seroma. In the follow-up studies, no signs of tumor growth or posttraumatic syringomyelia were detected (Blanquer et al., 2010). The authors did not observe any severe transplant-related adverse events, but there were 43 nonsevere events. Twenty-two (51%) resolved in $\leq$2 weeks and only hypoestesia and constipation were still present at the end of follow-up (Blanquer et al., 2012). The necropsy samples did not show any damage due to surgery. CD90+ cells were found around the MNs in the transplanted segments of the spinal cord. This suggests that the cells grafted into the anterior horns of the spinal cord survive a long time (Blanquer et al., 2012).

In another study, BM-derived hematopoietic progenitor stem cells were injected directly into the brainstem and in the upper spinal cord of 13 ALS patients with severe bulbar involvement. Stem cells (0.1 mL) were injected using a 21-gauge needle attached to a 1-mL syringe into multiple different areas in the spinal cord. No severe adverse events and some benefits in nine patients are reported (Deda et al., 2009).

Improvement of the surgical procedures with evidence of technology capable of a safe, targeted, localized administration in humans has been provided by Riley et al. (2011). This neurosurgical team in Atlanta developed a spinal cord stabilizer and an associated injection device. The device is anchored to the patient's vertebrae, and the stem cell preparation is injected into the spinal cord using a controlled pump to maintain cell viability and reproducible delivery rates. Accurate targeting to the ventral horn is determined for each patient by use of MRI. The system had been validated in a large animal, the minipig (Riley et al., 2011), and the FDA approved a clinical trial of Human Spinal Stem Cells injections into the spinal cord for ALS treatment. The study is currently underway. Twelve patients received either five unilateral or five bilateral (10 total) injections into the lumbar spinal cord at a dose of 100,000 cells/injection. All patients tolerated the treatment without any long-term complications related to either the surgical procedure or the implantation of stem cells (Glass et al., 2012). Next patients will be transplanted in the cervical spinal cord.

The results of these phase 1 clinical trials suggest that the delivery of cellular therapies to the spinal cord of ALS patients with a surgical approach can be proposed without significant adverse events at the thoracic and lumbar levels. Although Deda et al. (2009) report no adverse events with stem cells injection directly into the brain stem and upper spinal cord, transplantation in the cervical spinal cord need further phase 1 studies to ascertain the safety and feasibility.

Another critical point is the volume of cell suspension that could be injected into the living human spinal cord without adverse effects (Guest et al., 2011). From the analysis of these pilot studies, we can conclude that at least up to 2 mL in two injection sites can be injected.

10 INTRAMUSCULAR GRAFTING

Another interesting approach consists in intramuscular grafting of MSCs expressing growth factors. In the SOD1G93A transgenic mice model, the intramuscular transplantation with hMSCs engineered to secrete glial cell line-derived neurotrophic factor resulted in muscle hyperinnervation by MNs and significant increase of the number of neuromuscular connections and motor neuron cell bodies in the spinal cord. Moreover, it delays disease onset, improves locomotor performance, and increases their lifespan (Suzuki et al., 2007).

Brainstorm recently gained approval for a phase I clinical trial in ALS patients using a new technology. The strategy will be to perform injections of MSC-NTF cells intramuscularly in early stage patients and intrathecally (via a standard lumbar puncture) in progressive ALS subjects (www.clinicaltrial.gov).

11 MSCs AS IMMUNOMODULATORY AGENTS: INTRAVENOUS AND INTRATHECAL DELIVERY

It should be emphasized that many reports support a "touch and go" mechanism for the therapeutic effects of MSCs that does not require long-term engraftment into the CNS or other tissues (Uccelli et al., 2008). MSCs have been shown to possess immunomodulating properties (Uccelli et al., 2008). They inhibit T cell proliferation both *in vitro* (Di Nicola et al., 2002) and *in vivo* (Bartholomew et al., 2002). In experimental autoimmune encephalomyelitis, i.v. MSCs ameliorate clinical course and decrease demyelination, immune infiltrates, and axonal loss (Zappia et al., 2005).

Surprisingly, these effects do not require full CNS engraftment by MSCs, but rely on the property of MSCs to inhibit pathogenic immune responses and release neuroprotective and pro-oligodendrogenic molecules favoring tissue repair. Thus, it is reasonable that the main immunomodulatory activity of MSCs is exerted in the secondary lymphoid organs where MSCs migrate following i.v. administration and inhibit T cells homing in the CNS (Kassis et al., 2008; Zappia et al., 2005); systemic effects are observed after i.v. infusion of as little as 1 million MSCs that are largely trapped in the afferent vessels of the lung and apparently degraded there (Lee et al., 2009).

An immediate immunomodulatory effect induced by i.v. administration of MSCs has been shown in five ALS patients by Karussis et al. (2010). These included an increase in CD4(+)CD25(+) regulatory cells and a reduction in the proportion of activated dendritic cells and lymphocytes and of lymphocyte proliferation. The same results were obtained in a group of patients affected by multiple sclerosis. In this phase 1/2 clinical trial, the patients received both an intrathecal and an i.v. injection of autologous MSCs. The 6–25 months of follow-up did not reveal any significant immediate or late adverse effects, and the authors concluded that the procedure is feasible and safe. Moreover, the author conclude with a possible clinical benefit in terms of stabilization of the disease. However, the small sample size and the variability of the disease does not allow to draw any conclusions on the efficacy.

12 HOW MANY CELLS NEED TO BE INJECTED?

Future studies aimed at clinical translation should address the question of the number of cells to be transplanted in order to calculate a therapeutic and also a maximal tolerated "dose" of cells before toxicity becomes a limiting factor. Which is the suitable MSC dose in humans? One can hardly infer the cell dose from experimentation in animal models. One should aim to implant the largest possible number of viable therapeutic substrates (cells), so that the greatest local beneficial effect can be achieved. Also, a single dose might miss adverse events that might emerge in later trials or large effect. Mazzini et al. (2010, 2012) found no correlation between the number of transplanted cells and the incidence and severity of the side effects or the outcome.

Another controversy in stem cell therapy concerns cell aging, given that there is little evidence regarding the lifespan of transplanted cells. In preclinical experiments (Vercelli et al., 2008), hMSCs transplanted into the lumbar spinal cord survive for long periods (more than 10 weeks), without immunosuppression (in agreement with Liu and Martin, 2006). Allotransplantation seems to improve MSC survival even in the absence of immunosuppression (Xu et al., 2009). However, in the case of MSC transplantion, that issue is not as critical because the delivery of trophic factors and the immunomodulatory activity are the main outcome, independently of the longevity of the cells. Moreover, the capacity of MSCs to trigger proliferation, migration, and differentiation of endogenous NPs could explain some of the effects recorded in diseases in which few cells engrafted (Munoz et al., 2005).

Karussis et al. (2010) provided some evidence of immunomodulatory effects of MSCs within 24 h of intrathecal injection but claims of iron oxide-labeled MSCs persisting after 3–6 months was less persuasive. Blanquer et al. (2012) demonstrated by histopathological analysis that autologous BMNC grafted into the anterior horns of ALS spinal cord, survive a long time as perineuronal nets.

13 CLINICAL TRIAL DESIGN

One of the important hurdles in clinical study design for cell therapy trials is defining endpoints, as this is the measure of the trial's failure or success. This is particularly challenging given the degenerative nature of ALS and the complexity posed by the rate of progression and lack of validated surrogate markers of disease.

In 2008, the International Society for Stem Cell Research released a set of recommended guidelines for the development of stem cell-based treatments (Hyun et al., 2008). These recommendations include the use of experts in stem cell biology for peer review of research ranging from preclinical to clinical, emphasizing risks involved with stem cell-based therapies within the voluntary informed consent, new oversight criteria for medical innovative care that falls outside of the realm of a clinical trial, and the equality of benefits of stem cell treatments.

Clinical trial designs need to be debated focusing on critical questions. What do we expect the stem cells to do and what outcomes are predicted? How do we anticipate patients early and later in the disease course will respond to treatments?

The clinical protocol of the studies in humans using stem cells should be carefully designed so as to minimize unexpected patient-related factors that may have a negative impact on posttransplantation outcome.

The successful administration of stem cells will critically depend on their transplantation at the optimal stage of the disease course. Other characteristics should be carefully considered including age at the time of the procedure, disease duration, and disease severity. There is a tendency to enrol, in phase I trials with cell therapy, patients in the advanced phase of disease, in absence of any other viable options, because they may be more motivated and have a more acceptable risk/benefit profile than patients early in the disease course. (Table 1). However, the late stages of ALS are associated with significant MN damage that might create an inhospitable environment for cell therapy. Moreover, patients in the late stages of disease are more susceptible to surgical complications due to disease comorbidities. Considering the possible negative influence of aging on the spinal cord microenvironment, the survival and trophic activity of transplanted stem cells might be affected. Hence, we can speculate that younger patients might benefit most from stem cell transplantation. Furthermore, in the case of MSCs, the outcome can depend on the route of administration. The course of ALS and the clinical characteristics are extremely variable among patients; hence, emphasis should be placed on patient selection and stringent inclusion and exclusion criteria should be established, based on lessons from the pharmacological clinical trials.

Table 1 Clinical trials with adult stem cells in ALS patients

Stem cells	Study phase	Regulatory oversight	Route of delivery	Immunosuppression	Number of cells	Number of patients	Inclusion criteria	Outcome	Reference
Autologous BM-MSCs	I	Regional and Local Ethic Committees	Intraparenchymal transplantation (T4–T6)	No	Mean: 57×10^6 Range: 7.0–152×10^6	9	Age: 23–75 FVC >50% Severe functional impairment Duration of the disease: 8–80 months	Safe and well-tolerated even in long term (9 years)	Mazzini et al. (2003, 2006, 2008, 2012)
Autologous BM-MSCs	I	National Institute of Health Regional and Local Ethic Committees	Intraparenchymal transplantation (T7–T9)	No	Median: 75×10^6 Range: 11–120×10^6	10	Age: 20–61 Spinal onset Duration of the disease: <3 years FVC >50%	No serious transplant-related adverse events. MRI showed no structural changes (including tumor formation)	Mazzini et al. (2010, 2012)
Fetal olfactory ensheathing cells	Controlled pilot study	In accordance with guidelines issued by the Chinese Ministry of Health	Bilateral corona radiata	No	2×10^6	15	Age: 20–70	Safe The mean ALSFRS score remained stable in the first 4 months	Huang et al. (2008)
Autologous HSCs	Single-center pilot trial	Institutional Review Board	Intravenous infusion following total body irradiation; immunosuppression	Tacrolimus methotrexate	unknown	6	Age: 35–69 FVC >60% Duration of the disease: 5–30 months	Tolerated No clinical benefits HSCTs enter the human CNS at sites of motor neuron and engraft as immunomodulatory cells	Appel et al. (2008)
Autologous blood purified CD133(+)	Single-center pilot trial	Ethic and Research Committees of the Hospital	Frontal motor cortex	No	2.5–7.5×10^5	10	Age: 38–62 Duration of the disease: 18–42 months	Safe and well-tolerated (1 year follow-up) Patients survival significantly higher than control group ALS-FRS improvement	Martinez et al. (2009)
Autologous bone marrow (BM)-derived hematopoietic progenitors	Single-center pilot trial	Regional Ethics Board Ministry of Health Institutional Review Board	Intraparenchymal transplantation (C3–C4 level) CSF IV	No	4×10^6 15×10^6 5×10^6	13	Age: 34–71 Duration of the disease: 5 years Moderate–severe functional impairment	No complications Some degree of decline 1 year after transplantation	Deda et al. (2009)

Autologous Bone marrow (BM)-derived hematopoietic progenitors	Open single arm phase I trial	Clinical Trials Ethics Committee of the University Agencia Española de Medicamentos y Productos Sanitarios	Intraparenchymal transplantation (T3–T4)	No	462×10^6 range 138–603×10^6	11	Age: 33–61 FVC $>50\%$ Spinal onset	Safe and well tolerated (2 years follow-up) Histopathological exams showed a greater number of motoneurons in the treated segments and no degenerative ubiquitin deposits	Blanquer et al. (2012)
Autologous BM-MSCS	Phase 1/2 open-safety clinical trial	Ethics Committees of the Hospitals. Registered in the National Institutes of Health Database	Intrathecally and intravenously	No	54.7×10^6 CSF 24.5×10^6 i.v.	19	Age: 25–65	Feasible and safe. Immediate immunomodulatory effects The mean ALSFRS score remained stable during the first 6 months	Karussis et al. (2010)
Human spinal cord-derived stem cells (HSSC)	Phase I trial	University Institutional Review Board	Intraparenchymal transplantation (lumbar spinal cord)	Basiliximab Prednisolone Tacrolimus Mycophenolate	5–10 injections (100,000 cells/ injection)	12	Age: >18 years ALSFRS-R lower extremity subscore <1 (Group A) and >2 (Groups B, C) FVC $>60\%$	Safe and well tolerated One patient has shown improvement in his clinical status	Glass et al. (2012)

Abbreviations: BM, bone marrow; HSCs, hematopoietic stem cells; HSSCs, human spinal cord-derived stem cells; FVC, forced vital capacity (percentage of predicted); MSCs, mesenchymal stem cells; PBSCs, peripheral blood stem cells.

The profile of the ideal candidates should be separately defined for phase I to phase III studies, as criteria may change from one study type to another. Stem cell translational research by its nature entails a high degree of risk: the risk/benefit ratio of stem cells trials clearly seems to require that a phase I study is a mixed design examining safety and efficacy simultaneously. Moreover, in the field of cell therapy, contrary to drug phase I trials, pilot trials cannot be performed on normal, volunteers. This means that the team members should take serious measures to collect data regarding several kinds of side effects and data for general, unexpected adverse events have to be properly acquired in an effective manner, using, for example, the World Health Organization (WHO) severity scale.

At present, there is no clear consensus on how this studies should be performed in ALS (Table 1).

A "risk escalation" paradigm has been adopted for the recruitment of patients in the ongoing neuralstem phase I trial (Lunn et al., 2011). Under this paradigm, risk to patients receiving human spinal cord stem cell transplants escalates across the different cohorts (designated A–E, with cohort A being the lowest-risk and cohort E being the highest-risk group) according to disease severity and the number and placement of injections.

Intrathecal and intravenous transfer of autologous MSCs in patients with ALS was tested by Karussis et al. (2010) in a small phase 1/2 pilot study. No patients manifested severe adverse events.

Cell transplants may survive for several years in patients, or their effects may be irreversible. Therefore, stem cell therapy requires careful patient monitoring and extended follow-up. Long-term follow-up must consider the possibility of the development of a tumor, cyst, or syrinx at the site of transplantation. Advanced MRI, in particular diffusion tensor imaging (DTI), represents an important monitoring mean because it allows a satisfactory quantification of the iatrogenic damage. Mazzini et al. (2012) produced the first report of a very long-term follow-up (up to 9 years) of intraparenchymal transplantation of MSCs into the human CNS. The most relevant finding of this study consists in the complete absence of tumor formations or abnormal cell growth in neuroradiological scans. Assessment of the integrity and survival of the grafted cells at autopsy would be fundamental, but consent represents an ongoing challenge, since it depends on legal and cultural aspects which vary in different countries (Hyun et al., 2008).

Both studies with intraparenchymal (Mazzini et al., 2012) and systemic (Karussis et al., 2010) MSC delivery have shown a slight trend toward a delay in the decline of the performance in the functional rating scales. However, in the light of the small sample sizes, heterogeneity of disease severity within the cohorts and lack of adequate control groups in these pilot studies, the results must be interpreted with caution. Even though some indication about therapeutic efficacy could be derived from phase I trials, this is not a prerequisite for the onset of phase II studies, which specifically aim to evaluate efficacy.

Some open questions include the number of patients needed for phase II trials, how to quantify a response over a short-time frame. Clinical trial designs need to be debated owing to the importance of ethical challenges, including sham control groups to assess the efficacy of invasive therapy. Ideally, a transplanted group should be compared with one undergoing the same surgery but receiving the vehicle. Such a study is unlikely to be approved by institutional review boards in most European Countries. Although a randomized and blinded trial design is always preferable and should be undertaken wherever possible, an alternative approach is to carefully document the natural history of the disease and compare it with the outcome in transplanted patients in an open-label clinical trial. This trial design has been adopted in most pilot clinical trials conducted to date (Blanquer et al., 2012; Glass et al., 2012; Mazzini et al., 2012).

The chances of confirming or convincingly denying the presence of a beneficial effect of treatment are increased by studying patient populations with comparable disease severity; however, they may fail to identify susceptibility to adverse effects or alternatively to detect benefit in particular patient subpopulations not represented in the initial studies. Thus, the presence of some variation in participants in phases I–II stem cells trials may be appropriate.

The focus of phase II studies should include quantifiable clinical outcomes that result in a benefit for the patient. Examples of primary objectives for phase II studies include assessment of response rate (e.g., defined as improvement in neurologic function), time to disease progression, and overall survival. Unfortunately, the applicability and value of such endpoints may only be evident after an expensive and time-consuming trial is completed, with a long-term follow-up.

The cost-benefit analysis of the results must take into account as a major endpoint the quality of life of the patients and ethical issues should be addressed properly. Stable psychological status and social situation should be considered as a prerequisite in the choice of participants. Given the intense public interest and controversy surrounding stem cell "cures," it is essential to ensure that the participants are able to fully comprehend the potential benefits or the lack of benefit and risks associated with the procedure, and that the family and social supports are able to assist them.

14 CONCLUSIONS

MSC research and application is opening great opportunities in ALS treatment. Many trials of MSC treatment for ALS have been recently announced or started. This survey shows that preclinical and initial clinical data support the therapeutic potential of MSCs for ALS. But, while safety seems to be reliably demonstrated, particularly with autologous transplants, sustained curative benefit has not been consistently obtained. The scientific community and patients urgently need safety and efficacy to be addressed properly in the framework of rigorous controlled clinical trials. The distinction between stem cell trials approved by academic research ethic boards and the commercial delivery of stem cells is crucial for patients. Because MSCs are easily isolated and expanded and their safety consistently demonstrated, hundreds of patients with severe incurable neurological diseases such as ALS have been treated in uncontrolled conditions, encouraging so-called medical tourism. Numerous clinics around the world are exploiting patients' hopes by offering expensive, new stem cell therapies for ALS patients, which are claimed to be effective, but without credible scientific rationale, transparency, oversight, or patient protection. Recruitment and selection of appropriate patients for larger trials will be a challenge and will require national and/or international multi-center collaboration with multidisciplinary groups.

Nevertheless, encouraged by current progress, and especially by the terrific strides being made in preclinical studies, we envision a much more concerted effort toward translation that would make the process more accessible, integrated into academic and industry settings, and efficient, therefore improving the chance that the health benefits of research reach patients. Translation, by which we mean advancing scientific discoveries from the laboratory into practical applications for patient benefit, that is, "bench to bedside," requires a comprehensive collaborative team approach: research scientists and clinicians must work closely with regulatory agencies, patient advocacy groups, ethic bodies, cell manufacturing facilities, and industry to achieve the quality of studies and necessary funding to ensure success (Aboody et al., 2011).

References

Aboody, K., Capela, A., Niazi, N., Stern, J.H., Temple, S., 2011. Translating stem cell studies to the clinic for CNS repair: current state of the art and the need for a Rosetta Stone. Neuron 70, 597–613.

Appel, S.H., Engelhardt, J.I., Henkel, J.S., Siklos, L., Beers, D.R., Yen, A.A., Simpson, E.P., Luo, Y., Carrum, G., Heslop, H.E., Brenner, M.K., Popat, U., 2008. Hematopoietic stem cell transplantation in patients with sporadic amyotrophic lateral sclerosis. Neurology 71, 1326–1334.

Awano, T., Johnson, G.S., Wade, C.M., Katz, M.L., Johnson, G.C., Taylor, J.F., Perloski, M., Biagi, T., Baranowska, I., Long, S., March, P.A., Olby, N.J., Shelton, G.D., Khan, S., O'Brien, D.P., Lindblad-Toh, K., Coates, J.R., 2009. Genome-wide association analysis reveals a SOD1 mutation in canine degenerative myelopathy that resembles amyotrophic lateral sclerosis. Proc. Natl. Acad. Sci. U.S.A. 106, 2794–2799.

Bartholomew, A., Sturgeon, C., Siatskas, M., Ferrer, K., McIntosh, K., Patil, S., Hardy, W., Devine, S., Ucker, D., Deans, R., 2002. Mesenchymal stem cells suppress lymphocyte proliferation in vitro and prolong skin graft survival in vivo. Exp. Hematol. 30, 42–48.

Beers, D.R., Henkel, J.S., Xiao, Q., Zhao, W., Wang, J., Yen, A.A., Siklos, L., McKercher, S.R., Appel, H., 2006. Wild-type microglia extend survival in PU.1 knockout mice with familial amyotrophic lateral sclerosis. Proc. Natl. Acad. Sci. U.S.A. 103, 16021–16026.

Bensimon, G., Lacomblez, L., Meininger, V., The ALS/Riluzole Study Group, 1994. A controlled trial of riluzole in amyotrophic lateral sclerosis. N. Engl. J. Med. 330, 585–591.

Bernardo, M.E., Avanzini, M.A., Perotti, C., 2007. Optimization of in vitro expansion of human multipotent mesenchymal stromal cells for cell-therapy approaches: further insights in the search for fetal calf serum substitute. J. Cell. Physiol. 211, 121–130.

Berthòd, F., Gros-Louis, F., 2012. In vivo and in vitro models to study amyotrophic lateral sclerosis. In: Maurer, M.H. (Ed.), Amyotrophic Lateral Sclerosis. Intech Rijeka, Croatia, pp. 81–124.

Blanquer, M., Moraleda Jiménez, J.M., Iniesta Martínez, F., Gómez Espuch, J., Meca Lallana, J., Villaverde González, R., Pérez Espejo, M.A., Ruíz López, F.J., García Santos, J.M., Bleda Díaz, P., Izura Azanza, V., Sáez Gallego, M., de Mingo Casado, P., Vivancos Moreau, L., Carles Dies, R., Jiménez Veiga, J., Hernández Palazón, J., Guardiola Jiménez, J., Torres Del Rio, S., Antúnez Almagro, C., de la Rosa Jiménez, P., Majado Martínez, M.J., Sánchez Salinas, A., López, J., Martínez-Lage Sánchez, J.F., Martínez Pérez, S., 2012. Neurotrophic bone marrow cellular nests prevent spinal motoneuron degeneration in amyotrophic lateral sclerosis patients: a pilot safety study. Stem Cells 30, 1277–1285. doi:http://dx.doi.org/10.1002/stem.1080.

Blanquer, M., Pérez-Espejo, M.A., Martínez-Lage, J.F., Iniesta, F., Martinez, S., Moraleda, J.M., 2010. A surgical technique of spinal cord cell transplantation in amyotrophic lateral sclerosis. J. Neurosci. Methods 191, 255–257.

Bohl, D., Liu, S., Blanchard, S., Hocquemiller, M., Haase, G., Heard, J.M., 2008. Directed evolution of motor neurons from genetically engineered neural precursors. Stem Cells 26, 2564–2575.

Boido, M., Buschini, E., Piras, A., Spigolon, G., Valsecchi, V., Mazzini, L., Vercelli, A., 2012. Advantages and pitfalls in experimental models of ALS. In: Maurer, M.H. (Ed.), Amyotrophic Lateral Sclerosis. Intech Rijeka, Croatia, pp. 125–156.

Boillée, S., Koji, Y., Lobsiger, C.S., Copeland, N.G., Jenkins, N.A., Kassiots, G., Kollias, G., Cleveland, G., 2006. Onset and progression in inherited ALS determined by motor neurons and microglia. Science 312, 1389–1392.

Bongioanni, P., Reali, C., Sogos, V., 2004. Ciliary neurotrophic factor (CNTF) for amyotrophic lateral sclerosis/motor neuron disease. Cochrane Database Syst. Rev. (no. 3), pp. CD004302.

Boucherie, C., Caumont, A.S., Maloteaux, J.M., Hermans, E., 2008. In vitro evidence for impaired neuroprotective capacities of adult mesenchymal stem cells derived from a rat model of familial amyotrophic lateral sclerosis (hSOD1(G93A)). Exp. Neurol. 212, 557–561.

Boucherie, C., Schäfer, S., Lavand'homme, P., Maloteaux, J.M., Hermans, E., 2009. Chimerization of astroglial population in the lumbar spinal cord after mesenchymal stem cell transplantation prolongs survival in a rat model of amyotrophic lateral sclerosis. J. Neurosci. Res. 87, 2034–2046.

Caplan, A.I., 2009. Why are MSCs therapeutic? New data: new insight. J. Pathol. 217, 318–324.

Chaddah, M.R., Dickie, B.G., Lyall, D., Marshall, C.J., Sykes, J.B., Bruijn, L.I., 2011. Meeting report of the International Consortium of Stem Cell Networks' Workshop Towards Clinical Trials Using Stem Cells for Amyotrophic Lateral Sclerosis/Motor Neuron Disease. Amyotroph. Lateral Scler. 12, 315–317.

Chiò, A., Mora, G., Calvo, A., Mazzini, L., Bottacchi, E., Mutani, R., On behalf of the PARALS, 2009. Epidemiology of ALS in Italy: a 10-year prospective population-based study. Neurology 72, 725–731.

Clement, A.M., Nguyen, M.D., Roberts, E.A., Garcia, M.L., Boillée, S., Rule, M., McMahon, A.P., Doucette, W., Siwek, D., Ferrante, R.J., Brown Jr., R.H., Julien, J.P., Goldstein, L.S., Cleveland, D.W., 2003. Wild-type nonneuronal cells extend survival of SOD1 mutant motor neurons in ALS mice. Science 302, 113–117.

Corti, S., Locatelli, F., Papadimitriou, D., Del Bo, R., Nizzardo, M., Nardini, M., Donadoni, C., Salani, S., Fortunato, F., Strazzer, S., Bresolin, N., Comi, G.P., 2007. Neural stem cells LewisX+ CXCR4+ modify disease progression in an amyotrophic lateral sclerosis model. Brain 130, 1289–1305.

Dadon-Nachum, M., Sadan, O., Srugo, I., Melamed, E., Offen, D., 2011. Differentiated mesenchymal stem cells for sciatic nerve injury. Stem Cell Rev. Rep. 7, 664–671.

De Coppi, P., Bartsch, G.J., Siddiqui, M.M., Xu, T., Santos, C.C., Perin, L., 2007. Isolation of amniotic stem cell lines with potential for therapy. Nat. Biotechnol. 25, 10.

De Troyer, A., Kirkwood, P.A., Wilson, T.A., 2005. Respiratory action of the intercostal muscles. Physiol. Rev. 85, 717–756.

Deda, H., Inci, M.C., Kürekçi, A.E., Sav, A., Kayihan, K., Ozgün, E., Ustünsoy, G.E., Kocabay, S., 2009. Treatment of amyotrophic lateral sclerosis patients by autologous bone marrow-derived hematopoietic stem cell transplantation: a 1-year follow-up. Cytotherapy 11, 18–25.

Dejesus-Hernandez, M., Mackenzie, I.R., Boeve, B.F., Boxer, A.L., Baker, M., Rutherford, N.J., Nicholson, A.M., Finch, N.A., Flynn, H., Adamson, J., Kouri, N., Wojtas, A., Sengdy, P., Hsiung, G.Y., Karydas, A., Seeley, W.W., Josephs, K.A., Coppola, G., Geschwind, D.H., Wszolek, Z.K., Feldman, H., Knopman, D.S., Petersen, R.C., Miller, B.L., Dickson, D.W., Boylan, K.B., Graff-Radford, N.R., Rademakers, R., 2011. Expanded GGGGCC hexanucleotide repeat in noncoding region of C9ORF72 causes chromosome 9p-linked FTD and ALS. Neuron 20, 245–256.

Deshpande, D.M., Kim, Y.S., Martinez, T., Carmen, J., Dike, S., Shats, I., Rubin, L.L., Drummond, J., Krishnan, C., Hoke, A., Maragakis, N., Shefner, J., Rothstein, J.D., Kerr, D.A., 2006. Recovery from paralysis in adult rats using embryonic stem cells. Ann. Neurol. 60, 32–44.

Di Nicola, M., Carlo-Stella, C., Magni, M., Milanesi, M., Longoni, P.D., Matteucci, P., Grisanti, S., Gianni, A.M., 2002. Human bone marrow stromal cells suppress T-lymphocyte proliferation induced by cellular or nonscpecific mitogenic stimuli. Blood 99, 3838–3843.

Directive 2001/83/EC of European Parliament and of the Council of 6th November 2001 on the community code relating to medicinal products for human use.

Directive 2004/23/EC of the European Parliament and of the Council of 31 March 2004 on setting standards of quality and safety for the donation, procurement, testing, processing, preservation, storage and distribution of human tissues and cells.

Dominici, M., Le Blanc, K., Mueller, I., Slaper-Cortenbach, I., Marini, F., Krause, D., Deans, R., Keating, A., Prockop, Dj, Horwitz, E., 2006. Minimal criteria for defining multipotent mesenchymal stromal cells. The international society for cellular therapy position statement. Cytotherapy 8, 315–317.

Ericson, C., Georgievska, B., Lundberg, C., 2005. Ex vivo gene delivery of GDNF using primary astrocytes transduced with a lentiviral vector provides neuroprotection in a rat model of Parkinson's disease. Eur. J. Neurosci. 22, 2755–2764.

European Commission. The rules governing medical products in European Union. Good Manufacturing practices. (GMP). 1997; volume 4.

Ferraiuolo, L., Kirby, J., Grierson, A.J., Sendtner, M., Shaw, P.J., 2011. Molecular pathways of motor neuron injury in amyotrophic lateral sclerosis. Nat. Rev. Neurol. 7, 616–630.

Ferrero, I., Mazzini, L., Rustichelli, D., Gunetti, M., Mareschi, K., Testa, L., Nasuelli, N., Oggioni, G.D., Fagioli, F., 2008. Mesenchymal stem cells from healthy donors and sporadic amyotrophic lateral sclerosis patients. Cell Transplant. 17, 255–266.

Flemming, A., Schallmoser, K., Strunk, D., Stolk, M., Volk, H.D., Seifert, M., 2011. Immunomodulative efficacy of bone marrow-derived mesenchymal stem cells cultured in human platelet lysate. J. Clin. Immunol. 31, 1143–1156.

Forostyak, S., Jendelova, P., Kapcalova, M., Arboleda, D., Sykova, E., 2011. Mesenchymal stromal cells prolong the lifespan in a rat model of amyotrophic lateral sclerosis. Cytotherapy 13, 1036–1046.

Garbuzova-Davis, S., Willing, A.E., Milliken, M., Saporta, S., Sowerby, B., Cahill, D.W., Sanberg, P.R., 2001. Intraspinal implantation of hNT neurons into SOD1 mice with apparent motor deficit. Amyotroph. Lateral Scler. Other Motor Neuron Disord. 2, 175–180.

Garbuzova-Davis, S., Willing, A.E., Milliken, M., Saporta, S., Zigova, T., Cahill, D.W., Sanberg, P.R., 2002. Positive effect of transplantation of hNT neurons (NTera 2/D1 cell-line) in a model of familial amyotrophic lateral sclerosis. Exp. Neurol. 174, 169–180.

Glass, J.D., Boulis, N.M., Johe, K., Rutkove, S.B., Federici, T., Polak, M., Kelly, C., Feldman, E.L., 2012. Lumbar intraspinal injection of neural stem cells in patients with ALS: results of a phase I trial in 12 patients. Stem Cells 30, 1144–1151. doi:http://dx.doi.org/10.1002/stem.1079.

Gowing, G., Svendsen, C.N., 2011. Stem cell transplantation for motor neuron disease: current approaches and future perspectives. Neurotherapeutics 8, 591–606.

Guest, J., Benavides, F., Padgett, K., Mendez, E., Tovar, D., 2011. Technical aspects of spinal cord injections for cell transplantation. Clinical and translational considerations. Brain Res. Bull. 84, 267–279.

Gurney, M.E., Pu, H., Chiu, A.Y., Dal Canto, M.C., Polchow, C.Y., Alexander, D.D., Caliendo, J., Hentati, A., Kwon, Y.W., Deng, H.X., 1994. Motor neuron degeneration in mice that express a human Cu, Zn superoxide dismutase mutation. Science 264, 1772–1775.

Haidet-Phillips, A.M., Hester, M.E., Miranda, C.J., 2011. Astrocytes from familial and sporadic ALS patients are toxic to motor neurons. Nat. Biotechnol. 29, 824–828.

Hemendinger, R., Wang, J., Malik, S., Persinski, R., Copeland, J., Emerich, D., Gores, P., Halberstadt, C., Rosenfeld, J., 2005. Sertoli cells improve survival of motor neurons in SOD1 transgenic mice, a model of amyotrophic lateral sclerosis. Exp. Neurol. 196, 235–243.

Henriques, A., Pitzer, C., Schneider, A., 2010. Neurotrophic growth factors for the treatment of amyotrophic lateral sclerosis: where do we stand? Front. Neurosci. 4, 32.

Huang, H., Chen, L., Xi, H., Wang, H., Zhang, J., Zhang, F., Liu, Y., 2008. Fetal olfactory ensheathing cells transplantation in amyotrophic lateral sclerosis patients: a controlled pilot study. Clin. Transplant. 22, 710–718.

Hyun, I., Lindvall, O., Ahrlund-Richter, L., Cattaneo, E., Cavazzana-Calvo, M., Cossu, G., De Luca, M., Fox, I.J., Gerstle, C., Goldstein, R.A., Hermeren, G., High, K.A., Kim, H.O., Lee, H.P., Levy-Lahad, E., Li, L., Lo, B., Marshak, D.R., McNab, A., Munsie, M., Nakauchi, H., Rao, M., Rooke, H.M., Valles, C.S., Srivastava, A., Sugarman, J., Taylor, P.L., Veiga, A., Wong, A.L., Zoloth, L., Daley, G.Q., 2008. New ISSCR guidelines underscore major principles for responsible translational stem cell research. Cell Stem Cell 3, 607–609.

Ilieva, H., Polymenidou, M., Cleveland, D.W., 2009. Non-cell autonomous toxicity inneurodegenerative disorders: ALS and beyond. J. Cell Biol. 187, 761–772.

In'tAnker, P.S., Scherjon, S.A., Kleijburg-van der Keur, C., Noort, W.A., Claas, F.H., Willemzer, R., 2003. Amniotic fluid as a novel source of mesenchymal stem cells for therapeutic transplantation. Blood 102, 1548–1549.

Ince, P.G., Clark, B., Holton, J., Revesz, T., Wharton, S.B., 2008. Greenfield's Neuropathology. In: Love, S. et al., (Ed.), Hodder Arnold, London, pp. 947–971.

Jiang, Y., Vaessen, B., Lenvik, T., Blackstad, M., Reyes, M., Verfaillie, C.M., 2002. Multipotent progenitor cells can be isolated from postnatal murine bone marrow, muscle, and brain. Exp. Hematol. 30, 896–904.

Kaplan, L.M., Hollander, D., 1994. Respiratory dysfunction in amyotrophic lateral sclerosis. Clin. Chest Med. 15, 675–681.

Karp, J.M., Leng, P., Teo, G.S., 2009. Mesenchymal stem cell homing: the devil is in the details. Cell Stem Cell 4, 206–216.

Karussis, D., Karageorgiou, C., Vaknin-Dembinsky, A., Gowda-Kurkalli, B., Gomori, J.M., Kassis, I., Bulte, J.W., Petrou, P., Ben-Hur, T., Abramsky, O., Slavin, S., 2010. Safety and immunological effects of mesenchymal stem cell transplantation in patients with multiple sclerosis and amyotrophic lateral sclerosis. Arch. Neurol. 67, 1187–1194.

Kassis, I., Grigoriadis, N., Gowda-Kurkalli, B., Mizrachi-Kol, R., Ben- Hur, T., Slavin, S., Abramsky, O., Karussis, D., 2008. Neuroprotection and immunomodulation with mesenchymal stem cells in chronic experimental autoimmune encephalomyelitis. Arch. Neurol. 65, 753–761.

Kerr, D.A., Llado, J., Shamblott, M.J., Maragakis, N.J., Irani, D.N., Crawford, T.O., 2003. Human embryonic germ cell derivatives facilitate motor recovery of rats with diffuse motor neuron injury. J. Neurosci. 23, 5131–5140.

Kim, H., Kim, H.Y., Choi, M.R., Hwang, S., Nam, K.H., Kim, H.C., Han, J.S., Kim, K.S., Yoon, H.S., Kim, S.H., 2010. Dose-dependent efficacy of ALS-human mesenchymal stem cells transplantation into cisterna magna in SOD1-G93A ALS mice. Neurosci. Lett. 468, 190–194.

Klein, S.M., Behrstock, S., McHugh, J., Hoffmann, K., Wallace, K., Suzuki, M., Aebischer, P., Svedsen, C.N., 2005. GDNF delivery using human neural progenitor cells in a rat model of ALS. Hum. Gene Ther. 16, 509–521.

Koh, S.H., Baik, W., Noh, M.Y., Cho, G.W., Kim, H.Y., Kim, K.S., Kim, S.H., 2012. The functional deficiency of bone marrow mesenchymal stromal cells in ALS patients is proportional to disease progression rate. Exp. Neurol. 233, 472–480.

Lee, R.H., Pulin, A.A., Seo, M.J., Kota, D.J., Ylostalo, J., Larson, B.L., Semprun-Prieto, L., Delafontaine, P., Prockop, D.J., 2009. Intravenous hMSCs improve myocardial infarction in mice because cells embolized in lung are activated to secrete the anti-inflammatory protein TSG-6. Cell Stem Cell 5, 54–63.

Lepore, A.C., 2011. Human glial-restricted progenitor transplantation into cervical spinal cord of the SOD1G93A mouse model of ALS. PLoS One 6, 1–17.

Lepore, A.C., Rauck, B., Dejea, C., Pardo, A.C., Rao, M.S., Rothstein, J.D., Maragakis, N.J., 2008. Focal transplantation-based astrocyte replacement is neuroprotective in a model of motor neuron disease. Nat. Neurosci. 11, 1294–1301.

Liu, Z., Martin, L.J., 2006. The adult neural stem and progenitor cell niche is altered in amyotrophic lateral sclerosis mouse brain. J. Comp. Neurol. 497, 468–488.

Lunn, J.S., Sakowski, S.A., Federici, T., Glass, J.D., Boulis, N.M., Feldman, E.L., 2011. Stem cell technology for the study and treatment of motor neuron diseases. Regen. Med. 6, 201–213.

Mafi, P., Hindocha, S., Mafi, R., Griffin, M., Khan, W.S., 2011. Adult mesenchymal stem cells and cell surface characterization—A systematic review of the literature. Open Orthop. J. 5, 253–260.

Mareschi, K., Novara, M., Rustichelli, D., Ferrero, I., Guido, D., Carbone, E., Medico, E., Madon, E., Vercelli, A., Fagioli, F., 2006. Neural differentiation of human mesenchymal stem cells: evidence for expression of neural markers and eag K+ channel types. Exp. Hematol. 34, 1563–1572.

Mareschi, K., Rustichelli, D., Comunanza, V., De Fazio, R., Cravero, C., Morterra, G., Martinoglio, B., Medico, E., Carbone, E., Benedetto, C., Fagioli, F., 2009. Multipotent mesenchymal stem cells from amniotic fluid originate neural precursors with functional voltage-gated sodium channels. Cytotherapy 11, 534–547.

Mareschi, K., Rustichelli, D., Calabrese, R., Gunetti, M., Sanavio, F., Castiglia, S., Risso, A., Ferrero, I., Tarella, C., Fagioli, F., 2012. Multipotent mesenchymal stromal stem cell expansion by plating whole bone marrow at a low cellular density: a more advantageous method for clinical use. Stem Cells Int. doi:http://dx.doi.org/10.1155/2012/920581 Article ID 920581.

Martinez, H.R., Gonzalez-Garza, M.T., Moreno-Cuevas, J.E., Caro, E., Guttierez-Imenez, E., Segura, J.J., 2009. Stem call transplantation into the frontal motor cortex in amyotrophic Lateral Sclerosis patients. Cytotherapy 11, 26–34.

Mazzini, L., Fagioli, F., Boccaletti, R., Mareschi, K., Oliveri, G., Olivieri, C., Pastore, I., Marasso, R., Madon, E., 2003. Stem cell therapy in amyotrophic lateral sclerosis: a methodological approach in humans. Amyotroph. Lateral Scler. Other Motor Neuron Disord. 4, 158–161.

Mazzini, L., Mareschi, K., Ferrero, I., Vassallo, E., Oliveti, G., Boccaletti, R., Testa, L., Livigni, S., Fagioli, F., 2006. Autologous mesenchymal stem cells: clinical applications in amyotrophic lateral sclerosis. Neurol. Res. 28, 523–526.

Mazzini, L., Mareschi, K., Ferrero, I., Vassallo, E., Oliveri, G., Nasuelli, N., Oggioni, G.D., Testa, L., Fagioli, F., 2008. Stem cell treatment in Amyotrophic Lateral Sclerosis. J. Neurol. Sci. 265, 78–83.

Mazzini, L., Ferrero, I., Luparello, V., Rustichelli, D., Gunetti, M., Mareschi, K., Testa, L., Stecco, A., Tarletti, R., Miglioretti, M., Fava, E., Nasuelli, N., Cisari, C., Massara, M., Vercelli, R., Oggioni, G.D., Carriero, A., Cantello, R., Monaco, F., Fagioli, F., 2010. Mesenchymal stem cell transplantation in amyotrophic lateral sclerosis: a phase I clinical trial. Exp. Neurol. 223, 229–237.

Mazzini, L., Mareschi, K., Ferrero, I., Miglioretti, M., Stecco, A., Servo, S., Carriero, A., Monaco, F., Fagioli, F., 2012. Mesenchymal stromal cell transplantation in amyotrophic lateral sclerosis: a long-term safety study. Cytotherapy 14, 56–60.

Morita, E., Watanabe, Y., Ishimoto, M., Nakano, T., Kitayama, M., Yasui, K., Fukada, Y., Doi, K., Karunaratne, A., Murrell, W.G., Sutharsan, R., Mackay-Sim, A., Hata, Y., Nakashima, K., 2008. A novel cell transplantation protocol and its application to an ALS mouse model. Exp. Neurol. 213, 431–438.

Munoz, J.R., Stoutenger, B.R., Robinson, A.P., Spees, J.L., Prockop, D.J., 2005. Human stem/progenitor cells from bone marrow promote neurogenesis of endogenous neural stem cells in the hippocampus of mice. Proc. Natl. Acad. Sci. U.S.A. 102, 18171–18176.

Nagai, M., Aoki, M., Miyoshi, I., Kato, M., Pasinelli, P., Kasai, N., Brown Jr., R.H., Itoyama, Y., 2001. Rats expressing human cytosolic copper-zinc superoxide dismutase transgenes with amyotrophic lateral sclerosis: associated mutations develop motor neuron disease. J. Neurosci. 21, 9246–9254.

Nagai, M., Re, D.B., Nagata, T., Chalazonitis, A., Jessell, T.M., Wichterle, H., Przedborski, S., 2007. Astrocytes expressing ALS-linked mutated SOD1 release factors selectively toxic to motor neurons. Nat. Neurosci. 10, 615–622.

Neuhuber, B., Swanger, S.A., Howard, L., Mackay, A., Fischer, I., 2008. Effects of plating density and culture time on bone marrow stromal cell characteristics. Exp. Hematol. 36, 1176–1185.

Nicaise, C., Mitrecic, D., Pochet, R., 2011. Brain and spinal cord affected by amyotrophic lateral sclerosis induce differential growth factors expression in rat mesenchymal and neural stem cells. Neuropathol. Appl. Neurobiol. 37, 179–188.

Otto, W.R., Wright, N.A., 2011. Mesenchymal stem cells: from experiment to clinic. Fibrogenesis Tissue Repair 4, 20.

Phinney, D.G., Isakova, I., 2005. Plasticity and therapeutic potential of mesenchymal stem cells in the nervous system. Curr. Pharm. Des. 11, 1255–1265.

Pittenger, M.F., Mackay, A.M., Beck, S.C., Jaiswal, R.K., Douglas, R., Mosca, J.D., Moorman, M.A., Simonetti, D.W., Craig, S., Marshak, D.R., 1999. Multilineage potential of adult human mesenchymal stem cells. Science 284, 143–147.

Prockop, D.J., 2009. Repair of tissues by adult stem/progenitor cells (MSCs): controversies, myths, and changing paradigms. Mol. Ther. 17, 939–946.

Prockop, D.J., Brenner, M., Fibbe, W.E., Horwitz, E., Le Blanc, K., Phinney, D.G., Simmons, P.J., Sensebe, L., Keating, A., 2010. Defining the risks of mesenchymal stromal cell therapy. Cytotherapy 12, 576–578.

Regulation (EC) N. 1394/2007 of the European Parliament and of the Council of 13 November 2007 on advanced therapy medicinal products.

Renton, A.E., Majounie, E., Waite, A., Simon-Sanchez, J., Rollinson, S., Gibbs, J.R., Schymick, J.C., Laaksovirta, H., van Swieten, J.C., Myllykangas, L., Kalimo, H., Paetau, A., Abramzon, Y., Remes, A.M., Kaganovich, A., Scholz, S.W., Duckworth, J., Ding, J., Harmer, D.W., Hernandez, D.G., Johnson, J.O., Mok, K., Ryten, M., Trabzuni, D., Guerreiro, R.J., Orrell, R.W., Neal, J., Murray, A., Pearson, J., Jansen, I.E., Sondervan, D., Seelaar, H., Blake, D., Young, K., Halliwell, N., Callister, J.B., Toulson, G., Richardson, A., Gerhard, A., Snowden, J., Mann, D., Neary, D., Nall,s, M.A., Peuralinna, T., Jansson, L., Isoviita, V.M., Kaivorinne, A.L., Hölttä-Vuori, M., Ikonen, E., Sulkava, R., Benatar, M., Wuu, J., Chiò, A., Restagno, G., Borghero, G., Sabatelli, M., ITALSGEN Consortium, Heckerman, D., Rogaeva, E., Zinman, L., Rothstein, J.D., Sendtner, M., Drepper, C., Eichler, E.E., Alkan, C., Abdullaev, Z., Pack, S.D., Dutra, A., Pak, E., Hardy, J., Singleton, A., Williams, N.M., Heutink, P., Pickering-Brown, S., Morris, H.R., Tienari, P.J., Traynor, B.J., 2011. A hexanucleotide repeat expansion in C9ORF72 is the cause of chromosome 9p21-linked ALS-FTD. Neuron 72, 257–268.

Riley, J.P., Raore, B., Taub, J.S., Federici, T., Boulis, N.M., 2011. Platform and cannula design improvements for spinal cord therapeutics delivery. Neurosurgery 69 (no. 2 Suppl Operative), 147–155.

Sadan, O., Shemesh, N., Barzilay, R., Bahat-Stromza, M., Melamed, E., Cohen, Y., Offen, D., 2008. Migration of neurotrophic factors secreting mesenchymal stem cells toward a quinolinic acid lesion as viewed by magnetic resonance imaging. Stem Cells 6, 2542–2551.

Sadan, O., Melamed, E., Offen, D., 2009. Bone-marrow-derived mesenchymal stem cell therapy for neurodegenerative diseases. Expert Opin. Biol. Ther. 9, 1487–1497.

Scott, S., Kranz, J.E., Cole, J., Lincecum, J.M., Thompson, K., Kelly, N., Bostrom, A., Theodoss, J., Al-Nakhala, B.M., Vieira, F.G., Ramasubbu, J., Heywood, J.A., 2008. Design, power, and interpretation of studies in the standard murine model of ALS. Amyotroph. Lateral Scler. Other Motor Neuron Disord. 9, 4–15.

Sensebé, L., Bourin, P., Tarte, K., 2011. Good manufacturing practices production of mesenchymal stem/stromal cells. Hum. Gene Ther. 22, 19–26.

Sessarego, N., Parodi, A., Podestà, M., Benvenuto, F., Mogni, M., Raviolo, V., Lituania, M., Kunkl, A., Ferlazzo, G., Bricarelli, F.D., Uccelli, A., Frassoni, F., 2008. Multipotent mesenchymal stromal cells from amniotic fluid: solid perspectives for clinical application. Haematologica 93, 339–346.

Sorenson, E.J., Windbank, A.J., Mandrekar, J.N., Bamlet, W.R., Appel, S.H., Armon, C., Barkhaus, P.E., Bosch, P., Boylan, K., David, W.S., Feldman, E., Glass, J., Gutmann, L., Katz, J., King, W., Luciano, C.A., McCluskey, L.F., Nash, S., Newman, D.S., Pascuzzi, R.M., Pioro, E., Sams, L.J., Scelsa, S., Simpson, E.P., Subramony, S.H., Tiryaki, E., Thornton, C.A., 2008. Subcutaneous IGF-1 is not beneficial in 2-year ALS trial. Neurology 71, 1770–1775.

Suzuki, M., Svendsen, C.N., 2008. Combining growth factor and stem cell therapy for amyotrophic lateral sclerosis. Trends Neurosci. 31, 193–198.

Suzuki, M., McHugh, J., Tork, C., Shelley, B., Klein, S.M., Aebischer, P., Svendsen, C.N., 2007. GDNF secreting human neural progenitor cells protect dying motor neurons, but not their projection to muscle, in a rat model of familial ALS. PLoS One 2, e689.

Suzuki, M., McHugh, J., Tork, C., Shelley, B., Hayes, A., Bellantuono, I., Aebischer, P., Svendsen, C.N., 2008. Direct muscle delivery of GDNF with human mesenchymal stem cells improves motor neuron survival and function in a rat model of familial ALS. Mol. Ther. 16, 2002–2010.

The BDNF Study Group (Phase III), 1999. A controlled trial of recombinant methionyl human BDNF in ALS. Neurology 52, 1427–1433.

Uccelli, A., Moretta, L., Pistoia, V., 2008. Mesenchymal stem cells in health and disease. Nat. Rev. Immunol. 8, 726–736.

Uccelli, A., Laroni, A., Freedman, M.S., 2011. Mesenchymal stem cells for the treatment of multiple sclerosis and other neurological diseases. Lancet Neurol. 10, 649–656.

Uccelli, A., Milanese, M., Principato, M.C., Morando, S., Bonifacino, T., Vergani, L., Giunti, D., Voci, A., Carminati, E., Giribaldi, F., Caponnetto, C., Bonanno, G., 2012. Intravenous mesenchymal stem cells improve survival and motor function in experimental Amyotrophic Lateral Sclerosis. Mol. Med. 18, 794–804. doi:http://dx.doi.org/10.2119/molmed.2011.00498.

Vercelli, A., Mereuta, O.M., Garbossa, D., Muraca, G., Mareschi, K., Rustichelli, D., Ferrero, I., Mazzini, L., Madon, E., Fagioli, F., 2008. Human mesenchymal stem cell transplantation extends survival, improves motor performance and decreases neuroinflammation in mouse model of amyotrophic lateral sclerosis. Neurobiol. Dis. 31, 395–405.

Vercelli, A., Boido, M., Spigolon, G., Piras, A., Valsecchi, V., 2010. Motor behavior tests to detect the early symptoms in a mouse ALS model. SFN 2010 abstracts, San Diego, USA.

Wislet-Gendebien, S., Wautier, F., Leprince, P., Rogister, B., 2005. Astrocytic and neuronal fate of mesenchymal stem cells expressing nestin. Brain Res. Bull. 68, 95–102.

Xu, L., Ryugo, D.K., Pongstaporn, T., Johe, K., Koliatsos, V.E., 2009. Human neural stem cell grafts in the spinal cord of SOD1 transgenic rats: differentiation and structural integration into the segmental motor circuitry. J. Comp. Neurol. 514, 297–309.

Xu, L., Shen, P., Hazel, T., Johe, K., Koliatsos, V.E., 2011. Dual transplantation of human neural stem cells into cervical and lumbar cord ameliorates motor neuron disease in SOD1 transgenic rats. Neurosci. Lett. 494, 222–226.

Yan, J., Xu, L., Welsh, A.M., Hatfield, G., Hazel, T., Johe, K., Koliatsos, V.E., 2007. Extensive neuronal differentiation of human neural stem cell grafts in adult rat spinal cord. PLoS Med. 4, e39.

Zappia, E., Casazza, S., Pedemonte, E., Benvenuto, F., Bonanni, I., Gerdoni, E., Giunti, D., Ceravolo, A., Cazzanti, F., Frassoni, F., 2005. Mesenchymal stem cells ameliorate experimental autoimmune encephalomyelitis inducing T cell anergy. Blood 106, 1755–1761.

Zhao, C.P., Zhang, C., Zhou, S.N., Xie, Y.M., Wang, Y.H., Huang, H., Shang, Y.C., Lim, W.Y., Zhou, C., Yu, M.J., Feng, S.W., 2007. Human mesenchymal stromal cells ameliorate the phenotype of SOD1-G93A ALS mice. Cytotherapy 9, 414–426.

Combined Index

The combined index lists items in both volumes. Page number preceded by I: or II: refer to pages in Part 1 (*Progress in Brain Research*, vol, 200) or Part 2 (*Progress in Brain Research*, vol. 201), respectively.

H

I

J

K

L

O

P

Volume 149: Cortical Function: A View from the Thalamus, by V.A. Casagrande, R.W. Guillery and S.M. Sherman (Eds.) – 2005 ISBN 0-444-51679-4.
Volume 150: The Boundaries of Consciousness: Neurobiology and Neuropathology, by Steven Laureys (Ed.) – 2005, ISBN 0-444-51851-7.
Volume 151: Neuroanatomy of the Oculomotor System, by J.A. Büttner-Ennever (Ed.) – 2006, ISBN 0-444-51696-4.
Volume 152: Autonomic Dysfunction after Spinal Cord Injury, by L.C. Weaver and C. Polosa (Eds.) – 2006, ISBN 0-444-51925-4.
Volume 153: Hypothalamic Integration of Energy Metabolism, by A. Kalsbeek, E. Fliers, M.A. Hofman, D.F. Swaab, E.J.W. Van Someren and R.M. Buijs (Eds.) – 2006, ISBN 978-0-444-52261-0.
Volume 154: Visual Perception, Part 1, Fundamentals of Vision: Low and Mid-Level Processes in Perception, by S. Martinez-Conde, S.L. Macknik, L.M. Martinez, J.M. Alonso and P.U. Tse (Eds.) – 2006, ISBN 978-0-444-52966-4.
Volume 155: Visual Perception, Part 2, Fundamentals of Awareness, Multi-Sensory Integration and High-Order Perception, by S. Martinez-Conde, S.L. Macknik, L.M. Martinez, J.M. Alonso and P.U. Tse (Eds.) – 2006, ISBN 978-0-444-51927-6.
Volume 156: Understanding Emotions, by S. Anders, G. Ende, M. Junghofer, J. Kissler and D. Wildgruber (Eds.) – 2006, ISBN 978-0-444-52182-8.
Volume 157: Reprogramming of the Brain, by A.R. Møller (Ed.) – 2006, ISBN 978-0-444-51602-2.
Volume 158: Functional Genomics and Proteomics in the Clinical Neurosciences, by S.E. Hemby and S. Bahn (Eds.) – 2006, ISBN 978-0-444-51853-8.
Volume 159: Event-Related Dynamics of Brain Oscillations, by C. Neuper and W. Klimesch (Eds.) – 2006, ISBN 978-0-444-52183-5.
Volume 160: GABA and the Basal Ganglia: From Molecules to Systems, by J.M. Tepper, E.D. Abercrombie and J.P. Bolam (Eds.) – 2007, ISBN 978-0-444-52184-2.
Volume 161: Neurotrauma: New Insights into Pathology and Treatment, by J.T. Weber and A.I.R. Maas (Eds.) – 2007, ISBN 978-0-444-53017-2.
Volume 162: Neurobiology of Hyperthermia, by H.S. Sharma (Ed.) – 2007, ISBN 978-0-444-51926-9.
Volume 163: The Dentate Gyrus: A Comprehensive Guide to Structure, Function, and Clinical Implications, by H.E. Scharfman (Ed.) – 2007, ISBN 978-0-444-53015-8.
Volume 164: From Action to Cognition, by C. von Hofsten and K. Rosander (Eds.) – 2007, ISBN 978-0-444-53016-5.
Volume 165: Computational Neuroscience: Theoretical Insights into Brain Function, by P. Cisek, T. Drew and J.F. Kalaska (Eds.) – 2007, ISBN 978-0-444-52823-0.
Volume 166: Tinnitus: Pathophysiology and Treatment, by B. Langguth, G. Hajak, T. Kleinjung, A. Cacace and A.R. Møller (Eds.) – 2007, ISBN 978-0-444-53167-4.
Volume 167: Stress Hormones and Post Traumatic Stress Disorder: Basic Studies and Clinical Perspectives, by E.R. de Kloet, M.S. Oitzl and E. Vermetten (Eds.) – 2008, ISBN 978-0-444-53140-7.
Volume 168: Models of Brain and Mind: Physical, Computational and Psychological Approaches, by R. Banerjee and B.K. Chakrabarti (Eds.) – 2008, ISBN 978-0-444-53050-9.
Volume 169: Essence of Memory, by W.S. Sossin, J.-C. Lacaille, V.F. Castellucci and S. Belleville (Eds.) – 2008, ISBN 978-0-444-53164-3.
Volume 170: Advances in Vasopressin and Oxytocin – From Genes to Behaviour to Disease, by I.D. Neumann and R. Landgraf (Eds.) – 2008, ISBN 978-0-444-53201-5.
Volume 171: Using Eye Movements as an Experimental Probe of Brain Function—A Symposium in Honor of Jean Büttner-Ennever, by Christopher Kennard and R. John Leigh (Eds.) – 2008, ISBN 978-0-444-53163-6.
Volume 172: Serotonin–Dopamine Interaction: Experimental Evidence and Therapeutic Relevance, by Giuseppe Di Giovanni, Vincenzo Di Matteo and Ennio Esposito (Eds.) – 2008, ISBN 978-0-444-53235-0.
Volume 173: Glaucoma: An Open Window to Neurodegeneration and Neuroprotection, by Carlo Nucci, Neville N. Osborne, Giacinto Bagetta and Luciano Cerulli (Eds.) – 2008, ISBN 978-0-444-53256-5.

FSC
www.fsc.org
MIX
Papier | Fördert
gute Waldnutzung
FSC® C083411

Kolibri 360 GmbH
Ferdinand-Jühlke-Straße 7
99095 Erfurt, Deutschland
produktsicherheit@kolibri360.de